CREATIONISM IN TWENTIETH-CENTURY AMERICA

Volume 10

# CREATION AND EVOLUTION IN THE EARLY AMERICAN SCIENTIFIC AFFILIATION

# CREATION AND EVOLUTION IN THE EARLY AMERICAN SCIENTIFIC AFFILIATION

Edited by
MARK A. KALTHOFF

Routledge
Taylor & Francis Group
LONDON AND NEW YORK

First published in 1995 by Garland Publishing, Inc.

This edition first published in 2022
by Routledge
4 Park Square, Milton Park, Abingdon, Oxon OX14 4RN
605 Third Avenue, New York, NY 10017

*Routledge is an imprint of the Taylor & Francis Group, an informa business*



*British Library Cataloguing in Publication Data*
A catalogue record for this book is available from the British Library

ISBN: 978-0-367-43553-0 (Set)
ISBN: 978-1-00-314991-0 (Set) (ebk)
ISBN: 978-0-367-44061-9 (Volume 10) (hbk)
ISBN: 978-0-367-44064-0 (Volume 10) (pbk)
ISBN: 978-1-00-300735-7 (Volume 10) (ebk)

DOI: 10.4324/9781003007357

**Publisher's Note**
The publisher has gone to great lengths to ensure the quality of this reprint but points out that some imperfections in the original copies may be apparent.

**Disclaimer**
The publisher has made every effort to trace copyright holders and would welcome correspondence from those they have been unable to trace.

# New Preface to the Re-issue of 2021

This anthology of primary documents related to the early history of creationism in the United States first appeared a quarter century ago, in 1995. My interest in the topic had been aroused by my years of research on creationism, which resulted in *The Creationists* (New York: Alfred A. Knopf, 1992). In the meantime, a former student of mine, Edward J. Larson, had published an excellent legal survey, *Trial and Error: The American Controversy over Creation and Evolution* (New York: Oxford University Press, 1985). The philosopher of science Michael Ruse had published the edited volume *But Is It Science? The Philosophical Question in the Creation/Evolution Controversy* (Amherst, NY: Prometheus Books, 1988); and the anthropologist Christopher P. Toumey had just released *God's Own Scientists: Creationists in a Secular World* (New Brunswick, NJ: Rutgers University Press, 1994). Led by Willard B. Gatewood's *Preachers Pedagogues and Politicians: The Evolution Controversy in North Carolina, 1920–1927* (Chapel Hill: University of North Carolina Press, 1966), local studies had also begun to appear. Nevertheless, few, if any, research libraries had begun collecting creationist literature; and not one, to my knowledge, possessed even a complete run of the *Creation Research Society Quarterly*, launched in 1964.

During the past quarter century the landscape of creationism has changed dramatically. Since 1995 the institutional heart of creationism has shifted from the Institute for Creation Research, founded by Henry M. Morris in southern California in 1972, to Ken Ham's Answers in Genesis, headquartered in northern Kentucky. In 2007 the charismatic Australian-born Ham opened a $27-million Creation Museum in Petersburg, Kentucky, across the Ohio River from Cincinnati. Forty-five miles away, in Williamstown, Kentucky, Ham in July 2016 opened an Ark Encounter featuring a "life-size" replica of Noah's ark, at a projected cost of $150 million.

Such growth has attracted considerable attention, such as Susan L. Trollinger and William Vance Trollinger Jr., *Righting America at the Creation Museum* (Baltimore: Johns Hopkins University Press, 2016), and James S. Bielo, *Ark Encounter: The Making of a Creationist Theme Park* (New York: New York University Press, 2018).

The literature on the general history of creationism in the twentieth century has exploded, symbolized most dramatically by Edward J. Larson's Pulitzer Prize-winning volume *Summer for the Gods: The Scopes Trial and America's Continuing Debate over Science and Religion* (New York: Basic Books, 1997). Other significant contributions include Michael Lienesch, *In the Beginning: Fundamentalism, the Scopes Trial, and the Making of the Antievolution Movement* (Chapel Hill: University of North Carolina, 2007); Adam Laats, *Fundamentalism and Education in the Scopes Era: God, Darwin, and the Roots of America's Culture Wars* (New York: Palgrave Macmillan, 2010); Jeffrey P. Moran, *American Genesis: The Evolution Controversies from Scopes to Creation Science* (New York: Oxford University Press, 2012); and Adam R. Shapiro, *Trying Biology: The Scopes Trial, Textbooks, and the Antievolution Movement in American Schools* (Chicago: University of Chicago Press, 2013).

Still, access to creationist sources before the early 1960s remains patchy. To help remedy this condition, Routledge has agreed to reissue the 10-volume set of *Creationism in Twentieth-Century America.* I thank them for their continuing interest.

Ronald L. Numbers
April 2021

*Errata*

On page xxiv of the "Introduction" line 24 should not include the name "Henry Morris."

Line 25 should read: "The one-page article entitled "Comment of the 'Deluge Geology' Paper of J. L. Kulp" appeared anonymously in the June 1950 issue of the ASA *Journal.*"

On page xxxvi, the first line of note 59 should read, "Although impossible to establish, the following correspondence may suggest that Henry Morris was involved in the production of the anonymous article. . ."

*VOLUME*

# 10

# CREATION AND EVOLUTION IN THE EARLY AMERICAN SCIENTIFIC AFFILIATION

Edited with an introduction by
MARK A. KALTHOFF
*Hillsdale College*

**GARLAND PUBLISHING, Inc.**
New York & London
1995

**Library of Congress Cataloging-in-Publication Data**

Creation and evolution in the early American Scientific Affiliation / edited with an introduction by Mark A. Kalthoff.
p. cm. — (Creationism in twentieth-century America ; v. 10)
Includes bibliographical references.
ISBN 0-8153-1811-1 (alk. paper)
1. Bible and evolution. 2. Creationism. 3. Evolution (Biology)—Religious aspects—Christianity. 4. American Scientific Affiliation—History. I. Kalthoff, Mark A. II. Series.
BS659.C73 1995
231.7'65—dc20 94-45528
CIP

Printed on acid-free, 250-year-life paper
Manufactured in the United States of America

# Contents

# Series Introduction

In recent years creationism has enjoyed a stunning renaissance both in the United States and around the world. Public opinion polls show that 47 percent of Americans, including one quarter of college graduates, believe that "God created man pretty much in his present form at one time within the last 10,000 years." In the early 1980s two states, Arkansas and Louisiana, passed laws mandating the teaching of "creation science" whenever "evolution science" was taught in the public schools. The United States Supreme Court subsequently overturned these laws, but creationists actively—and often successfully—continue to promote their cause in local schools and churches.

Since the early 1960s creationism has become increasingly identified with a particular nonevolutionary belief known as "scientific creationism" or "creation science." Scientific creationists believe that all life on earth originated no more than 10,000 years ago, and some argue that the entire universe is equally young. To explain the appearance of age suggested by the fossil record, they typically invoke Noah's flood, which, they claim, deposited virtually the entire geological column in the span of a year or so.

Before the 1960s relatively few Americans, including religious fundamentalists, subscribed to such restrictive views of earth history. At the height of the antievolution controversies of the 1920s, for example, most creationists who expressed themselves on the subject embraced interpretations of the book of Genesis that allowed them to accept the evidence of historical geology for the antiquity of life on earth. They generally did so in one of two ways: either by assuming that the "days" of Genesis 1 really meant "ages" or by interposing a gap of perhaps hundreds of millions of years between the creation "in the beginning" and the relatively recent Edenic creation (or restoration, as some would call it) associated with Adam and Eve. Only a few fundamentalists at the time, mostly Seventh-day Adventists, insisted on the recent appearance of life and on the geological significance of the deluge. In recent years, however, through the influence of books such as John C. Whitcomb, Jr. and Henry M. Morris's *The Genesis Flood* (1961), organizations such as the Creation Research Society (1963) and institutions such as the Institute for Creation Research (1972), the so-called flood geologists, now known as scientific creationists, have co-opted the very name creationist for their once peculiar views.

Despite the undeniable importance of antievolutionism in American cultural history, few libraries, academic or otherwise, have collected more than the odd book or pamphlet on creationism, and early creationist periodicals are almost impossible to find. Whether the result of prejudice or indifference, such neglect has made it difficult for students and scholars to explore the development of creationist thought in the twentieth century. This collection of reprinted documents from the first six decades of the century makes available some of the most widely read works on creationism by such stalwarts as Arthur I. Brown, William Bell Riley, Harry Rimmer, Byron C. Nelson, George McCready Price, Harold W. Clark, and Frank Lewis Marsh. It also reprints, for the first time, three of the earliest and rarest creationist journals in America: the *Creationist*, the *Bulletin of Deluge Geology*, and the *Forum for the Correlation of Science and the Bible*.

# Introduction

The American Scientific Affiliation (ASA) officially dates from the first week of September 1941. Responding to an invitation from the president of Moody Bible Institute in Chicago, a group of five concerned, conservative Protestants with scientific training convened for four days on the Moody campus to assess perceived needs and outline plans for creating the proposed "Society for the Correlation of Science and the Bible."[1] Although the suggested name failed to win approval (as did fifteen others), the group gradually flourished, eventually becoming an organization with the appearance of a typical professional academic society. Annual meetings (beginning in 1946), a quarterly journal (beginning in 1949), and a bimonthly newsletter (beginning in 1959) were the ASA's most visible features. By 1961 (the year *The Genesis Flood* by John C. Whitcomb, Jr. and Henry M. Morris appeared) ASA membership stood at 806. It continued to increase until the late 1970s, when it peaked at nearly 2,700. Thus in the half-century following its inception, the ASA emerged as the primary forum for American evangelical scientists to engage with one another in order to explicate the relationship between science and their Christian faith.

Unlike earlier groups such as the Religion and Science Association (1936–1937) and the Deluge Geology Society (1938–1947) that had flared and fizzled after cantankerous in-house squabbling, the ASA managed to sustain its vitality despite divergent opinions.[2] This ability to escape institutional suicide did not issue from some kindly gentlemen's agreement to avoid debating the perennially volatile issues of creation and evolution. To the contrary, the story of the ASA is largely the story of twentieth-century evangelical scientists confronting questions of origins head-on, facing casualties, and living to tell about it. After fifty years of neither fully answering those questions nor self-destructing in the attempts to do so, the ASA still showed no signs of decay or willingness to abandon the origins debates.[3] Along the way, the organization and its members left a rich paper trail documenting their thoughts and struggles. These documents offer one lens through which to focus on the encounter between science and American evangelicalism in the twentieth century. This volume reproduces fifty-three selections (by thirty-one authors) that appeared during the ASA's first two decades.[4] They are best understood, however, when placed within the context of a narrative that begins before the ASA's 1941 founding.

By the late 1920s rigorous confrontations with modernism had dampened the fundamentalist vision of transforming America into a Christian nation.[5] Among other things, the Scopes trial symbolized the modernists' rejection of the conservative Protestant vision and legitimized their scorn for fundamentalists, who remained woefully out of step with received scientific opinion. Shunned by the dominant culture, evangelicals entered, during the decades prior to mid-century, a period of what historian George Marsden calls "exile." "The period from about 1920 to 1950 became a sort of academic dark age for conservative evangelical scholarship," suggests Marsden. "Biblical considerations had been ruled out of bounds in the sciences . . . [and] explicitly biblicist evangelical scholars virtually disappeared from those fields."[6]

Rather than ceasing and desisting, however, fundamentalists shifted their focus away from winning cultural hegemony and labored during the 1930's to establish their own institutional bases. Joel Carpenter has documented the dynamism that produced numerous evangelical voluntary associations, the most significant of which was the Bible institute, a phenomenon exemplified by Moody Bible Institute, "the national giant of institutional fundamentalism."[7] That the American Scientific Affiliation was born on the campus of Moody Bible Institute conjures various implications about the fundamentalist roots of the organization.

Bible institutes possessed little zeal, however, for incorporating science into their curricula. Instead, they generally eschewed secular "science falsely so-called" while championing their baptized Baconian conviction that the true "facts of science" could never conflict with God's word recorded in scripture. This outlook was rooted in a long heritage stemming from the thriving evangelical Baconianism of the antebellum period. Antebellum Protestants "nurtured the comfortable assumption," according to historian Theodore Dwight Bozeman, that "science and religion, Baconianism *and* the Bible, were harmonious enterprises cooperating toward the same ultimate ends."[8] If by the 1930s most American intellectuals had abandoned this nineteenth-century view, many conservative evangelicals did not know it. Fundamentalists, including the founders of the American Scientific Affiliation, persisted in proclaiming their archaic, but comfortable, philosophy of science. It is no coincidence that several authors whose writings appear in this volume possessed an inordinate fondness for the Baconian phrase "the real facts of science."[9]

It comes as no suprise, then, that fundamentalists in the 1930s had difficulty naming recognized credentialed scientists who embraced their unadorned harmonizations of science and Scripture.[10] Therefore, when young fundamentalists did actually learn science, it usually was not from instructors espousing sympathetic theological convictions. As Joel Carpenter has noted, "Fundamentalists who desired a Christian liberal arts education for their children in the 1930's sought it for the most part outside the movement proper."[11] These circumstances—the naive tendency of fundamentalists to believe science and the Bible could be harmonized according to a simplistic and dated Baconianism, the apparent paucity of recognized scientists with fundamentalist convictions, and the trend of young fundamentalists increasingly to expose themselves to the potentially deleterious spiritual effects of secular higher education—formed the catalyzing atmosphere in which the idea for an organization like the ASA could flourish. The source of that idea was a young southern California Baptist preacher in his late twenties.

Irwin A. Moon (1907–1986) had shelved his childhood love of science and rejected a Yale physics scholarship in order to attend Moody Bible Institute and pursue a career of Christian ministry. After completing his studies at the Bible Institute of Los Angeles and Los Angeles Baptist Seminary, he became pastor in 1929 of the Montecito Park Union Church in Los Angeles. Unable to suppress his youthful interest in the relation between science and his Christian faith, Moon reconsidered his assumption that the two must be antithetical. Over the next seven years he gradually developed a number of "gee whiz" scientific presentations in order to demonstrate the harmony of science and Christian belief. The tremendous successes of his amateur traveling science exhibition led him in 1937 to resign his pastorate and devote full-time effort to his burgeoning evangelistic road show, which he had christened "Sermons from Science."[12]

In the course of developing his scientific ministry Moon met increasing numbers of college-aged young people who questioned the viability of Christian faith in an age of science. He became more and more concerned that the materialistic philosophy of collegiate science courses was undermining the faith of many Christian college students. Moon had long entertained the "wistful dream" of an organization of Christian men of science. He believed that college students' faith would be buttressed against the assaults of atheistic science merely by knowing that such an organization existed. By late 1940 Moon had discussed his fears about the faith of college students and his idea for a remedy with two like-minded friends, Will Houghton (1887–1947) and F. Alton Everest (b. 1909). Houghton had left his pastorate at Calvary Baptist Church in New York in 1934 to become president of Moody Bible Institute, where, since 1938, he had supported Moon's activities by employing him through Moody's Extension Department. Everest, a Conservative Baptist with electrical engineering degrees from Oregon State College and Stanford, was an assistant professor of engineering at Oregon State College. Everest's wife, a former classmate of Moon's at the Bible Institute of Los Angeles, had introduced Everest to Moon a few years earlier at a "Sermons from Science" presentation in Corvallis, and the two had since corresponded from time to time.[13]

Not only was Moon rhetorically pursuasive—Everest once quipped, "That man could sell a pogo stick to a legless man"—but in Houghton and Everest, Moon had found an audience of energetic, like-minded allies. Houghton not only wished to prevail against the disturbing prospect of college students losing their faith, but also presumed that a group of Christian men of science could serve as instructors and consultants for scientifically untrained pastors, Sunday-school teachers, and Bible-institute faculty. Everest, who had read widely if uncritically, had been impressed by the ideas of such creationists as George McCready Price, Harry Rimmer, and especially the physician Arthur I. Brown. This naturally disposed him favorably toward the notion of taking some positive steps in the form of a society of Christian scientists to counter the effects of science in the hands of non-Christians. His correspondence with Moon following a November 1940 meeting reveals genuine enthusiasm for what the two then referred to as "the plan."[14]

After almost a year of correspondence regarding "the plan" for a "society," a final strategy for an organizational meeting materialized. Henry Parsons Crowell, a wealthy industrialist, founder of the Crowell Trust, patron of Moody Bible Institute, and father of Moody vice-president Coleman Crowell, had agreed to donate the funds required to bring together a group of evangelical scientists to lay the groundwork for a new organization. In June 1941, several conservative evangelicals with scientific training received a letter from Will Houghton inviting them to attend an early-September meeting at Moody Bible Institute in Chicago. Houghton presented his conceived purpose for the meeting as follows:

> Some of us believe the time has come for a meeting of science teachers who are Christians and who feel that some

> scientific facts are not having proper recognition, while some hypotheses are being presented as laboratory truth. It might be that eventually an organization will come into being. . . . Our thought is of an annual meeting, which could be a kind of clearing house for ideas. . . .
>
> There are two special things that such an annual conference might do, namely, help ministers and evangelists who are dealing with scientific subjects, and make available certain phases of truth. . . . There might also develop from this a handbook for college students which would forewarn them as to what they might expect in some textbooks and classrooms, and would inform them of further facts to be considered.[15]

In addition to Everest four men responded to Houghton's invitation: John P. Van Haitsma (1884–1965), who with his Ph.D. from the University of Michigan served as head of the "organic science" department at the Christian Reformed Calvin College in Grand Rapids; Peter W. Stoner (1888–1980), who had completed all but his dissertation for a doctorate from the University of California, Berkeley, taught mathematics and astronomy at Pasadena City College, and attended Pasadena's Lake Avenue Congregational Church; Russell D. Sturgis (1897–1969), who served as deacon of the First Baptist Church of Norristown, had earned a Ph.D. from the University of Pennsylvania in physical chemistry, and had served as head of the chemistry department at Ursinus College since 1928; and Irving A. Cowperthwaite (b. 1904), another Baptist, who completed a Ph.D. in physical chemistry at Columbia University in 1931 and then remained on the faculty at Columbia until 1937, when he left academe to become chief engineer and metallurgist at Boston's Thompson Wire Company.[16] Although Houghton and Moon stood by as enthusiastic observers, from the outset the job of designing and implementing plans for the new society belonged entirely to the founding five—with Everest as the point man and chief motivator. By the close of their organizational meeting, the founders had explicated their mission by listing ten objectives the new association of evangelical scientists would pursue.[17]

The days of September 1941 brimmed with activity for conservative American evangelicals. While the ASA founders convened their Chicago organizational meeting, the militant fundamentalist Carl McIntire (b. 1906) was spearheading his opposition to the mainline Federal Council of Churches by founding his American Council of Churches. But unlike McIntire's group whose *raison d'etre* embodied the negative emphasis of an attack upon liberals, most members of the American Scientific Affiliation envisioned a positive role for their new circle. The following spring a conference of evangelicals met in St. Louis and formed the National Association of Evangelicals (NAE). This event demonstrated that ASA founders were ushering conservative Christian scientists into a national arena of lively evangelical activity. Joel Carpenter has identified the founding of the NAE as a sign that fundamentalists were forging a "new evangelical coalition." The ASA, like the NAE "a fundamentalist-initiated but genuinely inclusive fellowship," was also emerging as a component of that new coalition.[18] But how inclusive would the ASA be? Although the founders had put together a fairly specific roster of plans, some key assumptions remained implicit and unofficial, notably the group's views on creation, evolution, Genesis, and geology. Was this an oversight or part of an agenda to protect other interests? The answer to this question was not always clear, even in the minds of early ASA members. Their views were evolving.

Having been elected chairman at the organizational meeting, Everest went to work vigorously. The eve of the Japanese attack on Pearl Harbor found the young engineer laboring over his manual typewriter pecking out the penultimate draft of his six-page manuscript, "WHAT IS *The American Affiliation of Scientists?*" After circulating among the other four founders, Everest's edited essay appeared in June 1942 as part of the nascent science and religion association's first promotional brochure. It carried the title "The Story of the American Scientific Affiliation" and was predictably brief. Telling the "story" of a nine-month-old organization that could claim only five full members required little ink, especially if there were things the group wanted to remain unspecified. Still, there was no want for juicy rhetoric. If successful, the essay optimistically noted, the American Scientific Affiliation would be the vanguard of "the most important movement since the ~~renaissance~~ reformation [sic]."[19]

A sane vision of history cries for one to dismiss immediately as naive megalomania the notion that a half-dozen excited fundamentalists could do anything on par with the Renaissance or the Reformation. Yet the American Scientific Affiliation did gradually inaugurate departures from an old worn path and slowly breathe new life into discussions of the age-old questions. In this superficial sense, the comparison with the Renaissance and Reformation is not entirely off the mark. Arguably this promotional brochure, which appears as the first entry in this volume of ASA publications, stands as a symbolic publication in the history of the American evangelical encounter with modern science. This is not because it is especially well written or critically informed. It is neither. But it does reveal signs of the old and offers a harbinger of things yet to come. On one hand, it bears distinctive marks of twentieth-century evangelicalism's fundamentalist heritage. It also reveals the primitive Baconian vision of the scientific enterprise that animated conservative Protestants through the World War II era. It disparages scientific ideas that, after all, "are only theories," and it reminds that "there is no disagreement between the statements of the Bible and the true facts of science."[20] On the other hand, the brochure foreshadows what would become in the post-War era a renewed endeavor among evangelicals to take science and the life of the mind seriously and to eschew a simplistic retreat into provincialism and intellectual mediocrity.[21]

It was this tension—between the Baconian biblicism of their fundamentalist heritage and their desire to meet the standards of a scientific community that was revered for its objectivity—that manifested itself best in the early ASA's treatment of evolution. The founders were no fans of Darwinism. Each rejected evolution. Cowperthwaite later explained, "All five founders were thoroughgoing anti-evolutionists in outlook."[22] In Everest's original draft of "WHAT IS *The American Affiliation of Scientists?*" he wrote, sounding remarkably like William Jennings Bryan, "[M]any [college students] are also known whose faith in Christ and belief in the Bible has slipped away, due to campus attitudes, many traceable directly to the door of evolutionary teaching. This representation of an unproven theory as an established fact, though it has the approval of a great host of scientists, is popular because it discounts the hand of God in Man's affairs."[23] The following spring John Van Haitsma wrote to the other four founders endorsing one of George McCready Price's books. He explained, "I mention it because it is in line with the aim of our society."[24] Clearly, the founders thought of themselves and of their organization as antievolutionist.

Nevertheless, these personal antievolutionary dispositions of the founders failed to translate into the construction of an antievolutionary organization. Everest's Bryanesque passage was edited out of the published version. When, during the summer and fall of 1942, Everest found himself engaged in correspondence with Arthur Pierson Kelley, a prospective member who had failed in his own attempt to launch an antievolutionary organization, Everest explained,

> Of course we are in the "anti-evolution" fight in a way, but we do not intend to get ourselves discredited at the outset by waving red flags before the eyes of scientists just to see the sparks fly. We are firmly convinced that our group will be a

> more powerful tool in the hands of the Lord if we maintain our standards so high that the confidence of those in the other camp will be compelled. For this reason we have chosen not to adopt deluge geology, anti-evolutionism, or anything else as the basis of our group.[25]

Later the same week Everest wrote to the members of the executive council (as the founders called themselves) reporting on his exchange with Kelley: "[Kelley's] group was strictly 'anti-evolution' so he was not only riding a hobby, but was *against* something too—a lethal combination."[26] Herein lies a fascinating part of the story. From a group of ardent antievolutionists emerged an organization that always refused to assume an official antievolutionary position. If this was not an oxymoronic state of affairs, it was at least confusing to some prospective members.

Walter E. Lammerts (b. 1904), a member of Mt. Olive Lutheran Church in Pasadena, put his genetics Ph.D. from the University of California to work as a plant breeder and horticultural consultant.[27] When in January 1942 several ASA members gathered for a weekend meeting at the southern California homes of Irwin Moon and Peter Stoner, Lammerts "was there for the purpose of becoming familiar with the aims and methods of the group."[28] Although Lammerts did not officially join the ASA until August 1946, he remained in touch with its leaders, eventually contributing works to ASA publications (four of which appear in this volume). Lammerts was a Missouri-Synod Lutheran, a follower of George McCready Price, and an ardent antievolutionist. Not surprisingly, he believed he was joining an antievolutionary association. He later wrote that when originally approached by Everest about joining, the ASA chairman had "assured [him] that the stance of the society would be anti-evolutionary."[29] Given that the *personal* positions of Everest and the other founders *was* antievolutionary, one can understand Lammerts's belief that their group would be also. But he was mistaken. A look at the first ASA constitution provides a clue to the reason for this.

By May of 1942 the executive council had finished drawing up the organization's first constitution. If this document did little to delineate a specific evangelical position on how to relate science to scripture, it did even less to specify a stance on evolution. This was, in part, because the founders did not know where publicly to stand on the evolution issue. Everest agreed with Van Haitsma on the wisdom of "leaving evolution until we know what we ourselves believe, as a group, and want to stand for."[30] The founders wanted the evolution issue to be shelved so they could "present a solid, well thought out front" to outside observers.[31] Consequently, the only specified constitutional objectives were "(1) To integrate and organize the efforts of many individuals desiring to correlate the facts of science and the Holy Scriptures. (2) To promote and encourage the study of the relationship between the facts of science and the Holy Scriptures. (3) To promote the dissemination of the results of such studies."[32]

Nothing in the constitution indicated antievolutionism. Why? There were three reasons. First, as indicated, the founders were ambivalent about taking a public position of any sort on the issue. Second, and related, was what Everest later pointed out to Arthur Kelley and to the council, that a negative approach—to be "anti-" anything—was a serious liability for any organization. Third, and most important, all the founders were Bible-believing evangelicals and they insisted that all new members would be as well.[33] Each affirmed the statement, "God is the Author of this Book [Bible], as well as the Creator and Sustainer of the physical world about us, I cannot conceive of discrepancies between statements in the Bible and the real facts of science."[34] In short, they were all creationists. Although to be a fundamentalist creationist at this time certainly did not require assent to any particular scheme of harmonizing science with scripture, it always meant at least one thing: opposing evolution.[35] And to affirm the ASA Creed meant that one was a creationist. To conceive of evolutionary creationism (or theistic evolution) required a greater imaginative stretch than could command the founders' worries.[36] Why should they sacrifice their positive approach, saddle themselves with an "anti-" label, only to make explicit what seemed obvious? There was no reason, especially since "theistic evolution," from their perspective, was a contradiction in terms. As evangelical creationists, then, whatever ways of correlating God's words and works they might encounter, evolution was not figured to be among the options. They could conceive of and fear internal wrangling and dissention, perhaps, but not that some member would turn up who endorsed evolution. That would run counter to a direct (but unstated and *assumed)* corollary of their creationism. Yet this was the critical assumption. *They* could not conceive of any Bible-believing creationist affirming evolution. This comforted them in the belief that their society would be positive in approach while attracting antievolutionist members. But that the founders and early members had difficulty imagining an evangelical Bible-believing evolutionist was no guarantee that such a species of creationist would not someday evolve in their midst.

Their assumption explains how from the outset the founders of the ASA, despite their personal antievolutionism, constructed a society that could, at least in theory, welcome an evolutionist (if one who affirmed their creed ever appeared). This was something they did not worry much about happening; in part because they did not think it would, and in part because, if it did, they wanted to be open to investigating what such a person would be like. Maybe there was more to God's ways than they had dreamt of in their philosophies. The reality of the matter, of course, depended not upon the founders' assumptions, but upon how future ASA members would approach that ominous task of "correlating the facts of science with the Holy Scriptures." As ASA history and the selections in this volume show, it would be done in many ways. And evolution would figure among them.

The founders' assumption that creation and antievolutionism were synonymous also shows how little history they knew. Many late nineteenth-century evangelicals had wrestled hard to come to terms with Darwin.[37] By the time the ASA came about, the founders' fundamentalist heritage had erased any collective memory of nineteenth-century evangelicals who had struggled to make peace with evolution. Names like Asa Gray, George Frederick Wright, and James Dwight Dana meant little or nothing to the ASA founders. In this respect, the story of mid-twentieth-century evangelical scientists' discoveries of ways to accommodate evolution would be a case of what Yogi Berra called "Déjà vu all over again." Of course, history never really repeats itself. But the ASA did have to learn nineteenth-century lessons in the context of twentieth-century science.[38]

Another irony for a group wishing to be positive in outlook was the inherent negativism associated with ferreting out the errors of zealous, but scientifically ignorant, Christian apologists. Among the ten objectives outlined at the organizational meeting was "The reviewing of written material to eliminate scientific errors."[39] This was one of Peter Stoner's hobby horses. His writings routinely echoed his claim that "the literature of the Church on the scientific accuracy of the Bible is in a pitiable state."[40] Everest, ever wanting to be positive in approach, hoped the ASA could fulfill this objective by developing "a 'stamp of approval' for publishers' use in the flyleaf of approved books and pamphlets. . . . I keep thinking of the 'Good Housekeeping Institute Seal of Approval,'" wrote Everest to Cowperthwaite, "but I hope ours means more."[41] The trouble was finding works meriting a positive endorsement.

Stoner, second only to Everest among the founders in his participatory zeal for the new group, agreed. For him correlating science with scripture did not mean defending six-day creationism, flood geology, or any other position that necessitated recasting scientific data into a mold fashioned by biblical hyperliteralism. Rather, if students were losing their faith in the college science classroom, it must be the result of having been reared in churches and Sunday schools tainted by scientifically illiterate pastors and teachers. While still a young graduate student at Berkeley during the 1909–1910 academic year, Stoner had acquired what would remain a life-long conviction that God had chosen him for "a tremendous job." That was the "job of correcting the great errors that existed in the Church's literature which contradicted the sciences, got the student into trouble, and caused thousands of them who had become Christians in the church to lose their faith when they went to college." Stoner had long believed that "God would produce men" of a like mind with enthusiasm for this labor; he came to see the ASA as that divinely ordained group of men.[42]

This hope that the ASA would make strides in the direction of scientific apologetics did not prevent Stoner from engaging in that work on his own. Stoner had keen interest in demonstrating the inspiration of scriptures, in countering errors that undermine the faith of young Christians, and in working out an accommodation between science and scripture. Many of his writings never saw publication.[43] But a few did. These included occasional articles for the fundamentalist magazines *Moody Monthly* and *Sunday School Times*. Most notable, however, was his eclectic science-and-religion monograph, *From Science to Souls,* first published by Moody Press in 1944 and reissued in 1952 by VanKampen Press of Wheaton, Illinois, under a new title, *Science Speaks.* In this edition some 350,000 copies came off the presses.[44] The introduction, preface, and first chapter of *From Science to Souls* are reproduced in this volume. This selection is important for several reasons. First, the ASA wanted to show its positive emphasis by endorsing reputable publications. Although it found so few meriting approval that the ASA eventually took to writing its own material, Stoner's quaint little volume earned the ASA endorsement in the form of Irving Cowperthwaite's introduction. This glimpse of what the ASA founders believed worthy of endorsing gives a quantitative and qualitative picture of the ideas current in the early 1940s. In the selection, Stoner reviews various concordist schemes for Genesis including the day-age theory, the intermittent day theory, the gap theory, and the canopy theory of the flood. Finding each wanting, he advances his own view based on the premise that the earth came from a "dark nebula." Always apparent is Stoner's motivating faith in the progress of science to vindicate the accuracy of scripture. Along the way he employs probability to demonstrate the inspiration of Genesis, one of his favorite exercises. He also gets in a jab at those Christians who teach what he considers scientifically indefensible doctrines. Despite his eagerness to bring Christianity into line with modern science, about evolution he concludes, "All theories of evolution must take the creative acts of God into account if they are to agree with all of the established facts."[45]

Another piece by Stoner reproduced in this volume is his chapter "Astronomy and the First Chapter of Genesis" from the ASA's "Christian Student's Science Symposium." Because a concern of Moon, Houghton, and all the founders had been the preservation of college students' faith, the ASA had planned from the outset to produce a "Handbook For College Students."[46] After slow but steady organizational efforts, hampered by the restrictions of World War II, the ASA had grown by 1948 to seventy-three members. In that year, having never lost sight of its initial goal, the ASA proudly announced the publication by Van Kampen Press of the first 3,000 copies of *Modern Science and Christian Faith: Eleven Essays on the Relationship of the Bible to Modern Science.* The dust jacket boasted, "To our knowledge, this is the first successful attempt to present a concise, scientifically accurate treatment of the relationship of the Bible to the many fields of modern science in a readable and understandable form in a single volume." The ASA advertised the book as "a 289–page Christian Classic that took twelve men five years to produce."[47] Indeed, a lion's share of the ASA's effort during the 1940s had gone into production of the volume. Editor Everest, with typical zeal for a task, circulated to the membership an occasional "Handbook News Letter." He subjected each chapter to the scrutiny of multiple readers.[48] The founders had convinced themselves that the stakes were high. As Moon explained in 1945, "I'm convinced that if [the book] is released without the most careful handling and explanation and before conservative Christians are sold on the purpose of the ASA, it will cause an explosion that will make the battle of Iwo Jima look like a friendly get-together."[49] Despite such fears, the book was well received. A second printing produced 2,000 copies; and in 1950 a second "enlarged" edition appeared.[50] Everest's retrospective assessment of the book is appropriate: "The enthusiasm and dedication of the early ASA members were fully matched by their inexperience. It must also be remembered that they were products of a less sophisticated early 20th century era. Viewed in the light of today, *Modern Science and Christian Faith* is indeed quaint and inadequate."[51] This volume reproduces three essays from the book, chosen because, more than the others, these chapters treat topics pertinent to creation and evolution. In addition to Stoner's concordist scheme for astronomy and Genesis are Edwin Gedney's "Geology and the Bible" and "Biology and Creation," jointly authored by William Tinkle (1892–1981) and Walter Lammerts.

Edwin Gedney (1904–1980), with master's degrees from Brown and Harvard, was an Advent Christian on the faculty of Gordon College of Missions. His chapter defends his "day-age" harmonization of Genesis and geology. It is also important for what Gedney omitted. The only reference to the flood geology of Price is a dismissive footnote inserted by Everest in the interest of "fairness."[52] Ironically, the next chapter was authored by two of Price's friends and disciples, Tinkle and Lammerts, who would be among the original ten founders of the Creation Research Society. Tinkle, head of the biology department at Taylor University, had a Ph.D. from Ohio State and had authored occasional articles in *Moody Monthly,* as well as a creationist text, *The Fundamentals of Zoology.* This chapter presents their refutation of evolution with a standard antievolutionist conclusion that "the Biblical account of the creation is corroborated, and that God finished His creation, just as Genesis states."[53] That the ASA had not reached unanimity on the correlation of science and scripture is clear from this volume. Nevertheless, the authors all possessed an unfeigned skepticism for evolution. If being an ASA creationist could mean many things, it still ruled out any endorsement of evolution. Before *that* could happen, ASA members would have to accept the earth's antiquity and dismiss flood geology.

This was accomplished largely through the efforts of J. Laurence Kulp (b. 1921), a young Wheaton College alumnus who joined the ASA in 1945, while completing a Ph. D. in physical chemistry at Princeton. Kulp went on to complete all but the dissertation for a second doctorate, in geology, from Columbia. Along the way he worked on the Manhattan Project, completed pioneering work in radiometric dating, and published widely in professional scientific and popular Christian periodicals.[54] Although Kulp had not contributed an essay to *Modern Science and Christian Faith,* he had assisted Everest in the editing process. His first ASA publication was the fruit of a seminar held in early fall of 1947 at the Pasadena home of Peter Stoner on "The Age of the Earth by Radioactive Methods." The papers given at this meeting were gathered and edited by Kulp and appeared the following year as *A Symposium on "The Age of the Earth."* This booklet is reproduced in its entirety here. With the exception of the anomalous paper by Lammerts, the volume offered a confident endorsement of radiometric dating and the prospects for mainstream geology.

Having broken the ice, Kulp was poised to carry out his "mission to educate the fundamentalist world" in the ways of orthodox geology. He also would be instrumental in engineering what he called "the necessary destruction of flood geology."[55] He realized these goals partially by writing and speaking and partially by recruiting bright young Christians (sometimes Wheaton alumni) to pursue graduate training and then write themselves. By 1952 Kulp had offered six contributions to ASA conventions, including his crucial refutation of "Deluge Geology" (which appears in this volume) at the 1949 meeting in Los Angeles.[56] This volume reproduces selected *Journal* articles by some of Kulp's like-minded ASA colleagues: Cordelia Erdman (Wheaton College geologist), Delbert Eggenberger (research physicist for Armour and Co.), and Herbert Feely (Upsala College geologist).[57] By 1951 Kulp had convinced most ASA members that he was a reliable authority. Even the distinguished J. O. Buswell, Jr., president of Shelton College (whose "Creation Days" argument for a figurative understanding of the Genesis days appears in this volume), spoke up during a discussion session with the comment, "I'm going to look up Larry Kulp and I'm going to get him to teach me some geology."[58] Kulp willingly played the role of evangelical geology instructor. Most ASA members learned his lessons willingly. Among the notable exceptions (whose *Journal* articles are reprinted in this volume) were Henry Morris, William Tinkle, and Walter Lammerts. The single item by Morris reproduced in this volume is a one-page article entitled "Comment on the 'Deluge Geology' Paper of J. L. Kulp." It appeared anonymously in the June 1950 issue of the ASA *Journal*. His conclusion that Kulp had been overly influenced by "the orthodox geological viewpoint" did little to convince the ASA members to reconsider the merit of Kulp's arguments.[59] By the time the organization celebrated its tenth anniversary, many ASA members were increasingly comfortable with the conclusions of orthodox geology. Attitudes toward evolutionary theory were beginning to undergo a gradualist transmutation as well.

Although the founders had temporarily shelved evolution at the outset, they knew the issue would not go away. Thus Everest sponsored an evening "Round Table Discussion" at the 1948 annual convention on the subject, "What Shall Be the Role of ASA in the Study of Evolution?" Earlier the same day Edwin Monsma (Calvin College professor of organic science), Russell Mixter (Wheaton College professor of biology), and theologian Bernard Ramm had contributed papers on evolution. Each is reproduced in this volume.[60] Mixter, perhaps the most liberal of the lot, rejected macroevolution as a product of the imagination, and Monsma was even more rigid in his opposition to evolution. Throughout the discussions that followed each paper (reprinted in this volume), Kulp played the role of geology professor attempting to get a hearing for a more mainstream position. The day's deliberations proved inconclusive, beyond demonstrating that the ASA was not yet prepared to climb on the evolution bandwagon.

In preparation for the 1951 annual convention Alton Everest prepared a banquet address, which, in edited form, appeared in the *Journal* (and is included in this volume) as "The American Scientific Affiliation—The First Decade." Another obligation prohibited Everest from attending the meeting, so eleven days before the conference he dashed off a hastily scribbled letter to James O. Buswell III, an ASA member and anthropologist at New York's Shelton College, the site of the convention.[61] Everest asked Buswell to have Cowperthwaite read his address and called to Buswell's attention to the lengthy quotation in his paper from geologist Harley Barnes (see p. 36 of ASA *Journal* article). In Everest's original draft, the passage, which Everest believed "to be a fair expression of the prevailing attitude of the majority," included a statement that the Bible does not necessarily rule out "God's use of evolution as a method of creation." Everest solicited reaction to this statement from Buswell and other ASA leaders. Buswell interpreted the statement as an endorsement of theistic evolution. Quite certain that nobody in the ASA adhered to such a view, he urged Cowperthwaite to delete the passage. It was stricken. What did remain, however, was Barnes's statement that "the Bible does not give unequivocal grounds for being anti-evolutionary."[62] So, after ten years, ASA creationism had drifted away from the founders' antievolutionism, but it had not changed enough to embrace theistic evolution. That would happen in the second decade.

The 1950s were a decade of tremendous change within the ASA. Controversies over factions and constitutional revision, membership growth from 134 in 1950 to nearly 800 by 1960, and the appearance of two ASA monographs, *Creation and Evolution* (1950) by Russell Mixter and *The Eye as an Optical Instrument* (1952) by Frank Allen (1874–1965), constituted just a few highlights.[63] Although already in 1951 members hailed from forty-one states, most resided in California or in a state bordering one of the Great Lakes. With over two dozen denominations represented, the preponderance was Mennonite, Baptist, Presbyterian, or "Undenominational."[64]

The two monographs that gained widest circulation during the fifties are reprinted here. Allen's appeared originally in 1949 as an ASA *Journal* article and is reprinted in that original form. It provides a glimpse of the ASA's early implicit endorsement of the design argument for God's existence. Allen, a Canadian with a 1902 Ph.D. from Cornell, also contributed a chapter to *Modern Science and Christian Faith.*[65]

Russell Mixter (b. 1906) was a 1928 Wheaton College graduate and Wheaton faculty member with a master's degree in zoology from Michigan State and a Ph.D. in anatomy from the University of Illinois. He sat on the ASA executive council from 1949 through 1954, serving as president the last four years.[66] Mixter's monograph was actually a compilation of papers he had written in the late forties. Its original intent had been to present a critical response to A. F. Shull's influential textbook, *Evolution.* Convinced that "people who place their faith in Divine Creation as the explanation of the development of life . . . have as much justification for their belief as do [atheistic evolutionists for theirs]," Mixter remained unwilling to make a case for macroevolution. Still, he determined that "the flood was not world wide and may be disregarded in a study of animal distribution." Furthermore, he considered likely the evolutionary development of organisms "within the order." Not surprisingly, his conclusion that "a creationist may believe in the origin of species at different times, separated by millions of years, and in places continents apart," was bothersome to some within ASA's antievolutionary contingent. As had been the case with Kulp, a response to Mixter's monograph appeared anonymously in the March 1954 ASA *Journal*. Ironically, he was taken to task for not being more thoroughgoing in his evolutionism. Mixter justified his moderate approach explaining, "I was trying to convert some creationists to a limited form of evolution." Despite this forthright admission, the ASA was not quite ready to come unglued over the evolution question. Instead, Mixter's views resonated within a forum where divergence from anti-evolutionism was increasingly tolerated.[67]

The second decade of ASA history witnessed a gradual evolution of evangelical attitudes toward evolution.[68] This transmutation was punctuated by at least four noteworthy events: first, the twelfth annual convention, held in 1957 on the campus of Gordon College and Divinity School; second, the thirteenth annual convention, held in 1958 on the campus of Iowa State College; third, the appearance in 1959 of the ASA Darwinian centennial volume, *Evolution and Christian Thought Today,* edited by Mixter; and fourth, a symposium on the topic "Origins and Christian Thought

Today," held on the campus of Wheaton College in early 1961 and featuring several contributors to the 1959 Mixter volume.

Irving Cowperthwaite observed in 1960 that the 1957 and 1958 annual meetings were benchmarks of a sort: "There appeared to be a growing conviction that inexorable pressure of expanding knowledge is about to force us to accept some formulation of the theory of evolution, including the evolutionary origin of man, and that we must adjust our thinking in accordance with this eventuality."[69] This "inexorable pressure" was felt at the 1957 meeting by those attending such sessions as the "Symposium on Mechanistic Views of the Origin of Life." Here Walter R. Hearn (b. 1926), an energetic young biochemist who would play increasingly prominent roles in the ASA, delivered a paper on "The Formation of Living Organisms from Non-Living Systems" (reprinted in this volume). At the same meeting Mixter presented a paper, "Stepwise Creation" (reprinted here as "An Evaluation of the Fossil Record"), which precipitated a discussion during which the term "theistic evolution" received its first, albeit tentative, endorsement at an ASA meeting.[70]

The following annual meeting provided a forum for similar discourse with a first-day panel discussion of "What is a Kind?—The Species Problem," chaired by J. Frank Cassel (b. 1916), a Wheaton alumnus on the biology faculty of North Dakota Agricultural College. The symposium, based on biologist Ernst Mayr's AAAS symposium, opened with a presentation by Cassel (reprinted in this volume).[71] Clearly, evolution was in the air at these meetings. Not everyone felt comfortable breathing deeply, however. This volume reprints, as an example of that fact, the 1958 convention contribution, "The Genesis of Kinds in Our Modern World," by the conservative Seventh-day Adventist biologist Frank L. Marsh (1899–1992).

In the following year appeared the ASA Darwinian centennial volume, *Evolution and Christian Thought Today,* which *Eternity Magazine* judged to be "The Most Significant Book of the Year." Over a decade in the plans, the book had originated in a suggestion from evangelical theologian Carl F. H. Henry that the ASA produce a "closely-knit refutation of evolution" for publication in 1959. The book read more like an endorsement of evolution than a refutation, however. The chapter on "The Origin of Life" by Walter Hearn and his student Richard Hendry was especially significant. Affirming the likelihood that "a complete metabolic machine may have appeared only after long periods of 'chemical evolution,'" they concluded that, "the expressions in Scripture regarding the creation of life [are] sufficiently figurative to imply little or no limitation on possible mechanism." It seemed that within the ASA evolution had evolved from being "a dirty word" into something like an official position. Thus Frank Cassel observed in his 1959 article, "The Evolution of Evangelical Thinking on Evolution" (reprinted here), that "in fifteen years we have seen develop within A. S. A. a spectrum of belief in evolution that would have shocked all of us at the inception of our organization."[72]

Some became worried that the ASA had taken an official position on the evolution issue or, at least, that certain members were trying to give the impression that it had.[73] That it had not was a matter of practice and policy. Various ASA *Journal* articles reprinted here do not endorse an evolutionary position.[74] Rather they demonstrate what one member later called the "fuzziness" of a group of people with an "exploring attitude."[75] Whether everyone possessed an attitude of openness is easily disputed, but the ASA *Journal* editor, Delbert Eggenberger, did his best to keep the ASA periodical an open forum. Circumstances compelled him to clarify this policy on a few occasions: "It is emphasized that there is <u>not</u> a uniform or official ASA interpretation. . . . To publish only articles of a particular theological system would defeat the very purpose for which the A.S.A. was founded." Later he wrote, "It would be easy to establish a 'party line' . . . [but] the A.S.A. has a purpose . . . of open-minded study that precludes such restrictions."[76]

Gradually, however, many of the more conservative antievolutionists came to feel that the ASA was emerging as a forum for vocal liberals who were open to evolution. The Wheaton College science symposium of 1961, "Origins and Christian Thought Today," was held, perhaps symbolically, the same month that Whitcomb and Morris's *The Genesis Flood* appeared in print. This volume reprints the presentations from that symposium (which the ASA *Journal* published) by Hearn, Cassel, and Buswell III. The Wheaton symposium triggered an avalanche of controversy for the fundamentalist college over the question "Is Evolution at Wheaton College?"[77] No such fireworks would have been unleashed had the symposium featured Whitcomb and Morris rather than liberal ASA members.

When Irving Cowperthwaite observed in 1960 that current trends might require ASA members to "adjust our thinking in accordance with" some concept of evolution, he was not merely prophesying for the American Scientific Affiliation. He was also proving how close history actually can come to repeating itself. In his account of Protestant intellectuals and organic evolution in the last half of the nineteenth century, historian Jon H. Roberts explains that by 1875 "the scientific community had defected to the evolutionists' camp." Upon perceiving this move, notes Roberts, American Protestants became convinced "during the last quarter of the nineteenth century that it was necessary to accept the theory and to accommodate Christian theology to its dictates."[78] Historically attuned evangelical scientists (had there been any) reading Cowperthwaite's 1960 conclusion would have quivered with a feeling of déjà vu. He wrote, "If evolution has become established beyond a reasonable doubt, we must accept it even though it means that we have to make an 'agonizing reappraisal' of our interpretation of Scripture."[79] Some ASA members were willing to follow their nineteenth-century forbearers and make that "agonizing reappraisal." Others were not. Of those who were not, some fled the ASA to form the Creation Research Society in 1963, but some stayed with the ASA. That some of these antievolutionists stayed with the ASA is proof that the organization had no official position. But it was equally clear that the antievolutionism of the founders had become an endangered species within this organization of evangelical scientists.

## Notes

1 F. Alton Everest to Irwin Moon, 29 March 1941, American Scientific Affiliation Collection (SC-26), Box 5, File 4, and accompanying "Proposed Constitution of the Proposed *Society for the Correlation of Science and the Bible,*" Box 1, File 3, Wheaton College Archives and Special Collections, Buswell Memorial Library, Wheaton College (hereafter ASA Papers).

2 For accounts of the Religion and Science Association and the Deluge Geology Society, see Ronald L. Numbers, *The Creationists* (New York: Alfred A. Knopf, 1992), pp. 102–157.

3 This is best illustrated by the appearance in 1986 of *Teaching Science in a Climate of Controversy: A View from the American Scientific Affiliation,* which stirred up considerable criticism from both creationists and evolutionists. By 1993, with over 100,000 copies in print, the booklet was released in its fourth revised printing and included "classroom exercises."

4 Determining which items to include in this volume was not easy. From 1949 through 1961 the *Journal of the American Scientific Affiliation* alone published over one hundred selections treating creation or evolution. This, of course, was not the only source of

material for consideration. Entries were chosen if they possessed special historical significance or if they helped to provide a cross-section of the type of material published by ASA members. Naturally, one could have justified including some items which do not appear here.

5 For the purposes of this discussion, the terms "fundamentalist" and "evangelical" are used largely interchangeably in reference to the nebulous transdenominational body of conservative American Protestants who during the period from the mid 1920s to the mid-1950s stressed evangelism and missions and adhered to such beliefs as "(1) The deity of our Lord Jesus Christ. (2) The substitutionary and sacrificial death of Jesus Christ for our sins and the historical fact of His bodily resurrection. (3) The presence and power of the Holy Spirit in the work of regeneration. (4) The unique inspiration, integrity, and authority of the Bible, as the Word of God. (5) The consummation of the Kingdom in the 'glorious appearing of the great God and our Saviour Jesus Christ.'" (Quotation from ASA Constitution Adopted August 1950, Article II—DOCTRINAL POSITION, Box 1, File 4, ASA Papers). From the mid 1940s onward the term "fundamentalist" increasingly referred to the militant, separatist, right wing of the evangelical community.

For helpful discussions of the defining nuances, see George M. Marsden, "Introduction: The Evangelical Denomination," in *Evangelicalism and Modern America*, ed. George Marsden (Grand Rapids, MI: William B. Eerdmans Publishing Co., 1984), pp. vii–xix; George M. Marsden, *Understanding Fundamentalism and Evangelicalism* (Grand Rapids, MI: William B. Eerdmans Publishing Co., 1991); Donald W. Dayton and Robert K. Johnston, eds. *The Variety of American Evangelicalism* (Downers Grove, IL: InterVarsity Press, 1991); Timothy Smith, "The Evangelical Kaleidoscope and the Call to Christian Unity," *Christian Scholar's Review* 15 (1986): 125–140; Leonard I. Sweet, *The Evangelical Tradition in America* (Macon, GA: Mercer University Press, 1984); George M. Marsden, *Fundamentalism and American Culture: The Shaping of Twentieth-Century Evangelicalism, 1870–1925* (New York: Oxford University Press, 1980); Ernest R. Sandeen, *The Roots of Fundamentalism: British and American Millenarianism, 1800–1930* (Chicago: University of Chicago Press, 1970); Ferenc Morton Szasz, *The Divided Mind of Protestant America, 1880–1930* (University: University of Alabama Press, 1982); Norman F. Furniss, *The Fundamentalist Controversy, 1918–1931* (New Haven: Yale University Press, 1954); and Willard B. Gatewood, ed., *Controversy in the Twenties: Fundamentalism, Modernism, and Evolution* (Nashville: Vanderbilt University Press, 1969).

6 Marsden, *Understanding Fundamentalism and Evangelicalism*, p. 148. On the Scopes trial, see Ferenc M. Szasz, "The Scopes Trial in Perspective," *Tennessee Historical Quarterly* 30 (Fall, 1971):

288–98; Szasz, *The Divided Mind of Protestant America,* pp. 117–25; and Judith V. Grabiner and Peter D. Miller, "Effects of the Scopes Trial: Was It a Victory for Evolutionists?" *Science* 185 (1974): 832–37.

7 See the following by Joel Carpenter: "Fundamentalist Institutions and the Rise of Evangelical Protestantism, 1929–1942," *Church History* 49 (1980): 62–75; "From Fundamentalism to the New Evangelical Coalition," in *Evangelicalism and Modern America,* George Marsden, ed.; and "The Renewal of American Fundamentalism, 1930–1945," Ph.D. diss., Johns Hopkins University, 1984. Also of significance for interpreting the state of evangelicals in this period is George Marsden, *Reforming Fundamentalism: Fuller Seminary and the New Evangelicalism* (Grand Rapids, MI: William B. Eerdmans Publishing Co., 1987). On Moody Bible Institute, see Virginia Lieson Brereton, *Training God's Army: The American Bible School, 1880–1940* (Bloomington: Indiana University Press, 1990), especially Chapter 10, "The Bible School and the Fundamentalist Movement"; and Gene A. Getz, *MBI: The Story of Moody Bible Institute* (Chicago: Moody Press, 1969).

8 Theodore Dwight Bozeman, *Protestants in an Age of Science: The Baconian Ideal and Antebellum American Religious Thought* (Chapel Hill: University of North Carolina Press, 1977), xv. See also Herbert Hovenkamp, *Science and Religion in America: 1800–1860* (University of Pennsylvania Press, 1978); and George Marsden, "The Evangelical Love Affair with Enlightenment Science," Chapter 5 in *Understanding Fundamentalism and Evangelicalism.*

9 See the section under the heading "The 'Real Facts of Science'" in Mark A. Kalthoff, "The Harmonious Dissonance of Evangelical Scientists: Rhetoric and Reality in the Early Decades of the American Scientific Affiliation," *Perspectives on Science and Christian Faith* 43 (December 1991): 260–62.

10 For example, much has been made of the fact that at the Scopes trial, William Jennings Bryan could name only the deceased G. F. Wright and the Adventist flood geologist G. M. Price as proponents of fundamentalist views. See Ronald L. Numbers, "The Creationists," in *God and Nature: Historical Essays on the Encounter between Christianity and Science,* ed. David C. Lindberg and Ronald L. Numbers (Berkeley and Los Angeles: University of California Press, 1986), p. 402.

11 Carpenter, "Fundamentalist Institutions and the Rise of Evangelical Protestantism," p. 68.

12 As "Sermons from Science" grew, Moon was increasingly booked by the United Service Organization and his message reached thousands. His addresses, which included presentations on sound, light, animals, the telescope, the microscope, etc., usually concluded when "he stands on a huge transformer and allows 1,000,000 volts of electricity to flash through his body." (Quoted from "'1,000,000-Volt' Sermons Daily by Irwin Moon," *Atlanta Journal,* 11 October

1941.) See also F. Alton Everest, "Chronology of Sermons From Science," 27 February 1963, Irwin Moon Papers, Moody Bible Institute Archives, Chicago, Illinois; Anon., "Biographical Material on DR. IRWIN A. MOON," Irwin Moon Papers; and F. Alton Everest, "The American Scientific Affiliation: Its Growth and Early Development," pp. 29–32, Box 20, File 14, ASA Papers.

13 "Events and Circumstances Leading up to the Founding of ASA," unsigned manuscript, n.d., ASA Papers (wistful dream); J. W. Haas, Jr., "Irwin A. Moon, Everest and Will H. Houghton: Early Links Between the Moody Bible Institute and the American Scientific Affiliation," *Perspectives on Science and Christian Faith* 43 (December 1991): 249–58; Everest, "The American Scientific Affiliation," pp. 29–32, ASA Papers; on Houghton see Wilbur M. Smith, *A Watchman on the Wall: The Life Story of Will H. Houghton* (Grand Rapids, MI: William B. Eerdmans Publishing Co., 1951).

14 F. A. Everest to A. Salter, 12 January 1943, Box 5, File 26, (pogo stick); Will H. Houghton to F. A. Everest et al., June 1941; Irwin A. Moon to F. Alton Everest, 31 December 1940, Box 5, File 1; and F. A. Everest to Irwin A. Moon, January 1941, Box 5, File 2, all in ASA Papers. For material by and about Brown, Rimmer, and Price, see volumes 3, 6, and 7 of this series, *Creationism in Twentieth-Century America,* and Numbers, *The Creationists,* pp. 57–101.

15 For Moon-Everest correspondence, see ASA Papers. On H. P. Crowell, see Richard Ellsworth Day, *Breakfast Table Autocrat: The Life Story of Henry Parsons Crowell* (Chicago: Moody Press, 1946). Will Houghton's June 1941 letter to F. Alton Everest and other prospective founding members has been referred to as "the birth certificate" of the ASA. The unedited text of the letter appears in Smith, *Watchman on the Wall,* pp. 142–144, and in Everest, "The American Scientific Affiliation," pp. 12–15. Extensive archival investigation and interviews with ASA founders has failed to produce a master list of recipients of Houghton's letter.

16 Of the founders, Everest, Stoner, and Cowperthwaite would remain active ASA members. Van Haitsma revealed his enthusiasm for science and religion when he published in 1941 his argument that Genesis corroborates modern genetics, in a little book *The Supplanter Undeceived or Jacob's Divine Instruction in Heredity* (Grand Rapids, MI: Privately Printed, 1941). He also supervised the Calvin College "Nature Scripture Study Club." But poor health led him to withdraw from the ASA in 1945. Sturgis, overburdened with other commitments (mostly a large teaching load), also faded from participation by 1945.

17 The list appeared in the minutes (Box 5, File 10, ASA Papers) as follows: "1. The compilation and publication of material to strengthen the faith of students. 2. The distribution of material of scientific accuracy to ministers and christian workers. 3. The reviewing of written material to eliminate scientific errors. 4. The establish-

ment of a summer school for educators and christian workers. 5. The establishment of a speakers bureau featuring scientists who would speak on biblical subjects. 6. The establishment of a correspondence school for the dissemination of scientific—biblical material. 7. The canvassing of scientists for christians in sympathy with the objectives of such an organization who would be interested in joining it. 8. The calling of an annual conference to which representatives of accredited religious schools would be invited to present for discussion and review the teachings of their schools which have scientific implications for the purpose of aiding them in working out all scientific errors. 9. The publication of a biblical-science news letter to be sent regularly to subscribers. 10. The preparation of a reference manual."

18 Carpenter, "From Fundamentalism to the New Evangelical Coalition," p. 12.

19 Typed manuscript, "WHAT IS *The American Affiliation of Scientists?*" p. 6, Box 20, File 14, ASA Papers. Everest had typed "renaissance" and then crossed it out and written "reformation" above it.

20 *The Story of the American Scientific Affiliation,* p.3

21 The degree to which evangelicals realized this objective is a subject of spririted historical discussion. See, for example, Mark Noll, *The Scandal of the Evangelical Mind* (Grand Rapids, MI: William B. Eerdmans Publishing Co., 1994).

22 Interview with Irving A. Cowperthwaite, July 27, 1991.

23 "WHAT IS *The American Affiliation of Scientists?*" p. 2, Box 20, File 14, ASA Papers. To compare with Bryan's argument about the harmful effects of evolution upon the souls of young people, see William Jennings Bryan, *The Last Message of William Jennings Bryan* (New York: Fleming H. Revell Co., 1925).

24 John P. Van Haitsma to F. A. Everest, et al., March 23, 1942, Box 5, File 16, ASA Papers.

25 F. Alton Everest to Arthur P. Kelley, October 4, 1942, Box 5, File 23, ASA Papers.

26 F. Alton Everest to Cowperthwaite, et al., October 7, 1942, Box 5, File 23, ASA Papers.

27 For discussion of Lammerts, see Henry Morris, *History of Modern Creationism,* 2nd edition (Santee, CA: Institute for Creation Research, 1993), passim; and Numbers, *The Creationists,* passim.

28 F. Alton Everest to Cowperthwaite, et al., January 26, 1942, Box 5, File 14, ASA Papers.

29 Walter E. Lammerts, "The Creationist Movement in the United States: A Personal Account," *Journal of Christian Reconstruction* 1 (Summer, 1974): 54.

30 F. Alton Everest to J. P. Van Haitsma, July 13, 1942, Box 5, File 20, ASA Papers.

31 F. Alton Everest to J. P. Van Haitsma, January 28, 1942, Box 5, File 14, ASA Papers.

32 "The Constitution of the American Scientific Affiliation," Article I—Objects, Box 1, File 3, ASA Papers.

33 See "The Constitution of the American Scientific Affiliation," Article II—Creed, p. 9 of the first promotional brochure, reproduced in this volume.

34 Ibid.

35 As Dudley Joseph Whitney of the Religion and Science Association lamented in the mid-1930s, fundamentalists were "all mixed up between geological ages, Flood geology and ruin, believing all at once, endorsing all at once." Quoted in Numbers, *The Creationists,* p. 102.

36 This is not to say they didn't believe that someone might self-identify as a theistic evolutionist. They just didn't envision such a person among their ranks. Van Haitsma even rhetorically asked the group at one point, "Does not the implied teaching of the Bible that the races of man all descended from Adam and Eve support a theistic organic evolution?" Van Haitsma to Council, March 23, 1942, Box 5, File 16, ASA Papers. But the question raised no substantive discussion.

37 The stories of evangelical accommodation of evolutionary theory were over half a century old by the time the ASA founders got together. For the best review of these people and their ideas see James R. Moore, *The Post-Darwinian Controversies: A Study of the Protestant struggle to come to terms with Darwin in Great Britain and America, 1870–1900* (Cambridge: Cambridge University Press, 1979); Jon H. Roberts, *Darwinism and the Divine in America: Protestant Intellectuals and Organic Evolution, 1859–1900* (Madison: University of Wisconsin Press, 1988); and David N. Livingstone, *Darwin's Forgotten Defenders: The Encounter Between Evangelical Theology and Evolutionary Thought* (Grand Rapids, MI: William B. Eerdmans Publishing Co., 1987).

38 This would include, among other developments, new techniques of geological age determination, the modern evolutionary synthesis, the Watson-Crick model for the structure of DNA (1953), key experiments by Miller and Urey (1953), advances in biochemistry, plate tectonics, etc.

39 See reference 17 above.

40 Peter W. Stoner, "Faith Lost at College," *Moody Monthly* (August 1944), p. 660. This article carries a subtitle that reflects well the conviction and ambition of the ASA founders: "A much needed exposé of the folly of spiritual pseudoscience—with a remedy for the evil."

41 See the section under the heading "Blackballing Fundamentalist Pseudoscience: Ironies in the Construction of an ASA Imprimatur" in Kalthoff, "The Harmonious Dissonance of Evangelical Scientists," pp. 262–64.

42 Peter W. Stoner, "God's Dealing with One Man," unpublished MS, n.d., pp. 55–56, and 95–99, Box 22, Files 4 and 5, ASA Papers.

43 His early unpublished works dating from the 1930s are devoted mostly to the inspiration of scripture. Titles include "Is the First Chapter of Genesis Inspired?" "Geographical Prophecies Prove the Bible," "Prophecy and the Inspiration of the Bible" (each about five single-space typed pages). A longer MS (38 pp.), "Come Let us Reason Together OR New Evidence of The Inspiration of God's Word," was devoted to similar themes. These I obtained courtesy of Irving A. Cowperthwaite.

44 Publication statistic is from Everest, "The American Scientific Affiliation," p. 19. In written correspondence, Moody Press was unable to provide specific data regarding the copies of *From Science to Souls* that appeared.

45 Stoner, *From Science to Souls,* p. 46. About the same time he also writes, "I wish we could change the teachings of the colleges on such subjects as the theory of evolution and have them show that the sequence of the fossils in the earth could be explained just as well by the creative acts of God that are recorded in Genesis 1." Quoted from "Faith Lost at College," p. 661.

46 See description in first promotional brochure, p. 5, reproduced in this collection.

47 Typed promotional brochure, ASA Papers.

48 Everest, "The American Scientific Affiliation," p. 90–94.

49 Irwin Moon to Will Houghton, February 24, 1945, Irwin Moon Papers.

50 On the favorable reception of the book, see review by Fuller Seminary's Edward J. Carnell, which appeared anonymously under the title "For Scientists by Scientists," *HIS* (February, 1949), 25–26. In 1954, *Modern Science and Christian Faith* was still judged by some [*Moody Monthly* (October), p. 60] to be "the best volume published in America" on science and religion between 1943 and 1952. No accurate publication data regarding the second edition have been found. Certainly, however, circulation was less than 10,000. The second edition was twenty-seven pages longer than the first edition, although it had only ten chapters compared with the eleven of the first edition. The second edition dropped chapters on "Mathematics and Prophecy" and anthropology from the first edition, and added a new "Christian View of Anthropology."

51 F. Alton Everest, "The American Scientific Affiliation," p. 93.

52 See page 58, note 28 of Gedney's chapter, reproduced in this

volume. On Gedney, see J. W. Haas, Jr. "Science and Mathematics at Gordon College—A Perspective on Attitudes, Programs and People: 1889–1989," in Ann Ferguson, ed. *Shaping a Heritage: Celebrating the Centennial* (Wenham, MA: Gordon College, 1989), pp. 122–136 and passim.

53 On Tinkle, see Numbers, *The Creationists,* passim; and Morris, *History of Modern Creationism,* passim. Typical of Tinkle's popular stance was his pronouncement that "the former idea that a species of plant or animal changes slowly but steadily into another type has been abandoned." See "Is Life a Continuous Flux?" *Moody Monthly* (July, 1945), p. 573.

54 Details of Kulp's life and ASA career from an interview with J. Laurence Kulp, November 7, 1990. See also Kalthoff, "The Faith and Science of a Faithful Scientist: The Case of J. Laurence Kulp," paper presented to The Pascal Centre International Conference on Science and Belief, Ancaster, Ontario, August 13, 1992; and Kalthoff, "The Harmonious Dissonance of Evangelical Scientists," 263–67.

55 Interview with Kulp (mission to educate); J. Laurence Kulp to F. Alton Everest, April 29, 1948, Box 8, File 14, ASA Papers (necessary destruction).

56 Kulp's ASA presentations also included "Antiquity of Hominoid Fossils" (1948); "The Carbon 14 Method of Dating" (1949); "Recent Developments in the Carbon 14 Method of Dating Fossils" (1950); "Age Determination and Dating Methods" (1951); and "Recent Development in Carbon 14 Age Measurements"(1952). Of the six, only "Deluge Geology" saw publication in the *Journal of the American Scientific Affiliation.* Interestingly, Kulp had been unable to attend the Los Angeles meeting, so the "Deluge Geology" paper had been read to an audience that included Price by Marie Fetzer, a Wheaton anthropology graduate.

57 Karl Turekian and Edwin Olson were also friends and colleagues of Kulp who published in the ASA journal on radiometric dating. See Turekian, "Carbon 14 and Cosmology," *Journal of the American Scientific Affiliation* 5 (June 1953): 21; and Olson, "Radiocarbon Dating" *Journal of the American Scientific Affiliation* 11 (March 1959): 2–11. Cordelia Erdman was one of a small group of influential ASA women. See J. L. Spradley and D. F. Chappell, "Wheaton Women in the Early ASA," *Perspectives on Science and Christian Faith* 44 (March 1992): 17–24.

58 "Discussion," *Journal of the American Scientific Affiliation* 4 (March 1952): 13.

59 The following correspondence establishes Morris as the author of the article: F. Alton Everest to Henry M. Morris, March 1, 1950, Box 10, File 5; Russell Mixter to Marion Barnes, April 18, 1950, Box 10, File 6; Allan A. MacRae to Russell Mixter, April 26, 1950, Box 10, File 6, all in ASA Papers.

60 Mixter's paper, titled "The Mechanisms of Evolution," was printed in the journal under the title "The Science of Heredity and the Source of Species." Ramm was not present at the meeting, so Irving Cowperthwaite read his paper on "The Scientifico-Logical Structure of the Theory of Evolution." Monsma contributed "Some Basic Presuppositions in Evolutionary Thinking."

Bernard Ramm would later prove to be highly influential in the ASA with the publication of his watershed volume, *The Christian View of Science and Scripture* (Grand Rapids, MI: William B. Eerdmans Publishing Co., 1954). For treatment of his roles for the ASA, see "A Bernard Ramm Festschrift," *Journal of the American Scientific Affiliation* 31 (December 1979); and Joseph L. Spradley, "Changing Views of Science and Scripture: Bernard Ramm and the ASA," *Perspectives on Science and Christian Faith* 44 (March 1992): 2–9.

61 See Buswell's article, "The Origin of Man, and the Bio-cultural Gap," reprinted in this volume. Ironically, the phrase "scientific creationism" originated with Buswell in his attempt to suggest an umbrella term that would embrace the spectrum of views from progressive creationism to theistic evolution. Of course, the term was hijacked by proponents of flood geology. See James O. Buswell, III, "A Creationist Interpretation of Prehistoric Man," in Russell Mixter, ed., *Evolution and Christian Thought Today* (Grand Rapids, MI: William B. Eerdmans Publishing Co., 1959), p. 188.

62 F. Alton Everest to James O. Buswell, III, August 17, 1951; James O. Buswell, III, to Irving Cowperthwaite, August 22, 1951; F. Alton Everest, typed MS, *THE AMERICAN SCIENTIFIC AFFILIATION—THE FIRST DECADE,* all from Irving A. Cowperthwaite Papers, courtesy Irving Cowperthwaite.

63 The ASA published three short monographs for public distribution. In addition to the two by Mixter and Allen was "Christian Theism and the Empirical Sciences" (1947) by Cornelius Jaarsma.

64 Report of membership data presented at the Sixth Annual Convention, *Journal of the American Scientific Affiliation* 3 (September, 1951): vii.

65 For a biographical sketch of Allen, see Brian P. Sutherland's obituary, "Dr. Frank Allen," *Journal of the American Scientific Affiliation* 18 (June 1966): 51.

66 For an accurate and brief discussion of Mixter, see Numbers, *The Creationists,* pp. 172–83. For further biographical reflections on Mixter, see D. F. Chappell's hagiographic article, "Christian, Teacher, Scientist, Mentor: Dr. Russell Mixter: Visionary for the Role of Science in Christian Higher Education," *Perspectives on Science and Christian Faith* 44 (March 1992): 10–16.

67 A. F. Shull, *Evolution* (New York: McGraw-Hill Book Co., 1936); on Mixter's intent, see Russell L. Mixter to H. Harold Hartzler,

December 18, 1963, H. Harold Hartzler Papers, courtesy H. Harold Hartzler; quotations from the monograph are from pp. 16, 17, 18, and 30; R. L. Mixter, "Reply to Criticism of 'Creation and Evolution,'" *Journal of the American Scientific Affiliation* 6 (March 1954): 28. The criticism of Mixter's monograph and the ensuing exchange from the ASA journal are reprinted in this volume.

68 For more thorough accounts than space here permits, see Numbers, Chap. 9, "Evangelicals and Evolution in North America," in *The Creationists,* pp. 158–83; and the section under the heading "Coming to Terms with the 'E-Word'" in Kalthoff, "The Harmonious Dissonance of Evangelical Scientists," pp. 267–69.

69 Irving A. Cowperthwaite, "Some Implications of Evolution for A.S.A." *Journal of the American Scientific Affiliation* 12 (June 1960): 12. This article is reprinted here.

70 Official Program of the Twelfth Annual Convention of the American Scientific Affiliation, obtained courtesy of H. Harold Hartzler. See Wilbur L. Bullock, "Comments On Dr. Mixter's Paper," *Journal of the American Scientific Affiliation* 11 (December 1959): 26.

71 Official Program of the Thirteenth Annual Convention of the American Scientific Affiliation, obtained courtesy of H. Harold Hartzler. Ernst Mayr, ed., *The Species Problem: A Symposium Presented at the Atlanta Meeting of the American Association for the Advancement of Science, December 28–29, 1955* (Washington, D.C.: American Association For the Advancement of Science, Publication No. 50).

72 *Eternity,* December 1960, p. 46; Carl F. H. Henry to F. Alton Everest, March 1, 1948, Box 8, File 13, ASA Papers; Walter R. Hearn and Richard A. Hendry, "The Origin of Life," in *Evolution and Christian Thought Today,* pp. 67, 69; Frank Cassel, "The Evolution of Evangelical Thinking on Evolution," *Journal of the American Scientific Affiliation* 11 (December 1959): 27, (dirty word).

73 Walter Lammerts to R. Laird Harris, November 12, 1962, Walter Lammerts Papers, Carton 1, Collection No. 84/149, Bancroft Library, University of California, Berkeley.

74 See, for example, the papers by Lammerts and Tinkle which appeared in the March 1961 issue of the journal.

75 Interview with Walter Hearn, August 7, 1990.

76 Delbert N. Eggenberger, "Journal Publications," *Journal of the American Scientific Affiliation* 8 (December 1956): 2. Delbert N. Eggenberger, "The Editor's Views," *Journal of the American Scientific Affiliation* 11 (December 1959): 2.

77 Vernon Charles Lyons, "Is Evolution at Wheaton College?" *Sword of the Lord* 28, No. 42, October 19, 1962, cover story. See also, V. R. Edman Papers, Box 3, Files titled "Evolution at Wheaton" and "Evolution at Wheaton (Mixter)," Wheaton College Archives and

Special Collections. This saga is also recounted briefly in Numbers, *The Creationists,* pp. 181–83.

78 Roberts, *Darwinism and the Divine in America,* p. x.

79 Cowperthwaite, "Some Implications of Evolution for A.S.A.," p. 12.

# asa

AMERICAN SCIENTIFIC AFFILIATION

◆

*Nineteenth and twentieth century science has brought forth some remarkable evidence substantiating the reliability of the Holy Scriptures. This same science has also brought in a great materialistic philosophy which is leaving a scar upon our civilization. An excellent method of combating the latter is to make known the former, and to this task this new organization of Christian men of science brings its rigorous thinking, specialized training, and humble faith.*

◆

*A brief explanation of the origin of the new group and its constitution follow.*

*Be ready always to give an answer to every man that asketh you a reason of the hope that is in you with meekness and fear.*—I Peter 3:15

—*asa*—

# THE *Story* OF THE AMERICAN SCIENTIFIC AFFILIATION

Throughout the week of September 2, 1941, there met in Chicago a group of five men of standing in their respective branches of science. Four of them are professors in recognized institutions of higher learning, and the fifth was on the faculty of the largest university of the country until recently, when he entered the industrial field. Two of them are heads of their respective departments. They are typical college professors; in addition, they are wholehearted Christians. They met to discuss some problems common to them all.

## CONCERNED WITH STUDENTS' FAITH

As college professors, it is natural that their first concern was for college students. While many sincere Christians are found among college students, and much fine evangelistic work is being done on the campus, many instances of students losing their faith in Christ and belief in the Bible through materialistic campus attitudes are known. It is common for scientists to develop theories as working aids, and for teachers to use them as teaching aids, but sometimes it is forgotten that they are only theories. There is no disagreement between the statements of the Bible and the true facts of science. Any disagreement results from man's attempt to interpret things he observes. Because the student often has access to only one side of the story, he cannot maintain an even spiritual keel while being educated in a world of facts if the facts which would place foundations under his faith are denied him. These men recognized this situation among students and were eager to alleviate it.

## CONCERNED WITH UNSCIENTIFIC DEFENSE OF THE BIBLE

These men had all read books, pamphlets, and articles by Bible teachers which deal with scientific matters. Although the biblical truths are presented in a commendable manner, some of the attempts to demonstrate the reliability of the Bible are shot through with inaccuracies, mistakes, and unscientific conclusions. Relatively few of these writings are by men who deliberately try to mislead readers. Most of the authors are genuine Christians, but are either unprepared to write in the scientific field selected, or are definitely misinformed. In their zeal to make God's Word known, they have done inestimable damage, for those reading these errors who are informed on scientific matters, often discount not only the errors but the whole work.

Some ministers and evangelists use scientific illustrations in an effective manner. Others, however, do violence to the thinking of any informed persons, in similar attempts. While to many of the lay public this may give a modern touch to the sermon and leave the impression that the speaker is well informed, to any college freshman a barrier is raised, which grows higher and higher with each succeeding similar incident. Usually the minister or evangelist has intentions of the highest type, that is, of driving home some precious gospel truth; but here again, misinformation or lack of accurate information is at fault, doing great damage in the minds of many. Thus a reputation has been built up over a period of years that religious speakers and writers are frequently loose in their use of illustrations from the fields of science.

How is it possible to help these authors, ministers, Bible teachers, evangelists, and college students? This group of five men spent a week considering that question, and the list of things that could be done demanded a permanent organization devoted to the carrying out of these plans. Thus the American Scientific Affiliation was formed.

**PLANS—Handbook For College Students**

A handbook for college students is being considered. This book would have a section for each field of science likely to be encountered on the campus, such as biology, geology, and perhaps even inferential fields such as psychology. In each of these sections the difficulties likely to be encountered by the student would be listed, the biblical position stated, and a list of the actual facts known, given. This book would be written on such a high academic level that the respect of the professors teaching the courses would be compelled without compromise to the fundamental position. The student so forewarned would be far better prepared to enter the classroom in any university than he is today.

**PLANS—Review Christian Books On Science**

Another phase of the proposed program is to review books, articles, and tracts of a biblical-scientific nature before publication. It was felt that publishers would welcome the opportunity to submit manuscripts, before sending them to press, to a group of earnest Christian scientific men who are specialists in the particular field involved. If the manuscripts are found worthy, authorization to use the name of the Affiliation and its stamp of approval would be given, which will eventually become a symbol of quality to readers. This would eliminate much of the loose parading of science for imagined gospel gains.

**PLANS—Scientific Summer School**

The establishment of a summer school for Christian workers and educators, dealing with the light that many phases of modern science has thrown on the Bible, was suggested. A correspondence course of this nature would also be valuable. Reviewing the material used by Bible schools and institutes before a body of scientists would insure that the most modern discoveries are used to their best advantage to prove the authenticity, historicity, and accuracy of the Bible on the one hand, and to discover gross scientific errors and fallacies which may have crept in through casual interpretation on the other.

#### PLANS—Writing Literature

The actual writing of books, pamphlets, tracts, etc., which bring together the most up-to-the-minute and most accurate scientific information available on the subject, would also be a part of the program. There is a wealth of material made available through recent discoveries in several fields of science which could be of great value to the expositor, illuminating many obscure Bible passages. Bringing these facts together into usable form would be a valuable contribution. At the annual meeting, scientific papers on the relationship of science and the Bible would be presented and discussed before experts in many fields. It would be quite unlikely that serious errors would remain after such a group had combed thoroughly the material presented.

#### PLANS—Spreading Accurate Literature

The dissemination of such material, once it was prepared, would be a job of major proportions. There are many exceedingly effective methods other than the printed page for distributing such information. Realizing the effectiveness of visual education, still and motion picture projectors could be used. Carefully prepared films as well as projectors could be distributed through local Bible bookstores, etc., to young people's societies, Bible classes, and churches. A bureau of doctrinally sound speakers well versed in scientific fields could be made available for conferences on campuses, in high schools, and in summer camps. Local members of the organization could help in this manner.

And so the list could be carried on and on. The field is ripe unto harvest and the Lord of the harvest is calling for laborers, in this case scientific men who love the Lord and His Book which reveals Him to us. There are many organizations which are doing an acceptable job of proving to the world that being a Christian is a red-blooded man's duty and privilege. This group of scientists can prove to the world that the principles of our Christian faith welcome investigation and that the Bible, being the Word of God and thus infallible, will withstand any encounter with science which might be proposed. Although it is not primarily a book of science, wherever it touches upon scientific or historical things it is accurate and true. Being a Christian does not mean mediocrity in science, as our roster will prove. Putting oneself in partnership with the Creator and Maintainer of the universe should be a prime requisite for any budding scientist. The search for knowledge should lead to God, for He is the seat of all knowledge.

One man said, in discussing this organization, "To my mind, this can become the most important movement since the Reformation." Whether or not this be true, if you are trained in some branch of pure or applied science and if you know Jesus Christ as your personal Saviour, you should make known your interest. You may be admirably suited for a portion of this task in some very unique manner. Teachers of scientific subjects in institutions of higher learning are in a particularly advantageous position to help the cause. Communicate with the chairman, or with one of the members of the executive council living in your locality.

*The works of the Lord are great, sought out of all them that have pleasure therein.*—Psalms 111:2

# THE Constitution

## OF THE AMERICAN SCIENTIFIC AFFILIATION

### Article I—OBJECTS

The objects of the Affiliation shall be:

(1) To integrate and organize the efforts of many individuals desiring to correlate the facts of science and the Holy Scriptures.

(2) To promote and encourage the study of the relationship between the facts of science and the Holy Scriptures.

(3) To promote the dissemination of the results of such studies.

### Article II—CREED

I believe in the whole Bible as originally given, to be the inspired work of God, the only unerring guide of faith and conduct. Since God is the Author of this Book, as well as the Creator and Sustainer of the physical world about us, I cannot conceive of discrepancies between statements in the Bible and the real facts of science. Accordingly, trusting in the Lord Jesus Christ, the Son of God, my Saviour, for strength, I pledge myself as a member of this organization to the furtherance of its task.

Signed ..................................................

### Article III—MEMBERSHIP

*Section 1.* Only those persons shall be eligible to membership in the Affiliation: (1) who subscribe to its creed; and (2) who have shown noteworthy achievement in original investigations in some branch of pure or applied science, or who by training and/or experience in scientific or biblical fields are qualified to further its aims.

*Section 2.* Upon the written recommendation of a member of the American Scientific Affiliation, a person may be elected to membership by a vote of the Executive Council.

**Article IV—OFFICERS**

*Section 1.* The affairs of the Affiliation shall be guided by an Executive Council of five members.

*Section 2.* The members of the Executive Council shall be elected for a term of five years, one member to retire each year, except in special cases, such as starting the rotation and filling unexpired terms.

*Section 3.* The officers of the Affiliation shall be elected by the Executive Council from its own ranks and shall consist of a president, a vice-president, and a secretary-treasurer. The term of office shall be one year. The President so elected shall be ex-officio chairman of the Council.

*Section 4.* Every reasonable effort shall be made, consistent with leadership ability, to select council members from the different fields of science as well as geographical areas.

**Article V—MEETINGS**

*Section 1.* The Affiliation shall hold an annual meeting at the time and place designated by the Executive Council. Other meetings may be authorized by the Executive Council.

*Section 2.* At this annual meeting the membership convened shall fill all vacancies on the Executive Council.

*Section 3.* The Executive Council shall hold at least one meeting during the year.

**Article VI—PUBLICATIONS**

*Section 1.* At the discretion of the Executive Council, the Affiliation shall publish such brochures as may appear desirable. The Executive Council shall have full power to enter into contract and agreement with publishing agencies in the name of the Affiliation.

*Section 2.* The Executive Council may review manuscripts prepared by persons outside the membership of the Affiliation, and authorize the use of the name of the Affiliation and the stamp of approval on manuscripts which are judged to be worthy from the standpoint of content and style. The Executive Council shall encourage writers before publishing their works, to avail themselves of this critical review by experts in the field.

#### Article VII—FUNDS

*Section 1.* The Executive Council shall fix the dues for members.

*Section 2.* The Executive Council shall receive bequests and gifts in the name of the Affiliation.

*Section 3.* The Executive Council shall expend its funds in accordance with the objects of the Affiliation.

*Section 4.* The activities of the Affiliation shall not be carried on for the purpose of deriving a profit. Should any profit result from the activities of the Affiliation, it shall not be credited to surplus and shall not inure to the benefit of any private member or individual.

#### Article VIII—AMENDMENTS

This constitution can be amended by a two-thirds vote at an annual meeting, provided the proposed change has been brought to the attention of the constituency at least ninety days previously.

# *Executive council*

## OF THE AMERICAN SCIENTIFIC AFFILIATION

*Chairman:*

F. Alton Everest, E.E.
1011 Encino Row
Coronado, Calif.

*Secretary-Treasurer:*

Irving A. Cowperthwaite, Ph.D.
10 Willoughby Road
Milton, Mass.

Peter W. Stoner, M.S.
56 N. Hill Avenue, No. 10
Pasadena, Calif.

John P. Van Haitsma, Ph.D.
1027 Benjamin Avenue
Grand Rapids, Mich.

Russell D. Sturgis, Ph.D.
26 Sixth Avenue
Collegeville, Pa.

# FROM SCIENCE TO SOULS

PETER W. STONER, M.S.
Professor of Mathematics and Astronomy;
Executive Council, American Scientific Affiliation

MOODY PRESS
153 Institute Place Chicago, Illinois

Printed in the United States of America

## CONTENTS

## INTRODUCTION

The American Scientific Affiliation, a group of Christian men of science, has carefully examined this manuscript and, to the best of its information, has found it to be accurate in its scientific content in the light of the accepted science of today.

Although the author is a member of the American Scientific Affiliation, the thesis here presented is not necessarily that of the group.

The author has made a laudable attempt to harmonize science and the Genesis account. His attitude is that any apparent lack of harmony is due to current misinterpretation of either scientific evidences or the scriptural statement. The author believes all truth is of God, and the real truth of science will agree wholly with God's own book, the Bible.

Irving A. Cowperthwaite, Ph.D., *Secretary*
The American Scientific Affiliation
10 Willoughby Road, Milton, Mass.

## PREFACE TO CHAPTER ONE

More than 25 years ago, there were Chinese government students in the University of California at Berkeley. Twelve of these students went to Dr. McAfee, pastor of the First Presbyterian Church, and asked that they be given a special Sunday school class so that they might obtain information about the Christian religion. These students were sent to this country to become engineers, doctors, etc., and to take this training back to China for the improvement of their country. They said that they considered that the religion of any country had much to do with its development, so they wished to learn what our religion was in order to interpret its effect on our culture.

Dr. McAfee came to me and asked me to take the class. I did not realize what was involved until I had started the work. Then I realized that they had no foundation whatsoever in our religion, nor had they any faith in our God or Jesus Christ.

After much prayer, I felt led to start the study of the Bible from its scientific connections, later to take enough prophecy to show the Bible's accuracy, and finally, to present Christ.

We spent the first winter studying the Bible's relation to science. The students took references to the University library and reported on them as faithfully as they would have, had it been a university course.

When we had completed the scientific part of the plan, the whole group went to the pastor and told him that they were convinced that our religion was the true religion, for no other religion had such proof of its having God as its author. They all accepted Christ as their Saviour and joined the Church.

I have carefully followed the development of science and its relation to the Bible in the years since that time. Today, developments of science have greatly increased the evidence of the inspiration of the Bible. I have attempted to set forth this evidence.

The references given in this chapter are only those that happen to be on my desk or most easily accessible to me. The reader can find equivalent material in a very great number of different books. These or corresponding references should certainly be read in order to obtain a satisfactory understanding of the subject under consideration.

In Part III are treated a few of the theories given by the Church long ago, and which are still in general use. These have tended to open the door to the destruction of the faith of our young people, when they have learned that scientific facts drastically disagree with their earlier teaching in the Church.

It should be clearly understood that Chapter One deals with the correlation between Genesis and science, as science stands today. Tomorrow science may discover new facts and this chapter may need revision. Every development of science in recent years has shown more clearly the truth of the Bible. We may be confident that future developments in science will only add to the evidence here produced.

Fig. 1—Spiral Nebula on Edge.
*Photograph, courtesy Mount Wilson Observatory.*

Fig. 2—Spiral Nebula Flat.

Fig. 3—Whirl-pool Nebula.

*Photographs, courtesy Mount Wilson Observatory.*

# CHAPTER ONE

## Part I

## GOD ON THE WITNESS STAND

### Let's Be Scientific

I HAVE A very high esteem for true scientists. I have worked with them for many years and have found them to be searchers for truth in any physical place where it could be found. I have found them to be men of integrity, so much so that they would as soon supply evidence to prove their earlier work false as to prove it true. Let us as Christian people be as open as the scientist in our search for the interpretation of our Bible. Let us consider all of the evidence and then draw unbiased conclusions regardless of our earlier beliefs or teachings.

### Scientific Theories

A scientific theory is only made to fit the known facts. The theory may or may not be true, it may not even be thought to be true by its authors. It helps the student to organize the facts in his mind and it often enables the scientist or engineer to predict future behavior or happenings. When additional facts are secured the theory is often revised to cover the new facts or is replaced by a new theory.

A few years ago we said the central 4,000 miles of the earth were solid steel. This was the theory based on the transmission of earthquake vibrations through the center of the earth. According to observations, at that time, the vibrations went through the center of the earth as they would if the center were solid nickel steel. Later investigations showed that the vibrations did not travel in all directions as they would in solid steel but rather were restricted in their travel as they would be in molten steel. So, our present theory is that the center of the earth is molten instead of solid steel.

Years ago our theory about the composition of matter was that the atom was the smallest particle of material. Now we have broken the atom down so that its nucleus is made up of protons and electrons, with electrons revolving about the nucleus. This has been the state of the theory of the smallest units of matter for some time. We may now be in the process of breaking these down into still smaller units.

Let us regard scientific theories in their true light; they account only for the known facts. We have two theories as to what light is; both are taught in all Physics classes. One states that light is only a wave, the other that light is real matter. One theory accounts for a part of the known facts about light but it takes the other theory to account for the remaining facts. Most scientists do not believe that both are true, many may doubt the truth of either, but the two theories together make it possible for the scientist to predict the behavior of light. The theories make it possible for him to design new optical instruments and know exactly what the instrument will do before it is constructed. Scientific theories are very useful, true, partly true or false.

**The Theory of Evolution**

The theory of evolution is based largely upon the fossils found in the various geological layers of the earth.

There are evidences of a very elementary form of life existing at the beginning of life on the earth. Later, more complex forms of plants and animals appeared, involving fish, reptiles, birds, mammals, and man.

The theory of evolution then accounts for the known geological facts when it states that life started, on the earth, in the most rudimentary forms and later appeared in more and more complex forms, including elementary forms of plants capable of living in sand and in the water, shell fish, fish, plants similar to the present flora, great reptiles, birds, mammals, and man.

There are many variations of the theory of evolution in some of its details dealing with the origin of the first life and the forces which caused the changes and developments. Some say that the origin of life and its changes were all automatic and influenced by environment. There is nothing in the fossils to indicate that this theory is not true, so from these facts alone it would be considered satisfactory.

Other scientists consider that the first life on the earth was created and that God took a hand in many of the later changes or development. This, too, finds no contradiction in the fossils in the earth's layers and must also be considered to account satisfactorily for all of the facts as found in Geology.

Most of the scientists seldom weigh these two positions against each other. They are little concerned about the difference. (The changes may come through mutations or through acts of God. Since either fits the facts equally well we should not expect most scientists to be greatly concerned.)

To many adherents of the Church it is a very important difference. It hits at the very heart of the question of the inspiration of the Bible, for the Bible claims that God did create.

Let us carefully consider the claims of the Bible and the evidences for their truth. Science has made great advances in the last few years and we should now be able to determine something of the scientific accuracy of the Bible's claims in its opening chapters.

### Genesis Is Evidence

This question of creation cannot be scientifically considered or settled without first hearing God's own testimony and claim. If we do not hear His testimony, we will not be considering all of the evidence before drawing our conclusions. No court of our land will convict a man without first hearing his defense and the evidence he has to offer. Let us in like manner consider the evidence that God has to present in favor of His claim to the creation of this universe and the life upon the earth before we decide that He did not create.

### An Old Battle Ground

The first chapter of Genesis has for generations been the ground for criticism of the Bible and of attacks on its authenticity. Let us, however, turn to it and carefully examine its record, for it is in this chapter that the main claim is made by God to His acts of creation. It will be necessary to read carefully and to analyze the account just as it is given, neither reading into the account things that are not there nor omitting things which are definitely stated.

There are thirteen acts very definitely named in this chapter. God claims in some places to have "created," in other places to have "made," and in other places he uses the word "let." God also claims to have done these things in a very definite order. We shall be very much concerned with this order for this account was written long before scientific information was available to guide the writer.

There is often disagreement regarding the date of writing this first chapter of Genesis. I shall not enter into a discussion of this subject, for while I have very definite ideas on it, the subject can in no way affect the present consideration. None of the scientific evidence which I shall use was established by science even at the latest date ever proposed for the writing of Genesis and some of the scientific information was not known 25 years ago. It is evident therefore that this material in Genesis did not come from scientific knowledge merely taken at an earlier date than the present. References to the teachings of Egypt, Babylon and other ancient civilizations regarding their theories of creation will be found on page thirty-nine.

**Thirteen Acts of God**

The following thirteen things will be found in the first chapter of Genesis as having been accomplished by God in the order given.

1. In the beginning God created the heaven and the earth (v. 1).
2. And the earth was without form, and void; and darkness was upon the face of the deep (v. 2).
3. And the Spirit of God moved upon the face of the waters (v. 2).
4. And God said, Let there be light (v. 3).
5. And God made the firmament, and divided the waters which were under the firmament from the waters which were above the firmament (v. 7).
6. And God said, Let the waters under the heaven be gathered together unto one place (v. 9).
7. And let the dry land appear (v. 9).
8. And God said, Let the earth bring forth grass, the herb yielding seed, and the fruit tree yielding fruit after his kind (v. 11).
9. And God made two great lights (v. 16).
10. And God created great whales, and every

living creature that moveth, which the waters brought forth abundantly, after their kind (v. 21).

11. —and every winged fowl after his kind (v. 21).
12. And God said, Let the earth bring forth the living creature after his kind, cattle, and creeping thing, and beast of the earth after his kind (v. 24).
13. So God created man in His own image (v. 27).

These are the thirteen items named in the first chapter of Genesis. In each one God claims some part. In Genesis 2:4, God claims that they are in the correct order. "These are the generations of the heavens and of the earth when they were created, in the day that the Lord God made the earth and the heavens." Let us examine these thirteen items and see if they are in the correct order. Let us also examine the items themselves and see if they are true.

### 1. The Origin of Matter

You may search, if you wish, every book of science, new or old, and you will not find a single theory as to the source of the material from which the universe was made. Every theory (Nebular Hypothesis, Planetesimal Theory, Tidal Theory, etc.) starts with the material universe already existing in some form or other and then deals with the steps which formed that material into its present condition in the solar system. (See *General Astronomy*,

Young, 1898, pp. 567-576; *Astronomy,* Duncan, Harpers, 1930, pp. 290-300, especially noting the last paragraph, p. 291; also, *Astronomy,* Skilling and Richardson, Holt, pp. 310-316.) No one challenges God's claim to the creation of the physical material of the universe.

There are now many strong reasons for dating the time of creation as somewhere in the neighborhood of 2,000,000,000 years ago. The age of the earth as determined by samples of radio active elements, such as Uranium, found in the igneous rocks, must be placed near 2,000,000,000 years. The oldest meteorites, both from the solar system and from outer space, are nearly 2,000,000,000 years old. The tidal forces between the earth and moon would require about 2,000,000,000 years for the moon to move from an early position near the earth to its present distance. The measured velocities of the island universes are such as would require about 2,000,000,000 years for them to have reached their present position if they had started from a common origin. Thus it would appear that 2,000,000,000 years ago, more or less, God created the heavens and the earth and set them moving at varying velocities from that place of creation. (See *The Solar System and its Origin,* Russell.)

### 2. The Nebular Hypothesis

"The earth was without form, and void." This statement has not always agreed with science. The theories of science about the formation of the solar system, in both Astronomy and

Geology, have a long and interesting history. The Nebular Hypothesis for something like a century held absolute sway as being practically a proven fact and not a mere theory. The theory went something like this: any system like the solar system (the sun with its planets revolving about it) started as a gaseous body, shaped like a flat disk, very hot and rotating slowly; then as the mass cooled off, it would contract and rotate more rapidly. Finally, small parts would be left behind and form planets, rotating and revolving around the central body in the some direction that it rotated.

This theory seemed plausible, it seemed to agree perfectly with the laws of Physics as we understood them. A glance at the solar system showed that it filled the conditions of the theory. The sun rotated in the same direction in which the planets revolved about it; nearly all of the planets rotated on their axes in the same direction; and practically all of the satellites revolved about their planets in this same direction and turned on their axes, still in this same direction. This certainly seems to indicate that this system developed just as the Nebular Hypothesis indicated. (See figure above.)

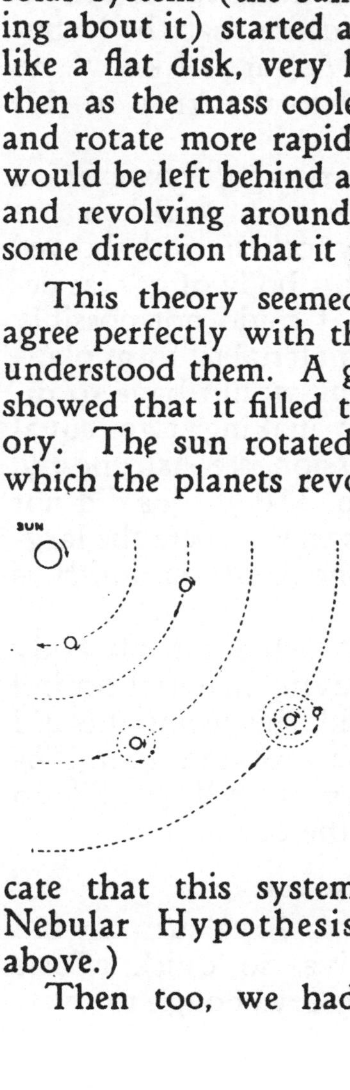


Then too, we had photographic evidence.

Photographs of spiral nebulae shown on edge (figure 1) certainly were flat disks. Others as seen from the side were quite round (figure 2). Others definitely showed pieces of material breaking off and forming planets (figure 3).

These photographs, with hundreds of others, furnished the clinching evidence that the Nebular Hypothesis was true. As far as I know, no scientist had any doubt as to the truth of this theory.

Approximately 25 years ago, however, a mathematical physicist went to work on the physics of the Nebular Hypothesis and showed that if we should have such a body of gas, as described in the hypothesis, it could not possibly throw off small bodies of material to form planets to the central body, but would have to divide nearly in the center making two equal bodies. (For a brief discussion, see *Astronomy*, Skilling and Richardson, p. 310.) This did not at all agree with the solar system where the largest planet is only about one one-thousandth as large as the sun.

It was also shown that such a gaseous body would have difficulty in leaving material behind when it shrank, and if it did, this material could not have collected to make a planet. (See *The Solar System and its Origin*, Russell, pp. 93-96 for other difficulties with the theory.)

**Island Universes**

It was very disturbing to have such an old hypothesis proven defective, so quick efforts were made to change the theory enough to overcome the difficulties. But they were of little avail, for at about this same time Dr. Hubble, working with the 100 inch telescope of Mt. Wilson, found that the spiral nebulae were not nebulous bodies as had always been supposed, but island universes, great aggregations of stars. These were shown to be similar to our own galactic system, but the nearest one was nearly 1,000,000 light years away, that is, the distance light will travel in 1,000,000 years going 186,000 miles each second. (See *The Realm of the Nebulae,* Hubble, Yale University Press.)

Our own galactic system, of which our sun is just one of the stars, contains something like 30,000,000,000 stars averaging as large or larger than the sun. Our galactic system seen from the edge, would look like figure 1; and seen from a side view it would look quite like figure 2. The sun with the earth would only be one of the 30,000,000,000 stars and would be located more than halfway from the center to the outer edge. The edge of the disk as seen from the earth is the Milky Way. The largest diameter of our system is about 100,000 light years. Each galactic system, or island universe, or spiral nebula, as you may choose to call them, is thought to be a very similar system to our own, to contain about the same number of stars, and to be about the same size.

### Nebular Hypothesis Unsatisfactory

These findings completely removed the Nebular Hypothesis from being a satisfactory description of the manner in which our solar system, or any other planetary system, could have been formed; for the photographic evidence was shown to consist of photographs of island universes instead of nebulous bodies and the theory was proven physically impossible. Today the Nebular Hypothesis is mentioned in books of science only for its historical interest.

Until the Nebular Hypothesis was discarded, science did not agree with the second verse of Genesis, in fact it contradicted it in every statement. The Nebular Hypothesis said that in the earlier age the earth was a disk-shaped body of gas, dense and very hot, giving off a great amount of light. Genesis says it was without form, void (very rare) and dark. Thus the two were exact opposites in every statement. During the century of belief in the Nebular Hypothesis the second verse of Genesis was one of the main points of attack against the Bible. It did not agree with science and furthermore its description did not fit any known physical object or astronomical type of body. As a Church we could only say that we did not understand, or try to translate the verse in a different manner so that we could give it a different interpretation. None of these produced any convincing argument of the accuracy of the biblical account.

### Every Theory Faulty

Following the Nebular Hypothesis, there sprang up other theories, the most prominent ones being the Planetesimal Theory and the

Tidal Theory. These were short-lived as they were both shown to be defective by Henry Norris Russell in his book, *The Solar System and its Origin.* Today the scientific world finds itself without a single theory of the formation of the solar system which has not been proven defective. The account in the first chapter of Genesis is the only one which has not been proven faulty and shown to violate the laws of physics.

### A Dark Nebula

But let us come back to this second verse. What does it refer to? It has been quoted by at least one nationally known astronomer as being the best description of a dark nebula that has ever been written. It is within the last 25 years that we have known of the existence of a dark nebula. All dark nebulae were considered merely spots in the sky where there were no stars. Figure 4 is the photograph which first definitely proved to the astronomer that these were clouds of dark material. On the right side of the photograph there are large numbers of small stars. These stars extend clear across the picture, but are covered on the left side by the dark cloud. You can practically see them shining around and illuminating the edge of the nebula. In the center of the photograph, a projection of the nebula extends out from the main body. A small section of the dark cloud is lighted up by a star in its neighborhood. There can no longer be any doubt about the existence of the dark nebulae. They are perhaps more prevalent than any other type of astronomical object except stars. Figures 5 and 6 show other dark nebulae in multitudes of shapes. Any recent Astronomy text will show you many other shapes of dark nebulae.

The suggestion has been made that dark nebulae can easily be the source of new stars. Practically every theory of the formation of our solar system supposes that our system started from some kind of a diffuse nebula. (See *The Evolution of the Earth and its Inhabitants,* p. 8, The Nebulae.) A bright or a dark nebula has the same composition. One has a star close enough to it to excite it to radiate light and the other has no such star and remains dark. The sequence of stars, accepted by Astronomy, starts the star as a rare gas, only hot enough to give off a faint red light. The difference between such a star and a nebula is slight, less in both density and temperature than the difference between many stars.

A more complete translation of this passage in Genesis would be: And the earth was shapeless, empty and darkness dwelt upon its face. It was of unmeasurable magnitude and in great commotion.

With this translation, which carries more completely the basic meaning of the Hebrew words, it seems totally impossible for this statement to refer to anything except a diffuse nebula.

We have been able to weigh astronomically the diffuse nebulae, and they have been found to be extremely rare (void or empty), rarer, in fact, than a good vacuum produced in our laboratories. They are dark, extremely rare and have been found to be in commotion. Now look at the photographs of dark nebulae and tell what shape they are. You can not help seeing that they fill perfectly the description of the second verse of Genesis. They are without form, very rare and dark. So today we have an astronomical body which perfectly fills the description of Genesis 1:2. We need no longer hang our head when the critic points at this verse. It certainly can not be considered absurd to think the earth came from the dark nebulae.

### The Earth Was Once Quite Hot

A theory involving the steps through which a planet could be developed from a dark nebula has not as yet been given by any scientist, but when it is it must involve two necessary steps. (1) Any consideration of the earlier layers of the earth or of its interior proves that at an earlier time the earth was hot and the surface has since cooled off. Therefore, any theory must include the earth as a fairly hot body, and then its cooling off to its present condition. (2) This theory must involve a diffuse nebula contracting into a more dense body and heating up; then cooling down again into a solid body like our planet. (This is in no way an attempt to make a new theory regarding the formation of our solar system. It has nothing to do with that problem.

Henry Norris Russell in his book, *The Solar System and its Origin*, pp. 123-126, briefly discusses the possibility of our solar system coming from a diffuse nebula. He states no obstacles to such a beginning.)

**3. Earliest Life**

"And the spirit of God moved upon the face of the waters." When this is translated so as to carry the basic meaning of the Hebrew words it will read something like this: And the Spirit of God was gently brooding on the face of the waters, *imparting life.*

Science claims that life first appeared in the sea in its earliest primitive forms. Nothing could agree better with the last half of this second verse of Genesis 1. The life spoken of here in this verse must be considered to be most elementary for the more complex types are all named at later periods.

**4. Light**

"And God said, Let there be light." The evidence in the astronomical world is very strong and well accepted that stars go through a cycle, coming from a diffuse nebula and first becoming visible as low temperature red stars, large in size, low in density. They then heat up to very hot stars, smaller and more dense than before. Later, after using up their energy, they gradually cool off and become dead stars or solid bodies like planets. (See *Astronomy*, Duncan, pp. 363-366, especially p. 364; and *Astronomy*, Baker, 3rd edition, pp. 420-423.)

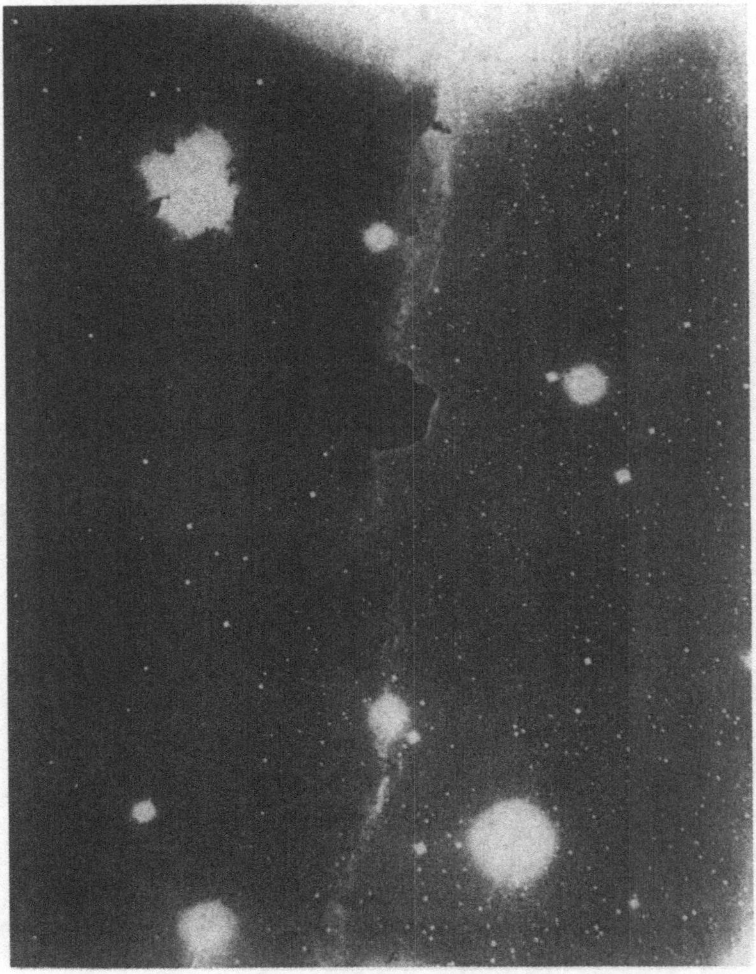

*Photograph, courtesy Mount Wilson Observatory.*

Fig. 4—Dark Bay Nebula.

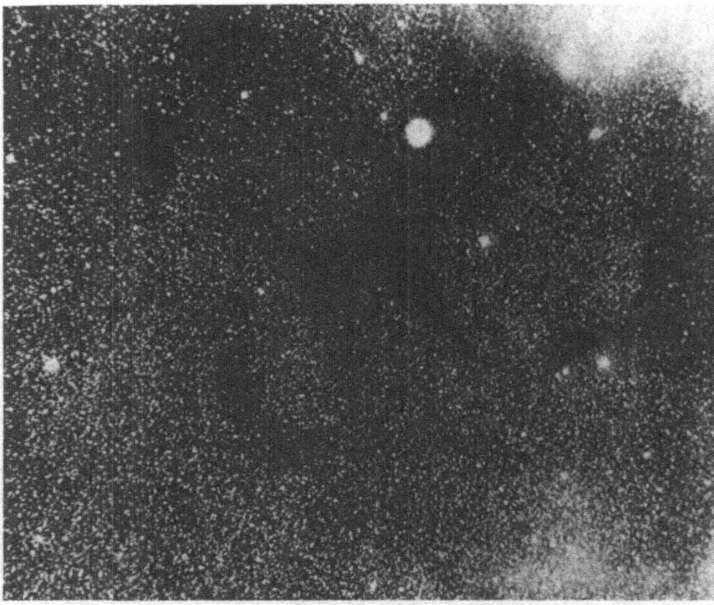

Fig. 5—S-shaped Nebula.

Fig. 6—Pelican Nebula.
*Photographs, courtesy Mount Wilson Observatory.*

According to this sequence the star apparently first becomes visible when very rare, large and just hot enough to glow red. It gradually rises in temperature to 14,000 centigrade surface temperature or even much higher. After its mass has been seriously reduced by radiation it slowly cools down and becomes a solid cold body.

In order to pass through this complete sequence a star's mass must not be less than one one-hundredth the mass of the sun. Since the mass of the earth is only 1/333,000 the mass of the sun it could not have reached such high temperatures. The light described in the third verse therefore may best be considered to come from the sun, as it still does, and not light coming from the earth at temperatures high enough to radiate light. With its small mass the earth probably never became hot enough to radiate light.

### 5. The Earth Cools Off

"And God made the firmament, and divided the waters which were under the firmament from the waters which were above the firmament." A firmament means a space. So the verse means that there were deposits of water on the earth, then a distance or space of air and dense clouds above. It is easy to show that as a planet (such as the earth) cools off it must go through this stage. When there was enough heat in the earth, there could have been no deposits of water upon it, for the heat would have evaporated them, and all of the water at one time would have had to be covering the earth as extremely dense clouds extending from the surface of the earth upwards. When the earth got cool enough, some of this water would have condensed upon the surface of the earth. The cooler it got, the more water we would find deposited on the earth, and the smaller would be the amount suspended in clouds. Yes, this condition of the earth was necessary, and had to follow the condition when the earth was much hotter. (See *The Evolution of the Earth and its Inhabitants*, Lull, Yale University Press, pp. 36-38.)

### 6. The Earth Was Covered With Water

"And God said, Let the waters under the heaven be gathered together into one place." This certainly indicates that at an early time the earth was covered with water. Geology very well attests to this early condition. (See *The Evolution of the Earth and its Inhabitants*, Lull, Yale University Press, p. 38; *Surface History of the Earth*, Joly, Oxford University Press, Preface and the Building of the Mountains, p. 107.)

### 7. The Continents Were Formed

"And let the dry land appear." Different theories are given by Geology for the rise and fall of continents which has gone on for great lengths of time; but I think all agree that in its earliest geological stages the surface was quite smooth and of a uniform height. When forces such as those produced by shrinkage raised parts of the planet above others, the continents were born. (See *The Evolution of the Earth and its Inhabitants,* p. 43.)

### 8. Vegetation

"And God said, Let the earth bring forth grass," etc. Chemistry has established the fact that it is impossible for any planet to have any amount of oxygen in its atmosphere without having large amounts of plants to continually replenish the supply, otherwise the oxygen would rapidly combine with the rocks and metals of the surface and deplete the atmosphere. Basically, plants are the food of animals. Some animals eat other animals but in the main must eat plants or they would exterminate each other. Therefore, we know that plants had to precede animals. I am not discussing very primitive forms of plants and animals, but plants and animals as described in Genesis. Perhaps this account in Genesis parallels the Pre-Cambrian period, but we can not be certain. We can, however, be certain that plants came before the animals. (See *Historical Geology,* Miller, Van Nostand Co., p. 70.)

### 9. Breaks Appear in the Clouded Sky

"And God said, Let there be lights in the firmament of the heaven to divide the day from the night; and let them be for signs, and for seasons, and for days, and years: . . . . And God made two great lights; the greater light to rule the day, and the lesser light to rule the night."

This passage in Genesis has also been one of the great points of attack of the critic against the Bible. He has said that the Bible could not be true, for this passage says that the sun was created after vegetation; and that would be impossible, for vegetation could not grow without the heat and light of the sun. Now it is perfectly true that vegetation could not grow without the heat and light of the sun, but Genesis does not claim that vegetation was created before the sun. The word used in this passage is not created but made. Scofield translates this word as "made to function," and it is a legitimate translation. The account of the creation of the sun and moon is given in verse 1, while the account of making the sun to shine on the earth is given in verses 14 to 18.

In item number 5 above, I described the condition of the earth at this early stage as being completely covered with clouds. If you should evaporate much of the water in the oceans and put it in the sky as clouds, the clouds would be so dense that you could not tell the time of day, the season of the year, or any other sign that we read from the sun today. So when God made the sun to function for signs, for seasons, for days, and for years, He accomplished this by breaking the clouds and letting the light shine through.

Suppose today were a very dark cloudy day and I should tell you that tomorrow I would *make* the sun shine on the earth. Would you think I intended to create a new sun and put it in the sky to shine on the earth? Certainly not. You would know that I intended to scatter the clouds so that the sun, which had been there for millions of years, could shine through. Let us be just as reasonable in interpreting our Bibles.

There is then no contradiction in vegetation appearing on the earth before the clouds were broken away so that the sun could shine through direct. Before the clouds were broken, there was still considerable internal heat in the earth; otherwise, nearly all of the water in the clouds would have already precipitated upon the earth, and the clouds would have broken. It is heat that keeps the water suspended in the atmosphere. This means that just before the clouds were broken we had a hothouse condition on the earth. The clouds above scattered the rays of light from the sun so they would not burn the vegetation, just as does the semi-opaque glass over a hothouse; and the interior heat within the earth kept everything quite warm and vegetation grew at its best. You cannot make a better setup for the advent of vegetation than that given in Genesis.

### 10. Fish Were Created

"And God created great whales (sea monsters) and every living creature that moveth, which the waters brought forth abundantly." If we turn to Geology we find the Devonian period designated as that of fish. This may easily be the period ushered in by the creation of fish. I am, however, not trying to say that these acts of God are connected to any particular geological period or layer, I am only concerned in their order. The creation of fish, certainly, could not have come after the Devonian period. Some authorities might put it earlier. (See *Historical Geology,* Miller, pp. 137-142.)

**11. Birds**

"—and every winged fowl after his kind." We might suggest the Jurassic period as being the most likely one in Geology to be ushered in by the creation of birds, for birds are abundant in this period. Again, however, some might choose to place the origin of birds at an earlier time, but definitely after fish. (See *Historical Geology,* Miller, p. 234.)

**12. Mammals**

"And God said, Let the earth bring forth the living creature after his kind, cattle, and creeping thing, and beast of the earth." Here I would suggest that the Tertiary period best fits the advent of these animals; however, someone may wish to place it earlier. The appearance of mammals will certainly be later than that of birds. (See *Historical Geology,* Miller, pp. 319-327.)

**13. Man Was Created**

And "God created man in his own image." This, all agree, is the last, or Quaternary period. No remains of man are found at an earlier date. (See *Historical Geology,* Miller, pp. 372-381 for the first appearance of man and what little is known about him.)

We then find that the thirteen things named in Genesis are in the correct order. We must therefore ask the question, from what source did Moses obtain this order? I think there are five possible answers. a. The information came from the schools of Egypt where Moses was educated. b. Perhaps you wish to say Moses did not write Genesis; that it was much later and came in the Babylonian period of culture. c. Perhaps it came from some other civilization. d. Perhaps he just made up the story, or sat down and reasoned it out. e. Perhaps it came by inspiration of God.

If these are the only possibilities, one of these five things must be correct; and if we can prove four to be impossible, we will then have established the fifth. Let us consider them in order.

a. Could this account have come from the schools of Egypt? The answer is no. We have, in our libraries, accounts taken from tablets of stone, a record of creation as taught by the Egyptians; and while it is similar to the account of Genesis in a few respects, it does not agree in most of its items or in the general description of events. It does not agree with present scientific information. Neither Moses nor any other person could have obtained this account in Genesis from Egypt. (See *Mythology of All Races*, Marshall Jones Co., Vol. XII; *Egyptian*, by W. Max Muller, pp. 33-127, 173-183 and 212-245.)

b. Could the account have been taken from the schools of Babylon? The answer must be emphatically no. We also have a record of the theories of creation as believed by the people of Babylon. One is something like this: two gods fought, one slew the other and out of his flesh he made the earth, out of his bones and teeth he made the rocks, and out of his blood he made the oceans and the rivers. No, this certainly is not the source. Other theories from Babylon are as fantastic. (See *Mythology of all Races,* Vol. V; *Semetic,* Stephen Herbert Langdon, pp. 277-325.) Note how very different these stories are throughout from Genesis!

c. We may search every possible source in our libraries but none give an account approaching that in Genesis 1.

d. Could Moses have guessed the proper order? If we say yes, the chance is extremely poor. The order of these 13 items would also be considered satisfactory if items 3 and 4 occupied any place from 3 to 7 inclusive. (Science has not yet advanced far enough to tell exactly where these two items must be.) The probability of Moses guessing and getting the correct order for these 13 items is

$$\frac{\text{The possible number of satisfactory orders}}{\text{The total possible number of orders}}$$

In symbols this is $\frac{{}_5P_2}{{}_{13}P_{13}}$. This is $\frac{20}{6{,}227{,}020{,}800}$ or $\frac{1}{311{,}351{,}040}$. Yes, we could say that Moses just guessed and while he had only one chance in 311,351,040 he did get it right. But that is not very logical.

Perhaps you say that is not likely, but he may have reasoned it out. This is just as unlikely as the other. Suppose you eliminate from your mind the account in Genesis and absolutely all literature on the subject and sit down to reason it out. What would you write? I do not know what you would write, but I think I should write something like this: In the beginning man was with God in heaven, but he was a physical being. Heaven was made for spiritual beings so man did not get along very well. Man and God talked things over and decided that the thing to do was to make a place which was better suited to human needs. Together they planned the earth so that it would have just the right kind of an atmosphere, etc. Certainly I would not put man down at the very end of the list, no matter what else I did. He must have a place in making this earth the wonderful place that it is. Certainly anyone who watches animals and fish, knows that animals learn to swim, but fish do not learn to walk, so we would all put land animals first and have some of them learn to swim so well that they just stayed in the water and became fish. Other animals kept trying to fly until the front pair of legs developed into wings. Perhaps you say that I am trying to be ridiculous. I am not. I believe this is a more logical way to reason it out than the account in Genesis. Yes, of course, we would all have God doing some special things for man;

such as creating cows and horses to make the work easier for him, and thus place them after man. I do not believe any man can logically reason out the order of creation, or even make a reasonable start at it.

Now this number of 311,351,040 does not tell the story at all. This is only the number of ways in which the 13 things can be satisfactorily arranged. Where did Moses get the thirteen things to arrange? Did Moses know all about dark nebulae so he could write a perfect description of one in the second verse? That is absurd, for the greatest of the scientists having many photographs of dark nebulae never guessed one existed until about 25 years ago. What chance had Moses, as a man, of writing a description of an object to be discovered nearly 4,000 years later? About 100,000,000,000 people lived from the day of Moses before anyone knew what his description meant. We should say, then, that he did not have more than one chance in 100 billion of being able to describe a dark nebula. Let's be conservative and say we will place our estimate at one chance in one billion. What chance did Moses have of knowing that the earth was once covered completely with water? I fear that there is no very definite way in which we can answer this question. Surely Moses had a very poor chance of guessing this early condition. Suppose we be conservative and say we will consider that he had one chance in 100. What chance did Moses have of guessing that the earth in its earlier stage was completely covered with clouds? Let's consider that his chance this time is one in 1,000. One in one million, or more, would perhaps be nearer accuracy.

Now what chance did Moses have when writing this first chapter of getting the thirteen items all accurate and in a satisfactory order? Using the estimates that we have just suggested, we must multiply them all together, and the result we find is that Moses had one chance in 31,135,104,000,000,000,000,000 of getting both the items and the order accurate.

Now this is a very small chance. Let us try to visualize it. Suppose we decide to have a drawing and have this number of tickets printed and numbered from 1 to 31,135,104,000,000,000,000,000. In order to get them printed, we should have to engage more than 8,000,000 presses, each capable of printing 2,000 tickets per minute; and then they would have to run day and night for 5,000,000 years to print this number of tickets. Now let one ticket be marked and the whole mass thoroughly stirred. Then, we will blindfold you and let you draw one ticket. Will you get the right one? Your chance is the same that Moses' chance would have been of writing this one chapter from the information which he had.

Again suppose we try to visualize this chance. Suppose these tickets are 1¼ inches square and they are printed from stock which requires 100 to make a thickness of an inch. How big would the pile be? It could hardly be called a pile for it would cover the whole United States from Canada to Mexico and from the Atlantic to the Pacific one mile deep (that is, piled evenly). Now mark one of these tickets, stir the whole mass thoroughly, and they will reach several miles high. Put on your blindfold and start out, dig as deep as you wish, travel as many days as you wish to reach the chosen spot and come up with a ticket and say, "this is it."

Now is it reasonable for us to say we think Moses took that kind of a chance and got it right? Certainly this is preposterous.

e. We have now disposed of the other four items as impossible or completely outside the realm of reasonableness. We must therefore accept our fifth supposition, that God inspired Moses in the writing of this chapter, as true. We have the claim, "All scripture is given by inspiration of God," and there is no evidence to contradict it.

Let us put the argument in another way. Suppose that I claim that I built a certain old house, but I have no witnesses, and people will not believe that I was the builder. How can I prove that I built it? I can prove that I built it if I can tell you some things which are hidden in the interior construction which no one can see by just looking at the house. If the house is torn open enough to reveal the things which I described, and the things are found, I will have established my claim.

Suppose in such a case I should say: "When the cement was being poured I dropped an old nailset in the cement in the north-east corner; in the south-east corner you will find a lot of old wire packed in with the concrete; you will find in the framing that 2 x 8 floor joists were used instead of 2 x 6 as is common; the south-west corner is framed by spiking together four studs instead of the common three." And so I go on and tell you 13 things that went into that house, running from the foundation to the roof. Then suppose you tear down that house and in the process you find the 13 items placed in agreement with my description, what conclusion will you draw about my claim as the builder? No one in the world would question my claim.

We are in exactly the same position with respect to the verification of these 13 items of Genesis. Science has proven these 13 things to be true, to the best information it can give today. God wrote such an account in Genesis so that in these latter days, when science has greatly developed, we would be able to verify His account and know for a certainty that God created this planet and the life on it. That He wrote this Book we call the Bible, we cannot doubt.

### The Theory of Evolution

Let us go back for a minute to the theory of evolution. Every act of God affecting life on this planet corresponds to a change in the series of fossils. Are these changes most logically accounted for by some freak or unusual act of nature, or are they more logically accounted for as creative acts of God? From the record written in the layers of the earth one explanation would suffice quite as well as the other. In the light of the evidence we have just presented, which proves the Genesis account to be true, we must accept the creative acts of God, and accept at the same time the authenticity and inspiration of God's Word.

This evidence is so strong that science cannot well ignore it. All theories of evolution must take the creative acts of God into account if they are to agree with all of the established facts.

By the recent advancements in science, Genesis has changed from the victorious battleground of the critic to the battleground where the victory lies with God.

"Be ready always to give an answer to every man that asketh you a reason of the hope that is in you" (I Pet. 3:15).

"The heavens declare the glory of God; and the firmament sheweth his handiwork" (Ps. 19:1).

## Part II

# DISCUSSION

### The Days of Genesis

The Hebrew word translated "day" in Genesis can just as well be translated "period of time." The words translated "morning" and "evening" may also be translated "beginning" and "ending." Thus, "And the evening and the morning were the first day," may also be translated, "And the beginning and ending of this work was the first period of time." (See the comments in the Scofield Bible about day, morning and evening, under Genesis 1.)

These words (*day, morning* and *evening*), in the Hebrew are practically the exact equivalents of the same English words. If you look them up in your dictionary you will find that day can be a period of time, that morning and evening can be the beginning and ending of any period of time as well as the beginning and ending of the 24-hour day.

This period of time may have been a 24-hour period or it may have been any other period of time, even a geological period. Furthermore the periods of time in Genesis may have been separated by other long periods of time. God is counting periods of time in which He was doing work on this earth of ours. If after the first act of God, the first period of time, a million years (or any other period of time) elapsed before He again acted, this second act would occupy the second period of time of God's work. This interpretation is also borne out in prophecy where God counts time against Israel. If He predicts that any event will happen to Israel in a certain length of time, that time is counted only while Israel dwells in Palestine as a nation. Are most of us not yet looking for the 70th week of Daniel? (See *The Time of the End,* Dean, Evangelical Publishers, pp. 19-23.)

Two different interpretations of the days of Genesis are possible and in accord with science and the Hebrew language. One is that the days are the different geological periods in which these particular developments took place. The other is that the days are the breaks between the geological periods. If God acted at the time of one of these breaks, then took no hand until the end of the next geological period and acted again in the end of that period (in the break between it and the next) then the breaks must be the days. If on the other hand His acts were the causing of new forms of life to develop from earlier forms (this is possible in the use of the word "made" which appears at certain places) then the geological periods would naturally become the days.

It is not at all necessary for one of these explanations to be true and the other false. Some of the days may be one type, and some the other type. The word "made" implies that it was brought into its present form from something already in existence while the word "created" implies that it was formed from nothing pre-existent. In other words, it is not unreasonable to suppose that wherever the word "made" is used, the day is a geological period, and wherever the word "created" is used, the day is the break between two geological periods. I think it is quite impossible to prove, at this time, which of the interpretations is correct.

### Criticism 1

The criticism is often made that the first and second chapters of Genesis each give an account of the creation and they do not agree. After finishing the account of creation, which extends to the fourth verse of the second chapter, God says, "These are the generations of the heavens and of the earth when they were created." In other words, this is the order in which all of these acts of God were brought about. In the remainder of the second chapter, God enlarges on the things affecting man. There is no contradiction here, merely an expansion.

The second account makes no claim as to order. Many items are left out and the discussion of some of the items occurs in a different order. God mentions man, then animals, in the second chapter. This is logical because He is dealing with man's control of the animals. I must point out that the order here makes no difference. Suppose I should tell you that on a certain day I did a list of things and tell you the order in which I did them. Afterwards if

I wish to go back and enlarge on any of the items, must I take them up in the same order in which I enumerated them? Certainly not. You would make no objection to my enlarging on them in any order I might choose. Should we not, then, give God the same right?

**Criticism 2**

The criticism is made that the date of the writing of Genesis is set too early and that Moses did not write it, but that it was written later by another author, perhaps Joshua. If you should take references of other Bible authors regarding the Pentateuch as the writings of Moses you would find that the evidence was very strong that Moses wrote these books. Take a good concordance, look up the word Moses and see how many books of the Bible credit Moses with the writing of the Pentateuch. Note how early this credit is given and that Christ also gives Moses credit (Matt. 19:7,8 and Deut. 24:1-4). There are other evidences which indicate that the book was first written much earlier by different men and then collected by Moses. (See *Messiah: His First Coming Scheduled,* David L. Cooper, Biblical Research Society.) It makes very little difference as to what explanation you choose regarding the writing of Genesis. Every argument that I have given would hold just as well if the book were written by Joshua, at a little later date. Joshua had no better way of gathering the information given in this first chapter than Moses.

**Criticism 3**

The criticism is also made that the second chapter was written long before the first chapter. I do not believe that this is true for the two chapters fit perfectly together and read as a continuous story from the first to the second chapter. But again what difference does it make? It certainly would be quite all right for God to have first inspired the account of His acts of creation dealing with man and later to have given a complete survey of the whole creation.

**Criticism 4**

The criticism is made that I have ignored the evidence that man came from the ape or both from a common ancestor as claimed by the theory of evolution. There are a large number of bones and organs nearly identical in the two. It is claimed that this must mean that one came from the other. No, it is not even evidence of it.

I am also a carpenter, by trade, and at my home you will find a house, a playhouse, a garage, a chicken coop. If you examine these very closely you will find many things in common. The studding will be found nearly identical in all of the buildings, the rafters will be found to be cut in exactly the same manner and put together in the same fashion in all four buildings. Does that mean that one developed from the other? Does it mean that the chicken coop came first, then the playhouse gradually evolved, and later the garage, then the house? Certainly not. It means that the same man built them all. Now let us be honest. Does a similarity of bones in the ape and man necessarily prove that man came from the ape, or the ape from man; or could it not just as well be taken as evidence that the same God created both? God made the ape with all of his marvelous bones, muscles, and organisms; then when he came to man who has many similar properties to the ape, God certainly used similar bones, muscles, organs, etc., in His creation of him. Where similar parts would not properly fit into the building of man in the image of God, God showed his originality in using different ones. No, this is no more an argument for evolution than it is for creation.

**Criticism 5**

The remains of man are found and dated by some back as far as 1,000,000 years ago. Should not this prove that the account in Genesis of the creation of man dating back about 6,000 years is false?

There are two variable quantities in this picture. In the first place we may be faulty in our time of 6,000 years to Adam. The period of 6,000 years has been quite carefully computed from the biblical accounts and may be fairly accurate; however, it is possible that there may be some gaps in genealogies which have not been detected, and the time back to Adam would then be more than 6,000 years. The other possibility is that the dating of the earlier remains of man are in error. We must remember that all of the remains of man occur in the most recent geological layers. Since these layers are sedimentary, much of them may have been washed from one place to another by streams of water and floods, making it quite impossible to be certain about the age of any single, small deposit. These deposits do not always stay where originally deposited and though a general age may be assigned to a certain depth, parts of the deposit at that depth may be much newer and other parts may be much older. Since only a very few remains of man have been found in surroundings which would indicate greater age than 6,000 years, the ages assigned to them may be quite in error. Especially is this true when only very small parts of a skeleton are found in one place, and that often in a stream bed, for this certainly indicates a moving of the original remains by water; and when that occurs there is no telling from what surroundings it came or with what other remains it may be redeposited. Ages of deposits by streams are particularly uncertain.

Some Hebrew scholars claim that the word "begat" only implies a descendant, a son, grandson, etc. If this is true it could extend the 6,000 years of Ussher's chronology to nearly any length of time. The period of time back to the first appearance of man is too indefinite and we are too uncertain about the age of man as given in the Bible to reach any definite conclusion about their agreement or disagreement, at the present time.

**Criticism 6**

The shape of some of these early skulls indicates a primitive man, ape-like in some facial features. That may be true, and yet I have never visited a museum which displayed these primitive forms and walked out on the street without being able to pick out, on the shoulders of living people, close counterparts to every exhibit. We are apt to classify any skull as primitive if it has a sloping forehead or other variation from the more common form. Dr. Wm. J. Tinkle, Prof. of Biology at LaVerne College, says, "Cromagnon man was very ancient, yet superior to modern man in cranial capacity, height and artistic ability."

**Criticism 7**

Six thousand years is not long enough to reproduce the present population of the earth. This is not true. If the flood occurred only 4,300 years ago, and if only the people in the ark survived on the whole earth, and if the population then doubled each 100 years (this could easily happen and more), the population to date would be more than 10,000,000,000,000. This is 5,000 times the present population of 2,000,000,000.

## Part III

# OTHER INTERPRETATIONS OF GENESIS I

Let us consider the explanation which places all of the geological ages between the first and second verses, declares that the earth was created perfect and was wrecked by the casting of Satan down to the earth, and claims that the days of Genesis involve only the reformation of the earth and the recreation of the life upon it in days 24 hours long, covering a total length of time of one week.

### The Earth Created Perfect

I think God did create the earth perfect, but perfect for His own immediate purpose, not necessarily perfect in the meaning which I would attach to the word. No two of us would likely agree as to what a perfect earth would be. Let us be careful of our interpretations. Not so long ago men were persecuted because they announced that the orbits of the planets were elliptical instead of circular. People then thought that the circle was the perfect curve and therefore God created the planets to move in perfect circles. Today we know that the planets all move in elliptical orbits and we also recognize the ellipse as just as perfect a curve as the circle. Let us be careful about reading into the Bible something that is not there, but which we think ought to be there. Our thoughts are not God's thoughts. We are extremely likely to be wrong.

Our Bible does not say that God created the earth perfect. Therefore, we dare not teach children so.

### The Fall of Satan

What do we know about an early destruction of the earth by the fall of Satan? Nothing. That is also read into the biblical account. The only effect on this earth caused by the casting of Satan out of heaven seems to be still future. (See Rev. 12:7-12 and Rev. 20:1-3.) These will certainly occur in the Great Tribulation period. Other references which are usually cited are Jer. 4:23-26; Isa. 24:1,5,21-23; Isa. 45:18; Ezek. 28:11-19 and Isa. 14:1,4,9-17. A careful study of all of these passages will show that they have nothing to say about a prehistoric time but refer to the Revelation passage, the casting of Satan out of heaven during the Great Tribulation period or to judgments against Jerusalem. In fact, we are clearly taught that Satan still has access to heaven, and is our accuser there.

Yes, there is one passage which is not as clear as the others, that is Luke 10:17, 18. Here, after the disciples report that the devils are subject to them, Christ says that he saw Satan fall from heaven like a star. The past tense is used, but so is the past tense used in many of the prophecies throughout the Bible. (In Isaiah 53 you will find several prophecies referring to Christ. Some of these are in the future tense, some in the present but most of them are in the past tense. This passage alone quite conclusively shows that the tense used in prophecies does not necessarily indicate the time when the event will happen or whether it has happened in the past. The number of such prophecies is great but I will mention only two more: Ps. 22:18 fulfilled in John 19:23,24; Ps. 34:20 fulfilled in John 19:32-36.) I think this is most likely prophetic, referring to the same event recorded in Revelation. We are told that Satan lost his position as the covering cherub, but we are not told that he was cast down to the earth or that anything on the earth was destroyed at that time.

I shall not further pursue this evidence, for in reality it has little to do with the subject at hand. Even if the earth had been wrecked by the fall of Satan between the first and second verses, there is an abundance of scientific evidence which shows that it could have no relation to the Genesis account.

### The Geological Ages

For sake of argument, let us ignore the evidence just cited and see what happens if we place the geological ages between the first two verses. Verse 2 says that the earth *was* without form, void and dark. Translate it *became* if you prefer. In either case, it perfectly describes a dark nebula: the physical condition of a nebula is that it contains no particle larger than the smallest particle of dust and its main composition is separate atoms with many electrons stripped off. Now suppose the geological layers were laid down between these two verses and then the earth was reduced to isolated atoms and reassembled again, would there be any fossils remaining? Not the slightest trace of one. No, according to this translation, the geological ages can not be placed between the first and second verses. Everything on this earth, in rock layers, fossils or anything else, happened after the second verse. Again, suppose that the second verse is changed in its translation so that it does not mean a separation of particles as described above. Let us see what is involved.

This explanation of Genesis is always given to make the days of Genesis 24-hour days. This then places all of the acts of God, described after the second verse as happening in recent times, approximately, say, in the last 6,000 years, certainly in the last geological period. This introduces several unsurmountable difficulties.

**Contradictions**

1. This would make light first to appear on this planet 6,000 years ago. We could not convince any geologist that light first appeared either 6,000 years or 60,000,000 years ago. This explanation would make the verses 3 to 5 definitely contradict known facts of science.
2. Verses 6 and 7 would tell us that the earth was completely covered with clouds and there were no oceans until 6,000 years ago.

Again, this drastically contradicts known information of science, as the covering of the earth solidly with clouds pre-dates nearly all geological eras and periods of time.

3. Verse 9 describes the earth as being completely covered with water, and according to this theory no continent rose from the water before some 6,000 years ago. Any geologist can produce definite evidence that it was hundreds of millions of years ago when our continents were first formed. So this explanation becomes absurd. Geology has very reliable means of setting approximate dates for many of its periods, best of which is the rate of change of Uranium to lead. This rate cannot be altered by any known means; the percentage of Uranium that changes to lead in a century is well known. In any layer where Uranium is found, the per cent which has changed to lead gives an authentic date to that deposit.
4. Verses 14-18 would, according to this explanation, tell us that the sun never shone through the heavy clouds until 6,000 years ago. This is quite as false as any of the other results above mentioned. (Note: the 6,000 years above mentioned is the time usually taught. Any other comparable age would give the same difficulty.)

I trust I have made it clear that this attempt to explain the first chapter of Genesis contradicts science at many points. These may easily be verified by referring to any available historical geology. We cannot hope to teach our young people this theory and then expect that their faith will not be shaken when they study science.

We accuse the colleges of destroying the faith of our young people. It is often true, but too often it is true because we have given our young people false teachings which are sure to lead them into dangers. Let us make sure that what we teach our children is true to the best information obtainable both from science and from the Bible, then we will not have them slipping away from the faith.

### The Canopy Theory

The canopy theory, as often thought, would place spherical shells of various kinds of material around the earth, then have the shells, one after the other, break down and fall upon the earth, forming the different layers of the earth, and the breaking down of the ice layer cause the flood and the ice age.

So far as I know this theory was never accepted by science. I have not been able to find it, or any reference to it, in any science text or encyclopedia. Generations ago it was known to be contrary to the laws of force which act throughout the physical universe. The rings of Saturn have supplied a field for research which has a definite relation to the canopy theory. As early as 1750 the principles involved in a canopy over the earth were quite well understood and by 1850, such a canopy was well known to be impossible. (See Young's *General Astronomy*, pp. 390-395, for a brief discussion of these principles.) I can think of nothing more dangerous than to try to teach such a theory to our younger generation.

May Bible students become more scientific and scientists become better Bible students; then we will be able to overcome the wrecking of the faith of our young people in the colleges.

CHAPTER III

# GEOLOGY AND THE BIBLE

*by*

*Edwin K. Gedney, M.A., Sc.M.**

## INTRODUCTION

The group of Earth Sciences designated by the general term 'Geology' presents many problems and has produced many a conflict in the mind of the student in his effort to correlate geological teaching with his Biblical background. The writer is keenly aware of the variety and intensity of these conflicts by reason of his own experiences of the past as a student of geology and related sciences, and his contacts with later generations of students in this field. Out of these experiences he is confident in stating that none of these difficulties is necessarily insoluble, and that it seems quite feasible to derive a rational and satisfying correlation between the facts of geology and the records of the Bible.

The difficulties that arise may be classed in two groups as major or minor problems. Because of the complexity of geology there arise throughout the Bible a host of problems which may be classed as minor. Some of these are rather important but not basic to an interpretation of either the Scripture or the earth, and are generally quite unrelated to each other. In this group we may place the influence of faulting action on the structure of the Jordan Valley and its environs, the possibility of the bearing of seismology upon the destruction of Jericho, and the likelihood of volcanism associated with the Jordan Rift Valley formation having

*Professor, Gordon College of Theology and Missions, Boston, Massachusetts.

played a part in the obliteration of Sodom. Here also a study of meteorology should give us much light on such weather factors in Bible history as hail, snow, floods, and the distribution of rainfall and desert areas. The peculiar ground water conditions of the Sinai Peninsula and the distribution of mineral resources of the Near East also merit some study. All of these problems and many others, however, may be classed as minor. Because of the limits of this chapter it will not be possible for us to engage in any extended discussion of these phases of the subject.

The major difficulties that arise in the relations between geology and the Bible have to do with such fundamental problems as the origin and development of the physical earth and the light that both paleontology and Scripture shed upon the development of life upon it. In dealing with these major problems we must carefully define the limits of our study. Geology properly begins with the earth as a discrete unit in the cosmos, hence, the discussion as to how this unit came to exist was covered under the heading of astronomy in Chapter II. The bearing of geology on the hypothesis of evolution consists in the historical record of the past as observed in the strata and cannot well include much discussion of the proposed mechanisms for evolution, nor of the frequently cited evidences for evolution derived from the fields of morphology, embryology and chorology which will be more adequately covered in the chapter on biology. In any case such evidences are descriptive of end-products and can be shown to be in harmony with evolution if that theory can be proved by some other means, but are not themselves necessarily proofs of evolution. It is only in the fields of paleontology and historical geology that a sufficient time-lapse is demonstrable to provide evidence for validating or invalidating the evolutionary hypothesis.

Limiting our general discussion therefore to the earth as a discrete cosmic body, and carrying its development down to historic times, we may now enter into a discussion of modern geological thought concerning this development and its relation to the Bible.

THE GEOLOGICAL CONCEPTION OF ORIGINS

Because the very beginning of the earth so far antedates the first observed facts of geology, speculation must be resorted to as a means of reaching an explanation of the arrival in space and time of this particular cosmic body. This speculation is stated and discussed fully in Chapter II. It need only be added here that the origin of the earth is so remote from any observed terrestrial fact that an increasing number of astronomers and geologists are becoming frankly agnostic with regard to the various theories of cosmic evolution. However it is necessary to start somewhere, and most speculative geologists begin with one of two hypotheses which are diverse at first, but eventually lead to the same end.

Professor Chamberlin,[1] and many others, think in terms of the planetesimal hypothesis and conceive of the earth as originating in a grouping of minute particles ejected from the sun under the influence of a passing star. This rudimentary earth was small, cold, and airless. As other minute particles of cosmic material were attracted to this body it grew by accretion and eventually became large enough and gained sufficient mass to attract from space and hold the rapidly moving molecules of water, nitrogen, oxygen, and carbon dioxide. As it grew the presence of increasingly heavy outer layers forced the inner particles to flow together to a considerable extent causing local

[1]For the development of the Planetesimal Hypothesis see:
T. C. Chamberlin, *The Origin of the Earth*, (University of Chicago Press, 1916).
——. *The Two Solar Families*, (University of Chicago Press, 1928).
T. C. Chamberlin and R. D. Salisbury, *College Textbook of Geology*, (New York: Holt, 1930), chap. 10.
F. R. Moulton, *Introduction to Astronomy*, (New York: Macmillian, 1925).
——. "The Planetesimal Hypothesis," *Science*, Vol. 68, No. 1771, (Dec. 7, 1928), pp. 549-559.

melting and the extrusion of the molten material through volcanic vents. The presence of water and atmosphere acting upon the rising and falling land surface began the processes of weathering which initiated soil development, the deposition of sediments of various kinds, and the molding of scenery more or less as we know it today.

Professor R. A. Daly[2] of the Harvard geological faculty, and others who follow the lines of the gas-nebula hypothesis (or any of the variants of the tidal theory), begin with a very different earth; an earth that is about the size of the present one, but gaseous and incandescent with a temperature of at least 3000° centigrade. Jeffreys calculates that this earth crustified relatively rapidly, possibly in a very few thousand years, and subsequently cooled much more slowly over a long period to the present time. This earth was so hot that the entire mass of the oceans, which now cover three-quarters of the terrestrial surface to an average depth of over two miles, was volatilized as steam, expanded fourteen hundred times in volume, and therefore surrounded the hot, naked rocks with an unbroken envelope of thick cloud. Most of the carbon now found in the great limestone and coal beds of the earth is said to have been present at that time as carbon dioxide in the atmosphere. As the temperature of the earth declined, water accumulated in bodies and the water cycle between clouds above and oceans beneath began producing weathering, soils, and sediments. Both of these speculations[3] bring us by differ-

[2]For the development of the Gas-Nebula Hypothesis see:
Sir James Jeans, *Astronomy and Cosmogony*, (Cambridge, England: Cambridge University Press, 1928), chap. 16.
——. *The Universe Around Us*, (New York: Macmillan, 1945), chap. 6.
H. Jeffreys, *The Earth, Its Origin, History and Physical Constitution*, (Cambridge, England: Cambridge University Press, 1929), chap. 2.
R. A. Daly, *Our Mobile Earth*, (New York: Scribner's, 1926), pp. 300-306.

[3]H. N. Russell, *The Solar System and Its Origin*, (New York: Macmillan, 1935), chap. 3. Professor Russell has introduced a variant of the disruption theories in which he suggests that the sun was originally a double star. A passing star caused a disruption of the companion to the sun, part of the material of which has been left behind to form some or all of the planets.

ent routes to the earth as we see it first in the rock records—an earth with atmosphere, water, and land masses upon which erosional agencies are already acting. This is our *first observed geological fact.*

Into this earth life was introduced. Here again it becomes necessary for the geologist to revert to speculation as to the origin of life. There have been many suggestions as to the processes that produced life. Some have suggested that life was brought to earth in meteoric fragments from some other cosmic body. This does not explain the origin of life, but simply places its origin somewhere else than upon earth. Dr. A. I. Oparin,[4] of the Biochemical Institute of the Soviet Academy of Science, has developed the thought of another group of speculators who hold that protoplasm, or living substance, was derived by long natural synthesis of primitive hydrocarbons in the molten or superheated earth. These are said to have been acted upon by the hot waters of the primitive seas to form complex colloids and eventually bacteria-like life. Many paleontologists think of life as coming about by the action of actinic radiation from the sun upon colloidal substances in the sands along the seashore or even in the open sea. Professor Shimer, of the Massachusetts Institute of Technology, summarizes this thought by suggesting that, "The characteristics of life may have made their appearance when a colloid of the requisite chemical structure came into existence."[5] Such a colloid of requisite chemical and energy composition is said by the non-creationist to have become a living organism and to have developed chemical catalysts and later nerve impulses and channels the integration of which produced mind.

---

[4]A. I. Oparin, *The Origin of Life,* translated with annotations by Sergius Morgulis, (New York: Macmillan, 1938).

[5]H. W. Shimer, *An Introduction to the Study of Fossils,* (New York: Macmillan, 1929), p. 17.

This life differentiated into plant and animal forms and, by processes now unknown, is said to have evolved through the invertebrate forms, the fish, the amphibia, the reptiles, and the mammals to man himself. This speculation brings us to the point where theory integrates with the *second great observed geological fact*—the occurrence of living forms as fossil traces in the strata.

These living forms were apparently abundant in the first important fossiliferous period, the Cambrian. About two thousand species of life have been found in the Cambrian strata in which all the phyla of animal life except the chordates or vertebrates, and most of the great classes are represented. This constitutes a *third great geological fact,* that all the invertebrate phyla appear contemporaneously with marked suddenness in the Cambrian differentiated into phyla, classes, and orders, and with no clear indication as to how they developed into this condition if they did develop at all. Dr. Austin H. Clark (Smithsonian Institute) says,[6]

> "So we see that the fossil record, the actual history of the animal life on the earth, bears out the assumption that at its very first appearance animal life in its broader features was in essentially the same form as that in which we now know it. . . . Thus, so far as concerns the major groups of animals, the creationists seem to have the better of the argument. There is not the slightest evidence that any of the major groups arose from any other."

In a later book Dr. Clark writes,[7]

> "The fauna of the Cambrian period were singularly similar to the animals of the present day," ". . . the facts are that all of the fossils, even the very earliest of them, fall into existing major groups. This is indisputable."

W. K. Brooks says of the Cambrian life,[8]

> "Far from showing us the simple unspecialized ancestors of modern animals, they are most intensely modern themselves in the zoological

[6] A. H. Clark, *Quarterly Review of Biology*, Dec. 1928, p. 539.
[7] A. H. Clark, *The New Evolution, Zoogenesis*, (Baltimore: Williams and Wilkins, 1930), pp. 101, 105.
[8] W. K. Brooks, *The Foundation of Zoology*, p. 216.

sense, and they belong to the same order of nature as that which prevails at the present day. . . . Nothing brings home more vividly to the zoologist a picture of the diversity of the lower Cambrian fauna and of its intimate relation to the fauna on the bottom of the modern ocean than the thought that he would have found on the old Cambrian shore the same opportunity to study the embryology and anatomy of pteropods and gastropods and lamellibranchs, of crustacea and medusae, echinoderms and brachipods, that he now has at a marine laboratory."

A. H. Cook writes,[9]

"The first undisputed traces of animal life, which appear in the Cambrian epoch, exhibit the same phyletic distinctions as now exist. Sponges, echinoderms, mollusca and worms, formed already, in those immeasurably remote ages, groups apparently as generally distinct from one another as they are at the present time."

This is a truly great fact that has been a source of marked perplexity to those who hold the view of evolution. Several noted geologists have spent a considerable part of their professional careers seeking some evidence of living forms in the Pre-Cambrian strata. Professor Percy E. Raymond writes:[10]

"Pre-Cambrian times were the real 'dark ages.' Many industrious and brilliant geologists have devoted their lives to the study of the rocks then formed, but though they have learned a great deal about them, they have been unable to establish a satisfactory chronology. Authentic history begins with the oldest really fossiliferous strata."

After some eighty years of search this effort must be said to show a singular lack of success. A considerable number of objects have been found and noted in the literature as fossils, but these have been exceedingly difficult to corroborate. Professor R. S. Lull, among American paleontologists one of the foremost proponents of evolution, states that over one-half of the total evolution of life had probably

[9]A. H. Cook, *Cambridge Natural History*, p. 5.
[10]P. E. Raymond, *Prehistoric Life*, (Cambridge: Harvard University Press, 1939), p. 39.

taken place before the Cambrian period began.[11] His statement makes the fact of the abruptness of appearance and completeness of differentiation of Cambrian fossils most remarkable in the light of the paucity of traces of their ancestry.

Professor Percy E. Raymond of Harvard University, in his Presidential address to the Paleontological Society of America entitled "Pre-Cambrian Life," states that the life of the Pre-Cambrian eras as represented by actual fossils cannot be shown to comprise more than four groups such as the various algae, sponges, and worm-like creatures. He further states that "If paleontologists were called upon for strictly scientific evidence, they could not prove their determination of any one of these groups is correct."[12] In the latter part of his very frank paper, Prof. Raymond reasons backward from the nature of Cambrian life and by inference describes what the Pre-Cambrian life might have been like. This, however, accounts for the first half of evolutionary development by theory and not by observation of evidence. In this paper Prof. Raymond also notes and discusses seven different theories propounded to account for this absence of Pre-Cambrian fossils, including a very interesting contribution of his own.

This conjointness of appearance of invertebrate life forms is in marked contrast to the *fourth great geological observation* – that subsequent to the Cambrian period the new forms do not appear at the same time, but that there is a succession of appearances of progressively advanced forms set apart from each other by rather sharp breaks. We find the fish, the amphibia, the walking, flying, and feathered reptiles, the birds, and lastly the mammals. During each of

---

[11]R. S. Lull, *The Ways of Life*, (New York: Macmillan, 1929), Rev. ed. Professor Lull gives a good discussion of the relative duration of the geologic eras in relation to the forms of life evidenced by fossils in each.
——. *Organic Evolution*, (New York: Harper, 1925).

[12]P. E. Raymond, "Pre-Cambrian Life," *Bulletin of the Geological Society of America*, (1935), Vol. 46, p. 375.

the so-called 'Ages' in which these creatures successively predominated there is a development of type and genera over a considerable time with an abundance of fossil representatives available. It is at the crucial transition points, where one of these classes is said to have changed into another by some evolutionary process that evidence is wholly lacking. Dr. A. H. Clark writes:[13]

"No matter how far back we go in the fossil records of previous animal life upon the earth we find no trace of any animal forms which are intermediate between the various major groups or phyla. . . . . The greatest groups of animal life do not merge into one another. They are and have been fixed from the beginning. . . . No animals are known even from the earliest rocks which cannot at once be assigned to their proper phylum or major group. . . . There can be only one interpretation of this entire lack of any intermediates between the major groups of animals. . . . If we are willing to accept the facts at their face value, we must believe that there were never such intermediates, or, in other words, that these major groups from the very first bore the same relation to each other that they do at the present day."

(Clark quotes Dr. Gerritt Miller of Smithsonian Institute), "The complete absence of any intermediate forms between the major groups of animals, which is one of the most striking and most significant phenomena brought out by zoology, has hitherto been overlooked, or at least ignored."

The writer has friends who have been searching without success for many years in the Devonian strata for some evidence of the evolution of the crossopterygian fish, or of the lung-fish, into amphibia.[14] The remarkable appearance and

[13]A. H. Clark, *The New Evolution*, pp. 189ff.

[14]Fruitless efforts have been made to trace the development of *Osteolepis* and other crossopterygian fish into amphibia. Raymond discusses these efforts in his book *Prehistoric Life*, pages 99 and 100. He finds that the *only* kind of fish that had a limb capable of modification into an amphibian leg was the crossopterygian. He then notes that the *only* kind of fish with swim bladders adapted to breathing were the dipnoi or lung fish. He observes that Osteolepis had a cavity in the skull with passages to the mouth cavity which may have been olfactory in function (although whether air or water was taken in would be difficult to establish). From this he *assumes* that it must have had a lung. He says, "We are justified therefore in postulating (sic) the existence in early or Mid-Devonian times of a ganoid with fins somewhat like those of *Sauripterus* and lungs like the dipnoans. This was the material from which the amphibia

disappearance of the reptiles succeeded by highly differentiated land, sea, and air inhabiting mammals and birds, without suitable graduations between, has long been a problem to the palentologists. The fact of the progressive appearance of these living forms seems definitely clear from geological observation. The gradual evolution of these forms from one to the other is equally clearly not demonstrated by the testimony of the strata, but must be inferred, hypothesized, or supplied by faith.

This constitutes our *fifth important Geological observattion* – the absence from the strata of transition fossils to link together the important forms. Many end-products of a supposed evolution are found, but the connectives, of which we should reasonably expect to have an immense number, are missing. For example, during the Mesozoic Era the land, sea, and air are occupied by reptiles as the dominant forms. There are running, walking, creeping, swimming and flying reptiles. Some thirty or more species of feathered creatures sometimes classed as ancestral birds have also been recovered. These creatures are, aside from the feathers, closer in their resemblances to reptiles than to the birds, and inasmuch as they experience a degenerative change during the era, losing the power of flight, and to a considerable degree that of walking as well, cannot and are not to be regarded as intermediates between the reptiles and birds but are generally considered as a separate development.[15]

At the end of the Mesozoic Era a tremendous change took place in living forms. The new fauna of the succeeding

might (sic) evolve." In other words he finds evidence of a nose in one kind of fish, a lung in another kind, and "*postulates*" that there must have been a form of fish, not yet discovered, combining these features, which "might" have evolved into amphibia. He further notes the discoveries of Lange and Koch in Greenland, of late-Devonian amphibia and says of it, "unfortunately they are typical four-footed creatures only slightly more fishlike than some later members of their group" (page 109).

[15]See discussion in Raymond's *Prehistoric Life* pages 185-188. He discusses the three common theories for the evolution of the birds and recognizes the lack of evidence for them.

Cenozoic Era is markedly different. Of some 21 great classes of life common in the Mesozoic, eight become totally extinct, six continue but are greatly limited in spread, four continue and increase markedly, and three appear for the first time and expand. Of the nine orders of the class of reptiles that flourished in the Mesozoic, all but four die out suddenly and completely and one of the four continues greatly limited. The class of birds, exclusive of the feathered reptile forms mentioned that do not seem to belong here, begins widely differentiated into its orders and many of its genera. The eleven orders of the class of mammals appear in the earliest period of the Cenozoic.

Efforts have been made, with very doubtful success, to find mammal remains before the Cenozoic. Certain very small forms have been found in the Mesozoic that have been allocated to Mammalian class, but no connectives between these and the highly differentiated land, sea, and air-adapted mammals of the Cenozoic can be shown. Professor Cleland, in his textbook of Geology, says:[16]

> "In the present imperfect state of our knowledge of the life at the beginning of the Cenozoic, it is perhaps even more difficult to account for the presence of highly developed mammals, as soon as the reptiles become extinct, than to account for the disappearance of the latter. It is improbable that the mammals on the earth today were descended from any of the mammals whose remains have been found in the later Mesozoic rocks."

The fact of these gaps in connections between the important forms can only be incorporated into the hypothesis of evolution by an exercise of faith. There are many such gaps.

A *sixth important geological observation* is the directionalism of the progressive appearance of living forms. Observed successions of forms in the strata are obviously in determined lines and do not follow fortuitous variations as selection or mutation would be expected to produce. This

16H. F. Cleland, *Geology, Physical and Historical*, (New York: American Book Co., 1916), Part II, p. 590.

has been recognized by many scientists with the result that Theistic evolution, implying a guiding 'Life-Principle,' often called God, has been introduced. The God of theistic evolution is, however, usually an intangible something introduced as a sort of trigger effect where evidence is lacking. The "Mystical principle of progressive development" of Nageli and others falls into this category.[17]

The very fact that such efforts have been made to provide a power to promote directionalism of development is strong evidence of the failure of the evolution concept to explain successfully this important observed fact.

### *The Test of the Geological Interpretation*

The foregoing survey of the generally accredited geological interpretation of the earth's origin and development, and the relation of that interpretation to some of the great geological facts permit us now to arrive at a general assessment of the validity and value of this current scientific view. Three things are at once apparent from the preceding discussion.

First, with the possible exception of the rather vague theistic evolution variant, it is not possible to reconcile the interpretation with the Scripture. This is due to the fact that the two are essentially irreconcilable. The scientific interpretation is basically mechanical in character, blind in its operation, and quite void of the supernatural. The Scripture is based primarily upon the supernatural as an effective cause. This is indeed a most profound difference, and definitely separates the two views. Few of the proponents of the evolution concept, although they frequently claim the existence of a sort of vague, actionless Deity, recognize the personal, omnipotent God indicated in nearly every verse of the Genesis account.

---

17H. H. Newman, *Evolution, Genetics and Eugenics*, (University of Chicago Press, 1925), p. 34. Prof. Newman has an extensive summary on preceding and following pages, of various forms of evolution theory. The particular citation summarizes Nageli's and other orthogenetic views.

Second, we note the tenuousness of the chain of facts by which the thread of the interpretation is carried along. The interpretation consists of a group of theoretical bridges spanning the yawning gulfs between the observed facts. It does not fit upon these factual buttresses too well. It implies a gradual unfolding of forms, legato in character, whereas the strata indicate a progression that is staccato or interjectory in character.

Third, the element of faith is as essential, or even more essential, than in the rival interpretation based upon the supernatural. The only way to bridge the gaps in the progression of life is by faith. Obviously any theory of Origins must involve some faith, for no human observers attended the generation of the earth, nor were the events of this beginning recorded in observed strata. However, for a group prone to criticize the exercise of faith on the part of those who hold to the supernatural view, and to claim the necessity for factual bases for scientific thinking, the proponents of this mechanical interpretation manifest an astonishing amount of pure faith and muster an unusually small array of facts in proportion. Dr. A. H. Clark says:[18]

> "Biologists are, of course, confessedly evolutionists, but it is really remarkable how little evidence they advance in support of their position."

Dr. E. L. G. Watson (British Zoologist) writes:[19]

> "The vast body of the tree of evolution is entirely imaginary, and no material creatures have been found to correspond to it."

In the first place the entirety of cosmic evolution and the early phases of terrestrial generation must be taken on faith. Lull, in the book previously referred to,[20] indicated that the first half of organic evolution probably took place

[18]Clark, *Op. Cit.*, p. 533.
[19]E. L. G. Watson, *Transactions of the Victoria Institute*, (1938), p. 11.
[20]Lull, *Op. Cit.*

before we find fossil remains, and this means that one-half of the next phase of terrestrial development, the origin and differentiation of life, must be taken on faith. The hiatuses between the great vertebrate classes, the fish, amphibia, reptiles, birds, and mammals, must be leaped in faith. Some time ago the writer asked one of his students to prepare a chart of the evolution of man from the Cretaceous to the present using various recent texts as sources and putting in the known connections of the family tree in solid lines, and the hypothetical connections and branchings in dotted lines. His surprise was intense when, after his chart was complete, his attention was called to the fact that although in other respects it resembled standard charts it had the peculiar character that *all* the connecting lines on it were dotted (See Fig. 5). In other words, the evolution of the human being from the reptile, the various branchings from the new and old world monkeys, the orthograde primates, and the great anthropoid stem, all the various splits from the main line, can be accepted only by the exercise of faith. Professor E. A. Hooton of Harvard University, writes concerning the many variant restorations of prehistoric men as follows:[21]

"Some anatomists model reconstructions of fossil skulls by building up the soft parts of the head and face upon a skull case, and thus produce a bust purporting to represent the appearance of the fossil man in life. When, however, we recall the fragmentary condition of most of the skulls, the faces usually being missing, we can readily see that even the reconstruction of the facial skeleton leaves room for a good deal of doubt as to details. The various reconstructions of the skull of Piltdown Man by Smith-Woodward, Keith, and other experts, differ widely one from another. To attempt to restore the soft parts is an even more hazardous undertaking. The lips, the eyes, the ears, and the nasal tip, leave no clue on the underlying bony parts. You can with equal facility model on a Neanderthal skull the features of a chimpanzee or the lineaments of a philosopher. These alleged restorations of ancient types of man have very little, if any, scientific value, and are likely only to mislead the public. To model a bust

21E. A. Hooton, *Up from the Ape*, (New York: Macmillan, 1931), p. 332.

of Pithecanthropus erectus from the skull cap and the two or three teeth is a palpable absurdity. We do not know anything of the minutiae of the appearance of the Pithecanthropus, Heidelberg, Piltdown, or Neanderthal types. We have no knowledge of their hair form, hair distribution, pigmentation and the detail of such features as I have mentioned. So put not your trust in reconstructions."

There is a most interesting and excellent article by a noted student of ancient anthropoid forms showing the extremely tenuous nature of evidence for Pithecanthropus and other supposed early men.[22] The writer shows an unusual ability to assume that truly objective attitude typical of a true scholar, but so frequently lacking in publications on both sides of the question of Evolution. It is not possible to quote this extensive article here, but it is well worth reading in its entirety by any young Christian student disturbed by the evident authoritativeness with which the validity of Pithecanthropus and other doubtful forms are presented in textbooks and college lectures. Dr. Miller writes, in summary,

"We do not hesitate to confess that in place of demonstrable links between man and other mammals, we now possess nothing more than some fossils so fragmentary that they are susceptible of being interpreted as something else."

A great number of highly intelligent persons, however, exercise this necessary element of faith. Raymond,[23] after showing that there were no absolute or unquestionable evidences of life in the Pre-Cambrian, exercised faith, and reasoning backward described the life that "must have existed there." Henry Fairfield Osborn acknowledges frankly that after nearly a century the quest for causes to implement evolution has failed, yet remains strong in his faith in the hypothesis.

---

[22]G. Miller, "The Controversy over Human Missing Links," *Smithsonian Institute Annual Report,* (1928), pp. 413-466.
[23]Raymond, *Op. Cit.*, see last third of the paper.

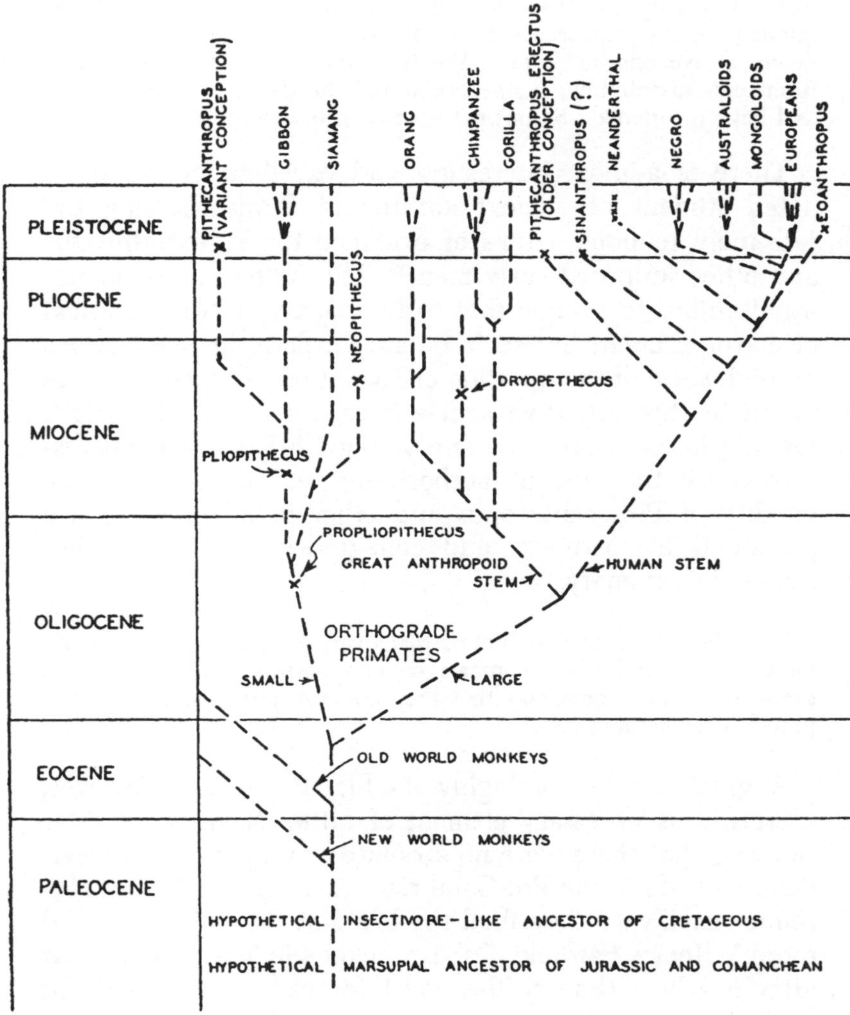


*Fig.* 5—The Family Tree of Humanity (modified after Keith). The broken lines indicate *conjectured* relationships. *Note*: The family trees designed by various anthropologists and paleontologists differ markedly from one another.

Dr. Osborn, from his position as President of the American Museum of Natural History in New York City, had come to view the various methods suggested by investigators from a really objective viewpoint. He was therefore, better able than the inventors of theory, to assess their relative merits and weaknesses. In the introduction to his book he writes:[24]

"In constrast to the unity of opinion on the law of evolution is the wide diversity of opinion on the causes of evolution. In fact, the causes of the evolution of life are as mysterious as the law of evolution is certain. . . . Between the appearance of *The Origin of Species* in 1859, and the present time there have been great waves of faith in one explanation and then in another: each of these waves of confidence has ended in disappointment, until finally we have reached a state of very general scepticism. Thus the long period of evolution, experiment, and reasoning which began with the French natural philosopher Buffon, one hundred and fifty years ago, ends in 1916 with the general feeling that our search for causes, far from being near completion, has only just begun. . . . It is best frankly to acknowledge that the chief causes of the orderly evolution of the germ are still entirely unknown, and that our search must take an entirely fresh start."

Dr. Osborn's "entirely fresh start" consisted in combining mutationism and selection into his Tetrakinetic theory, but neither this nor any later combination has proved generally satisfactory as a mechanism in evolution.

Many able scientists recognize the weaknesses and difficulties associated with the problem of substantiating and implementing the theory, yet accept the basic principle of evolution faithfully, have no alternative that seems to them to be acceptable. For the intelligent mind that does not acknowledge the activity and reality of the supernatural it is the only alternative, and such a person is forced to look to the hypothesis as the only rational interpretation of the world and of life.

---

[24]H. F. Osborn, *The Origin and Evolution of Life*, (New York: Scribner's 1918).

Dr. W. R. Thompson of the Imperial Institute of Entomology says,[25]

> "The concept of organic evolution is very highly prized by biologists, for many of whom, it is an object of genuinely religious devotion, because they regard it as a supreme integrative principle. This probably is the reason why severe methodological criticism employed in other departments of Biology has not yet been brought to bear against evolutionary speculations. There are, however, indications that this criticism will not now be long delayed."

Professor W. B. Scott writes,[26]

> "My experience with graduate students of Biology has shown me that usually their traning has so completely taken for granted the truth of the evolutionary doctrine, that they have but a vague conception of the testimony by which that doctrine is supported."

The writer's association with graduate students during the past twenty years in strongly corroborative of this statement.

Dr. Paul Shorey (University of Chicago) writes,[27]

> "There is, in fact, no cause that is so immune from criticism . . . as evolution. . . . An ambitious young professor may safely assail Christianity, or the constitution of the United States, or George Washington, or female chastity, or marriage, or private property . . . but he must not apologize for Bryan. It is not done."

For the Christian, however, who does acknowledge an active, real, supernatural power, the blind, mechanical, poorly supported concept of evolution may seem highly inadequate and he prefers to think in terms of the second main division of this chapter.

### The Biblical Interpretation of Origins

The Biblical interpretation of Origins is essentially based upon the supernatural. It has long been the opinion of the writer that the most difficult verse in the Bible is the

---

[25]W. R. Thompson, *Science and Common Sense,* (New York: Longmans, Green, 1937), p. 229.
[26]W. B. Scott, *The Theory of Evolution,* (New York: Macmillan, 1917), p. 10.
[27]P. Shorey, *Atlantic Monthly,* Oct. 1928, p. 478.

first one. If this verse can be received in its full significance, all the miracles, prophecies and other evidences of the supernatural will hold nothing but interpretive exegetical difficulty for the reader. If it is not received, there will be a stumbling block in nearly every chapter of the book.

Like the scientists, the conservative Biblical interpreters are not wholly agreed among themselves as to just how to interpret the first chapters of the Bible. While all of them agree as to the inspiration of the text there is wide disagreement as to the rendering and significance of the Hebrew original. These are not differences of bases, however, but rather differences of interpretation. The principal theories of interpretation will be studied and an attempt made to relate them to the observed geological facts in the succeeding paragraphs.

### Instantaneous Creation

The oldest interpretation of the passage is undoubtedly the Instantaneous Creation Theory. The proponents of this view think of the earth and its inhabitants and the universe as having been brought forth by a series of six direct creative acts accomplished in a period of six days of twenty-four hours each or a total of one hundred forty-four hours. In other words, the entire universe was produced out of nothing in less than a week. This view was generally held by the Church down through the later middle ages of history, but not universally. There were a few, such as Augustine and Thomas Aquinas who dissented from the general opinion. On the basis of this view the earth was regarded as between five and six thousand years old, and very justly so, for no one had as yet discovered any reason for thinking otherwise. It is obviously impossible to correlate the data of geology with the Genesis account on the basis of this interpretation, for its very nature obviates the testimony of geology. The influence of this view has diminished greatly in the last century, but it is still held by many devout Christians.

*The Influence of New Discoveries*

About a century and a half ago there began a great acceleration in the exploration of the world from an investigative or scientific point of view. This was not as remarkable as some writers seek to have it appear; the remarkable thing is that for thousands of years few had done it. The secrets of chemistry, physics and biology began to be brought to light. With the invention of the spinning jenny and the steam engine to operate it the industrial age with its tremendous development of knowledge through discovery and research began. Among other things men discovered the stratification of the sedimentary rocks and the significance of the fossils contained in them.[28] Others went down into the coal mines to obtain the new fuel which was then just coming into general use. There they found that the coal beds were composed of the carbonized remains of pre-historic plants. Many successive coal beds were sometimes found with intervening strata of various rocks apparently deposited when the surface was depressed beneath the sea between the forested or coal forming epochs. A tremendous time interval was indicated in the growth and carbonization of these forests and the deposition of intervening sea deposits.

---

[28]Editor's Note: In all fairness it should be mentioned that there are those who believe the various geological strata do not have any time significance, but merely indicate ecological zones, i.e., a segregation of former life more sharply on the basis of ecological factors than we now know it to be. The adherents to this view believe modern paleontologists are guilty of "reasoning in a circle" by dating strata by fossils included, and, in turn, dating fossils by the strata in which they are found. Prof. George McCready Price, author of *The New Geology* (1923), is foremost champion of this thesis.

The author of this chapter, however, believes that the above view is a result of an abbreviated experience with the practical use of fossils and a failure to comprehend the history of age correlation. He also points out that physical relations constitute the ultimate basis for determining the age of strata, and that only the high positive correlation which has been found between the position of the strata and the fossils contained therein gives geologists confidence in using fossils for identification of layers.

Still others visted such places as Specimen Ridge in Yellowstone Park. Here they found that forests had grown to maturity; then had been destroyed and buried by considerable thicknesses of volcanic ash ejected by some local volcano. On this ash bed a new forest grew and the trees reached large size only to be destroyed by a new eruption. In the side of the two thousand foot cliff can be discovered a dozen successive forests representing intervals of great length between ash effusions—the trees now completely petrified, many of them in vertical position. (See Fig. 6). Not only this, but after the destruction of the last forest the ash bed became lithified and a river canyon about two thousand feet deep was cut in it exposing the forests in cross section. (See Fig. 7). In addition to the time required for this action, beneath the lowest forest were found various rocks that were ancient long before the first forests grew upon them. Obviously a tremendously long time factor was represented here.

In the Grand Canyon of the Colorado, geological explorers found that in the very bottom of the mile deep gorge were exposed highly contorted rocks that had apparently been deposited in immense thicknesses in early seas, solidified by lithification into rock, then elevated and folded into mountains. Subsequently weathering agencies had worn down the mountains and upon their upturned and eroded rock strata new and great thicknesses of sediments had been deposited in another prolonged submergence beneath the sea. This series of strata had also been lithified into rock and in turn uplifted and broken into block-like mountains. These had been eroded to a level plain by weathering. Then a new submergence took place and in the new sea over three thousand feet of strata were deposited. These were lithified and the whole region elevated high above sea level. Since then the Colorado river had had time to cut a gorge five to fifteen miles wide and a mile deep through these rocks. (See Fig. 8). A time period far longer than anyone had ever conceived before must have passed since the beginning of these events.

Evidences such as the above came in from all parts of the earth, and two important and highly controversial problems were engendered. These proved to be exceedingly difficult to harmonize with the Special Creation interpretation of Genesis. *First,* was the problem of the fossils. Why were the fossils present in the strata? What did they represent? Apparently God must have created the fossils *in situ* and as fossils, and the natural question arose as to why this seemingly illogical procedure should have taken place. *Second,* was the obviously very great age of the earth as implied by the evidences of the rocks. Geologists differed widely among themselves as to how old the earth was, but all agreed that the five to six thousand years admitted by the theologians of the day were hopelessly inadequate to admit the production of the observed facts. The controversy over these problems was further complicated by the introduction of the non-supernatural, mechanical hypothesis of evolution.

A great divergence of opinion came to exist between the scientists of the day and the theologians. The three "Swords of Satan," sarcasm, caustic speech, and ridicule, were wielded with deadly skill. In this the advocates of the Biblical interpretation, unfortunately, proved themselves as able as their scientific opponents. The natural consequences emerged from the bitter debates of the last decades of the past century; feelings were hurt, personalities exchanged, and the great widening gulf between the Biblical interpreter and the scientific mechanist came to exist.

Many sincere and intelligent Bible students and not a few devout scientists, notably James D. Dana,[29] the greatest

[29] J. D. Dana, *Manual of Geology,* (New York: Ivison, Blakeman, Taylor, 1880), 3rd ed., p. 604.

of our early American geologists, greatly regretted this unfortunate situation and cast about for some means of rapprochement between the two fields of learning. These men felt that the special creation hypothesis failed to meet the situation. They believed in a God who was able to create the fossils *in situ* and the earth with the appearance of great age, but felt that it put too great a strain on the credulity to think that this was the way in which He actually worked.

They reasoned that we have two records of God's handiwork. One of these is the Bible or the Book of Words. It is an inspired book and authoritative. It has an object or purpose, which is to reveal the great plan of God for the Universe and His relations with mankind at various times in the unfolding of that plan. On the other hand every act leaves its trace. The pressure of the thumb upon the desk leaves its scarcely visible print, while the blows of an axe would demolish it. We have then in the broken pages of the rock strata the traces of God's activity in what we might call the Book of Works. Hugh Miller,[30] the founder of geology, had this thought in mind when he wrote his geological treatise entitled, "*Footprints of the Creator.*" The Book of Works is not inspired nor is it prepared for a purpose. Many of its pages are missing or damaged. It is like the fragmentary and disordered manuscript left by an author in the preparation of a finished work.

These books, being the works of the same author, should agree where they record the same events. We do not regard the Bible as a textbook of Egyptology. When we desire to know something about the history of Egypt we visit the library or museum and consult the Egyptologists and their works. However, we may expect that where the Bible and Egyptian history overlap there should be good

[30]H. Miller, *The Footprints of the Creator* or the Asterolepis of Stromness, with memoir of the author by Louis Agassiz, (Boston: Gould and Lincoln, 1851).

agreement with the proved findings of the Egyptologist.

In like manner, the Bible is not a textbook of geology, astronomy, or any other science. We would not consult it to find out the successive epochs of the Carboniferous periods nor for data concerning the periodicity of volcanic or glacial phenomena. However, where the Bible discusses matters of geology, as it does briefly in Genesis, we have every reason to expect agreement between it and the proved facts of Geology, bearing in mind that its purpose is not to convey scientific truth, but to satisfy the minds of men who knew little of scientific truth, as to their origin and the origin of the earth on which they lived in connection with the revelation of God's plan for them.

Where there is disagreement, therefore, there must be something wrong, either with our observation of the facts of nature, or in our interpretation of the Words of the Book. The evidences of the last century for the great age of the earth seemed so overwhelming that the latter possible source of error seemed more likely to those sincere investigators and they began a resurvey of the early chapters of Genesis. This resurvey resulted in the rise of two alternate theories of interpretation.

*The Interval or Reconstitution Hypothesis*

This interpretation is called variously the interval, reconstitution, or cataclysmic hypothesis, the choice of terms depending upon whether the emphasis in the mind of the writer is placed upon the time element, the character of the Divine work implied, or the nature of the event involved.

The proponents of this view postulate a time interval of indefinite and unascertainable duration between the first and second verses of Genesis. They would read the first two verses of the chapter about as follows:

"In the beginning God created the heaven and the earth.

And (after an indefinite period of time suitable for the geological ages) the earth became waste and empty."

This interpretation implies then that the universe existed for a period of time of possibly very great length; then at least the surface of the earth experienced a purgative catastrophe in which most or all of its life perished. Subsequent to this catastrophe God recreated the earth in a period of six days of twenty-four hours each, or a total of one hundred forty-four hours.

The basis for this new rendering of the second verse of Genesis was said to be in a new significance given to the conjunction "and," and a new translation of the verb "was." The conjunction was said to imply a break or departure from the preceding words and thought, and the verb was retranslated to signify "becoming." Support for the idea of a previous creation and testing of life upon the earth has been adduced from Isaiah 24:1; 45:18; Jeremiah 4:23-26; and Ezekiel 28:12-19. These passages are regarded by many as implying a previous testing and fall of angelic beings, terminated by a cataclysmic destruction and subsequent reconstitution of the earth.

A number of favorable and unfavorable criticisms were immediately brought to bear upon this interpretation. From the favorable viewpoint it was acknowledged to be an excellent compromise, and helped to solve the great difficulties involved in the fossils and the age of the earth. It obviated the creation of the fossils *in situ* and it gave unlimited opportunity for any time interval between the verses. When the radiologists announced that the lead-thorium and lead-uranium ratios in the oldest known rocks indicated an age of the order of 1800 million years, and that the earth was undoubtedly much older than these rocks, the new fact was easily integrated into the interval implied by this theory of interpretation. These favorable aspects still hold good today.

There were also a number of unfavorable criticisms raised which are also still valid today. One of these is that the interpretation still places a great and perhaps unnecessary strain on the credulity by implying the creation of thousands of species instantly and the entire recreation of the earth in less than a week. From the geological viewpoint it fails to explain the orderly progression of the fossils, but merely relegates them to an earlier creation. We are still left with the same old problem of whether the orderly appearance of the fossils is a product of evolution, or whether they are products of successive creations. Furthermore it is impossible to correlate the great facts of geology and the observed minor facts of that science with the Genesis account on the basis of the interpretation. This is due to the fact that these phenomena are said to have occurred during the interval and therefore no mention of these geological events is made in the Bible, for the events from verse three onward have to do with historic times only.

From the viewpoint of the Hebrew scholar several criticisms were levelled at the theory.[31] It was pointed out that the word "and" at the beginning of verse two is simply the ordinary Hebrew "Waw conjunctive," a simple connective which *in itself* cannot indicate such a break as that asserted in the proposed interpretation. It occurs at the very beginning of both Leviticus and Numbers with nothing preceding it, and has no special significance as of a break between the verses. Many Hebrew scholars also disagree as to the justification for giving the verb of verse two the sense of "becoming." There is a Hebrew expression that has this sense, but this is not used in the verse. The word used is the simple verb "to be" and appears to signify

[31]For a critical study of the Hebrew the author has sought the opinions of Hebrew scholars. He is particularly indebted to Dr. Burton L. Goddard, Dean of Gordon College of Theology and Missions for reviewing portions of this chapter in which word-studies are involved. The Brown, Driver and Briggs lexicon has been used in preparing the material.

that the earth was at the time in an unformed, unfinished state, much like the condition of clay in the potter's hands before he molds it into a definite shape. Furthermore the exegesis of Ezekiel 28, and the other references mentioned as implying a creation previous to the cataclysm, was and is regarded as very questionable by a large number of very able Bible students.

The interpretation, however, found a ready acceptance by numbers of Christians who sought reconciliation of their belief and the findings of science, and it is held today by a large part of the conservative Christian Church. It is the most prominent of the views set forth in the footnotes on Genesis in the widely used Scofield Reference Edition of the Bible.[32]

## Progressive Creation

Many of the students of the last century adopted a third view which came to be known as the Progressive Creation Theory. The proponents of this view interpreted the word "day" of the first chapter of Genesis as signifying a period of time rather than as a twenty-four hour day. The hypothesis implied the origin of the earth by six or seven stages, or periods, of indefinite and not necessarily equal duration, in which the present world conditions were produced by a series of creations or creative acts separated by considerable time intervals, with intervening development and change of species.

Inasmuch as this interpretation is apparently the only one whereby it is possible to arrive at a specific harmony between the modern geological record and the verses of Genesis, it will be necessary to enter into a little longer explanation of various matters that bear upon the development of the theory. The first of these preliminary discussions has to do with the use of the Hebrew word im-

[32]*The Holy Bible*, Edited by C. I. Scofield, (New York: Oxford Unversity Press, American Branch, 1917), footnotes 2, 3 and 4, page 3; footnote 3, page 4.

plying the idea of creation, which may be transliterated "Bara." It is used only four times in the first chapter of Genesis; once in connection with the creation of the heavens and earth (Gen. 1:1), once in connection with animal life (Gen. 1:21), and twice concerning human beings (Gen. 1:26,27).[33] The other words used for the introduction of light, the luminaries, the fish and the other phenomena and objects of the chapter are words that signify "making," "causing to appear," "bringing forth" and other ideas that do not connote creation, forming, shaping, in the sense of "bringing into being out of nothing." Only three creative acts are expressly mentioned, although others may have occurred. It is presumed, for example, by the proponents of this view that plant life was created, but the Hebrew of the text does not expressly state it so. A literal rendering of the first five Hebrew words of the eleventh verse might be given as follows:

"And God said, Let the earth sprout sprouts . . ."

The second preliminary discussion has to do with the phrase used as a sort of terminus to the activities of the various days, "And the evening and the morning were the . . . day." It is natural for us to think of this as meaning ending and beginning, for so it is in our culture. This was not true of the Hebrew age. The beginning of the day was the evening and it is logical to think that the writer intended it to signify that here. The first day started in darkness, the subsequent days began with the close or darkening of the light of the day before. Evening and morning then have the sense of "beginning" and "ending" in the passage.[34]

---

[33]In the entire Old Testament the simple stem of this verb is used some thirty times, but in every case it refers to the activity of God, and in no case is there any suggestion of pre-existing materials being used in the creative process.

[34]In a personal comment Dr. Allan A. MacRae, President of Faith Theological Seminary suggests that there would hardly be a regular evening and morning before the sun appeared and consequently these terms must be used figuratively as suggested above.

*Fig.* 6—Petrified tree trunks at Specimen Ridge, Yellowstone National Park. More than a dozen such petrified forests occur at this location, one above the other. (*Photo by Mr. J. P. Iddings, U.S. Geological Survey*)

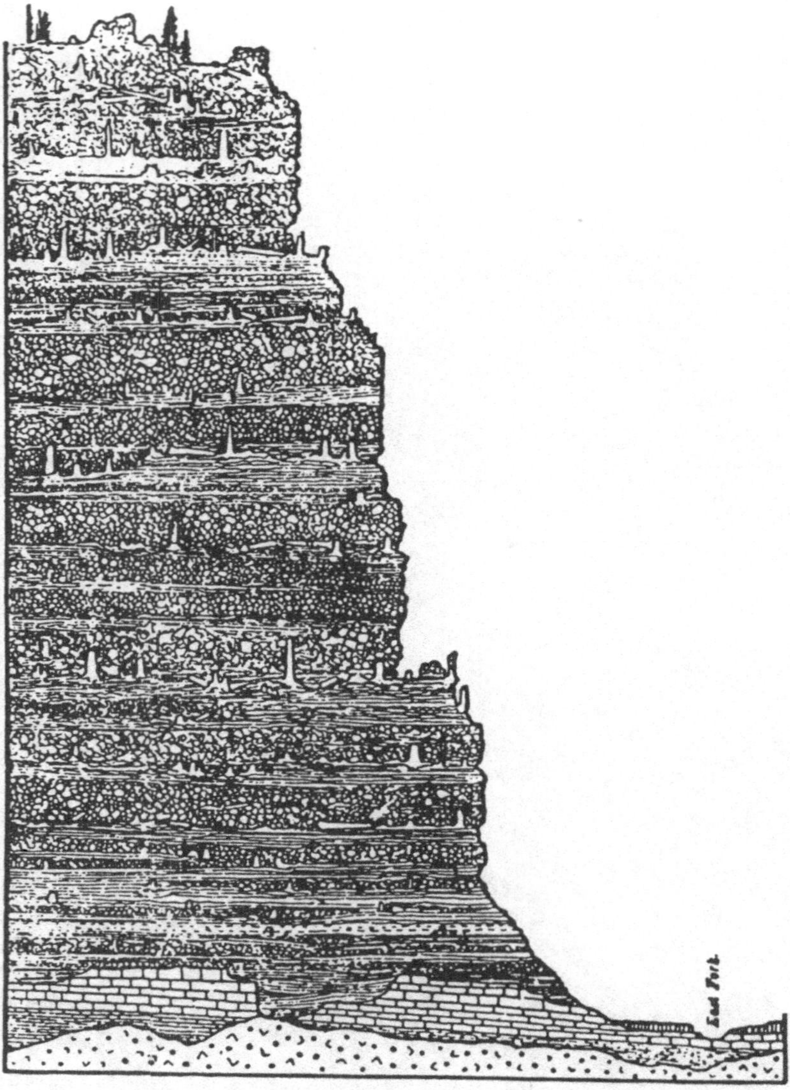

*Fig.* 7—A cross-section sketch of Specimen Ridge, Yellowstone National Park. Long periods of time must have been necessary for the successive growth to maturity and destruction by volcanic eruption of the dozen layers of forests which constitute this ridge. Further time was necessary for the river to cut its way down through this two thousand foot cliff. (*After Holmes, U.S. Geological Survey; taken from Miller's "Historical Geology," Van Nostrand*)

The third preliminary discussion is of the use of the Hebrew word for day—transliterated "yom." The students of the last century put much study upon the uses of the word, for it was the basis for the chief difficulty in the controversy between the Biblical and scientific accounts. They quickly discovered that the word may be interpreted in a number of ways.[35] For example, it is translated six times as "age" (in the plural), sixty-seven times as "time," and also has had several other scattered equivalents in our language that have been given it in the various translations. When it is translated by our word "day," as it usually is, it may have a number of implications.

It is used to designate:

a. The twenty-four hour period from evening to evening;
b. That portion of the above period that is light, (Gen. 1:14,16,18);
c. A prophetic year, (Ezek. 4:6);
d. A day set apart with a distinctive purpose as the Day of Atonement, or the Sabbath Day, (Gen. 2:3);
e. A period of time of indefinite duration during which God begins and ends some specific work, (Gen. 2:4; Isa. 2:12; 22:5).

All but the third of these usages will be found in the first chapters of Genesis.

With this orientation we may proceed to suggest a harmony of Genesis with geologic facts and with recent geological speculation where facts are not to be found. It must be borne in mind in connection with the preparation of such a harmony that the Book of Words presents a comprehensive outline with few details. Great events are passed over in a word. Some are omitted entirely. On the

[35]Although the Scofield Reference Edition of the Bible stresses the Reconstitution view, it also indicates usages of the word "day" and the phrase "evening-morning" that are quite in harmony with the suggestions of the Progressive Creation view. See footnotes 1 and 2 on page 4 of the Scofield Bible.

# A SUGGESTED HARMONY

## Days of Genesis

### Day I

"In the beginning God created the heaven and the earth. And the earth was without form, and void; and darkness was upon the face of the deep. And the Spirit of God moved upon the face of the waters. And God said, Let there be light: and there was light. And God saw the light, that it was good: and God divided the light from the darkness. And God called the light Day, and the darkness he called Night. And the evening and the morning were the first day" (Gen. 1:1-5).

### Day II

"And God said, Let there be a firmament in the midst of the waters, and let it divide the waters from the waters. And God made the firmament, and divided the waters which were under the firmament from the waters which were above the firmament: and it was so. And God called the firmament Heaven. And the evening and the morning were the second day" (Gen. 1:6-8).

### Day III

"And God said, Let the waters under the heaven be gathered together unto one place, and let the dry land appear: and it was so. And God called the dry land Earth; and the gathering together of the waters called the Seas: and God saw that it was good. And God said, Let the earth bring forth grass, the herb yielding seed, and the fruit tree yielding fruit after his kind, whose seed is in itself, upon the earth: and it was was so. And the earth brought forth grass, and herb yielding seed after his kind, and the tree yielding fruit, whose seed was in itself, after his kind: and God saw that it was good. And the evening and the morning were the third day" (Gen. 1:9-13).

### Day IV

"And God said, Let there be lights in the firmament of the heaven to divide the day from the night; and let them be for signs, and for seasons, and for days, and years: And let them be for lights in the firmament of the heaven to give light upon the earth: and it was so. And God made two great lights; the greater light to rule the day, and the lesser light to rule the night: he made the stars also. And God set them in the firmament of the heaven to give light upon the earth, And to rule over the day and over the night, and to divide the light from the darkness: and God saw that it was good. And the evening and the morning were the fourth day" (Gen. 1:14-19).

### Day V

"And God said, Let the waters bring forth abundantly the moving creature that hath life, and fowl that may fly above the earth in the open firmament of heaven. And God created great whales and every living creature that moveth, which the waters brought forth abundantly, after their kind, and every winged fowl after his kind: and God saw that it was good. And God blessed them, saying, Be fruitful, and multiply, and fill the waters in the seas, and let fowl multiply in the earth. And the evening and the morning were the fifth day" (Gen. 1:20-23).

### Day VI

"And God said, Let the earth bring forth the living creature after his kind, cattle, and creeping thing, and beast of the earth after his kind: and it was so. And God made the beast of the earth after his kind, and cattle after their kind, and every thing that creepeth upon the earth after his kind: and God saw that it was good. And God said, Let us make man in our image, after our likeness: and let them have dominion over the fish of the sea, and over the fowl of the air, and over the cattle, and over all the earth, and over every creeping thing that creepeth upon the earth. So God created man in his own image, in the image of God created he him; male and female created he them. And God blessed them, and God said unto them, Be fruitful, and multiply, and replenish the earth, and subdue it: and have dominion over the fish of the sea, and over the fowl of the air, and over every living thing that moveth upon the earth. And God said, Behold, I have given you every herb bearing seed, which is upon the face of all the earth, and every tree, in the which is the fruit of a tree yielding seed; to you it shall be for meat. And to every beast of the earth, and to every fowl of the air, and to every thing that creepeth upon the earth, wherein there is life, I have given every green herb for meat: and it was so. And God saw every thing that he had made, and, behold, it was very good. And the evening and the morning were the sixth day" (Gen. 1:24-31).

### Day VII

"Thus the heavens and the earth were finished, and all the host of them. And on the seventh day God ended his work which He had made; and He rested on the seventh day from all his work which he had made. And God blessed the seventh day, and sanctified it: because that in it he had rested from all his work which God created and made" (Gen. 2:1-3).

OF GENESIS AND GEOLOGY

| EVENTS NOTED IN SCRIPTURE | GEOLOGICAL DEVELOPMENT | GEOLOGICAL TIMETABLE |
|---|---|---|
| The heavens and earth<br>Light | Matter, the universe,<br>Earth and other planets. | Cosmic Era |
| Succession of day and night | Mists surrounding earth.<br>All oceans in atmosphere. | Azoic Era |
| Firmament (expanse)[36] | Waters begin to collect.<br>Clouds and oceans. | |
| The earth out of the sea | Continental shields and beginning of graduation. | Archaeozoic Era |
| The earth brings forth plant life | Algae and other plant forms. | Proterozoic Era |
| Luminaries 'made to appear' | By condensation the cloud envelope breaks so that the sources of light become visible at the surface. | |
| Sea brings forth:<br>Moving creatures[36]<br>Flying creatures (fowl)[36] | Invertebrate life of all kinds.<br>Fish of various types.<br>Amphibia and insects. | Paleozoic Era<br>Cambrian<br>Devonian<br>Carboniferous |
| Sea monsters[36]<br>Animal life of all kinds.[36] | Reptiles of land, sea and air. | |
| Cattle[36]<br>Creepers[36]<br>Beasts of the earth[36] | Mammals of the land, sea and air. True birds. | Cenozoic Era |
| Man[36] | | |
| The Sabbath rest. | No notable new forms. | Psychozoic Era |

other hand the Book of Works is rich in detail but the comprehensive outline must be supplied by theory. The two are therefore complementary to each other. The Bible may be regarded as a complement to geology by supplying the comprehensive plan and causation. Geology serves as a complement to the Bible by providing a wealth of detail to amplify the outline of Genesis. The following table suggests such harmony and is given as an aid to the student in his own research rather than constituting a complete and perfect result in itself.

[36]For the convenience of the student all Hebrew words have been transliterated:

*Firmament* – Hebrew "raqia." It has the sense of something spread or stretched out, therefore an expanse. It can hardly be anything hard and solid as the English "firmament" would imply for in Genesis 1:20 we find the birds flying through it. Apparently signifies the open atmosphere between the heavy unbroken cloud layers above and the waters beneath.

*Moving Creatures* – Hebrew "nephesh chayyah." Identical in the Hebrew with "living creature" in verses 21 and 24, and with "living soul" in chapter 2, verse 7. The idea of movement is derived from the Hebrew "sherets" which has the significance of swarms or an active mass of moving animals. A more literal translation of the first phrase of verse 20 might be given as follows: "And God said, Let the seas swarm (swarms of) living creatures." This is in excellent harmony with the nature of life in the Cambrian and other earlier Paleozoic periods.

*Fowl* – Hebrew " 'oph." This has the significance of "That which covereth with wings," hence, a flying creature. The flying creatures with which the Hebrew commonly had to do were the birds, but the word could also signify any other flyer such as reptiles or mammals.

*Whales* – Literally sea monsters. Hebrew "hattanninim." Descriptive of the great reptilian and elasmobranch forms in the sea.

*Living Creatures* – See "Moving Creatures" above. Signifies breathing creatures or that which has breath or life.

*Cattle* – Hebrew "behemah." Any large quadrupedal animal principally of mammalian type. Sometimes in the plural, specifically used of the hippopotamus.

*Creepers* – The idea of creepers in verses 24 and 25 is derived from the Hebrew "ramas" which signifies "to creep, crawl, or glide swiftly." It can be used of a reptile, or any other animal that crawls or moves rapidly with very short steps.

*Beasts of the earth* – This included in the mind of the Hebrew many mammalian organisms of the non-domestic type. The carnivores, for example, would belong here.

*Man* was formed of the material elements of the earth and into this material the immaterial breath or spirit of life was breathed. How God organized this material part of man, how long it took Him, and just when the formation took place is not clear from the passage.

## Conclusion

We have surveyed the major problems that involve the science of geology and the Bible. Our brief survey brings us to the conclusion that the vital difference between these two areas of knowledge in their interpretations of the earth and the life upon it lies in one's attitude toward the supernatural. To the mechanist the supernatural is an evanescent, intangible concept that is to him wholly unsatisfactory and not suited to a place in his view of the universe. To the supernaturalist the blind, mechanical groping of evolution seems hopelessly inadequate as an interpretation of the earth and the progressions of its life.

One's attitude toward these rival views will vary with his placement of faith, for both views require faith for their acceptance. Many able men who do not seem to find a place in their cosmos for an active, personal, supernatural God have placed their faith in an evolutionary materialism. On the other hand, many highly intelligent and able thinkers, who have come to have a personal experience with the God described in Genesis, find no difficulty in placing their faith in the supernatural view, or rather in the interpretation that includes the supernatural as a guiding and impelling force. Our attitude then will depend to a considerable degree on our attitude toward the supernatural. The realm of the supernatural is essentially the realm of the Spirit. Spiritual things must be spiritually discerned, and are opaque to any other sort of inspection.

The modern scientific age challenges us as Christians to be receptive to new truth as it is discovered. The leaders of our faith counsel us to make certain that a new idea is the truth before we hasten to adopt it. Our reason tells us to examine with great care any new teaching that appears to displace an old established line of thought in which we have been trained, until we have weighed it and found it satisfactory. We must avoid the two rocks of gullibility on the one hand and the ostrich-like attitude of some who deliberately close their eyes to new truth. Hear in closing the counsel of one of the wisest men in the history of the Christian Church in these matters. This is wisdom of fifteen hundred years ago but it is singularly applicable today and in any age as all true wisdom is. Augustine writes to the students of his day as follows:[37]

"It often happens that one who is not a Christian hath some knowledge, derived from the clearest arguments or from the evidence of his senses about the earth, about the heavens, about the other elements of this world, about the movements and revolutions, or about the size and distance of the stars, about certain eclipses of the sun and moon, about the course of the years and the seasons, about the nature of animals, plants and minerals, and about other things of a like nature.

"Now it is an unseemly and mischievious thing, and greatly to be avoided, that a Christian man speaking on such matters, as if according to the authority of Christian Scripture, should talk so foolishly that the unbeliever on hearing him, and observing the extravagance of his error, should hardly be able to refrain from laughing. And the great mischief is, not so much that the man himself is laughed at for his errors, but that our authors are believed, by people without the Church, to have taught such things, and are so condemned as unlearned, and cast aside, to the great loss of those for whose salvation we are so much concerned.

"For when they find one belonging to the Christian body, falling into error on a subject with which they themselves are thoroughly conversant, and when they see him moreover enforcing his groundless opinion by the authority of our Sacred Books, how are they likely to put trust in these Books about the resurrection of the dead, and the hope of eternal life, and the kingdom of heaven, having already come to regard them as fallacious about those things they had themselves learned from observation, or from unquestionable evidence?

"And indeed it were not easy to tell what trouble and sorrow some rash and presumptuous men bring upon their prudent brethren, who, when they are charged with a perverse and false opinion, by those who do not accept the authority of our Books, attempt to put forward these same Holy Books in defence of that which they have lightly

[37]Augustine, *De Genesi ad Litteram*, Lib. i, Cap. xix, n. 39.

and falsely asserted: sometimes even quoting from memory what they think will suit their purpose, and putting forth many words, without well understanding either what they say, or what they are talking about."

The danger of abrupt decisions in some of these matters is also discussed by Augustine in these words:[38]

". . . if we find anything in Divine Scripture which may be variously explained without injury to faith we should not rush headlong, by positive assertion, either to one opinion or the other; lest, if perchance the opinion we have adopted should afterwards turn out to be false, our faith should fall with; and we should be found contending, not so much for the doctrine of the Sacred Scriptures as for our own; endeavouring to make our doctrine to be that of the Scriptures, instead of taking the doctrine of the Scriptures to be ours."

Another great Christian thinker of the Middle Ages, Thomas Aquinas, in discussing these very problems of the first chapter of Genesis writes as follows:[39]

". . . in questions of this sort there are two things to be observed. First, that the truth of Scripture be inviolably maintained. Secondly, since Scripture doth admit of diverse interpretations, that no one cling to any particular exposition with such pertinacity that, if what he supposed to be the teaching of Scripture, should afterwards turn out to be clearly false, he should nevertheless still presume to put it forward, lest thereby the Sacred Scripture should be exposed to the derision of unbelievers, and the way of salvation should be closed to them."

---

[38]*Ibid*, Cap. xviii, n. 37.
[39]Thomas Aquinas, *Summa Theologica*, Pars Prima, Quaest. lxviii, art. 1.

CHAPTER IV

# BIOLOGY AND CREATION

*by*

*William J. Tinkle, Ph.D.**

*and*

*Walter E. Lammerts, Ph.D.***

Let us consider now the animate realm, the realm of organic nature. In His arrangement of the cells of plants and animals the Creator has written a record which the trained person can read. Likewise His inspired Word has made certain statements about plants and animals, although it deals primarily with man's salvation. Since God is the author of both the Bible and nature, it is evident that there can be no discrepancies between them. When discrepancies seem to arise we must dig more deeply into the facts.

God reveals Himself in the intricate structure and beauty of plant and animal tissue. If one strips off the outer layer from a leaf and looks at it under a powerful lens the leaf tissue is magnified, revealing its minute arrangement. It is a layer of cells fitted nicely together, all having similar size and shape, and within each cell there is a smaller body called a nucleus. If one takes a thin slice of a root tip and magnifies it still more, using a compound microscope, he observes that the nucleus is made up of regularly shaped parts, including the rod-shaped chromosomes. *The more we magnify, the more order is revealed to us.*

*Professor and Head, Department of Biology, Taylor University, Upland, Indiana.

**Horticultural Consultant to Manchester Boddy, Rancho del Descanso, La Canada, California, formerly of University of California, Los Angeles, Calif.

However, this intricate structure is not characteristic of the work of man. When magnified, the edge of a very sharp razor blade looks like a jagged and very irregular saw; as if a nervous man had cut out saw teeth with his eyes shut. Or if we look at fine printing we see irregular blobs of ink on criss-cross fibers of paper; and shifting to the highest magnification of the microscope, we can not read the printing. Man has learned to make minute machinery, such as a watch, that performs its function reliably. But the Master Mechanic is not cramped by small spaces, nor does He do His work carelessly where it may never be seen.

The facts discovered by modern scientists afford no basis for rejecting the account of creation given in Genesis. The Bible states the broad, central facts of beginnings in simple dignified language. No human observer was present to explain the creative process but it is narrated by Moses, one of the ablest and most learned men of ancient times. There was ample opportunity for God to reveal these facts to Moses while the latter was on Mount Sinai forty days, receiving the law directly from God.[1]

It is well known that the above attitude is not shared by all biologists; some have even denied parts of the Bible in the name of science. Instead of a planned formation of organisms, which must needs have been the case in creation, they speak of the formation of organisms by the interaction of the forces of the universe. This controversy has raged so long that it is necessary to devote to it nearly all of this chapter. In certain books[2] evolution is called the philosophy of biology, and some may even go so far as to say that it is the religion of the biologist. Thus the subject is too important to be dismissed with a few words.

This unfortunate and needless conflict probably never would have arisen if the church of the Middle Ages had

[1]Exodus 34:28.
[2]Parker and Haswell, *Textbook of Zoology*, Vol. 2, 3rd ed., p. 623.

been true to her divine mission. Since she was not, intellectual men broke away from her tyranny, and then went farther in their views than the facts justified.

Lest we be too severe in our criticism of medieval theologians we must note that the world had sustained terrific losses and most institutions sank lower than the church. There was little education; ignorance, superstition, poverty and war prevailed. To a certain extent the church restrained the evil and soothed the ills of humanity; and no doubt conditions would have been much worse without it. But the priests tried to control even what men thought and all but strangled science by their autocracy.

The learned men of the Middle Ages, called schoolmen, claimed to believe the Bible but did not truly represent it. They studied their philosophy and ancient authorities more than the Word of God, and made no scientific observations. Living organisms were thought to be entirely static and not influenced by chemical and physical laws, although the Bible states nothing to this effect.

The biological break with church authority came in the middle of the nineteenth century, which was late. Considering the progress of "the enlightenment" it might have been expected a century and a half sooner. The theory of evolution had been advocated by Buffon and Lamarck but it was so manifestly at variance with Genesis that people hesitated to accept it.

We wish to make it clear that modern scientific creationists have but little in common with the medieval schoolmen. We disclaim also the mistakes of those theologians who rant against science without understanding it. While agreeing with the evangelical school of Christianity, the authors are scientists and make their living by teaching it or by adding to it through research.

Creationists agree with some statements that have been written to support evolution; for example, modern views of migration and adaptation. However we consider such discussions neutral to the argument; in other words, they support the viewpoint of the creationist quite as well as that of the evolutionist. Unlike some who have claimed to be champions of the Bible, we recognize that changes occur. When the first wingless fruit fly emerged from its pupa case and looked with wondering eyes at the perfect wings of its parents, this was change with a vengeance. But such occurrences as this multiplied by a million would in no way take the place of creation. Such changes give no hint as to how our animals became large, complex, and well adapted to their environments.

Informed workers appreciate now, as no one could appreciate fifty years ago, that there are limits of possible variation; that any change beyond this limit leads to crippling and extinction. Now we know that an organism is governed by its chromosome pattern which provides for small changes but precludes those beyond a certain degree. A fruit fly without wings has had a small change in a chromosome, yet the change in its anatomy is such as to handicap it. A large germinal change, such as loss of a chromosome, would cause its death at an early stage. Thus it now can be shown that it is not good science to say that if a new variety is seen to arise, a new order may arise just as easily as the variety.

This brief mention of chromosomal changes is merely introductory. Those who wish to follow the proof will find it in technical terms in a later section. We cannot state our meaning definitely nor satisfy critical scholars without using technical terms.

It is in the field of genetics along with that of geology that the proof or disproof of evolution lies. Genetics is based upon careful observation and experiment carried on by hundreds of trained investigators. If genetics had been well understood a hundred years earlier the theory of evolution might have been much different or not accepted at all.

### Meaning of Evolution

The original meaning of the word evolution was "unfolding." In this sense the development of embryos was evolution but biologists no longer use the term in this sense. Like many English words it is used loosely and with different meanings, so that definitions are highly important. One rather common view is that organic evolution is the sum of all the changes, good and bad, that are observed in species or races of plants and animals. This definition does not postulate much of a theory or doctrine. It gives no ideas about the direction or extent of changes or of the frequency of change. If one limited his concept of evolution to the changes that have actually been observed without theories to account for them, and conclusions as to their effects, he could believe in organic evolution without endangering his belief in the Bible. But those who have formulated the doctrine of evolution have gone far beyond this definition in their claims.

When a biologist mentions the doctrine of evolution without any qualifications he usually means Charles Darwin's conception of evolution, which amounts to this: All the species of plants and animals now in the world, including men, have arisen by a series of small changes from one or a few very simple forms. Darwin did not account for the origin of these first forms of life. This conception involves a belief in changes which are predominantly in the direction of increased size and complexity since the original germ of life is supposed to have been minute and extremely simple. The motive forces of this development, it is claimed, are the natural forces which scientists can study at the present time. It is this definition the authors of this book have in mind whenever the term evolution is used without qualification. No doubt evolutionists are sincere, but creationists interpret the facts of nature differently.

### Not Alternative to Creation

Before Charles Darwin wrote the "Origin of Species" there was very general belief in the creation of kinds of animals and plants much as we find them today. Now we hear the doctrine of evolution advocated as an alternative to the doctrine of creation. But it is not and can not be a complete substitute for creation, for it fails to account for the origin of life, which was a very important part of creation. Some theory as to the origin of life such as spontaneous generation[3] must be added to the doctrine of evolution by those who advocate it. But after an enormous amount of research no one has been able to produce life. Evolutionists state frankly that they do not know how the primordial cells of protoplasm came into being. The doctrine of evolution is a theory, not of primary origin, but of elaboration; it attempts to explain how increased size and complexity were attained in organisms. Thus one can not rid his mind of the problem of creation (if he finds it a problem) by believing in evolution.

One scientist said in the presence of one of the authors that he believed in evolution so that he would not have to accept anything supernatural. However, as noted above, he can not thus get away from his difficulty. He is trusting a theory that does not attempt to explain his chief perplexity.

There are a number of good people who say they believe Darwin's doctrine is correct but they believe God guided the process and made every change come to pass. This is one way to settle the discrepancy, but it is not founded either upon the Bible or upon science. Moreover this entire-

[3]R. W. Hegner, *College Zoology*, (New York: Macmillian, 1942), 5th ed., p. 721. W. J. Tinkle, *Fundamentals of Zoology*, (Grand Rapids: Zondervan, 1939), p 79.

ly ignores the creation of man in God's image and the fall of man, and in order to make this compromise one must explain much of the Bible figuratively; not only poetic writings which should be so interpreted but also realistic prose. History teaches us that compromises may be useful for a time but that they are only temporary arrangements. Note also the following comments from a modern philosopher:[4]

"In view of the current desire for intellectual prestige on the part of religion and for moral respectability on the part of science, any compromise is likely to make a strong appeal. . . . Peace on this basis is a peace of diplomacy, which merely prepares the way for intenser antagonism."

As stated above there are some passages of the Bible that will not stand a figurative interpretation. Consider Genesis 1:27, "God created man in his own image, in the image of God created he him." There must be an essential likeness between God and man, else the passage has no meaning at all; a likeness that is essentially spiritual, since God is a spirit (John 4:24). This creation involved a plan, a mental pattern, and the employment of means to bring about an end. Such purposeful endeavor, called teleology, is out of fashion at present in evolutionary circles, but fashions come and go.

At this point it will clarify the situation if we consider a case of real evolution, although not organic evolution and not upward in direction as organic evolution is supposed to be. Go to a cliff that is being washed away by the waves of the sea and you will find pebbles being formed as the cliff is broken down. No pebble is planned but each is what it is in size and shape because of the grinding action of rolling versus the resistance offered by the cohesion of the particles of the pebble. At the corners the stresses due to the wearing action are greatest, resulting in a tendency for

[4]M. C. Otto, *Things and Ideals*, (New York: Holt, 1924), p. 208.

the pebble to become a sphere if perfectly homogeneous inside; but if an axis of hard material traverses it the pebble tends to become a spindle. The small and soft pebbles are ground to fine sand, and no pebble remains the same in size or shape.

Man could hardly have attained the likeness of God by such interaction of blind and purposeless forces; nor could each species accrue a definite pattern of chromosomes, the same number and shape in each member of the species.

### Why It Is Believed

One reason why so many people believe in evolution is because so much has been written on the subject. While much has been written, very little that substantiates the theory has appeared during the last quarter of a century. During this time, however, notable discoveries have been made in genetics some of which have created difficulties for the evolutionists. These discoveries will be discussed later. The situation is well illustrated by an announcement of a new book in 1931.[5]

> "Professor Lindsey's new book promises to set up several trains of activity in the scientific world. It is a challenge to the *present lethargic state of evolutionary thinking*. . . . You will certainly admire the clarity of his thought on a subject so *susceptible of befuddlement* and you will find interest in his program for further investigation of the *evolutionary riddle*."

A look at one of the large volumes on evolution impresses one that surely the subject is proved; but on reading it one finds data on genetic experiments, cytological studies, fossils from far and near, comparisons of skeletons, etc.; showing slight changes in some species and great stability in others. The result is that before one is through one has gone off into a number of interesting side lanes and has lost sight of the usual definition of evolution. But the tendency on the

[5]A. W. Lindsey, *Problems of Evolution*, (New York: Macmillan, 1931). The italics have been added.

*Fig.* 8—Bright Angel Canyon, Grand Canyon National Park. A long time was required for the river to cut this mile-deep canyon through solid rock. (*Courtesy U.S. Department of the Interior*)

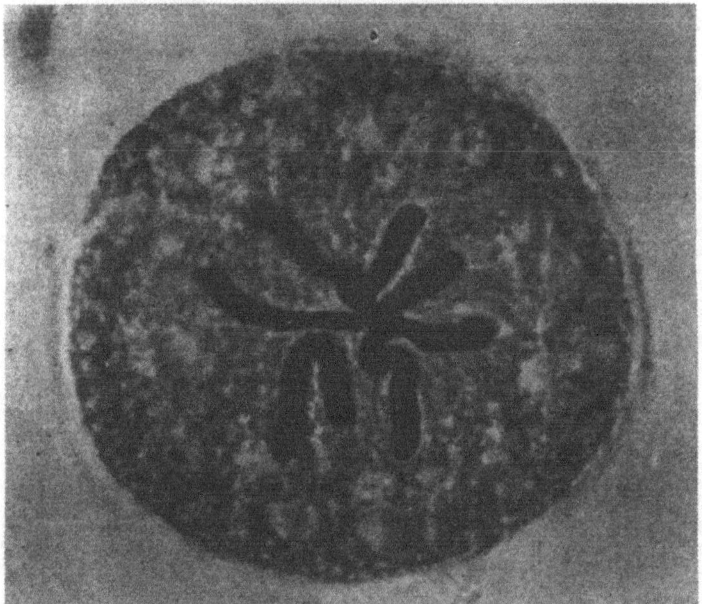

*Fig.* 9–Chromosomes in cell of Ascaris (a worm). In these rod-shaped bodies the traits are passed on to future generations. (*Courtesy General Biological Supply house, Chicago*)

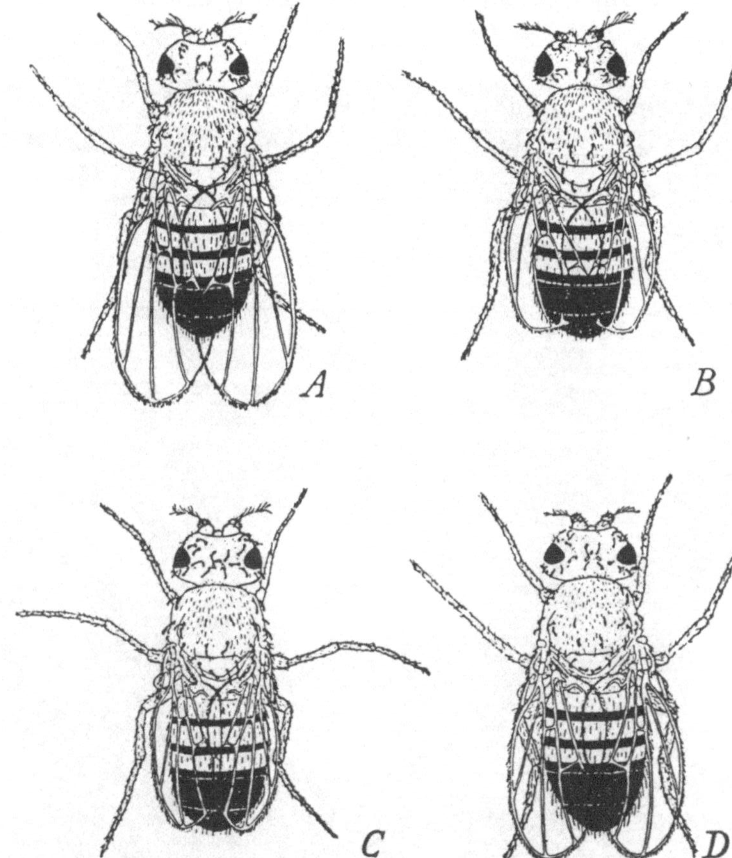


*Fig.* 11–Mutations in Drosophila resulting from radium irradiation. (A) Normal; (B) Rudimentary wing; (C) Miniature wing; (D) Small wing. These are true mutations and they are inherited in regular Mendelian fashion. (*From Hanson in the Journal of Heredity*)

part of the reader is to feel that a man who can cite so much that actually has been observed must be correct in his main thesis; forgetting the fact that the data do not all support his thesis.

A reader's confusion is lessened if he knows what type of evolution the author is attempting to establish. If we say that changes in plants and animals have occurred, resulting in an increase in the number of sorts, we are not in disagreement with the Bible. But if on the other hand we say that the impersonal forces of the environment have constructed our present species from a tiny germ of life in the ocean to new forms, ever larger, more complex, more beautiful, and better adapted to succeed, we are saying quite a different thing. It is this latter viewpoint of evolution which has not been proved for it is quite apparent that the changes observed in nature are not predominantly upward.

Another reason why some people believe in evolution is because it seems to be consistent with the progress that we observe in civilization. For instance, a man plans a machine and builds it by his intelligent toil. He leaves it to his son, who keeps it and improves it. This is progress. Yet this progress is due to planning and effort, while evolution is supposed to trend upward because of inherent forces. This is quite a different thing. Nor does the son inherit a higher intelligence because his father designed the new machine.

### Struggle For Existence

It may surprise some readers to learn that the author of the idea of universal struggle was not a biologist but a sociologist, Thomas Robert Malthus, Professor of history and political economy at the college of the East India Company (1805-1834). Malthus wrote "An Essay on Population" in 1798 in which he stated that life tends to increase much faster than food, resulting in overpopulation and a struggle for existence. He stated that life tends to increase in geometrical ratio, while food tends to increase only in arithmetical ratio. He should have recognized, however, that there can be no distinction between life and food since food is living things or substances produced by living things and that plants and small animals, serving as food for larger animals, reproduce faster than the large animals. He certainly would have been greatly surprised if he could have known that under other circumstances cotton was to be plowed under and little pigs killed to get rid of troublesome surpluses.

Charles Darwin borrowed the idea of struggle from Malthus and made it one of the main principles of his book, "The Origin of the Species." He coined the well-phrased slogans, "struggle for existence" and "survival of the fittest"; which catch one's attention whether the supporting theory is understood or not.

There is a struggle in a forest where young trees, overtopped by large ones, grow very slowly until the old trees fall down. It is a struggle, however, only in a metaphorical sense. It requires effort but not always struggle for a creature to keep alive.

Selection either by man or nature cannot create anything new but only makes more of a certain type already existing.[6] Moreover, natural selection does not tend to favor the more complex forms, the "higher species" of those who classify organisms. Yet it is hard to see how man and other complex species could develop from a tiny bit of protoplasm unless the more complex animal did have the advantage. Man is very much afraid of disease-causing bacteria which are among the simplest of plants. They are the victors in competition in spite of simplicity. Other examples of survival of simple forms are not hard to find. One can put the water flea, Daphnia, under a hand lens and see its in-

[6]T. H. Morgan, *Evolution and Genetics,* (Princeton University Press, 1925), p. 126.

testine, its beating heart, and its young within, ready to be born alive; then put it into a glass of water with Hydra and see it devoured by that simple, two-layered little Coelenterate that has none of these structural features. On the abandoned hills of Pennsylvania, jack pines come up and shade out the shorter vegetation, including plants with a much more complex fruiting structure than those of the pines. Thus natural selection is not making species more complex.

In theory, the organisms that are most fit will survive, but those of greater complexity, greater size, or even greater strength are not necessarily the more fit. Tennyson in "Maud" describes a very fit little shell:

> "Slight, to be crushed with a tap
> Of my finger-nail on the sand,
> Small, but a work divine,
> Frail, but of force to withstand
> Year upon year, the shock
> Of cataract seas that snap
> The three-decker's oaken spine
> Athwart the ledges of rock
> Here on the Breton strand."

It is evident that the shell was fit to survive because it was small enough to fall into the cracks between the rocks. Other shells are fit to survive because of thickness, others because of tough muscles that hold them together, while still others have tissues that are not suited for food. In the end, what kind of animal is a fit one? It is one that does survive. But this is reasoning in a circle, and reveals that Darwin's slogan is little more than a truism. Selection caused by struggle does not account for more complex types.

The idea that nature tends to produce too many little plants and baby animals is true in enough cases that we cannot deny it. It is true that in the resulting competition not all of these organisms will grow large enough to leave offspring. The struggle for existence followed by natural selection does tend to eliminate abnormal individuals, the cripples, including most mutants, and so tends to maintain a standard within a species. It does not raise the standard of a new generation above that of the parental generation. Selection contributes nothing new, for it is negative in its action. And since it does not tend to replace simple species with more complex ones, it is not an alternative to creation as narrated in the Bible.

In the following sections we shall see that evolution cannot be explained by the inheritance of "acquired characters," by recombination of characters, nor by chromosomal aberration. This will leave only mutation to be seriously considered. In the following discussion one must be mindful of the definition of evolution as given above, viz., that all present species of plants and animals are supposed to have come about by upward changes from very simple forms.

### Inheritance of Acquired Characters

Since J. B. Lamarck was the most noted proponent of this doctrine it sometimes is called Lamarckianism. An "acquired character" is a change in a plant or animal due to the environment, to use, or to disuse. No one doubts the occurrence of such characters but authorities agree that they are not inherited by the next generation. If a man toils until he develops callouses on his hands, tans his face, and solidifies his muscles, his children will be no different in these respects than if he had not undergone such discipline. Chinese children, as another instance, still are born with normal sized feet after many generations of foot binding by their mothers.

In order to inherit an acquired modification there would have to be some way for the body cells to influence the eggs or sperms that are formed. Charles Darwin thought of these germ cells as being made up of particles from all over the body, and assembled by the blood stream. This hypothesis, however, is incorrect since when blood was transfused from one animal to another by Galton there was no change in transmitting qualities.[7]

Since some scientists had explained evolution by the inheritance of acquired characters, they were loath to give it up. Herbert Spencer was so convinced that he wrote, "Close contemplation of the facts impresses me more strongly than ever with the two alternatives—either there has been inheritance of acquired characters or there has been no evolution."[8] And Ernest Haeckel writes in the same vein:[9]

> "Belief in the inheritance of acquired characters is a necessary axiom of the monistic creed. . . . Rather than agree with Weissmann in denying the inheritance of acquired characters, it would be better to accept a mysterious creation of all the species as described in the Mosaic account."

The Bible makes a few references to acquired characters but it does not say that they are inherited. Ezekiel reproves the people for quoting the proverb, "The fathers have eaten sour grapes, and the children's teeth are set on edge."[10] In giving the ten commandments[11] God does give warning that He visits the iniquities of the fathers upon the children unto the third and fourth generation. The inheritance of acquired characters is one method by which this warning might be carried out, but true inheritance

[7]Castle and Phillips grafted ovaries from a pure black rabbit into a white rabbit, whose ovaries had been removed. After this white female recovered she was mated three times with a white male rabbit. All the offspring, six in number, were black. This proves that the color of the young animals was determined by the genes in the ovaries and testes, and that these hereditary factors were not influenced by the body color of the female. It is clear that the eggs came from the ovary of the black rabbit and were not influenced by the whiteness of the mother's fur. Similarly an acquired character does not change the cells that will give rise to the next generation. H. E. Walter, *Genetics*, (New York: Macmillan, 1938), 4th ed., p. 24.

[8]J. A. Thompson, *Heredity*, (New York: Coleman, 1926) 5th ed., p. 189.

[9]*Ibid.*

[10]Ezekiel 18:2.

[11]Exodus 20:5.

would not stop with the fourth generation. There are other ways in which parental sins influence children, for instance, the unwholesome or immoral environment in which they are brought up, or infection at birth as in syphillis.

It seems that Jacob[12] shared the belief of his time about prenatal influences, that a mother's experiences affect the unborn progeny. Since the contract with his uncle was that spotted kids should belong to him, he placed spotted rods before the parent goats. Van Haitsma[13] has written a careful discussion of this Biblical account in the light of modern science and finds no discrepancy between the Bible and recent scientific discoveries.

In the herd mentioned above it is true that many spotted kids were born, but not because of Jacob's clever scheme. God revealed to Jacob in a dream the true reason:[14] "All the he-goats which leap upon the flock are ring-streaked, speckled, and grizzled; for I have seen all that Laban doeth unto thee." The rams which impregnated the ewes were spotted, therefore the kids were spotted in large numbers. To the casual observer they were of solid colors, for all the spotted goats had been removed; but their hereditary factors or genes for color were mixed; a condition which the geneticist calls heterozygous. At this point modern genetics helps us, for breeding tests have shown that spotting is recessive to solid color in goats; making it possible for a goat to have spots that can be transmitted, although they do not appear to the eye.

We need not go into further detail on acquired characters, a subject on which all leading modern geneticists agree.[15]

12Genesis 30:37ff.

13J. P. Van Haitsma, *The Supplanter Undeceived*, Privately printed, 1941, sold by H. Kuizema and Son, Grand Rapids, Michigan.

14Genesis 31:12, A.S.V.

15A. H. Sturtevant and G. W. Beadle, *Introduction to Genetics*, (Philadelphia: W. B. Saunders, 1939), p. 48. T. H. Morgan, *Evolution and Genetics*, (Princeton University Press, 1925), chap. 12.

### Recombination of Characters

The second type of change recognized by geneticists is recombination of characters. While these are real and actually occur, they can not account for the alleged building up of the organic world through evolution. The principle of these variations was described by Gregor Mendel after he crossed two varieties of plants. Characters which were formerly scattered in different individuals were combined in one offspring. For example, if a pea plant bearing yellow round seeds is pollinated with pollen from one bearing green wrinkled seeds, some peas in the second filial generation will be of the new combinations, yellow wrinkled, and green round. Along with these will be peas of the same combinations that were used to start the experiment, namely, yellow round and green wrinkled. The recombinations are new varieties in a way, but in another sense there is nothing new in them, since their characters were in existence before. It is clear that this process has little or nothing to do with evolution; since if a complex animal were to be built up from a mass of protoplasm, new characters, unknown before, would have to be added from time to time.

At this point many a person has been misled, for it has been said that natural evolution works like the improvement of farm plants and animals.[16] This improvement is largely due to the crossing of varieties followed by selection. Often a plant or animal having a desirable character will be imported from a distant region for the purpose of crossing it with our varieties, so as to get a type having the good traits of both. Our modern strawberries have not been developed simply by the improvement of the tiny, soft, wild strawberry, *Fragaria virginiana,* but by crossing it with a large, firm strawberry of South America.[17] The

---

[16]For a more extended discussion, see Wm. J. Tinkle, *Popular Misconceptions of Variety Origins,* School Science & Mathematics, June 1944.

[17]W. Popenoe, *Journal of Heredity,* Vol. XII, No. 10.

American lard type hogs are not a transformation of the long-legged, sharp-backed boars of Northern Europe, but a result of crosses with the wild hogs of China,[18] which are always fat or in fair flesh. In 1904 the regal lily was imported directly from Western China, where it grew in the wild state[19] When civilized man first discovered corn, potatoes, and tomatoes, there were varieties of all three very much like our leading modern varieties.

There is no evidence that man ever has produced a new character through selection. It is true that some characters may be improved in this way, as for instance the production of butterfat by cows or of sugar in beets. But as we continue to select, the rate of improvement slows down until after a time there is no more change, no matter how much we select. The result of such breeding is a group of plants or animals, called a pure line, in which each individual has hereditary factors just like those of every other individual. Then, to be sure, it does not matter which ones are chosen as breeding stock; the progeny will be the same. Johannsen of Denmark found that it was very easy to develop a pure line of beans by planting one bean and using the seeds from this one planting for subsequent plantings. He then selected the small beans and planted them in one plot, and planted the large beans of this pure line in another plot; but the two plots produced beans of the same average size. Strange as this result may seem, it has been amply verified by other scientists,[20] and it is only logical to conclude that the beans were alike in their hereditary factors for size, even though their actual volume was different. The big beans were big simply because of better nourishment or a better position in the pod. This added size is a good example of an acquired character.

---

[18]C. Keller, European Domestic Animals, report of Smithsonian Institute, 1912.
[19]U. S. Department of Agriculture, Yearbook, 1937, p. 951.
[20]L. H. Snyder, *Principles of Heredity*, (Boston: Heath), 2nd. ed., p. 306.

We see then that agriculturists have imported, crossed, and selected in order to make improvements. If the same processes occur in nature they increase the number of varieties without making new or advanced characters. It makes little difference whether we call the results of these combinations new species or something else, the fact is that nature can not make advancement simply by arranging new combinations of that which she already has. In a population having mixed genes—heterozygous—selection is effective, but its effectiveness slows down and eventually stops when the population has become homozygous. New characters must be developed before evolution can occur. We must therefore consider those changes known as chromosomal aberrations to see if they are of the kind that could lead to evolutionary progress.

### Chromosomal Aberration

Chromosomes are found in each cell of a plant or animal during nuclear division and are readily seen under the microscope if the material has been stained, since they absorb stain more readily than surrounding portions. Fig. 9 shows the chromosomes in the cell of a worm, Ascaris. By extensive breeding experiments and cytological studies the hereditary factors called genes have been located in the chromosomes of the fruit fly *Drosophila melanogaster* and in corn *Zea mays,* and maps have been constructed showing their location.

Chromosomes are found regularly in pairs, a constant number for each species; for example, four pairs in *Drosophila melanogaster* and twenty-four pairs in man. One member of each pair has come from the father, while its mate came from the mother. In ordinary cell division each chromosome divides exactly in half along its entire length. In the maturation of egg or sperm in the gonads of the animal there is reduction division, at which time the chromosomes line up in pairs, then the members of the pair separate and the cell divides; the result is that the chromosomes are reduced to one-half the regular number. When egg and sperm unite in fertilization, the normal number of chromosomes is restored.

Rarely do we find a plant in which the chromosome number is doubled or halved. The effects of these different chromosome numbers in the case of the Jimson weed, *Datura*, are shown in Fig. 10.

If one species were to arise from another it would of necessity involve changes either in chromosome number and shape; or in the arrangement, number, and kind of factors within the chromosomes. Yet the nature and causation of these chromosome variations have proved to be most elusive. If a species with a normal allotment of eight chromosomes in its eggs should produce an egg having only seven chromosomes (such as have indeed been observed) and this egg were to be fertilized it would not be the ancestor of a seven-chromosome species, but it would die. If a nine-chromosome egg is formed and fertilized it may grow to maturity but the resulting individual will be unbalanced and lacking in vigor. If it lives long enough to reproduce, the progeny will not have the same peculiarities of this abnormal individual. All plants or animals with extra chromosomes have so far proved to be unstable due to relatively low transmission of the extra chromosome through the pollen or egg, and they are soon eliminated from any population left to reproduce naturally. Only a few plants or animals with two extra chromosomes occur, and these are quite infertile. It is clear then that differences in chromosome number among the species can not be accounted for by past deficiencies or duplications.

Now let us attempt to visualize what would be involved to account for the production of one species from another that has the same number of chromosomes. If this proves to be difficult, we can be assured that any other type of evolutionary change would be still more difficult. The gene arrangement of apparently very similar species such as the fruit flies, *Drosophila miranda* and *Drosophila pseudoobscura* is radically different.[21] The differences in these species are so profund that the chromomeres (small bead-like structures in which the genes are presumably located) fail to pair entirely or else form very complex pairing configurations. Genes which *D. miranda* are far apart, lie next to each other in *D. pseudoobscura.* From this it is concluded by evolutionists that inversions[22] have taken place in the phylogenetic development of these species from a common ancestor. Again some genes which in one species are located in the same chromosome are borne in different chromosomes in the other. Such differences are accounted for by translocations.[23] However, homologues[24] of many sections of *D. pseudoobscura* have not been found in *D. miranda.* Many of these sections in which homologous parts have not been detected in the other species are quite long. The nature of these sections is admitted by Dobzhansky and Tan, 1936, to be quite puzzling, and the explanation favored by Dobzhansky[25] is that "some chromosome sections have been so thoroughly rebuilt by repeated inversions and translocations that their disc patterns in the salivary chromosomes no longer resemble each other and no pairing of the homologous genes takes place." Dobzhansky and Tan estimate that aside from chromosome sections present

[21]T. Dobzhansky and C. C. Tan, *Studies on Hybrid Sterility* III, Zeit. fur Ind. Abst. und Vererb. 72; 88-114; 1936.

[22]*Inversions* are sections in the chromosome in which the normal order of the factors is inverted. Thus abcdefghijk has become abihgfedcjk.

[23]A *translocation* is a section of one chromosome transferred to a different non-homologous one. Most translocations are reciprocal, i.e., a section of one chromosome is exchanged for a section in a different or non-homologous one.

[24]*Homologous* chromosomes are those in which the genes and their order of arrangement is identical. Usually, of course, one homologue comes from the female, the other from the male parent.

[25]T. Dobzhansky, *Genetics and the Origin of Species*, (Columbia University Press, 1941).

in only one species, a minimum of 49 breakages and rebuilding by inversion and translocation are needed to account for the derivation of the chromosomes of one species from those of the other, or both from a common ancestor. Very many more would be needed to account for the sections without homologous parts found in closely related species of plants such as indicated by hybrids of *Nicotiana glutinosa* and *N. tomentosa*.[26]

Though these investigators believe that the gradual accumulation of inversions and translocations would account for these differences it would seem that more evidence is needed to make this explanation plausible. It is well known for example, that translocations and inversions are frequently lethal when homozygous in Drosophila or else produce visible changes in the form of the flies, with which is associated reduced viability, i.e., ability to maintain life. Though some young individuals homozygous[27] for a translocation or inversion are viable, it is only logical to expect that the same ratio of viable to inviable individuals has occurred in the past as occurs now. This makes the presumed accumulation of inversion and translocations difficult to accept. Furthermore, translocation heterozygotes[28] usually produce 50% or more of inviable gametes[29] due to the presence of duplications and deficiencies. Hence the reproductive potential[30] of translocation heterozygotes is lowered. They are then subject to selection pressure[31] which

[26]I. Elvers, *Interspecific Hybridization in Nicotiana XIV*, University of California Publications in Bot. 17, pp. 341-354.

[27]*Homozygous* is the condition of an individual (called homozygote) in which any given genetic factor (e.g., in this case a translocation or inversion) is doubly present, due usually to the fact that the two gametes which give rise to this individual were alike with respect to the factor in question.

[28]*Heterozygous* is the condition of an individual (called heterozygote), in which any given genetic factor has been derived from only one of the two parents.

[29]A *gamete* is a mature male or female sex cell; an egg or sperm.

[30]*Reproductive potential* is the potential number of living offspring an individual plant or animal may theoretically produce by random mating.

[31]*Selection pressure* is the pressure for or against survival of a particular type or variety due to natural selection.

tends to decrease their frequency and eventually eliminate them altogether. As soon as a translocation becomes homozygous, however, the negative pressure ceases. The difficulty which a translocation undergoes in becoming established is greatest during the time when it is in a minority, which occurs mostly in heterozygous form.

In becoming established in a population a translocation is subject to the same laws of scattering of variability and of slow accumulation by selection as is the case in mutations. According to Patau,[32] a mutation with a selective advantage[33] of 1% (or a translocation homozygote with a similar advantage) would increase in frequency from .01% to .1% of the population only after 900,230 generations. The most difficult feature of the whole picture is that no inversions or translocations have been shown experimentally to confer even a 1% selective advantage. Accordingly it would seem very unlikely that the postulated accumulation of inversions and translocations needed to build up such differences as are found in *D. psuedoobscura* and *D. miranda* ever took place. Then how could they have been transformed from a common ancestor?

Much has been written about doubling of chromosome number as a method of formation of new species. One of the most famous cases on record is that of the intergeneric Raphano-Brassica double-diploid hybrid, first reported by Karpechenko in 1924 and 1928. *Raphanus sativus,* the common radish, is characterized by having 9 pairs of chromosomes, and *Brassica cleracea,* the cabbage, has 9 pairs. The $F_1$ hybrid according to Karpechenko showed no pairing, 18 univalent chromosomes occuring at the first metaphase of

---

[32]*Die mathematische analyse der Evolution-vorgange.* Zeitschrift fur Induktive Abstammungs und Vererbungslehre, 76:220-228 (1938).

[33]*Selective advantage* is the advantage in favor of survival conferred on a variety by natural selection because of some advantageous character or physiological behavior.

the reduction division. Only a few seeds were set, and these arose from the union of unreduced eggs with unreduced sperms, giving rise to double diploid or amphidiploids in the $F_2$ generation. Karpechenko states that 60% of these $F_2$ plants were fully fertile, 25% had reduced fertility and 15% were sterile. Two of his fertile $F_2$ plants gave 1067 and 892 seeds respectively per 100 fruits on open pollination as compared to 1200-1300 for cabbage and 627 per hundred for the radish. No quadrivalents were reported, only regular chromosome pairing or bivalent formation occuring, though he does state that "doubling does occur in the amphidiploid hybrids which *as a rule* have quite normal reduction division." As so reported this intergeneric hybrid is a perfect experimental demonstration of the formation of a polyploid species because it showed (1) regular formation of bivalents and (2) was fully fertile. The bivalent formation would follow naturally from the lack of pairing reported in the $F_1$ plants. However, Howard[34] who, together with Richharia,[35] reinvestigated this interesting hybrid, points out that there actually must have been at least a low frequency of bivalent formation in Karpechenko's $F_1$ hybrids for some of his published metaphase figures (see Karpechenko, 1924, p. 389) showed only 16 or 17 chromosome bodies at diakinesis. These, Howard believes, led to irregularities in chromosome distribution with resultant formation of gametes deficient in chromosomes or chromosome segments and so may account for the 40% of Karpechenko's amphidiploid plants which were partially sterile.

Richharia and Howard in their exhaustive reinvestigation of this hybrid report results quite different from those obtained by Karpechenko as follows: (1) The $F_1$ hybrids

---

[34]H. W. Howard, *Fertility of Amphidiploids,* Journal of Genetics 16:239-273.
[35]R. H. Richharia, *Cytological Investigation of Hybrids,* Journal of Genetics 34, 19-44, (1937).

formed a variable number of bivalents at the metaphase usually averaging 2-3 per cell. (2) Many of the $F_2$ plants had fewer than 36 chromosomes and those with 36 chromosomes showed variable pairing at first metaphase, univalents, bivalents, and quadrivalents occurring. At second metaphase, nuclei with 17 and 19 chromosomes were found and accordingly even in the $F_4$ generation the plants had from 33 to 37 chromosomes. (3) The $F_2$ plants were only partially fertile. In fact the fertility even in the $F_4$ varied from 5 to 42%. Howard believes that this formation of quadrivalents in the $F_2$ and resulting sterility in the $F_2$, $F_3$, and $F_4$ generations follows simply from the fact that bivalents are formed in the $F_1$. The difference between the results obtained by Richharia and Howard and those of Karpechenko are basically due to the greater amount of pairing in Richharia and Howard's $F_1$ hybrid. This difference has not been accounted for but Howard states that a similar difference in frequency of bivalent formation has been observed by Kihara in *Triticum Aegilops* $F_1$ hybrids in preparations from the same plant fixed on different dates. The work of Howard is very thorough and was done later than that of Karpechenko when more was known about the pairing possibilities of homologous parts in chromosomes of even distantly related species, with resultant bivalent formation in the $F_1$ and quadrivalent formation in the $F_2$.

In any event, *one of the most famous examples of the origin of polyploid species has not been verified on reinvestigation,* and until some one uses the same strains as Karpechenko and shows that his results can be duplicated, we cannot accept the Raphano-Brassica amphidiploid as a good example of the origin of a polyploid species; for one of the fundamental rules of science is that experiments, to be universally accepted, must be capable of complete verification.

### Mutation

Having reviewed the role of acquired characters, recombination of characters, and chromosomal aberration in the supposed gradual upbuilding of the organic world, let us now turn our attention to mutations. If the gene is a self-reproducing unit in the chromosome it undoubtedly must have a definite chemical structure, for that is true of all known physical bodies. Alterations in the gene which do not prohibit it from reproducing itself may be the source of the variations which we call mutations. Unfortunately no methods are available for making a comparative analysis of the original and of the presumably mutated gene. The only evidence for the occurrence of a change in the gene is the appearance of a variant called a mutant, which follows Mendel's law in inheritance (i.e., gives a 3:1 or a 1:2:1 ratio in the second filial generation, when crossed with its ancestral type.)

Mutations have been studied in greater detail in the fruit fly, *Drosophila melanogaster*, than in any other organism (see Fig. 11). As far as studied however, the kinds of mutations which occur in other animals and plants are similar. The first mutation to be studied changes the eye color from red to colorless and reduces the viability. Three mutations reduce wing size in different degrees and a fourth eliminates the wings entirely. Another mutation changes the balancers into wings, giving a fly with four wings, a characteristic which insect specialists consider less specialized. Other mutant characters are black body color, hairless, eyes reduced to bars, eyeless, and various colors of eye. Careful tests by Muller have shown that a majority of mutations are lethal, causing death in early stages of development.

A majority of all mutations observed are defective changes or neutral types of changes which have associated physiolo-

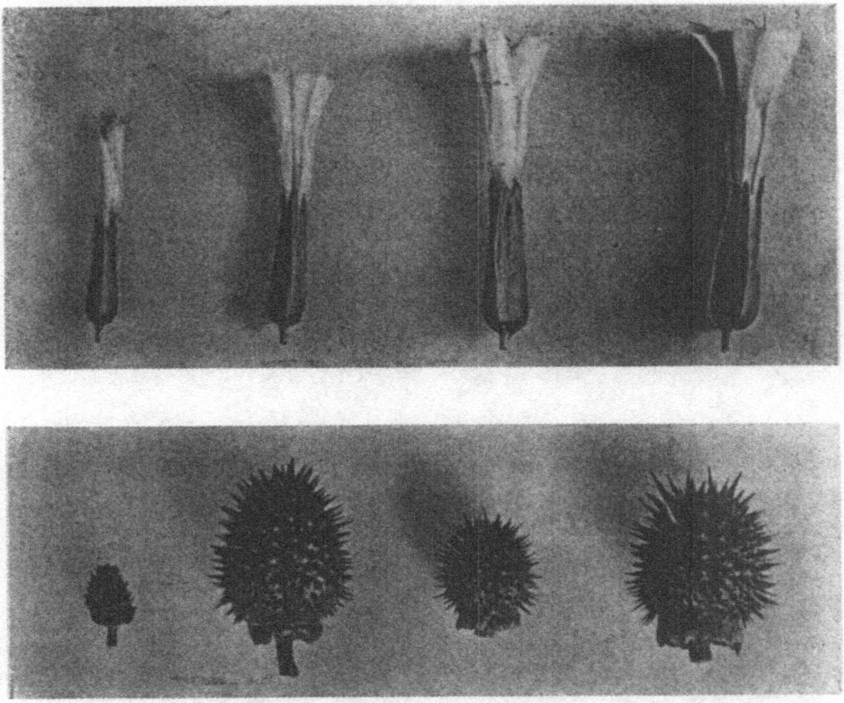

Normal

A B C D

*Fig.* 10—Flowers and capsules of Jimson weed *Datura* from haploid, diploid, triploid, and tetraploid plants, respectively. The diploid type B is normal and bares more seeds than the other. The haploid type has half the normal number of chromosomes, the triploid one and one-half the normal number, and the tetraploid twice as many chromosomes as the normal type. (*Courtesy Dr. A. F. Blakeslee*)

*Fig.* 12—Restoration of Old Man of Cro-Magnon by J. H. McGregor. (*Courtesy American Museum of Natural History New York*)

gical effects which decrease the ability of the organism to maintain life.[36]

"Mutations are usually harmful to the organism. This is to be expected, as the organism is a delicately adjusted mechanism, and a random change would more likely be deleterious than beneficial. The vast majority of mutations have been found to be detrimental to the continued existence of the individual or the species."[37]

These, if accumulated, would not seem to give us the sort of changes needed for species formation; especially not the larger and more complex species required to enable evolution to progress from amoeba to man.

Certain mutations of fruit flies under a particular set of environmental conditions are superior to the wild type in viability according to Timofeef-Ressovsky. Thus the mutation "eversae," which causes a change in wing venation has only 98.3% of the viability of the wild type at 59°-60° F., 98.5% at 82°-86° F., but 104% of the viability at 75°-77° F. In accordance with accepted scientific procedure, this rare exception should be verified by other investigators before being accepted as final. However, according to the results of Timofeef-Ressovsky when the mutation eversae is combined with one called abnormes, the combination has only 83.7% of the viability, while eversae bobbed has only 85.5% of the viability of the normal type. Eversae singed is still 103.1% of the viability of normal, a decrease of only 1% from that of eversae alone. Again the combination of venus abnormes and lozenge, each of which decreases viability somewhat, produces a summation of deleterious effects.

[36]Horlacher and Killough reported in American Naturalist, Vol. 67, pp. 532-538, a mutation in cotton, *Gossipium hirsutum,* which changed the leaves from deeply lobed to more nearly entire. While this mutation increases the combined surface of a given number of leaves, it may not be an advantage to the plant, since the upper leaves of the mutant shade the lower leaves more than in a normal plant; and this shading is a detriment. Outside of a few doubtful examples like this one, no mutations favorable to the organism have been found.

[37]Snyder, *Op. Cit.*, p. 220.

Now it may readily be seen that the combination of five or six of even the most viable mutations so greatly reduces the viability of the fruit fly as to make its continued survival even under ideal conditions quite difficult. And only by the accumulation of many mutations could species as they occur in nature be presumed to originate, for they obviously differ from one another in many genes.

One of the principal tenets of the evolution theory is that the presumed changes occurred from causes which now continue to be in operation and can therefore be observed and studied experimentally. Now the fruit fly has about 26 generations per year, which allows for the observation of over 780 generations since 1910 when it was first carefully studied by Morgan and associates. Furthermore, Muller[38] demonstrated that by X-rays the rate of occurrence of mutations in *Drosophlia melanogaster* can be increased 59 to 143 times that of untreated flies. These mutations according to Muller and others are of the same kind as those which occur "spontaneously." The speeding up process makes it possible to study in one year the kinds of mutation which normally would occur only during a period of from 59 to 143 years. Again, in *Drosophlia pseudoobscura* Mampert[39] has discovered a "Mutator" gene which when homozygous increases the mutation rate about 70 times. In such a case the equivalent of 1820 generations instead of the usual 26 may be studied in the course of one year. Again, in this species as in *D. melanogaster* the many mutations which have occurred are all deleterious or at best neutral in their effect. It may thus be concluded that careful observation during a period comparable to from 900 to 1820 years of study of the larger mammals with breeding cycles of one year, has yielded little or no evidence of the

[38]H. J. Muller, *Artificial Transmutation of the Gene*, Science LXVI. 84. (1927).
[39]K. Mampert, *High Mutation Frequency in Drosophila Pseudoobscura, race B*, Nat. Acad. Sciences; 29 (1943), pp. 137-144.

accumulation of mutations which either substantially increase the viability of the mutated form or make it more likely to survive in nature.

A frank summary of the situation with respect to the search for methods was made by John M. Coulter in 1923;[40] the same situation exists today.

> "There is a much more serious problem of evolution, however, which is still baffling. The variations observed which result in new species as tested by genetics, and for which the cytological machinery has been observed, produce species either laterally or retrogressively; that is, species of the same phylogenetic level, or of declining rank. There is as yet no adequate explanation of progressive evolution, the advance from one great phylum to another."

*Thus we have considered all that genetics has to offer*—acquired characters, recombination, chromosomal aberration, and mutation—*and find in none of them a method for the building up of our complex species from simple units of protoplasm.* But we maintain that if such a method existed some hint of it would have been revealed by the enormous amount of biological research that is carried on.

### Evidence From Fossil Records

There are many who claim that transformation of species is proved by the fossil records of plants and animals. However, the facts on which this theory is based can be explained in more valid ways. As some geological problems have been discussed in Chapter III, only a few plant and animal records will be mentioned here.

The buried animals and plants, known as fossils, are valuable records and some of them represent species that have

[40]*Science Remaking the World,* (Doubleday Page, 1923), p. 177. While he does not agree with Coulter in the use of the word "species," Goldschmidt of the University of California reaches a similar conclusion. In his "Material Basis of Evolution" he proves in detailed fashion that small mutations can not add up to new species; that the varieties so formed are not incipient species. The latter half of his book however, in which he speculates how evolution might come about by large mutations, the very kind we have found to be most destructive of viability, is not impressive.

not changed since very early times if at all. A classical example of this is Lingula, a genus of brachiopod (a shell-fish) living today in the Pacific Ocean and found in the same form in the Silurian rocks, which are among the deeper formations of the sedimentary rocks.[41] If all the buried species had changed into modern species it would be a remarkable demonstration of evolution in reverse at least so far as size is concerned; for among these fossils we find the giant beaver which was nine feet long, the imperial elephant which was as tall as the giraffe and well proportioned, the reptile, Brachiosaurus, eighty feet long and forty-five feet tall, and horsetails twelve inches in diameter as contrasted with the largest modern species of one inch.

Some one will object that size does not indicate an advanced type of organism; but the evolutionists claim that life began as a tiny mass of protoplasm and changed into larger forms, so that increase in size is part of the theory. Many animals and plants of early geological time were not only large but well proportioned and highly ornamented,[42] showing that the history of life on our planet has not always been a gradual, progressive development.

Perhaps more has been written on the geological history of the horse than of any other species. *Eohippus* is supposed to have changed to the horse, *Equus,* by the loss of three toes. It is a well-known fact that there is no place where the remains of modern *Equus* and its supposed ancestors *Eohippus* and *Mesohippus* are found one above the other in the supposed evolutionary order, but they are found in widely scattered locations. While it is frequently claimed that *Equus* developed from *Eohippus,* some scholars believe each fossil type represents a co-ordinate genus, no genus of the group having developed from another.

---

[41]Hegner, *Op. Cit.,* p. 204.
[42]Cf. Ammonites; H. F. Cleland, *Geology, Physical and Historical,* (New York: American Book), p. 527.

### Evidence of Early Man

There has been much comparison of remains of early man with the Biblical account of creation and geneology. The supposed Pre-Adamic race seems to have no support either in history or in paleontology. Perhaps some have theorized upon such a race not related to Adam in order to account for Cain's wife. This is not necessary for Genesis 5 states that Adam lived 930 years and begat sons and daughters; during such a long life, perhaps a number of them. Cain or some one else must have married a sister. Although such marriages are frowned upon in our day modern genetics teaches us that the offspring of relatives is sound where there are no inferior traits, expressed or latent, in either parent. At that early day very few mutations had occurred, so that there was no reason why relatives should not marry.

Several skeletal remains of man have been discovered that are supposed to be ancient, but evidence for their great antiquity is not satisfactory. Again, types like those from Java, Piltdown, and Heidelberg, of which only one specimen is known, may not be typical of any race. Java man, *Pithecanthropus erectus,* consists of a skull cap, a femur, and three teeth; Piltdown man about half the skull bones; and Heidelberg "man" is a single lower jaw. There are two types of which a large number of specimens have been found, namely Neanderthal and Cro-Magnon.

The Neanderthal was a rather peculiar chap. Only five to five and one-half feet tall, he nevertheless had large muscles (which we learn from the attachments on the bones) and a firm jaw. The brow was low and appeared even lower because of the massive ridges above the eyes. However, when the cranial capacity is measured in a typical specimen we find it to be 1723 cubic centimeters. Modern man ranges from 950 cc. to 2020 cc.; so that there is every reason to believe that Neanderthal man had a brain and an intelligence quotient comparable to that of modern man. He buried his dead with ceremonial care, placing flint implements in the grave, indicating a belief in the future life.[43] Some lived, perhaps, contemporaneously with Cro-Magnon man.

Before the writer is a plastic reproduction of a skull of Cro-Magnon man alongside a modern skull. It is interesting to consider this superior race (See Fig. 12), with a cranial capacity 150 cc. above the average of modern man, averaging six feet in height and well proportioned, living before the metals were employed. H. F. Osborn states that at least some of them lived at the same time as Neanderthal man. They skillfully chipped stone implements, and made paintings on the walls of caves that are so good that at first it could not be believed that prehistoric man had made them. These painting reveal that Cro-Magnon man had great power of observation as well as manual skill. About eighty skeletons of this type have been discovered, and these along with the Neanderthal skeletons indicate that the *average* structure of early man was about the same as that of modern man.

### Evidence from Comparison

Lastly let us consider the comparison of species and the light it sheds upon creation and evolution. That there is considerable likeness or homology in the animal and plant kingdom is readily seen. Some contend that this indicates descent from a common ancestor, while others maintain with equally good logic that we should expect the works of one God to be alike. However the likeness is far from complete in many cases, and it seems that creation by an intelligent Being would account for these differences more readily than the theory of common descent; especially since, as mentioned above, it is so hard to find progressive, hereditary changes taking place at present.

[43]H. F. Osborn, *Men of the Old Stone Age*, (New York: Scribner's, 1936).

The theory of recapitulation, or "biogenetic law" as Ernst Haeckel called it, states that the embryos of some species of animal resemble the adults of other species; e.g. the embryo chick is supposed to resemble the fish; therefore a fish-like organism is supposed to be the ancestor of the chicken. In fact, however, the resemblance is among embryos themselves, not between an embryo and an adult of another species. This was the fourth conclusion of Von Baer, which Haeckel did not quote. We can expect one embryo to resemble another since both start from a fertilized egg and have considerable likeness also at the end of their development.

Haeckel at first claimed that the embryo went through *all* the stages of its phylogenetic or evolutionary development, in other words, "climbed its own ancestral tree." But since many supposed stages came too early, some too late, many supposed stages were missing, many structures in embryos had no reference to an ancestor, the result was that the theory was reduced and made less definite. Kebel gives an instance of this confusion: Teeth in vertebrates, he points out, are supposed to be much older than the tongue, and yet in the embryo mammal the tongue develops before the teeth.

At the present time there is claimed to be a resemblance between the embryo and adults of other species in certain *selected organs;* while organs of the two species that show no resemblance are ignored. But it is hardly scientific thus to select only the data that fit one's preconceived theory; science is supposed to weigh all evidence impartially.

There are alternating arches and furrows on the neck of the human embryo which are supposed to resemble the gill arches and clefts of a fish. But their relationship is thrown into question by the fact that the first arch develops into the lower jaw (which it does not do in the fish) and the second forms the middle and outer ear (which the fish does not have). None of them form gills.

Much is made of the fact that three kinds of excretory organs develop successively in the mammalian embryo, the fore-, middle-, and hind-kidney. The fore-kidney is found in adult hag fishes, the middle-kidney is functional in amphiba, consequently this is the best example that can be found for recapitulation. Courville[44] points out difficulties, in that all three kidneys use the same excretory duct; and that the fore-kidney in man is by far the best developed within the group of mammals. "In other words, the primitive kidney is better developed in man than in birds, and is equally as well developed as in the reptiles. On this basis the reptiles evolved from the birds instead of vice versa, and insofar as the fore-kidney is concerned, man has not evolved further than the reptile stage. The better way of accounting for these organs in man is to recognize their temporary place as simple excretory organs at a stage when no complex organ for this function has yet had time to develop."

Indeed this seems to be the true principle of development; that organs develop in the most direct manner, considering the circumstances and needs. The nervous system begins to develop early in life, not because the supposed ancestors had over-sized brains like the embryo, but because the nervous system is so complex that it takes it a long time to develop.

As noted above, if recapitulation be true it should be found in all the systems; yet nature has preserved no traces of the supposed changes from gills to lungs.

It has been shown that the theory of recapitulation is not a necessary explanation in that the experimental embryologists, a modern and very active group, do not use it in

[44]C. B. Courville, *The Recapitulation Theory*, Bulletin of Deluge Geology, Vol. I, No. 2, August 1941.

their research. And since Haeckel had to devote so many pages to the explanation of exceptions, it is not a lucid interpretation of the observed facts.

### Summary and Conclusion

The intricate and orderly structure of living things indicates that they were built according to a plan, and the Bible states they were formed by a Creator. However, the tyranny of church leaders in the Middle Ages alienated intellectual men and induced them to look elsewhere for an interpretation.

Some students of nature have proposed theories which they sincerely thought accounted for the origin of organisms without recourse to the Biblical record. The present century has been marked by experimentation rather than by speculation, consequently earlier theories have been subjected to an acid test. It has been demonstrated that neither "acquired characters" nor recombination after crossing furnish any new characters that are stable. Only mutation and chromosomal aberration give new characters, but these are not progressive in an evolutionary sense. A method which could produce evolution admittedly is unknown.

The fossil records, early human remains, homology, and recapitulation are taken as evidences by some, but when we scrutinize the data we find in them evidence for creation. The best modern research corroborates the Bible and strengthens our faith. The first chapter of Genesis relates that God created plants and animals "after their kind" and we observe at present a strong tendency for them to remain within certain well defined limits. If we eliminate from biology the wishful thinking and the cleverly worded slogans, the residue—which includes all the essential and useful parts of the great science—does not conflict with the Bible. We conclude that the Biblical account of the creation is corroborated, and that God finished His creation, just as Genesis states.

## Chapter II

# ASTRONOMY AND THE FIRST CHAPTER OF GENESIS

*by*

*Peter W. Stoner, M. S.**

Astronomy is one of the oldest sciences, yet most of the progress in this field has taken place in comparatively recent times. About thirty years ago our knowledge of the size of the sun, its planets, and their satellites as well as the distances involved was fairly complete. However, we were still teaching the Nebular Hypothesis which has now been completely abandoned. The source of the sun's heat was a complete mystery. Dark nebulae were thought to be holes in the heavens. All objects were thought to be parts of our own galaxy. The spiral nebulae were described as small gaseous bodies in the process of forming new solar systems.

In the last thirty years we have found that the sun's heat is a manifestation of atomic energy. We have quite well determined the size, shape, content, and period of rotation of our galaxy. We have stretched our measurements of distance from a few hundred light years to about 600,000,000 light years. We have found each spiral nebula to be a galaxy similar to our own, and have measured their velocities and their distances from us. We have fairly well determined the ages of nearly all astronomical objects.

Except for information on the solar system, a textbook

*Chairman, Department of Mathematics and Astronomy, Pasadena City College, California.

of astronomy fifty years old is of little scientific value today. This modernity of astronomy is in direct contrast to the antiquity of the Genesis account. Conservative scholars familiar with the Hebrew date the writing of Genesis at about 1500 B.C. Others place this date as late as 750 B.C., but as to the general antiquity of the book there is universal agreement.

If Genesis is only a book of human origin and its allusions to astronomy reflect the knowledge extant at the time of writing, we would expect it to be full of gross scientific errors. If, on the other hand, we find Genesis to be in agreement with the latest factual developments of modern astronomy, we must conclude that God supplied the information as the book claims.

Today there appears to be complete harmony between astronomy and Genesis. This does not mean that all astronomers believe the Genesis account, although it does seem that astronomers as a group frequently possess a more reverent attitude toward God than other scientists. It means that a careful study of the facts and well-established hypotheses of astronomy reveals a striking consistency with the outline of origins found in the first chapter of Genesis.

This agreement has not always existed. In fact, 35 years ago astronomy and the account of creation in the first chapter of Genesis differed in many points, making it quite impossible to correlate the two. Year after year advancements were made in science resulting in an improved agreement between Genesis and astronomy. Within the past generation not a single instance is known where astronomy that has once agreed with Genesis has later reversed itself so as to disagree. There are many instances, however, where subsequent developments have greatly strengthened earlier partial confirmations.

It is possible that future developments in astronomical theory or interpretation of astronomical observations may again result in disagreement with the Genesis account. If this should occur, we would have to wait patiently for further developments of the science or re-examine our interpretations of Genesis with the firm conviction that when astronomical knowledge is advanced enough, and full understanding of the Biblical text is at hand, agreement is sure to come. It is the purpose of this chapter to present some examples which should be greatly strengthened as new discoveries are made and scientific knowledge becomes more complete.

### The Time of Beginning

"In the beginning God created the heavens and the earth."[1]

Science formerly taught the "conservation of matter," that matter was indestructible, that we could change its state but we could neither add to nor subtract from the total amount of matter. Today it is taught that matter can be changed into radiant energy such as light or heat and cease to exist as matter. Under the heading, "Mass Changed to Heat," Dr. Robert S. Richardson of Mount Wilson Observatory writes:[2]

"Implied in this theory of the source of the sun's heat is the idea that mass can be changed into heat. Scientists used to regard matter as indestructible. 'Conservation of Matter' was the formula expressing this belief. Now it is known that a given amount of matter can be changed into a given calculable amount of energy (heat). The relationship between mass and energy is expressed by the following equation formulated by Einstein: $E=mc^2$, where E is the energy in ergs, m the mass in grams, and c the velocity of light in centimeters per second. . . . We find that the sun loses by its radiation $4.2 \times 10^{12}$ grams per second or 4,200,000 metric tons, which would be equivalent to 4,620,000 of our ordinary tons per second."

[1]Genesis 1:1, A.S.V. "The heavens and the earth" probably constitute the whole physical universe.

[2]W. T. Skilling and R. S. Richardson, *Astronomy*, (New York: Holt, 1941), pp.192-194. See also: R. H. Baker, *Astronomy*, (New York: Van Nostrand, 1938), pp. 301-303; J. C. Duncan, *Astronomy*, (New York: Harper, 1926); H. N. Russell, *The Solar System and its Origin*, (New York: Macmillan, 1935), p. 26.

In this day of atomic bombs and atomic energy it is not necessary to pile up evidence that matter can be changed to energy. The man in the street knows that one tenth of one percent of the mass concerned was changed into energy in the explosion of the first atomic bombs.

This suggests the concept that the matter of the physical universe may not have always existed. A few years ago astronomy had nothing to say about the origin of the physical universe either as to time or place.[8] Recent findings, however, indicate a definite beginning at a definite time in the past.

Studies of (1) the age of the earth, (2) the age of the moon, (3) the age of meteorites, and (4) the age of the galaxies all indicate that each had a time of beginning and that that time was not infinitely remote in the past. The evidence is worthy of serious consideration.

*Age of the Earth*

The element uranium is radioactive, and after several changes breaks down into lead which is stable. Under all producible physical conditions it has been found that approximately 1/637th part of uranium changes to lead each ten million years. When a deposit of uranium is found sealed in igneous rock, the crust of which has not been cracked to allow the entrance of any foreign substance, it is possible to determine the percentage of uranium which has changed to lead and thus determine the approximate time elapsed since the rock crystalized about the uranium. The oldest samples of uranium thus found in the earth in-

[8]All of the theories of the formation of the solar system: the nebular hypothesis, the planetesimal theory, the tidal theory, etc; all begin with the material from which the solar system was formed as existing in some definite form. These theories only deal with the development of the solar system from some accepted or assumed point of departure. They say nothing about creation. See C. A. Young, *General Astronomy*, (1898), pp. 567-576, for early concepts; J. C. Duncan, *Astronomy*, (New York: Harper, 1935), 3rd ed., pp. 290-300, especially notice the last paragraph, p. 291; W. T. Skilling and R. S. Richardson, *Op. Cit.*, 310-316; H. N. Russell, *Op. Cit.*

dicate ages up to 1,800 million years.[4] The earth of course, must be older than that. Based on this method, 2,000 or 3,000 million years might be a reasonable estimate for the minimum age of the earth, for it is believed that the earth went through several earlier stages before rock could crystallize.[5]

*Age of the Moon*

The tidal forces acting between the earth and the moon (*See Fig.* 1) gradually slow down the rate of rotation of both, and cause the moon gradually to recede from the earth. Physicists have tried to compute the time required by the tides to move the moon from a position relatively close to the earth to its present position. We do not know how close to the earth the moon may have been to start with, but by basing the computations on the idea that it was very close we obtain a result which is the limiting age of the moon. Since the depth of the oceans is an important factor in determining the effect of the tides, and since the depth of the oceans through the past ages is not accurately known, it has been impossible to arrive at a definite answer. Such computations, however, have yielded estimates of a few thousand million years.[6] This is compatible with what we know about the age of the earth itself.

*Age of Meteorites*

Meteorites which come to the earth such as the one shown in Fig. 2 are of two types. One type belongs to our solar system, revolving around the sun much as do the planets, and the other type comes to us from outer space.

[4]Editor's Note: Opponents of the radioactivity method of dating contend that we do not know that the rate of decompositioin has remained constant over this great length of time. In addition to this, difficulties relative to contamination and chemical instability are pointed out. Considerable credence is given the method by practically all workers in this field, however, for the amount of helium and various isotopes of lead give additional checks on the results. See C. Goodman and R. D. Evans, Bulletin of the Geological Society of America, Vol. 52, No. 4, pp. 491-544, (1941).

[5]H. N. Russell, *Op. Cit.*, pp. 19-22.

[6]*Ibid.*, pp. 32,33.

The meteorites, in the vicinity of the earth, traveling less than twenty-six miles per second will be turned back by the gravitation of the sun before they can get far from the solar system. The meteorites whose velocity is greater than twenty-six miles per second will recede to an unlimited distance from the sun and, therefore, do not belong to our solar system. Uranium specimens have been found in both of these classes of meteorites. The oldest meteorites found by different investigators vary from 2,000 to 6,000 million years.[7] Thus, the oldest material moving as meteorites appears to be of about the same general age as the earth and moon.

### *Age of the Galaxies*

The velocities of the galaxies or island universes, (*Figs.* 3 *and* 4) as measured by the red shift,[8] increase nearly proportionally to their distances from us. These velocities are so distributed that if the island universes were to be traced backward they would appear to have originated from one place at one time and that about 2,000 or 3,000 million years ago.[9]

The red shift indicates velocities for distant galaxies (island universes) as high as tens of thousands of miles per second. To many, this velocity has seemed to be unreason-

[7]*Ibid.*, pp. 22,23. The Harvard Books on Astronomy, F. G. Watson, *Between the Planets*, (Philadelphia: Blakiston, 1941) pp. 197,198.

[8]The spectrum is composed of light separated according to its wave length, running from the longer wave lengths of red at one end to violet, the shorter waves, at the other. Each element produces light of certain definite wave lengths, which appears in the spectrum as a definite pattern of lines, always at the same place in the spectrum unless the source of light is, relatively, moving toward or away from the observer. If the source is approaching, the pattern of lines for any element will appear further toward the violet. If the source is receding, the lines will appear further toward the red than they would if the source were stationary with respect to the observer. This shifting of the spectral lines is used extensively to determine the velocities of planets and stars with respect to the earth. Practically all spectra of island universes have their lines shifted toward the red. These shifts are very great for the more distant universes and less for the closer ones. This shift has become commonly known as the "red shift."

[9]Russell, *Op. Cit.*, pp. 23, 25.

able, so attempts have been made to explain the shift in some other way than a measure of velocity.

The idea was advanced that light might change its speed after traveling such great distances and consequently change the length of its waves. The velocity of light from distant galaxies has been measured and found to be the same as light from a local source, so that idea has been discarded.

Another theory involves the leaking of energy from photons of light. So far, this theory has no scientific background in atomic physics.

Dr. Edwin Hubble, in the annual Sigma Xi address before the American Association for the Advancement of Science, Dallas, Texas, December 30, 1941, states:[10]

> "It may be stated with confidence that red shifts either are velocity shifts or they must be referred to some hitherto unrecognized principle in nature. . . . The present distribution of red shifts could be adequately described on the assumption that all the nebulae were once jammed together in a small volume of space. Then, at a certain instant, about 1,800 million years ago, an explosion occurred, the nebulae rushing outward in all directions and with all velocities. Today, of course, we find the nebulae distributed according to their inititial velocities, those moving most rapidly have reached the greatest distances, while the laggards are still in our vicinity. Although this picture is over-simplified, it suggests the importance attached to the so called 'age of the Universe' –1,800 million years."

Dr. Hubble goes on in this article to point out that other explanations of the red shift may be forthcoming. He also points out some difficulties involved in the above explanation, including a greater concentration of nebulae at greater distances than at the lesser distances. The 200-inch telescope at Mt. Palomar is expected to shed light on the problem.

Another method of attack comes from Harvard University.[11]

---

[10]Science, Vol. 95, pp. 212-215.

[11]B. J. Bok and P. F. Bok, *The Milky Way*, (Philadelphia: Blakiston, 1941), one of the Harvard Books on Astronomy, pp. 185,188,189.

"Our galaxy has not been rotating sufficiently long for the interchange of energy between stars of different types to become effective. From our considerations of stellar encounters it would seem very unlikely that the stars would show so much individuality in their motions if our galaxy had existed in its present form for as many as . . . two billion solar years. . . .

"The very fact that we find a fair percentage of all known stars of spectral class A in clusters of various degrees of concentration is the best available proof that our galaxy cannot have existed in its present form for much longer than 10 billion solar years."

"These computations show that, if the red shift is explained as a velocity shift, the expansion of the universe of galaxies would probably have begun some 3 billions years ago."

The age of the galaxies is one of the new fields of research and from the quotations just given it is seen that the various methods of computing their ages give results varying from about 2 to 10 billion years. For such a new field this variation is surprisingly small, in fact only a few years ago these same results varied from 600 million years to 5 trillion years. We may expect that in a few more years these values will be in much better agreement.

## The Creation

Astronomy has thus pretty well bracketed the ultimate age of the whole physical universe, at least in its present state, as something between 2 and 10 billion years.

Much research work has been done dealing with the present state of the universe and the changes taking place. But the information is not yet sufficiently complete for the astronomer to reach any definite conclusion regarding the origin and development of the universe. One possible theory is that the physical composition of the universe originated at one place a few billion years ago after which forces drove various parts of the matter away in different directions and at various speeds, forming our known universe. Another possible theory[12] is that the physical universe expands and contracts in cycles. This theory would indicate that a few billion years ago the whole physical universe was drawn together into a small space and everything was completely reformed. Every astronomical body was reduced to a gaseous or nebulous state and its identity was lost in becoming a part of the great whole. This involved changes of the form of matter, even to the development of complex radioactive elements from the simpler ones. Then, as in the former theory, forces developed and drove the matter out into space eventually to form our present universe. This cyclic theory holds that the universe is now expanding but will later contract into one place and repeat the cycle.

If the first theory is true we can approximately date the creation of Genesis 1:1. If the second theory is true, the day of creation may have been earlier or the last time of compression may still be the recorded act of creation. The new elements brought forth and the birth of every astronomical body could possibly warrant the use of the scriptural term "create."[13]

If the ultimate age of the earth, the moon, the meteorites, the galaxies, and in fact, all known astronomical objects (at least in their present form) is between 2 and 10 billion years, or if the ultimate age limit is any other finite period of time, we are led to conclude that the universe was either created from nothing or that its state of existence was very drastically changed at a limited period of time in the past. Genesis states a creation of the universe. It does not state

[12]The discussion of still other possibilities may be found in the Harvard Books on Astronomy, H. Shapley, *Galaxies*, (Philadelphia: Blakiston, 1943), pp. 207-219.

[13]The Hebrew word for "create" (*bara*) means "to create from nothing." Even those Hebrew scholars who hold that a secondary meaning of "constructive work from existing materials" is valid, believe this word in Genesis 1:1 to mean "create from nothing" based upon a consideration of the form of the word, the context, and its association with other words meaning "construct." See discussion on pages 65, 66.

when or how that creation took place. Astronomy not only does not contradict a creation but gives all of the above evidence that such a creative act actually did take place.

*Another Argument for Creation*

The radiation of the sun is produced by the loss of about 4 million tons of its material every second. Through radiation absorbed from other stars and its collection of meteorites, it regains only about 1/2000th of this amount. The sun definitely appears to be running down. Every star is losing mass at a comparable rate and seems to be no more successful at recovering matter than is the sun. In other words, the universe is running down. If every star of the universe is running down, the stars must have been started at a time not infinitely distant. In other words, there must have been a creation regardless of what theory of the formation of the universe may be true.

*The Earth from a Diffuse Nebula?*

"And the earth was without form, and void; and darkness was upon the face of the deep."[14]

For about 150 years, and ending only 25 or 30 years ago, the nebular hypothesis held sway in the scientific world. According to this hypothesis, the whole solar system started from a single spiral nebula shaped like a flat disc, giving off great amounts of light and becoming increasingly denser as it contracted. Thus this hypothesis was considered completely to contradict the second verse, as it claimed a definite early shape, not "without form." It claimed that great amounts of light were being given off, not "dark." It claimed that in this early state it was a dense gas, not "void." Throughout the reign of the nebular hypothesis the second verse of Genesis was, perhaps, the most seriously attacked point in the whole Bible. The critic even asked, "What kind of an astronomical body was without form and

[14]Genesis 1:2a.

void and dark?" There was no answer for the critic. The devout scientist only believed that some day the Biblical statement would be better understood. Eventually this nebular hypothesis was rejected completely when it was understood that well-established physical laws would have to be broken for it to operate.

When dark nebulae were discovered, it became apparent that here, perhaps, was the stuff from which our earth originated as described in Genesis 1:2. Recently one of our astronomers said in the hearing of the author, that this second verse of Genesis 1 was the best description of a dark nebula[15] that man had yet written. Considering the earth as originating from a dark nebula would not necessarily mean that it did so directly. It may have, and again the whole solar system, or perhaps several systems may have come from one diffuse nebula.[16]

The concept that new stars came from nebulae[17] has been prominent in astronomy for two hundred years or more. The stellar sequence is very generally accepted by astronomers. According to this sequence the star starts as a rare gas, only hot enough to glow red. The star then contracts, becomes denser and hotter, gradually reaches a high maximum; the temperature then begins to decline and in sufficient time becomes a dead body, giving off no light. The first form of the star, a rare gas only glowing faintly red, differs (in absolute amount) much less from a diffuse nebula than many stars differ between themselves in both density and temperature. If this sequence is followed backward and the big red star becomes still rarer and of lower

---

[15]The Hebrew words rendered, "without from, and void," are compatible with terms suggestive of the dark nebula such as the following: "desolation," "primeval world-material," "primeval nebular state," "chaos," "confusion," "hollow or empty," or "without distinction of object or form."

[16]A bright nebula and dark nebula are the same in composition. The bright nebula simply has a bright B or O type star close enough to illuminate it while the dark nebula has no such star. See B. J. Bok, and P. F. Bok, *Op. Cit.*, pp. 110-126.

[17]See the subject of the stellar sequence in any modern astronomy text.

temperature, the state of the diffuse nebula will eventually be approached. Many authors of astronomy texts start the sequence with the diffuse nebula and have the rare, faint red star as the second step.[18] The diffuse nebula is today the most generally accepted starting place for stars. The planets may have had a common starting place with the sun. Thus we see that the possibility of the earth originating from a dark nebula is in full accord with both the Genesis account and modern astronomical knowledge.

### *The Sun's Early Light*

"And God said, Let there be light: . . . and God divided the light from the darkness. And God called the light Day, and the darkness he called Night. . . ."[19]

Where did the light come from? Certainly from the sun, for if the earth had become hot enough to give off the light there would have been no darkness. Astronomically, was the earth once hot enough to be a star? Geologically, we know that the earth was once warmer than it is now. It could not, however, have been hot enough to be a star, for theory says that the smallest body that can produce enough internal heat to become a star is about 1/100 part of the mass of the sun. The earth has only 1/333,000 the mass of the sun. It could, therefore, never have been a star. The light on the earth, therefore, according to this theory always came from the sun. Thus this scriptural account by the introduction of the words "Day" and "Night" has shown its minutest details to be accurate.

### *Clouds and the Sea*

"And God made the firmament, and divided the waters which were under the firmament from the waters which were above the firmament: . . ."[20]

---

[18]R. S. Lull, *The Evolution of the Earth and its Inhabitants*, "The Nebulae," (Yale University Press, 1918), p 8.

[19]Genesis 1:3-5.

[20]Genesis 1:7.

*Fig.* 3—This is what might be called an average galaxy, or island universe, containing 30 to 100 billion stars as well as many nebulae, star clusters, etc. (*Courtesy Mount Wilson Observatory*)

*Fig.* 4—This is the way our galaxy would appear from a great distance. The sun and its planets would be a little more than half way from the center to the outer edge. The ring of dark nebulae surrounding the galaxy seems to be present in all galaxies. (*Courtesy Mount Wilson Observatory*)

The word translated "firmament" means space.[21] In other words, God made a space between the oceans and the clouds. It is generally believed that in an early stage the earth was quite hot, possibly much hotter than the boiling point of water. If so, there could have been no oceans as the heat would have evaporated all the free water and the earth would have been completely covered with dense clouds coming right down to its surface. As the surface cooled this water would begin to condense upon the surface of the earth and form bodies of water, but still the heat of the earth would be sufficient to keep a large part of the water suspended as clouds. They would form a very dense covering right down to the surface of the water. When the earth cooled still further and approached its present temperature a space would develop between the oceans and the clouds. The clouds for a long time would still continue to be very dense and completely hide the sun, moon, and stars. Yes, the earth had to go through a stage in which a space formed between the oceans and the clouds.[22]

*The Clouds Break Up*

"And God said, Let there be lights in the firmament of the heaven to divide the day from the night; and let them be for signs, and for seasons, and for days, and years: . . . And God made two great lights; the greater light to rule the day, and the lesser light to rule the night: he made the stars also."[23]

The creation of all the heavenly bodies is recorded in the first verse of Genesis.

The word "created"[24] is not used here in the fourteenth verse, rather it is the word "made." Other translators give, "made to function," "made to appear," "made to shine," etc. But the fourteenth verse is really its own key to the meaning, for God is here described as making these heavenly

[21]See Footnote 36, p. 70.
[22]Lull, *Op. Cit.*, pp. 36–38.
[23]Genesis 1:14, 16.
[24]See pp. 66 and 134.

bodies so they can be used for signs, seasons, days, and years.

On a cloudy day the place of the sun in the sky cannot be located and the sun cannot be used to tell the time of day. It is also impossible to tell if the sun is high overhead at noon indicating summer, or low in the south indicating winter. When the clouds break, and only then, can the sun be used to determine signs, seasons, days, and years. Genesis, too, must refer to the breaking of the clouds.

Before the clouds were cleared, and when there was more heat in the surface of the earth than now, and when the clouds were thin enough to let much of the light and heat of the sun through, the earth would have been a great hothouse, perhaps from pole to pole. The condition of the earth was perfect for the introduction of vegetation as described in the previous verses.

Geology testifies that at about this geological period, seasons became evident on the earth, but before this time the whole earth was tropical. The thick layer of clouds and the warm earth would make a very general tropical condition diffusing the sun's rays. Breaking up of the clouds would allow the heat of the sun to become localized, developing different temperature zones[25] much as we find them today.

### Summary

Every type of astronomical object which is available for careful study has had a maximum probable age placed upon it by different and independent approaches. These indicate a beginning or a creation of the universe. Every star is losing energy and mass. They must therefore, have had a beginning.

Previous to thirty years ago, dark nebulae were not known to exist. Today astronomers believe new stars de-

[25]R. C. Moore, *Historical Geology*, "Comments on Climate," (New York: McGraw-Hill, 1933), pp. 155, 173 and 254.

velop from diffuse nebulae. The second verse of Genesis is consistent with this idea that the earth developed from a dark nebula.

Genesis 1:3-5 indicates that the first light on the earth did not come from the earth itself but from an external source. Astrophysics agrees by telling us that the earth is too small to have ever given off light of its own.

Genesis 1:6-10 and 14-19 give us a vivid account of the earth's various stages in its cooling off process. These steps agree with the most modern and up-to-date information known as to the astronomy, physics, and geology of the earth.

Thirty years ago there were serious discrepancies between Genesis and astronomy. Today there is excellent agreement brought about by development of the science. Further developments are certain yet to come. They will, no doubt, show still more emphatically that the Genesis account is entirely accurate.

Although Genesis was written thousands of years ago, every reference to astronomy in this first chapter is corroborated by the best of our present scientific information. And yet we note that books of astronomy, written 25 years ago or more, are full of serious errors and anything written more than a few hundred years ago would only make amusing material suitable for the entertainment of the reader.

It obviously would have been impossible thousands of years ago for men to have possessed all of the scientific knowledge which was required to write this first chapter of Genesis. Only the builder of a house can tell with complete accuracy the details of its hidden construction. Only the Creator of the universe, the earth, and the life upon it could have told thousands of years ago the minute details of that creation.

# A Symposium on "The AGE of the "

BY

MEMBERS

of the

# American Scientific Affiliation

1948

# TABLE OF CONTENTS

RADIOACTIVITY METHODS OF GEOLOGICAL DATING

## I. INTRODUCTION

In historical apologetics there are probably no more fundamental questions than the age of the earth and the age of specific rock strata. There is abundant evidence for the great age of the earth. A consideration of the physical and chemical processes involved in the deposition, metamorphism, and erosion of rock formations gives qualitative proof for its great antiquity. The superposition and apparent structure of the strata in any local area make it possible to determine relative ages of the units present. In order to determine the absolute age of any particular strata or to get some quantitative idea of the time since solidification of the earth's crust, other methods must be sought.

A method which will theoretically meet this need is based on the constancy of the rate of disintegration of certain radioactive isotopes which occur in the earth's crust. (There are several other methods of potential quantitative value for Pleistocene chronology only, i.e., for the past few hundred thousand years.)

Since these radioactive methods, at least in theory, can determine the absolute age of rock strata in a local sequence they make possible long distance correlations and therefore may ultimately produce an objective geological time scale into which paleontological discoveries may be fitted.

In view of this it is evident that Christians should evaluate the accuracy of these methods and be ready to accept pertinent implications.

An initial informal attempt at such an evaluation was made Sept. 26, 1947 at the home of Prof. P. W. Stoner, Pasadena, California by the Los Angeles section of the American Scientific Affiliation where a seminar was held on the subject "The Age of the Earth by Radioactive Methods." Various papers were presented and discussed. Most of those presented were written and sent in to the elected editor who has freely altered them in the interests of integration and accuracy. (In a number of cases more recent data is available which made the paper in error and hence revision was imperative.)

This colloquium is not presented as authoritative but it should be stimulating to ASA members. It will give some basic information on the subject which is probably unknown to many and will indicate the direction of modern researches in geological dating. It further gives an excellent illustration of the good that can come from informal sectional seminars.

The final paper in this compilation was appended after the editor reviewed the other papers. The reason for the appendage was twofold. First there is a considerable amount of more recent information that was not available to the contributors to the seminar. Second there was no general summary of all the presently usable radioactive methods.

The bibliography is suggestive only. It includes articles referred to by the writers of the succeeding papers. The editor has added others with notes on those which are most significant.

## II. THE BIBLICAL EVIDENCE OF THE AGE OF THE EARTH
(Professor Bernard Ramm)

In this paper we are not discussing the geological evidence for the great antiquity of the earth. Our problem is to discuss the near or remote antiquity of the earth from Biblical data.

The most perplexing problem for us to face in the interpretation of Genesis is the time element. There is on the one hand the Lightfoot-Usher tradition. Usher dated creation at 4004 B.C.; and Lightfoot went on to say that it was on the week of October 18-24th, with Adam created on October 23 at approximately nine A.M. forty-fifth meridian time. (Brewster, Creation, p. 109). Opposite to this is the current geological position which demands an astronomical figure for the age of the earth. We may leave the Usher tradition intact and deny the time element to geology, attributing it to a developmental or evolutionary bias, and insist that all its material can be interpreted satisfactorily otherwise. Or, we may accede to the geologists' claim for time and re-examine our interpretation of Genesis.

Suppose for a minute that the geologist is right. (That this is a very probable supposition can be easily demonstrated. The following articles present only a small fraction of the evidence that can be brought to bear. Ed.) Where do we add the time in Genesis? First of all, Usher's 4004 B.C. date may be pushed back to 10,000 B.C. (or so) as far as the origin of man is concerned. (Or further, depending on the interpretation of the Genesis account, yet maintaining the plenary view of inspiration. Ed.) This job of pushing man's creation back was done by the Princeton theologican of the past century who re-examined the genealogical records of Genesis and found that a simple father-son relationship was not at all adequate for an explanation of the lists. James Orr, the great British scholar, has concurred in this judgment and it must be remembered that he was the one scholar German criticism feared and respected among the conservatives. (James Orr, "WORLD," I. S. B. E., 5:3106-3108).

But this is a little "bite" of time. The long epochs of geology may be fitted into the Genesis record two ways. The first is by way of secondary-creation. There is an old Jewish interpretation that says Genesis I dealt with the remodeling of just Palestine, the original creation not even being touched upon in Genesis 1:1 and is, therefore, irrelevant to the Biblical record. We reject this interpretation because it, first, obviously misses the full force of the Genesis I account which depicts the creation of a universe, and secondly, it leaves the real cosmological-cosmogenical problem unsettled which would be a fatal mistake for a full-rounded system of Christian thought, and a dependable doctrine of inspiration.

Another form of secondary-creation is that Genesis 1:1 deals with the creation of the universe and world. Then this was put in the hands of Satan, then an unfallen angel. He fell and judgment was passed on the earth and it was destroyed. All the geological time then occurs during the second verse, or between verses one and two. The six following days of creation are limited only to the earth, and are as usually interpreted, twenty-four hour days. This view was first advocated, as far as we know, in Pembers, Earth's Earliest Ages; taken over by Scofield in his Reference Bible and currently defended in Rimmer, Modern Science and the Genesis Record, and Higley, Science and Truth. To our way of thinking there are several fatal objections to this interpretation. First, the Hebrew exegesis involved is most untrustworthy, the details of which we cannot enter into here. You simply cannot untangle the vast problems of geological science with forced Hebrew exegesis. Secondly, the reading into Genesis I of an elaborate angelology involving the ruination of creation is nowhere clearly and unquestionably supported in the Bible. It is gratuitous and arbitrary to bring in Isaiah 14 and Ezekiel 28 between Genesis 1:1 and 1:2. It is to be regretted that when this theory is defended by its proponents they do not clearly state that they are importing these passages into Genesis but present them as if they naturally and unquestionably belong there. It is also to be regretted that their interpretations of Isaiah 14 and Ezekiel 28 are not supported by an exhaustive and thorough study of these passages apart from what they have been told that these passages ought to say. Thirdly, the very evident design of Genesis 1:1 is the creation of the universe and world. It is incredible to think that the bulk of that great record is really "the second time around." On the one hand it puts all the agreement between geology and Genesis in the blank spaces of Genesis 1:1 and 1:2. It would be embarrassing to tell a geologist that the Bible agrees with his geology in the blank space between verses 1 and 2. Again, it would leave as a great mystery the particular facts of the geological record. The progress from the azoic era up to the appearance of man would be a profound enigma. (The facts of geology make this interpretation unacceptable. Instead of a real "blank" between Genesis 1:1 and 1:2, the "reconstructionist" usually infers that the geological record is one of judgment catastrophe. Such is unknown to the careful student of the earth's crust. Rather there is excellent evidence that from the oldest geological records, rocks were formed by the same physical and chemical process as are operative today and furthermore that volcanic, orogenic, and epeirogenic (broad warping) activity occurred sporadically throughout geologic history. Ed.)

The only real way in which I can see that the time demanded by geological science is to be found in what I call the geologic day theory of Genesis. The days are immense periods of time. They need not be of equal length. They need not correspond specifically with the various eras, periods, or epochs of a geological chart. But they are the six great days in which the creativity of God was at work bringing to pass what matter was helpless to do.

The opposition to this interpretation by insisting the sabbath week is seven literal days is not serious as seven periods could be readily typified by seven days. In reference to the objection that the seventh day would have had to be millions or at least thousands of years old it may be said: (1) That the expression "evening and morning" is absent meaning that as far as creation is concerned it is finished; (2) we may well be in the rest of creation now; and (3) that the days of creation are of unequal length depending specifically on the purpose to be accomplished. For example, according to Schuchert and Dunbar pre-Cambrian time occupies three-fourths of all geologic time! (Historical Geology, 4th edition, p. 70).

Furthermore, Gruber has pointed out that a strict chronological principle can not be used in Genesis 1. He points out that although the sun, moon, and stars were created in Genesis 1:1, their appearance is not given till the fourth day. There fore for purposes of clear delineation chronological features may either be sacrificed or distorted such as in the appearance of plant forms in the third day apart from animal forms. (Cf. Gruber, The Six Creative Days, p. 56 ff. This writer has never found anything yet so simple, clear, lucid, fruitful and intelligent on the subject under discussion as Gruber's work here cited).

In returning now to our original thought, this interpretation of Genesis establishes a working basis between Genesis and geology which pays full respect to the plenary inspiration of the Scriptures, and takes careful intelligent consideration of the mass of information and somewhat confirmed hypotheses of the science of geology. It furthermore gives to geology a genuine interpretive principle which the recreation-theory does not even profess to do, and which the deluge-theory endeavors to do but fails.

Frist, let it be said that this view is not new in interpretation. Gardiner (Evolution of Christian Doctrine) points out that this view has been held in all the great epochs of church history. Lange (Genesis), and Walshe, (The Principles of Catholic Apologetics) have pointed out that Augustine and Gregory of Nyssa, long before the days of modern geology, insisted that these seven days were "divine days" and hence could not be thought of as 24 hour days.

Secondly, this has been the view, for the most part, of the great German, British, and American Conservative scholars. The reason is that it alone gives the rightful place to both the works and word of God.

Thirdly, a critical examination of the terms clearly indicates the tremendous scope of the passage, and the utter helplessness of handling it in terms of 24 hour days. Taylor Lewis (in Lange, Genesis) very forcibly and remarkably deals with this, the details of which , cannot be entered into here. (Also see Gruber, The Six Days of Creation).

Fourthly, there are passages of Scripture that demand a great antiquity of the earth. This is brought out by Lewis in the only exposition of this kind we have ever seen in Lange's Genesis, entitled "Allusions to the Six Days in other Parts of the Bible" (p. 135 ff). He commences by showing that the objection to the geologic day theory of Genesis on the basis of the sabbath week is invalid. Then he comments on other Scriptures that indicate a vast antiquity for the age of the earth.

For example he refers to Proverbs 8:22 and translates it: "The Lord possessed me from the antiquities of the earth," and goes on to give exposition to show that the idea here is that the world is incredibly old. He refers to Hebrews 11:3 and paraphrases it: "By faith we understand that the ages, the eternities, the saecula, or great world-times were mediated . . . or put in order, by the Word of God.

In Micah 5:1 he indicates that the expression "from everlasting" is literally "from the days of eternity" and refers to creative times. The contrast then is between the brief earthly life of Christ compared to his previous immeasurable antiquity.

This point need not be drawn out any further save to mention that by similar exegesis the same point is made quite clear, and to us, convincing by reference to Psalm 104, Psalm 90, and several passages in the book of Job. We commend a careful examination of the evidence Lewis presents in the volumn here referred to. A cursory report loses the fullness and the pungency of the argument.

In conclusion we believe that the time element as a general principle can be granted to the geologist. We find the time element in the Bible by (1) interpreting Genesis I after the manner set down by Augustine, (2) by noting those passages of Scripture that can only be understood if the earth is of a great antiquity.

## III. HELIUM AND LEAD METHODS OF DATING
(Professor Roland N. Icke)

I would like to discuss these two subjects, basing my report almost entirely on the Chemical Review article (1933 p. 305) by Urry listed in the bibliography. (This reference is considerably out of date. The most recent summary is to be found in the G.S.A. Bulletin (1941, p. 491). Ed.) There are three different disintegration series which have been applied to the lead method of dating: the decomposition of $U^{238}$ to $Pb^{206}$ giving off eight helium atoms, $U^{235}$ to $Pb^{207}$ giving seven helium atoms, and $Th^{232}$ to $Pb^{208}$ giving six helium atoms. These series provide three independent ratios which can give indications of the age of a mineral sample if it contains both uranium and thorium. The disintegration constants for these series have been determined with sufficient accuracy so that they are not limiting factors. The determination of the isotopes can be done quite accurately by Aston's method using the mass spectrograph. The other theoretically possible cross checking measurements include (a) helium-uranium and helium-thorium ratios and (b) the ratio of the radiogenic Pb isotopes which vary with time for a uranium bearing mineral.

The helium method is applicable only when the samples have minute amounts of radioactivity and yield about $10^{-4}$ to $10^{-6}$ cc. of helium. In such cases the actual amount of lead produced is so small that it is not measureable.

There have been many experiments performed in an attempt to determine the constancy of the disintegration rate for radioactive isotopes. Temperatures up to 3000 degrees Centigrade, and pressures up to 20,000 atmospheres, extremely high electric and magnetic fields and high gravitational forces have been applied, with absolutely no change in the rate of decomposition. Jeans calculated it would take a temperature of $5.8 \times 10^9$ degrees to appreciably alter this rate. (The following paper will give additional direct evidence for the constancy of radioactive disintegration with time. Ed.)

In order for the lead method to be accurate, one must be certain that the mineral did not change by admixture from the outside or by losses from the inside, and that only radioactive disintegration took place. There are five factors which must be known.

1. There must be analytical data for the mass of the uranium, lead and thorium per unit mass of the mineral.

2. The atomic weight and also the amounts of each of the isotopes of lead produced by disintegration.

3. The disintegration constants of all of the disintegrating atoms.

4. The lineage of the lead isotopes.

5. There must be some way to ascertain the possibility or probability that lead was present at the time the formation was laid down. It is necessary to determine each isotope separately and to determine whether or not common lead was present at the time of formation. All of these factors must be determined for the lead method, in order that real accuracy may be certain.

(It might be pointed out here how the determination of primary lead is made. The lead isotopes observed in nature have the atomic weights 204, 206, 207, and 208. All of these are radiogenic except 204. The ratio of these isotopes to each other is well known for ordinary lead--such as obtained from the usual lead ore deposit. Therefore the amount of primary lead present in any uranium mineral will be given by the concentration of $Pb^{204}$. In many samples analyzed by Nier this is negligible. Ed.)

Helium determinations have been made on igneous rock and on many of the fine-grained minerals of such rocks. Urry believes the results on the separated minerals to be more reliable (This has been proved with later reseraches, Ed.)

In the helium method there are four things to be known; (1) The quantities of uranium, thorium and helium in a rock or mineral, (2) the disintegration constants of the isotopes involved, (3) the probability of primary helium, and (4) the probability of radiogenic helium leaking from the rock or mineral through out geologic time.

It seems improbable that helium could be trapped in a mass of molten material. As a free gas it would have very little opportunity to be trapped. The question now is whether helium was absorbed by the rock on cooling. That of course necessitates or presupposes that the helium in the atmosphere nearby must be present in very high concentration in order to be absorbed in sufficient quantities. In order to investigate this possibility, though, Urry carried out several experiments in which he heated different types of pulverized rocks at high temperatures that he could maintain in the laboratory, in an atmosphere of pure helium. He found that there was not a detectable amount of helium absorbed by the material and he stated that he could measure $10^{-10}$cc., so he could say that there was less than $10^{-10}$cc. absorbed in the samples. Urry had some evidence that the helium in these different radio-active minerals was held within the crystal structure and was not an absorption phenomenon. (A theoretical study of the probability of helium retention in perfect crystals has been carried out by Keevil, 1940 reference, which adds credence to Urry's conclusions. Ed.)

(Another item of interest is that the helium method has been used to determine the age of meteorites. Meteorites contain radioactive material, and in every case the age of these particles was found to be within the limits of age determinations on other types of rock taken from the earth's surface--the maximum value being 2600 million years. This finding could indicate one of two things, either that the meteorite was of solar system origin and thus the same age as the rest of the solar system.) The fact remains that the meteorite appeared to be the same age as the earth, and there is no way to determine which of the above possibilities is true. This question is somewhat outside our topic. Some meteorites have been found by the helium method to be older than the oldest mineral estimated by the lead method. (It should be noted that there are a number of recent researches on the age of meteorites. These are being carried out at the University of Chicago and the Geophysical Laboratory. Ed.)

(In summary, I would like to say that as far as the helium method is concerned, it seems to have fewer errors involved and fewer factors which are uncertain. The helium method is certainly not self-sufficient for geologic age determination but it is a powerful supplementary tool.)

## IV. PLEOCHROIC HALOS
(Professor F. Alton Everest)

Determination of the ages of rocks by radioactive methods calls for a knowledge of the rate of disintegration of the uranium or thorium into helium and lead. This disintegration rate as it now exists can be measured very accurately. Furthermore, observations on the rates of disintegration of the radio-active elements have been made under the highest attainable temperatures, pressures and magnetic fields with no noticeable effect. Since the minerals in which the radioactive clock elements reside would have been changed over under much less severe conditions than those indicated above, it is reasonable to assume that alteration of these rates has, in fact, not taken place in the geologic history of the crust of the earth.

However, a direct method for examining the constancy of these disintegration rates throughout geologic time is highly desirable. Can we be positively assured that the physical constants involved have not varied in the past?

Pleochroic halos provide us with information as to the constancy of radioactive rates of decomposition throughout geologic time. Flakes of brown mica contained in certain granites show a sprinkling of tiny spots under the microscope (100 to 400 X magnification). Some of these spots under careful examination show a beautifully developed pattern of concentric rings called "mica halos" or "pleochroic halos." It is found that the rings are actually concentric spherical shells or halospheres, shown in section along the cleavages of the mica.

At the center of each halosphere is found a minute radio-active crystal, the cause of the concentric rings. (This "radio-active crystal" is usually some rare earth mineral such as zircon which contains a small amount of uranium or thorium as substitution or impurity in the lattice. Ed.) In the disintegration of the radio-active atoms, alpha particles (helium atoms) are ejected. The alpha particles slows down in penetrating the mica, and finally comes to a stop within a few thousands of a millimeter. The alpha particle ionizes the mica, the greater ionization taking place, as in air, near the end of its travel where the velocity is low. The mica material is discolored by this ionization process resulting in the discolored sphere surrounding the tiny particle of radioactive matter. If the particle is composed of several radioactive elements, concentric halos of different characteristic radii will result. The rate of formation of the halos is a function of the amount of radioactive material at the center, but the radii are essentially independent of this factor.

G. H. Henderson has developed a special Halo Photometer for making accurate measurements of the radii of mica halos. This instrument has eliminated some of the earlier minor discrepancies of Joly and others whose measurements were affected by lighting, microscope focus, and various subjective factors. Henderson has shown that Pre-Cambrian halos are just as sharply defined as those in younger rocks. Further, the radii of these older halos are identical with those found in younger rocks. Halos can also be "grown" rapidly in the laboratory, and the radii of such artificially produced halos correspond to those found on the rocks of various ages. Since these radii are functions of the disintegration constants* of the radioactive elements giving rise to the halos, this means that the radioactive disintegration rate has not varied appreciably since early geologic time.

---

* This is due to the fact that the energy of the particle ejected from the nucleus of an atom is directly related to the disintegration constant and this energy determines the depth of penetration of the alpha particles in any given material. Ed.

Thus pleochroic halos provide independent and direct testimony to the constancy of the rate of disintegration of radioactive elements and give to the geologist great confidence in methods of dating the rocks based upon radioactive decomposition. The radii of pleochroic halos produced by a given radioactive element are the same whether produced in the laboratory with sources of greater intensity, or observed in pre-Cambrian to Cenozoic rocks. This relieves us of the necessity of assuming without experimental demonstration that rates of disintegration observed today have been the same throughout geologic time.

----------

## V. USE OF THE MASS SPECTROGRAPH IN AGE DETERMINATION
(Dr. Earl Rex)

We may, perhaps, say that the history of the mass spectrograph began with the discovery of the positive rays in a cathode ray tube by Goldstein in 1886. It was while he was studying a stream of electrons flowing from the cathode of a partially evacuated tube that he noticed a stream of particles flowing through some holes or "canals" in the cathode in the opposite direction. We now believe that these particles were positive and were atoms of the residual gas ionized by the bombardment of the electrons.

Beginning in 1907, J. J. Thompson investigated these rays by the so-called "parabola method." His results led him to suspect there were two isotopes of neon of atomic weights, 20 and 22. This was definitely confirmed in 1919 by Aston by the use of the first mass spectrograph. Others constructed mass spectrographs having ingenious features. The one used by Mattauch is of particular interest. It employs a radial electric field which deflects the positive ions through $31^{\circ}50$ and then deflects them in the reverse direction through a $90^{\circ}$ magnetic field. In a recent application, this field was 18,800 gauss across a 1/8 inch gap. Energy is given the ion source by a compact Tesla transformer wound on Lucite cylinders.

(The most significant paper on the application of the mass spectrograph to the measurement of geologic time is by A. O. Nier, The Isotopic Constitution of Radiogenic Leads and the Measruement of Geologic Time, Phys. Rev. vol. 55, p. 153-163 (1939). This work greatly improved the accuracy of the method. This paper also includes a systematic study of many ordinary and radiogenic lead samples. From this excellent work it is clear that these analytical techniques are not limiting factors in the evaluation of the primary lead in radio-active minerals.

The mass spectrograph will be used for a number of other important radioactive methods including the strontium-rubidium and strontium isotope ratio procedures which will be described in the last paper of this series.

The importance of the mass spectrograph is strikingly shown in the case of the uranium-lead method. Prior to Nier work there was some uncertainty in all of the uranium-lead measurements because of the question of the amount of primary lead present. Geochemical separation of uranium and lead takes place so efficiently that usually the amount of ordinary lead present in the crystal of a radioactive mineral is much less than that of radiogenic origin except in the case of very young minerals. Hence although there are minor variations in the isotopic constitution of ordinary lead the correction required in subtracting this from the radiogenic sample in a uranium-bearing mineral is negligible. It is clear, however, that isotopic analyses must be made for any satisfactory age result with the uranium-lead, rubidium-strontium, or strontium isotope ratio methods.

Further the determination of the proportions of $Pb^{207}$ to $Pb^{206}$ which are the daughter elements of $U^{235}$ and $U^{238}$ respectively, permits an independent check on the uranium-lead measurements. Because $U^{238}$ and $U^{235}$ have different rates of decay, the proportions of the lead isotopes will vary systematically with time. If the age as obtained from Uranium-lead ratio differs from that determined by the $Pb^{207}$: $Pb^{206}$ ratio it is clear that the specimen has undergone alteration since the first ratio would be affected to a much greater degree than the latter by chemical attack. Therefore where there is good agreement between these independent methods, the result may be accepted with reasonable assurance. The following table taken from Goodman and Evans (1941) is illustrative of the better results.

Comparison of U-Pb with $Pb^{207}/Pb^{206}$ Ages (in million years)

| Locality | U-Pb | $Pb^{207}/Pb^{206}$ |
|---|---|---|
| Connecticut (Uraninite) | 366 - 371 | 375 |
| Katanga (Pitchblende) | 580 - 995 | 610 - 665 |
| Ceylon (Thorianite) | 530 | 485 |
| Besner (Uraninite) | 765 | 825 |
| Pied des Monts (Cleveite) | 880 | 905 |
| Norwegian (Cleveite) | 1085 | 1090 |
| Wilberforce (Uraninite) | 1077 | 1035 |
| Great Bear Lake (Pitchblende) | 1251 | 1420 |

Hence the value of mass spectrographic analyses in age determination work is evident. Ed.)

Unless all the various disintegrations took place at varying rates in the past, and other methods of dating are in error by the same amount and in the same direction, we have succeeded in finding a method of dating parts of the world.

(See special bibliography to accompany this paper in the general Bibliography at the end of the papers.)

---

## VI. CRITIQUE OF RADIOACTIVITY ESTIMATES OF AGE OF THE EARTH
(Dr. W. E. Lammerts)

This discussion is based on Lane's paper on Measuring Geologic Time: Its Difficulties[1] & D. J. Whitney's paper on Radioactivity and the Earth's Age.[2] Lane is undoubtedly eminent in his field and has achieved a very fine reputation; Whitney though just a layman and student of the subject, nevertheless uses a good deal of common sense in studying the various problems involved. The basic principle involved in using radioactivity as a means of measuring geologic time may be stated by the following example. If you were to watch a popcorn machine and count the number of kernels popped in a second, and the number already popped, you could figure out how long the machine has been running, even thou you couldn't predict exactly when any given kernel of corn would pop. That is essentially our problem in trying to date the age of the rocks. That is, though radioactive elements, and rates of disintegration are known, the time of disintegration of each constituent atom is unpredictable. Yet ordinarily in any rock containing a radioactive element, there are many emissions per second per milligram and the number of emissions per hour is constant and

1. Lane, Alfred C., 1938. Measuring Geologic Time: Its Difficulties. Smithsonian Report 1937: 235-254.
2. Whitney, D. J., 1938. Some Incongruities of Radioactivity Estimates of the Earth's Age. Pan American Geologist 69: 15-22.

characteristic for each element regardless of any variation of pressure, temperature or electric charge, which is likely to occur in the earth's crust. The question is, how do we know that the elements presumably resulting from disintegration were not already there? Or to return to our illustration of the popcorn machine, how do you know that some of the popcorn hadn't been sold before you arrived? This question is answered by saying that each derived element i.e. derived lead, has a distinctive atomic weight. The problem, however, arises as how to find that weight, especially since mixtures of the various isotopes occur in such minute quantities. (This has been successfully carried out by Nier as discussed in the preceeding paper. Ed.)

Lane discusses the two methods of getting the ages of rocks and materials which have been mainly used; namely the amount of helium and of lead accumulated from the explosions of uranium, of radium, and thorium atoms. The amount of lead from radioactive disintegration is very small in comparison even with the few grams per ton of ordinary lead present and until recently it has been impossible even to attempt to get the proportion of isotopes in the rock lead, as it would involve getting about a gram of lead from a rock which contains less than 22 grams per ton. Of this gram only a small portion would be the lead derived from the explosion of uranium and thorium atoms, perhaps only 10%. For instance Dr. Urry has computed for the "whinsill," a well known sheet of intrusive basalt that occurs in the British Carboniferous, that the lead derived from the uranium and thorium would be about 0.049 grams per ton, whereas the actual amount of lead present was 3 to 5 grams per ton. For these reasons the uranium-lead method is only used on uranium minerals in which the concentration of radiogenic lead becomes appreciable while the uranium-helium method will work on these rocks and the minerals separated from them that are low in uranium content.

Lane goes on to say that in molten rocks the helium gas was mainly driven off at the time they were molten, and furthermore that helium derived from the atmosphere can be recognized by the neon which goes with it. Whitney claims that indirect evidence makes it appear that helium would not be driven from the magma when it intruded the upper rocks, but would remain, in part at least, to be counted as if it were all derived from radioactive decay after the rock formed. That is, Whitney mentions that when granite and certain other igneous rocks are heated to red glow noteworthy amounts of nitrogen and certain other gases are driven off.

Presumably the nitrogen was there from the time of formation, as no one claims it to be the result of radioactive disintegration. Similarly any helium present in the molten magma would not at all be driven off, therefore to consider it to be a factor which showed the age of the rock would certainly be untenable. I presume that the other argument that helium which is trapped from the air can always be detected by its contamination with neon is a sound one. (It should also be noted that experiments have shown that for rocks of the same age but different type the helium content goes up linearly with radioactivity content. Furthermore the uranium-helium ages are relatively in the same order as the uranium-lead ages. This could not happen in the large number of samples if primary helium were an appreciable factor. Ed.)

Regarding pleochroic halos there is this interesting statement in the Bulletin of the National Research Council #80, 1931, page 181-190, "The color of the halo before reversal or bleaching sets in, depends on the number of alpha particles expělled from the central inclusion, and this in turn depends on (a) radioactivity and (b) the age of inclusion. Natural halos are developed by a relatively low degree of radioactivity operating over a very long period of time. Halos can, however, be made artificially by using a high degree of radioactivity over a relatively short time. (These look very similar in appearance.) Suppose that a natural halo is formed in biotite with exactly the same depth of color as an artificial halo produced under controlled conditions in a flake of the same biotite.

If now the radioactivity of the nucleus can be estimated then the time during which it must have acted can easily be calculated. This period will correspond to the age of the halo, i.e. to the age of the nucleus and host mineral and this is the important assumption provided that the radioactivity of the nucleus can be readily estimated." And, of course, in view of the minute diameter of the whole pleochroic halo you can readily see how minute the nucleus is and how difficult such estimates must be. (Note that this criticism applies only to the use of pleochroic halos in absolute age determination -- not to the argument that the disintegration rate has remained constant throughout geologic time. Further modern techniques may well make the analysis of the radioactive content of the halo determinable. Ed.)

Whitney points out on page 40 of the 1936 report of the Committee for the Measurement of Geologic Time, five analyses which are far too low for the assumed ages of the formation from which they come. Moreover in nearly every list of analyses, serious inconsistencies may be found in the age indicated by the radioactive analysis and the place in the time scale from which the rock is believed to have come, that is the time scale based on the usual methods used in age determinations.

(These anomalies are not as bad as the author indicates. Firstly the data of the thirties showed considerable inconsistency due to a number of factors both theoretical and experimental. Age determination has passed through this stage of "growing pains" and more recent results are much more consistent. See tables of the major references by Goodman, Evans, Hurley et. al. Ed.)

Furthermore the results of various methods of estimating time by radioactive analysis frequently do not check with each other and with other methods of determining geologic time. Thus Lane makes the interesting statement that "in radio active minerals containing a considerable amount of a radio active element, the lead radio is the more significant, and, if there is considerably more helium in the mineral than in the surroundings, the helium diffuses. The helium methods are then likely to give ages that are too low, and the ratio of lead to the uranium and thorium, if the ordinary lead has not been determined, ages that are too high. Insofar as they check, we can be sure of not being far off."[3] Now obviously the geologic time table as determined by fossil evidence is the standard of comparison or check, and is apparently assumed by radio-activity investigators to be substantially objective and correct. If then a helium determination indicates an age significantly lower, or a lead determination an age signignificantly higher it is discarded or explained away as not being valid due to diffusion of helium, leaching of uranium, or other causes. It would seem very critical study is needed of this whole problem in order to avoid the danger of reasoning in a circle.

If helium is not retained 100% by the mineral containing a small amount of uranium or thorium, the ages will be lower than those determined by the uranium-lead method. If uranium happens to have been leached (the most ordinary form of uraninite alteration) from a uranium mineral, the U-Pb age will be high. Insofar as the lead and helium methods check with each other, with logical structural interpretation, and with a number of other independent age determinations, we can be reasonably sure of not being far off. (These results are therefore not considered as "checks" merely if they agree with the paleontological age. Actually the agreement of the best age determinations with the relative position of the geologic time table derived from physical and to a lesser extent paleontological grounds is amazing. Ed.)

---

3. Lane, Alfred C. Measuring Geologic Time, Its difficulties, page 250,

On page 554-555, Bull. of the G.S.A., vol. 52 we find the following "considering magnetite as unity, the above results indicate that the specific retentivity of feldspar is about 0.3 and of pyroxene about 0.8." Thus the only mineral so far that appears to have 100% retentivity and thus of most use for helium determinations seems to be magnetite and in Table 8, page 555, of the above Bulletin, age ratios of various magnetite deposits are given. Though substantial correlation between geologic age and helium age determination here exists, analysis of many more samples per given geologic age are needed before the data can delimit periods with accuracy. Thus even in the data presented in Table 8 the magnetite from late Jurassic which is a geologically younger period than the Triassic, actually has a higher age ratio in $10^6$ years than some specimens from the Triassic horizons, i.e. 130 as against 126-145. In general it may be said that though methods of age determination are becoming much more exact and refined, in order to establish the geologic time scale with precision, many more determinations per age are necessary for data to have the desired statistical significance.[4] Certainly many of the determinations previously made have little more than a rough qualitative value. (This is particularly true of determinations by the helium method prior to 1940 and U-Pb determinations prior to Nier's work in 1938-39. Ed.)

4. See also Burdick, C. L. The Radioactive Time Theory and Recent Trends in Methods of Reckoning Geologic Time. The Forum of Society for Study of Natural Science Vol.I pg 59-59.

## VII. GENERAL DISCUSSION OF LATEST RESULTS IN GEOLOGICAL DATING
(Prof. P. W. Stoner)

A few years ago I talked with Dr. Buwalda, chairman of the department of Geology at California Institute of Technology. Dr. Buwalda and Dr. Evans did much of the early work in determining geological time by the radioactivity of uranium. Dr. Buwalda is still at California Institute of Technology, while Dr. Evans has left and is working with Dr. Goodman at M. I. T. A report of the work of these latter two men is found in the bibliography (1941). I talked with Dr. Buwalda again yesterday. Dr. Buwalda selected the samples and used only those which appeared to be very definitely sealed in igneous rock and from which they considered the removal of uranium or lead unlikely. He believes that the present radioactive dating to be in a rough stage, just like any new scientific work is for the first few years or first several decades.

Dr. Buwalda said he was confident that the ages as now determined are of the right order of magnitude. When asked what he meant by the right order and what % of error he would think possible in the extreme, he said that he would like to put a factor of 2 as an absolute limit. [He said, normally the age of the earth was given as 2 billion years, in the extreme it might come out as a 1 billion but it certainly would never be shown to be 100 million years or anything like that. The ages are of the right order. He also spoke about the methods of helium and lead, that helium definitely gave a lower limit to the age while lead gave the upper limit.] He also stated that so far as he knew only one scientist working on the atomic ages was in any way doubtful of the results, all others were convinced that their results were of the right order.

I would like to review a few of the things which have been mentioned and add a few items here and there. Helium gives the lower limit because some of the gas migrates. When helium is placed in a container whose walls are composed of a certain rock and great pressure is placed on the helium, a certain amount of the helium has been found to pass through the rock walls in a given time. This amount differs with the type of rock used. The helium found sealed with the sample therefore is less than the total amount liberated from the uranium. (It should be observed here that it has been demonstrated that the leakage in the experiment described above is along major fractures and fissures in the rock which may be submicroscopic but are large compared to the interatomic spacings. Further it can be shown that in a reasonably fresh and unstrained rock the number of helium atoms that would be ejected into these channelways for easy exit are a negligible fraction of the total. Most of the helium atoms are trapped in the atomic lattice.Ed.) From efforts already underway (M. I. T. laboratories) this amount of migration may be fairly accurately determined and corrections made for it so that a good determination of the ages of all these samples may be made.

It was mentioned earlier that it was difficult to think of uranium liberating helium when it was in the molten state, and then retaining that helium until some later age when the uranium and the supporting rock solidified. In a less spectacular but almost as effective way Pb is separated from U during solidification by the greatly different geochemistry of the two elements.

The main danger with the lead date appears to be due to the possibility of water or some other agent entering the sample after everything has solidified and removing some of the lead and some of the uranium. Since most agents will remove more uranium than lead, we may expect all such samples to give an age for lead which has the possibility of being too great. (The isotope ratio method mentioned before takes care of this problem rather satisfactorily. Ed.)

For the above reasons, I think we should expect some inconsistencies. This does not mean, however, that the results need be thought of little value. When we compare the different geological ages given by one of these elements alone, such as helium there are a few striking discrepancies in that what we were sure was an older sample is given a younger age. If we average several determinations in this vicinity, the discrepancies mostly disappear. If we keep in mind that the real age lies some place between the age given by helium and that given by lead, we have practically no discrepancies. All that is claimed for the method is that the true age lies between these two limits.

I asked Dr. Buwalda what he considered to be the most dependable material now available on the work done in these determinations. He referred me to the Goodman and Evans article (1941) listed in the bibliography.

Following is a list of several of the ages taken from this report. In spots where several ages were listed for helium and only one or two for lead I have averaged them.

| He age | (million yrs.) | Lead Age | |
|---|---|---|---|
| 21 | | 34 | |
| 58 | | 70 | |
| 150 | | 220 | |
| 200 | | 232 | |
| 54 | | 69 | (an evident discrepancy) |
| 200 | | 349 | |
| 290 | | 580 | |
| 600 | | 765 | such close |
| 1000 | | 1000 | (agreement purely fortuitous) |
| 1830 | | 1850 | |

(The interested reader is advised to study the original article which gives most of the reliable values available pre-war)

The last two are pre-Cambrian. With the exception of just one discrepancy, and it is possible that the sample was wrongly classified, the order of the ages agree with the order of the geological strata as previously determined. They would indicate a very ancient earth. The age of 2 billion years does not appear to be far out of line.)

Various isotopes of lead have been identified as being derived from different disintegration series of uranium and thorium. It is now known just what ratio should exist between these isotopes. Several measurements have been made determining the age from the isotopes and they have been reasonably consistent with the straight lead values.

When these methods are perfected and many samples studied, we may expect very reliable results giving us the ages of many geological formations in the earth's historical sequence.

---

## VIII. PRESENT STATUS OF AGE DETERMINATION IN GEOLOGY

(Dr. J. Laurence Kulp)

Like most other fields of physical and chemical research which depend on the advance of our knowledge of radioactivity and other nuclear processes, the matter of radioactive age determination has made tremendous strides in the last decade. Most of the earlier references are out of date and some of the latest information is not yet available in print. Therefore it would seem desirable to briefly summarize the various radioactive methods that are presently recognized as having value in geologic age determination and to indicate their limitations.

There are a number of basic requirments which must be met in order that the method is valid. These are enumerated for each method followed by an estimate of the potential effective range, and finally the geologic conditions in which the particular method may be used.

1. Rubidium-Strontium Method

This method depends on the fact that $Rb^{87}$ is beta-ray active and decays, with a half life of about $10^{11}$ years to stable $Sr^{87}$.

Basic Requirements

a) Disintegration constant must be accurately known.

Although the half-life of $Rb^{87}$ was only known to $\pm$ 40% at the beginning of the war, a recent redetermination of this quantity by Eklund (Arkiv. Mat. Astron. Phys. A33, No. 14, p. 79 (1946)) gives $5.8 \times 10^{-10} \pm 15\%$. Private communication from Prof. Dodson of the Brookhaven National Laboratories says that this can certainly be reduced to $\pm 5\%$ with present techniques in thie country. Therefore this is no longer a limiting factor, As far as relative values for the ages are concerned this uncertainty is of no consequence.

(b) Accurate measurement and sampling.

Spectroscopic procedures have been worked out by Ahrens (1948) which yield sufficient accuracy except in the late Cenozoic.

(c) Primary strontium in crystal must be evaluated.

Fortunately geochemical processes tend to separate Rb and Sr and hence most Rb minerals contain only very minor amounts of "ordinary" or primary strontium.

However, this quantity of ordinary strontium may be subtracted out by measuring the strontium isotope ratios and knowning that "ordinary" strontium contains $Sr^{87}$ only to the extent of 7.5%.

(d) Addition or subtraction of Rb or Sr during life of mineral.

This method meets this requirement much better than the U-Pb method. The average pegmatite feldspar from which Rb is separated shows no effect of alteration even in pre-Cambrian rocks. Thus since the radiation is insufficient to shatter the mineral as in the case of the uranium minerals, and the chemical alteration appears negligible, it can be safely assumed for most cases that addition or subtraction of material has not occured. In this respect it is markedly superior to the helium method.

Effective Range

A maximum range for high (2-5%) Rb content minerals with the present means of analysis would be from the oldest known rocks to 10,000,000 years ago (Pliocene). For low Rb (less than 0.05%) as in numerous muscovites, or where the ordinary Sr concentration is high, the range has an upper limit closer to , 500,000,000 years (beginning of Cambrian). Its greatest usefulness will therefore be in application to pre-Cambrian rocks.

Area of Geology

Rb occurs appreciably in potassium bearing minerals in granite Pegmatites of all ages. Since these occurrences are very restricted in the earth's crust, the method might seem quite limited in its application. However, Ahrens' work clearly indicates the possibility of applying the method to many muscovites and hence to most old (Paleozoic or older) granites. Further it should be emphasized that Rb minerals occur in pegmatites in which uraninite has been found thus making an independent cross-check between these methods possible. If such check is produced with statistical numbers of samples, great credence can be placed on both the U-Pb and Rb-Sr determinations in other areas where they are not found together. Preliminary data indicates that this is the case.

The future development and application of this new method seems most promising.

2. Uranium-Lead Method (Includes Thorium-Lead and Lead Isotope Ratio Methods)

These methods have been treated in the previous papers so only brief reference will be made here. The method, of course, depends on the decay of uranium into lead through the various intermediate isotopes.

Basic Requirements:

(a) Half-life accurately determined. -- not a limiting factor.

(b) Analytical procedures for U, Th, and Pb well-known for some time.

(c) Primary lead may be determined by mass spectrographic analysis of the lead isotopes.

(d) The addition or subtraction of material may be detected by petrographic examination, chemical analysis of the mineral, and the presence of secondary minerals. A more critical evaluation of such subtraction may be made by studying the lead isotope ratios which is in effect another independent method of age determination.

Effective Range

Early Pre-Cambrian to 1,000,000. The upper limit is determined by the length of time for a measurable amount of radiogenic lead to accumulate. The lower limit is a result of the denominator becoming small in the age equation, i.e. uranium content gets so low that uncertainty increases.

Area of Geology

This method cannot be used for ordinary rocks for two reasons. (a) The concentration of radiogenic lead cannot be determined and (b) there are several orders of magnitude more of ordinary lead than radiogenic lead. Therefore the method is restricted to uranium minerals or to rare earth minerals that contain appreciable (greater than 0.001%) U or Th.

The geologic location for such minerals are usually granite pegmatites but may include certain types of igneous rocks and metallic veins.

3. Uranium-Helium Method

As previously described this method depends on the measurement of the helium accumulated from uranium and thorium disintegration.

Basic Requirements

(a) Half-life well known.

(b) Measurement of the helium evolved from a rock is only a limiting factor for very young rocks (Pleistocene).

(c) Primary helium is not important for extrusive igneous rocks (lavas etc.) and while it might contribute to the helium content of an intrusive rock under unusual circumstances, this can be detected from other data.

(d) Addition of helium during the history of the rock is no significant factor since any helium pressure from below would leak through the major fractures, or at least move around grain boundaries and hence not accumulate. The problem of retentivity, however, is a major one. Careful experimentation indicates that this can be conquered. It depends on finding minerals with 100% retentivity and then calibrating other minerals. A necessary requirement is that the minerals used and compared should be of the same mineralogical, textural, and metamorphic environment. Magnetite and certain common sulfides already seem potential applicants for the job.

Effective Range

For ordinary minerals it should be effective from early Pre-Cambrian (2,000,000,000?) to 1,000,000 yrs. For optimum measurement the early Mezozoic should be the upper limit.

Minerals with higher uranium content such as phene, zircon, etc. might be useful up to 1000,000 yrs. if the leakage is not prohibitive. With increasing age the leakage would probably become significant with these minerals.

Uranium minerals might be useful in this method for very young rocks to the present but older minerals would certainly be severely shattered and a large fraction of their helium lost.

Area of Geology

Potentially any extrusive or intrusive body may be used, possibly hydrothermal sulfides in ore deposits, clays and many minerals under most diverse conditions.

4. Strontium Isotope Ratio Method

Since it can be reasonably assumed that the strontium isotope ratio in ocean water is roughly in equilibrium with that of the earth's crust, the $Sr^{87}/Sr^{88}$ should consistently increase through geologic time from the continual addition of $Sr^{87}$ due to the disintegration of $Rb^{87}$.

Thus Wickman (1948) suggests on the basis of the estimated concentration of Rb and Sr in the earth's crust how this ratio should change with time. Since Sr is precipitated with calcium in the many vast anhydrite and gypsum beds in the stratigraphic column, a study of the Sr isotope ratios should make it possible to place each gypsum-anhydrite bed in its proper place in the time table.

The accuracy in this case is dependent mainly on the sensitivity of mass spectrographic analysis. Presently it is indicated that the geologic periods can certainly be set apart and with reasonable future development, the sensitivity of the method might be 1,000,000 years.

Basic Requirements

(a) For absolute ages the isotope ratios at any point in time must be known. Since this depends on both the disintegration constant of $Rb^{87}$ and the assumptions regarding the relative abundances of Rb and Sr in the earth's crust such an absolute age might be questioned. It could be calibrated by some other method, however, and since the factors mentioned above would in no way affect the relative ages, all ages would then be put on the absolute scale.

(b) The accurate determination of the Sr isotope ratios can now be carried out. Increased accuracy in this department will however greatly increase the sensitivity of the age measurements.

(c) The possibility of isotopic enrichment during deposition must also be investigated but this is not expected to be prohibitive.

Effective Range

Oldest sedimentary rocks to Pleistocene.

Area of Geology

Anyhydrite and gypsum beds are known throughout the stratigraphic column from the Cambrian to the Present. Certain fossils may concentrate Sr in their calcareous shells. In this case the fossils themselves might be used. This is almost the ideal for the paleontologist.

A great potential advantage of this method in contrast to the preceding ones is that the accuracy of the age determination is independent of the age of the deposit. It depends only on the accuracy with which the Sr isotope ratio may be determined.

This is a very promising experimental field that no one has yet investigated. It is probable that something will be done within the year.

## 5. Radioactive Inequilibria Method

This method depends on the observation that the fine clay which precipitates very slowly on the deep ocean floor absorbs radium and ionium preferentially. Thus the deep sea sediments are enriched in ionium over that ordinarily in equilibrium with uranium. It is possible to measure the approach to equilibrium (which is really the measurement of the decay of the excess ionium) with depth of sediment and thus for a given depth the age may be determined. This method has the greatest value in unraveling the story of the Pleistocene glaciations. Since the half-life of ionium is about 80,000 years and most radioactive sources are measureable over about 5 half-lives, the lower limit of usefulness of this method would be about 400,000 years. The accuracy is amazingly good in the upper part of the range, i.e. about 50,000 years the values are good to probably one or two thousand years. This method was applied by Urry (1941) and is being exploited presently.

### Basic Requirements

a) The disintegration constants are known accurately.

b) The amount of excess ionium is determined by a measurement of the radon which is in equilibrium with the radium product of ionium and can be done with high accuracy.

c) It is assumed that the excess of ionium precipitated for a given deep sea clay type remains constant for several hundred thousand years. A priori this is reasonable because a homogeneous process (clay precipitation for a single mineralogical type) is extracting ionium from a homogeneous environment (the ocean). Experimentally this assumption seems valid as smooth curves of ionium concentration vs. time are obtained when examining deep sea cores by this method.

### Effective Range

5000 - 400,000 yrs.

### Area of Geology

Deep sea sediments. Possibly recent continental clay deposits and glacial clays. A project is under way to obtain such deep sea cores on a grid system across the North Atlantic with subsequent age measurement by this radioactive inequilibria method.

## 6. The Carbon$^{14}$ Method

The method depends on the fact that cosmic radiation is constantly producing $C^{14}$ from $N^{14}$. This $C^{14}$ then enters the carbon dioxide cycle and hence all living things have an equilibrium concentration of radioactive $C^{14}$ in their tissues. When the plant or animal dies and is buried in such a way that it is essentially removed from the atmospheric-living carbon dioxide cycle, this equilibrium concentration of radioactive carbon starts to decrease according to the known half-life of the isotope. If the activity of any carbonaceous archaeological find can then be determined, it is possible to place a reasonable date as to its time of life.

### Basic Requirements

a) The half-life has recently been remeasured and further work is being pursued so that this should shortly be known with such accuracy that it will not be the limiting factor.

b) In order to measure the weak and very minute quantity of $C^{14}$ by its radioactivity, it must first be enriched by a pilot plant diffusion cascade such as was used in separating U235 from U238 in the Manhatten Project. This has been done successfully -- see Libby et. al. (1948) in bibliography.

c) The matter of addition or subtraction must be considered with each find as a special case.

Effective Range

1000 to 30,000 years.

Area of Geology

Latest glacial epoch, archaeology.

Thus there are a number of feasible radioactive methods for age determination of geological strata. Together they cover the span of geologic time. Since they overlap, they permit independent cross-checking of results.

Indeed the next decade should show the greatest advance in the calibration of the geologic time scale that geologists could hope for. We must be vigilant......

---

## BIBLIOGRAPHY

I. Most Significant Papers

1. Bulletin of National Research Council #80, Physics of the Earth; IV-The Age of the Earth. (This is an excellent summary of the work and problems up to the mid-thirties. In many ways it is only of historical interest now.)
2. Goodman and Evans, "Age Measurements by Radioactivity," Bull. Geol. Soc. Am., vol. 52, pp. 491-544 (1941) (Best general reference and best reference on details of lead and helium methods.)
3. Nier, A. O., "The Isotopic Constitution of Radiogenic Leads and the Measurement of Geologic Time," Phys. Rev. pp. 153-163 (1939). (This is the classic reference on the lead isotope work.)
4. Ahrens, L. H., "The Determination of Geological Age by Means of the Natural Radioactivity of Rb," Trans. of the Geol. Soc. of So. Africa. (in press for 1948) (Best reference for Rb-Sr Method)
5. Hurley, P. M. and Goodman, C., "Helium Retention in Common Rock Minerals," Bull. G. S. A. vol. 52, pp. 545-560 (1941) and "Helium Age Measurement I. Preliminary Magnetite Index." ibid. vol. 54 pp. 305-324. (These are latest and best references on helium work.)
6. C. S. Piggot and E. D. Urry, "Time Relations in Ocean Sediments," Bull. of G. S. A. vol. 53, pp. 1187-1210 (1942). (Best reference on radioactive inequilibria method. For other references see works by Urry 1938 through 1944 in Am. Jour. of Sci.)

7. Wickman, F. E., "Isotope Ratios, A Clue to the Age of Certain Marine Sediments," Journal of Geology p. 61 (1948) (Reference on Sr-isotope method).

8. Anderson, Libby, et. al., "Natural Radiocarbon from Cosmic Radiation" Phys. Rev. p. 931 (1947) (Nov. 15th issue). (Reference on $C^{14}$ method)

9. Henderson and Sparks, "A Quantitative Study of Pleochroic Halos," IV New Types of Halos, V. The Genesis of Halos, Roy, Soc. London Pa. Vol. 173 A, p. 238-264 (1939) (Latest on pleochroic halos)

II. Papers on Mass Spectrograph technique.

1. Goldstein, Berl, Ber. 39, 691, (1886)
2. J. J. Thompson, Phil. Mag. 13, 561 (1907)
3. Aston, Phil. Mag. 38, 709, (1919)
4. Aston and Fowler, Phil. Mag, 43, 514 (1922)
5. Dempster, Proc. Amer. Phil. Soc. 75, 760, (1935)
6. Bainbridge and Jordan, Phys. Rev. 50, 282 (1936)
7. Straus, Phys. Rev. 59, 102 (1941)
8. Mattauch, Phys. Rev. 50, 617, (1936)
9. Shaw and Rall, Rev. Sci, Instru., 18, 315 (1947)
10. Lane, Smithsonian Report, 235-54, (1937) (General on age determination).

III. Other Papers

1. Joly, J., "Pleochroic Halos of Various Geological Ages," Proc. Roy. Soc. Series A, vol. 102, pp. 682, 705, (1925)

2. Rose, J. L. and Stranathan, R. K., "Geological Time and Atomic Composition of Radiogenic Lead," Phys. Rev. Vol. 50, pp. 792-6.

3. Henderson, G. H., and Bateson, S., "A Quantitative Study of Pleochroic Halos.", Proc. Roy. Soc. Series V, vol. 145, Part I p. 563, Part II p. 582.

4. Lane, A. C. and Urry, W. D., "Ages by the Helium Method," Bull. of G.S.A. vol. 46, pp. 1101-1120 (1935)

5. Urry, W. D., Chem. Rev., vol. 1, pp. 306-343 (1933)

6. Schumb, W., Evans, R. D., and Leaders, W. M., S. Am. Chem, Soc. vol. 63, p. 1203, (1941).

## MONOGRAPH TWO

# Creation and Evolution

by

Russell L. Mixter, Ph.D.
Professor of Zoology
Wheaton College
Wheaton, Illinois

Published by The American Scientific Affiliation, a group of Christian Scientific men devoting themselves to the task of reviewing, preparing, and distributing information on the authenticity, historicity, and scientific aspects of the Holy Scriptures in order that the faith of many in Jesus Christ may be firmly established.

As the aim of the Affiliation is to encourage the interchange of ideas, the views expressed by this author are not necessarily subscribed by the group.

Additional copies may be obtained from:
Secretary-Treasurer
The American Scientific Affiliation
1006 North President St.
Wheaton, Illinois

To the authors and publishers whose works have been cited goes my deep appreciation for permission to use the quotations in this monograph.

To Dr. Roy M. Allen, Miss Cordelia Erdman, Miss Marie Fetzer, Dr. Paul Harrison, Dr. Emmett Herring, Mr. Robert Ogren, Mr. Paul Jewett, Dr. J. Laurence Kulp, to members of The American Scientific Affiliation who have reviewed parts of my manuscript, to my colleagues Drs. J. L. Leedy and C. O. Mack, Mr. C. E. Luckman and Miss Betty Urquhart, to students and friends, I am grateful for the letters or conversations whose helpful criticisms have contributed to this work.

The Alumni Association of Wheaton College made possible a leave of absence from teaching for a semester during which much of this material was obtained. For this grant I am particularly thankful.
Mar. 10, 1950.

# Creation and Evolution

## INTRODUCTION

Current books on evolution are taking for granted [1, 2, 3] that animals and plants are derived from other kinds of animals and plants. Abundant factual material is presented which is interpreted to mean that from one or a few simple forms have descended all the types now existing or that have lived in the past. No one denies facts: every intelligent man should try to see what the meaning of the facts is, but facts may have other meanings than those which are held by the majority of writers. The facts and interpretations of biologists and geologists are to be examined in these notes with the purpose of distinguishing the possible conclusions that may be drawn from observations and experiments regarding the origin of kinds of organisms. The evolutionary hypothesis and the creation viewpoint will be compared in their ability to harmonize the facts and their approach to unsolved problems.

Creationists of today are not in agreement concerning what was created according to Genesis. Some hold that God made the first forms of every species, if by species is meant those forms which produce offspring among themselves but will not have living or fertile offspring with other forms.[4] Others believe that the Creator endowed animals and plants with the ability to produce descendants as different from each other as are the members within a major group of animals.[5] To some, the house fly, the fruit fly and the horse fly were separate in their origin. Others may even believe that one species of fruit fly that will not produce fertile offspring with another species (Drosophila melanogaster and Drosophila simulans) had a distinct origin at the time of creation. But there are non-evolutionists who think the evidence favors the probability that all the species of flies may have had the same ancestors. This does not mean that flies and bees, which belong to different major groups of insects, had the same ancestry. Reasons why special creations are believed necessary to originate the basic forms of each major group will be presented in the chapters on Fossils and Heredity.

Brewster[6] describes very interestingly the varied uses of the word "create" and the ideas that have been held by Creationists. He holds that "create" and "make" are interchangeable and synonymous with the Biblical phrase "Let . . . bring forth." "But make means almost anything. We elders 'make' beds and the morning train . . . . So make means anything from 'shape' to 'do' and create is only a narrowed and strengthened 'make.' " He proposes that if anything living comes directly from inorganic matter or dead flesh, the occurrence be called spontaneous generation; if one sort arises from another living sort, that is evolution. If neither occurs and still there is life, "we may call it Special Creation." Accepting this definition, let Creation be implied whenever any group of plants or animals cannot be proved to have come from any other group by a series of transitions. The evolutionist does not know where verte-

(1)

brates came from. "The oldest ancestors of the vertebrates are unknown, and may always remain unknown."[7] Likewise, as Brewster remarks, " . . . God 'created' has always meant 'I don't know.' "[8] Because of his belief in revelation, a Creationist can say, "I don't know **how** it was done, but I know **Who** did it." One may judge whether logic favors relying on an unknown source or on a Person to whom much power is attributed in the Scriptures.

The "I don't know" of the Creationist is likely to soothe him into a complacency so that he does not search the natural sciences for explanations of "things as they are." Perhaps the phrase "God made it that way" is an excuse for lack of research. Adam, before the fall, was commanded to "replenish the earth, and subdue it." One of his first tasks was naming the animals and birds. During an eight year period, "Dr. Gertsch, of the American Museum, has described some 500 new species of North American spiders."[9] The work begun by Adam is continuing or some would say "being repeated" so that "a million and a half species of animals and plants combined is a conservative estimate"[10] of the total number to be expected. Adam had to learn in order to "subdue the earth," especially after his fall. Likewise, today's generation must learn from the past and add to its knowledge to fulfill its task assigned to it through its earliest parent.

The meaning of evolution in Shull's words is " . . . in general . . . the origin of species."[11] If the Savannah Sparrow of the mainland gives rise to the Ipswich Sparrow,[12] living on Sable Island, 100 miles off the coast of North America, evolution has occurred. But this is a very restricted use of the term. In this sense, Creationists can be called evolutionists. Any change at all of any amount, no matter how small, may be called evolution. So believing that the descendants of Adam and Eve are now members of different races is believing in evolution in this restricted sense.

In addition to its use as the "origin of species," the term evolution implies "from the beginning of organic life on the habitable planet, a gradual unfolding and branching out into all the varied forms of beings which constitute the animal and plant kingdoms.[13] From one sparrow to another and from protozoa to men are comprehended in the term "evolution." It is the intent of this discussion to determine how far this process has proceeded. It should not be concluded that because one accepts transitions of some kind into others that he necessarily accepts all the evolutionary changes that have been postulated.

In order to understand the problem of origins, the student needs courses in biology and geology and in addition should have an understanding of the hereditary mechanism. The textbook, **Evolution**, by A. F. Shull of the University of Michigan is helpful for those whose knowledge of heredity may be at a minimum when beginning their study of evolution, for this book contains a survey of our knowledge of genes and chromosomes, the bearers of inheritance. For more comprehensive studies, one should read **Genetics and the Origin of Species** by Theodosius Dobzhansky of Columbia University.

# THE ORIGIN OF VARIETY IN NATURE

There are about 800,000 kinds of animals and 250,000 kinds of plants.[14] "Kinds" means "species" as used here. Our problem is to decide whether each of these kinds was specially created or whether a number of basic types have descendants differing enough to account for the large number of species. One difficulty in answering is that those who classify living things disagree regarding what are genera, species, and sub-species. This lack of unanimity results from the great variation between and within types that can be set apart from other types by any structural or psysiological difference. Mayr states that "at least 94 of the listed 755 full species of North American birds will be considered by some authors to be merely sub-species of other species." However, "there is extremely little disagreement in well-worked taxonomic groups as to the limit of the species. Nearly all authors will agree as to what is a species and what is not, except for the border-line cases . . . . " We shall not be hindered to any great extent, then, by lack of agreement among biologists as to whether different animals are to be classified as different species or merely varieties within the same species. We shall be more concerned with finding whether different animals, no matter what they are called, could have arisen from common ancestry or whether they could have arisen only by a direct creative act and not from a previous form.

One should not insist that "kind" means species. The word "kind," as used in the Bible, may apply to any animal which may be distinguished in any way from another, or it may be applied to a large group of species distinguishable from another group, such as birds as distinct from reptiles. In Genesis 1:11, three kinds are mentioned: grass, herbs and fruit trees. Here groups, each containing a large number of species, are being discussed. In Genesis 30:32, the "kinds" are varieties of a species of sheep, that is, speckled and spotted ones as different from black sheep.

Three sugestions have been made regarding the use of the word "kind" in Genesis: (1) a breeding type, so that the descendants are of the same "kind" as the parents, no matter in how many respects offspring differ from parents; (2) the idea in God's mind of what each type of animal should be;[15] (3) the phrase, "after its kind," may mean "all kinds of" in the sense that it appears to be used in the account of the flood (Gen. 6:20).[16]

The species concept was originated by Linnaeus in 1758. He assumed that the species he could identify were the kinds created in Eden. Later in life, Linnaeus presumed the "genus" to be the "kind" of Genesis. Today "Linnaean species are sometimes nearer our genera, and Linnaean genera are almost like our families;"[17] yet, "a majority of the Linnaean species are still treated as species."[18] Instead of equating "kind" with any category of classification, it is wise to study living and fossil plants and animals to discover what variety is possible by descent from an original type (e.g. the races of mankind from the first pair) and what types cannot be related by descent but would require distinct creations. Orthodox Christians readily admit that present-day living forms are not the same as their ancestors when first created, because they easily conceive both the American Indians and the African Negroes to be descendants of Adam and Eve; likewise, Christians should believe that interbreeding animals with similarities and differences no greater than those separating human races may likewise have a common ancestry. An attempt will be made to decide where the line may be drawn between "specially created" forms and types which have descended from other types.

One reason for holding the idea that all species are not separately created is presented by Shull.[19] Several species have similarities of appearance so they are placed in the same genus. These are readily distinguished from another collection of species which are classified together in a second genus. The obvious reason why the members of a genus are similar is that they descended from a common ancestor. To illustrate, observe three groups of sparrows. In the genus Zonotrichia are three species possessing white wing bars. The White-crowned Sparrow has a gray throat and a black and white-striped head; the White-throated Sparrow has a white throat and a black and white head; and the Harris's Sparrow has a black crown and black throat. Other differences are found among them but these are the most obvious. This genus is easily distinguished from the genus Melospiza, which includes the Lincoln's, Swamp, and Song Sparrows, and readily told from the genus Spizella composed of the Tree, Chipping, Clay-colored, and Field Sparrows. The similar members of these groups are assumed to have a common origin. One population of White-throated Sparrows may have lost white throats and increased the white in the crowns in addition to other hereditary changes to become White-crowned Sparrows. That a change from White-throated to White-crowned Sparrows is possible by the alteration of hereditary determiners will be apparent in the section on the heredity. It appears that God made the White-throated Sparrow with the possibility of becoming the White-crowned Sparrow. This is comparable to Adam and Eve becoming the parents of the Negro and the white man.

Not all genera can be as easily distinguished as the ones used for illustration. A study of Mayr's[20] chapter on "The Higher Categories" reveals that the limits of the genus are subjectively determined. Some genera could be lumped together with others and a genus with many species could be split into several genera. No one doubts the similarity between species, but the specialists may differ on just which ones to put in one genus and which into another. The same confusion exists regarding some members of larger categories.

It is observed that some animals cannot be easily assigned to a group but appear to be intermediate between it and another group. The evolutionist says this is best explained by believing all groups have descended from a few forms of life or from one.[21] However, there is definitely a grouping within groups as arranged by the systematist. Species are gathered into genera, genera into families, families into orders, orders into classes and classes into phyla. Shull maintains that "this grouping within groups would not be expected if each group had arisen independently of other groups of the same rank . . . . The very obvious order which exists amid all the diversity of living things can scarcely be reconciled with a totally independent origin of each of the groups."[22] This argues against Fixity of species on naturalistic grounds. Hamilton has given an argument against Fixity of species on grounds of design; he writes ". . . . would not a world order in which every species was different from every other species be far harder to attribute to one God than the world order with its similarities such as we see around us?"[23]

Similarities between the species of a genus have at least two explanations: The species have remote ancestors in common or they have been made by the same Designer. Note some details which reflect on the explanations. The male and female Red-wing Blackbirds are quite unlike in color markings, yet they may have had the same parents. In shape and size and in the structure of all parts except the reproductive organs, however, they are nearly identical. Common heredity accounts for the similarities and permits the differences. Also, in many structures the Red-wing and the Baltimore Oriole are similar. Knowledge of their history does not go back far enough to tell whether they are related or not. One may infer relationship because of the many items of likeness. On the other hand, birds as different as the Oriole and Red-wing may have been separate creations. The idea that they are similar because God made them resemble each other is as logical as the assumption that they inherited their similarities from the same distant ancestors. But the Red-wing and the Oriole belong to the same order of birds, and because members of the same order have descended from the same remote ancestor, as the paleontological record will indicate, it may be concluded that birds no more widely separated structurally than are Orioles and Red-wings have arisen from one stock living in the past.

Some Orthodox theologians have seen that the facts of natural variability suggest the derivation of some species from others. F. E. Hamilton writes, "The theory of a common mind back of the similarities of the organic world seems far more logical than the theory of descent, especially if we hold, as the present writer does, that in many cases the species or even genera are descended from common original types created by God."[24] A comparable statement is, "All the common similarities between species can be accounted for on the theory that God created certain 'kinds,' and that these kinds broke up into the various species as we have them in the world today . . . ."[25]

Douglas Dewar, author of a series of articles in the Sunday School Times[26] does "not assert that every species, or every genus, or, even, every family has been especially created. I do not know what the units of creation are; it may be that they do not exactly coincide with any of the present systematic categories."[27] The intent of A. Rendle Short's passage on the "Creation-Narrative in Genesis," written for the Inter-Varsity Fellowship of Evangelical Unions, London, is that there is plenty of room for differences of opinion on what are the kinds of Genesis, how many there were, and how much their descendants have changed since creation. He concludes, "That which must be firmly held by Christians who honour the Bible as the Word of God is the fact that God is the Creator of the the heaven, the Earth, and all living things, whatever methods He may have used; that the creation-narrative of Genesis is a true account; and that man is a special creation of God, though this does not necessarily mean that God created him out of nothing at all."[28]

One student thought that if anyone ever produced "life" or a living thing in a laboratory, he would have done away with the idea of God as Creator. But why should one think so? If one finds how a plant makes starch, he has not done away with the plant. If one makes life, he has not done away with God, who made it similarly in earlier ages. The more mechanistic a thing is shown to be, the more need there is to presume that a machinist made it. The more we understand "life" and its activities from a mechanistic viewpoint, the more need there is to recognize a Divine Machinist.

## ADAPTATION

To believers in both evolution and creation ". . . the nice adjustment of the structural and physiological properties of organisms to the environment is truly remarkable."[29] Adaptations to living in water or in the air, to four-footed or upright posture, to herbivorous or carnivorous diets speak of forces that have cleverly channeled animals into their behavior or suggest a Creator, who by means of those forces, or directly, without them, has placed animals where they now live. Again it may be said that even if it is thoroughly understood how an adaptation is brought about, one need not discard the explanation that a Person is responsible for the structures and functions and the environment that work so harmoniously together. In fact, knowledge of how something occurs will increase the appreciation of the skill of the one who does it.

Note that "the characters of the higher ranks (classes, orders, families) are often, perhaps usually adaptive, those of genera less often or less certainly so, while specific distinctions are usually not adaptive."[20] This suggests that the adaptive nature of family, order, and class distinctions are the result of planning by a Creator. The species have few distinctions which are adaptive. One species of a genus probably can catch its prey as readily as any other. But the anatomical arrangements peculiar to bats and moles can be more readily imagined to be the result of creative activity than acquired by a slow accumulation of gradual changes from non-flying and non-digging animals.

## SUMMARY

Because the differences between the species within a genus or the genera within a family are comparable to the distinctions between human races, we are justified in believing that all the members of a genus or family have descended from the same ancestors. This does not mean, however, that all animals and plants have arisen from one or a few original types of life. A study of fossils and of heredity will indicate to what extent animals may be derived from others and what groups had their origin from direct creative activity.

## BIBLIOGRAPHY


[1] Dobzhansky, Th. 1941 *Genetics and the origin of species.* Second edition. Columbia Univ. Press, New York. p. 8.

[2] Goldschmidt, R. 1940 *The material basis of evolution.* Yale Univ. Press, New Haven. p. 4.

[3] Shull, A. F. 1936 *Evolution.* McGraw-Hill Book Co., New York. p. V.

[4] O'Toole, G. B. 1925 *The case against evolution.* The Macmillan Co., New York. p. 4.

[5] Hamilton, F. E. 1931 *The basis of evolutionary faith.* James Clarke & Co., London. p. 92.

[6] Brewster, E. T. 1927 *Creation.* The Bobbs-Merrill Co., Indianapolis. pp. 67-92.

[7] Romer, A. S. 1937 *Man and the vertebrates.* The University of Chicago Press, Chicago. p. 20.

[8] Brewster, *op. cit.,* p. 83.

[9] Mayr, E. 1942 *Systematics and the origin of species.* Columbia Univ. Press, New York. p. 5.

[10] Dobzhansky, *op. cit.,* p. 6.

[11] Shull, *op. cit.,* p. 10.

[12] Mayr, *op. cit.,* p. 164.

[13] Lull, R. S. 1940 *Organic evolution.* The Macmillan Company, New York. p. 6.

[14] Shull, *op. cit.*, p. 1.

[15] Clark, H. W. 1940 *Genes & Genesis.* Pacific Press Publishing Assn., Mountain View, Calif. pp. 97-103.

[16] Brewster, *op. cit.*, p. 65.

[17] *Ibid.*, p. 189.

[18] Dobzhansky, *op. cit.*, p. 365.

[19] Shull, *op. cit.*, p. 2.

[20] Mayr, *op. cit.*, Chapter 10.

[21] Darwin, C. R. 1872 *The origin of species.* Sixth edition. Merrill and Baker, New York. p. 34.

[22] Shull, *op. cit.*, p. 7.

[23] Hamilton, *op. cit.*, p. 150.

[24] *Ibid.*, p. 150

[25] *Ibid.*, p. 149.

[26] Dewar, D. 1946 "Some Recent Literature Concerning Evolution," *Sunday School Times.* 88:6, p. 107; 88:7, p. 127; 88:8, p. 153; 88:9, p. 176.

[27] Dewar, D. 1937 *More difficulties of the evolution theory.* Edward Arnold & Co., London. p. 10.

[28] Short, A. Rendle 1943 *Modern discovery and the Bible.* The Inter-Varsity Christian Fellowship, Chicago. pp. 69-71.

[29] Shull, op. cit., p. 12.

[30] *Ibid.*, p. 14.

# THE MEANING OF SIMILARITIES BETWEEN SPECIES

Within one family can be found children closely resembling their parents, and those which are less like their parents and more nearly resemble their grand parents. These likenesses to parents and differences from parents occur in persons known to have common inheritance. Likewise we concede common ancestry to Negroes and whites if both are traced back to prehistorical times. Orthodox Christians find the source of all races in Adam. All races have fundamental similarities such as muscular and nervous structure, and minor differences such as skin color, hair texture, and distribution of hair. Both the similarities and differences result from the interaction of hereditary determiners (genes) and their environment. A particular pair of genes may influence skin color, but other genes also do; and even if a gene has its major effect on skin color, it may likewise influence hair color or eye shape or nearly any part of the body. In other words, the determiners interact with each other to produce their result.

Any human race has many genes that are identical with those of other races. A Negro may have genes for dark skin and yet possess the same genes for a blood group as a white person. Within a race occur a large number of combinations of genes so that one member may be tall and blond, another short and dark. In fact, there is no "pure race." Essentially then, any human being is related to any other because he has a stock of genes common to every other member of the race. Man is one species, for he can and does readily combine his genes with those of any race and can have living, reproducing offspring.

The similarities between different species are sometimes the result of the same kind of genes. Two species of fruit flies, Drosophila melanogaster and Drosophila simulans, are very much alike in appearance. There is also a close resemblance between the germ cells of the two species. Hybrids reveal that many genes are the same in both kinds of flies. These genes are homologous (they have the same locations on the chromosomes, which are the bodies in the germ cells bearing the genes); the characters resulting from them are homologous (they have the same kinds of genes causing them and they look alike). Therefore the species are considered to be related.

"Where hybrids cannot be obtained (as in the case of the majority of comparisons of Drosophila species) the judgment regarding gene homology is of necessity based on the similarity of the phenotypic effects of the genes. Admittedly, this is a serious flaw in the method, which must, consequently, be applied with greatest caution. 'Mimic' mutants produced by nonallelic genes in the same species are not a rare occurrence"[1]. An illustration is found in eye colors. Two red-eyed flies of one species look alike. But one is the wild type red and the other is scarlet. The scarlet mimics the red and is caused by a pair of genes different from wild type genes. Here, then, is a case of similarities resulting not from descent from a common ancestor but from gene alterations.

Another type of similarity not resulting from descent is caused by parallel mutations.[2] Two species of Drosophila have white-eyed forms but both were observed to arise independently from the wild flies which were red-eyed in each species. Drosophila melanogaster had been red-eyed for many generations and then a white-eyed type appeared; a corresponding event happened in Drosophila simulans. Identical appearances are not from common parentage but by mutations (genic changes) which occurred in each of two distinct species. The white eyes of Drosophila simulans did not descend from the white eyes of Drosophila melanogaster. The white eyes are homologous. ". . . They are due to corresponding alterations in corresponding parts of the hereditary constitution."[3] "When these occur and are preserved in stocks which are already specifically distinct, the Darwinian concept of homology breaks down. For the homology, though perfectly real, no longer implies descent from a common ancestor showing the common feature."[4] It is concluded, then, that some similarities between different species may not be the result of their kinship.

The meaning of similarities usually held by biologists is as follows. Drosophila species resemble one another in external appearance and in chromosome pattern more than they resemble house flies. In the majority of features, species of Drosophila are similar; therefore they are considered to be related. The respects in which they resemble house flies are evidences of relationship to them, although the kinship is not as recent as among the Drosophila themselves. Although in particular features, as white eyes, a similarity may not indicate common ancestry, the sum total of all similarities is taken to indicate relationship. Huxley emphasizes that the arm of man, the wing of a bird and the flipper of the whale have a common structural plan and it cannot be conceived that they arose independently for there are so many steps necessary if arm, wing and flipper evolved from some previous structure not like themselves that "no biologist would venture to suggest" that they "could have been separately evolved in more than one stock."[5] That is, the dificulties in getting wings or arms by accumulating many changes suggest that wings were evolved from arms; they could not have come from several independent sources, so they must have descended from one ancestral structure. To illustrate this from human structures one would say that a white man's skin color shows he is not directly related to a Negro, but the muscles and bones of his arm show they have their remote ancestry in common. Likewise, the arms of men and wings of birds testify that both have come up from similar ancient ancestors.

As previously indicated it is usually held that homologous organs result from similar genes. Harland, on the contrary, considers that "organs such as the eye, which are common to all vertebrate animals, preserve their essential similarity in structure or function, though the genes responsible for this organ must have become wholly altered during the evolutionary process, since there is no reason now to suppose that homologous organs have anything genetically in common."[6] To see the meaning of this quotation think back to the time when there were neither birds nor mammals. A quadruped had arms, upper arms and hands determined by a group of cooperating genes. In later generations, it is presumed, the arm and hands were modified in the direction of a wing by mutations in some descendants and towards the mammalian arm in others. Perhaps after several mutations, the genes formerly concerned with the arm could cooperate with other genes which could not previously act on the arm. Eventually when wings were complete, genes were acting which had had no effect in the original quadruped and some genes helping to make the arm of the first quadruped were no longer affecting the development of the arm at all. We are now supposing "that homologous organs" do not "have anything genetically in common" and yet supposing at the same time that homologous organs are indicating genetic kinship. If there is reason to assume that a man's arm and a whale's flipper do not have the same genes, there is reason to believe they did not have the same ancestry. If similar characters, involving only the quality of eye color as in the example from Drosophila, do not necessarily mean descent from a common ancestor and if the complex organ in one class, such as birds, so similar to the corresponding organ in another class such as mammals, does not mean a common ancestry for both, it is doubtful whether any similarities can be taken to mean kinship. It is not certain that "similarity can be reasonably attributed to only one cause, namely, heredity from a common ancestor."[7]

One further consideration weakens the idea that similarity necessarily implies kinship. It is the "wide spread occurrence of convergence."[8] An example is used in birds. A tern has a bill shaped like a dagger; so do herons and kingfishers. All have bills which are obviously adapted to catching fish. Here are quite similar structures in different orders of birds. The similarity is explained as a result of independent acquisitions and not because kingfishers descended from terns.

Because we so readily conceive the common structures in human families to be proof of kinship and allow that variations are also possible in the same family, one might grant the assumption that similarities and differences between species and genera of animals are also best explained by the theory of descent with modification. In addition to common ancestry, convergence and Harland's explanation there is another. It is the Design of Creation. Students who were ". . . theoretically inclined, such as Goethe and Oken, regarded the existence of structural plans common to a large number of animals as evidence of some form of planning in the act of creation. In extreme form, this theoretical view found the basis of homology in the existence of a limited number of archetypal ideas in the mind of the creator."[9]

A university professor said, "Design of creation is a psychological explanation which can not be definitely disproved." But can the evolutionary assumption be proved? Do the fossil lineages show that the resemblances in arm and wing can be traced back to the same source? Dobzhansky holds that the paleontological series do show this[10] because classifications based on comparative anatomy correspond in general to those based on fossil pedigrees. Huxley says that "save for a few fossil lineages, we do not and cannot know the actual course of events in the evolution of a group."[11] A problem for the classifier is that successive stages in one lineage may be more like those of another lineage than like their predecessors in the same lineage.[12] This weakens the idea that species resembling other species are in the same ancestral stream. They may be undergoing parallel changes or have reached their present degree of likeness by convergence, that is, changes in unrelated groups tending to make them similar. Of course, when an evolutionist says a group is unrelated to another it is a relative matter, for all animals and plants are considered to be related if traced back far enough. In the discussion on paleonotology, it will be indicated that within orders or families there are some genera which evidently have changed into other genera, but that orders usually can not be traced back to parental stocks belonging also to other orders. In concluding that one order arose from another, evolutionists place their reliance on the principle we are examining, namely, that similarity of one group to another group suggests its source in that group.

To what extent has experimental genetics indicated that similarity is the result of kinship? Recall that the hybrids between Drosophila melanogaster and Drosophila simulans prove that much of their similarity is a consequence of their having the same genes. Presumably most of these genes are identical because they belonged to one species in the past. Some mutations have arisen after the stock separated into the two species, as in the example of white eyes. The variations of inheritance have formed two species from one. Whenever geneticists can show common genes, they are entitled to infer common ancestry. Genes common to two species can be revealed whenever hybrids can be produced; it is unnecessary that the hybrids be fertile. Hybrids are known between forms considered different species (Drosophila melanogaster and Drosophila simulans); between different genera (cabbage by radish); between different sub-families (ferret by skunk); and between different orders (Fundulus by mackerel). In the latter case, the embryos were abnormal but did show both maternal and paternal characters. Such results indicate that certain species have hereditary units compatible with those of other species. The body of a skunk is produced by the genes he received from his two parents; the body of the hybrid between the ferret and the skunk came from a combination of ferret and skunk genes. If ferret and skunk can be put together into one animal, both of them could have arisen from one of them. The ferret and skunk are akin because they can be crossed. They are similar because of their kinship. But animals which cannot be crossed can give no evidence of relationship; it must be inferred from the resemblances. If cats and beavers will not cross, we have no way of knowing how much genic material they have alike, if any. To decide whether they have ancestors in common we must consult paleontological evidence to see if they can be traced back to common ancestors.

## RESEMBLANCES BETWEEN EMBRYOS

In view of the resemblances between adult animals, one would expect their embryos to be even more similar. For development to occur at all, it must begin with a single cell, in most cases fertilized by a sperm. The processes leading to the differentiation of the germ layers are fundamentally the same in all animals as also are the in- and out- foldings needed to produce the organs.

All vertebrates are characterized by a dorsal nerve cord, an internal skeleton and "gill slits opening from the throat to the outside or touching the outside wall in at least the developmental stages."[13] In the human embryo, "the ectoderm of each groove and the entoderm of its complementary pouch then meet and unite; the thin plates thus formed only rarely rupture and complete the gill slit condition."[14] Associated with these pouches are blood vessels, muscles, cartilaginous arches and grooves entering between the arches from the outside. A study of the derivatives of the gill arches[15] will convince one that practically all of the material which becomes gills in a fish is used for the construction of certain of the structures of the throat region in man. A common plan is followed in all vertebrates and modified for the needs of the developing individual and the fully formed organism. This common plan may be attributed to an ancient form of vertebrate whose descendants have been modified into fishes and air breathers or to a Mind which uses the same fundamental process for all vertebrates but varies it at will for specific purposes.

A debate on the significance of the gill arches is carried on between Dewar's books[16] and the volume by Davies.[17] Dewar insists that the "fish plan" of arteries in the head region is necessary even for mammals to have an adequate early circulation. Davies replies that the modification of the "fish plan" suggests that air breathers are "an after thought." Such modification of a common plan in all embryos to the needs of the organism is the result of thought. Each class of vertebrate, whether fish, amphibian, reptile, bird or mammal, has an efficient circulation developed from this basic type. Surely creative activity is not bound to make an entirely different arrangement of blood vessels for each class. It has done well if it makes an efficient arrangement. If certain of the blood vessels are found only for a short time in the embryo and then degenerate, one need not say this is to a Creator's discredit. Rather let him notice that the Creator is more sparing of His plans than of material.

## VESTIGIAL ORGANS

An organ which appears to remain undeveloped in the adult is called vestigial. Creationists have felt that vestigial organs are evidence of degeneration in animals whose former condition was nearer perfection. ". . . Genesis iii.14 may be interpreted as perhaps implying that there was a time when the snake did not crawl in the dust but walked."[18] "The existence of true vestiges . . . is not inconsistent with the theory that every main organic type has been specially created and that some types have degenerated."[19] On the other hand vestigial structures are held by evolutionists to indicate relationship. Creationists believe that an alternative view is possible, that vestigial organs have a function or are genetic abnormalities.

Actually not all organs claimed to be vestigal are useless. ". . . It is not well to call an organ functionless prematurely and it has ever been the history of anatomy to discover new functions for so-called vestigial structures—witness the present interest in the endocrine glands"[20] "So many of the so-called vestiges of embryology may prove to play leading roles in the development of the individual."[21]

The appendix, generally regarded as vestigial, is a pouch ending blindly at the beginning of the large intestine. In the same position, an adult rabbit has a caecum nearly a foot long in which vegetable material ferments in order to be more completely digested. But the human appendix averages 3½ inches, although it may be absent or as long as 9 inches. Its cavity is too small to permit digestive activity within it. It produces a slight amount of digestive fluid. There are many lymphocytes in the connective tissue. A widely accepted opinion on the function of lymphocytes is that they are the stem cells from which arise all other types of blood cells, both red and white. It may be that a convenient supply of these cells at the beginning of the large intestine is helpful in combating the action of the numerous bacteria found there. To be sure, the appendix becomes inflamed, necessitating removal. In this, it shares the lot of nearly all the parts of the body, for scarcely any organ is free from the possibility of the destructive influences of degenerative processes.

The appendix may not even be a vestigial structure. In a review of the book, **"Up From The Ape"** by Hooten, William L. Strauss, Jr. has commented that, "There is no longer any justification for regarding the vermiform appendix as a vestigal structure."[22]

The coccyx ". . . . corresponds to the tail vertebrae of other mammals. It is much reduced in size."[23] To it are attached two muscles (levator ani and sphincter ani) which act as a cradle for the pelvic contents, constrict the anal opening and assist in the expulsion of the fecal mass. A portion of the heaviest muscle of the thigh is attached to the coccyx, as well as a rudimentary muscle (coccygeus) which assists one of the cradle muscles in upholding the pelvic floor.

"In rare instances the coccyx is overdeveloped and produces a tail-like projection."[24] "Specimens as long as 3 inches have been recorded in the newborn; most of these are soft and fleshy, but a few have contained skeletal elements."[25] In a motion picture made by Dr. P. Kenneth

Gieser, there was a Chinese baby with a very evident tail. But should it be presumed that because a part of the body has developed in a manner resembling some animal, that the body is therefore derived from that animal? There are abnormalities of the bodily development that cannot be taken to refer to any ancestral condition, but are merely disarrangements of the embryological process. To what ancestor does hair-lip point? Are we descended from an animal with an open-roofed skull and virtual absence of brain because the human newborn occasionally has the anomaly called cranioschisis? Because the limbs may fail to develop, were our ancestors limbless? There are no criteria for judging when an anomaly points to inheritance from a remote ancestor and when to a deranged development.

The external ear muscles are classified as muscles of the scalp and considered vestigial by evolutionists. One extends from the skull to the ear and covers the posterior auricular vessels and nerve; the other two go from the ear to the tough connective tissue of the scalp. These muscles and the epicranius muscle aid in keeping the scalp tense. This is necessary because where the scalp is not attached to the skull directly by connective tissue, it is held by muscle. Another suggested use for the external muscles is that the ear has an increased blood supply by way of them.[26]

"The slender red crescent at the inner corner of the human eye"[27] called the semilunar fold resembles the third eyelid found in both birds and some modern reptiles. Birds are not in our ancestry, and "we do not really know whether the paleozoic reptiles possessed this particular feature, or not. At any rate, the fold serves to regulate the flow of tears."[28]

The hind legs of whales are so reduced that in some species they do not appear on the surface. Regarding them Hamilton asks, "Suppose a whale originally had hind legs. Does that mean that it is a descendant of reptiles who walked on land? Perhaps, but is it not more reasonable to suppose that such rudimentary bones were used in a fin as a means of swimming and that such a fin was later on lost through a mutation? There is no reason for thinking that because the animal possessed such organs it was ever anything but a whale. As for the snake, the same thing can be said."[29] Examples are known of the loss of limbs by mutations. Snyder pictures a family in which the father and children have neither hands nor feet.[30] Observers of the fruitfly have worked with vestigial wings. Both hereditary abnormalities are the result of gene changes within single species. Vestigial-winged fruit flies are not descendants of a long series of species whose wings gradually became smaller; the transition occurred in one species by a germinal change that caused reduction of wings in one generation.

"The splint bones at the sides of the feet of horses recall the ancestral condition in which there were at least four toes."[31] The splints are vestiges, but also structures serving useful purposes. "These, as Hayes points out, (1) strengthen the leg, (2) serve as an attachment for certain muscles, (3) in conjunction with the cannon bone form a groove in which lies the upper part of the suspensary ligament,—an elastic brace supporting the fetlock and counteracting the effects of weight."[32]

## BIBLIOGRAPHY

[1] Dobzhansky, Th. 1941 *Genetics and the origin of species.* Second edition. Columbia Univ. Press, New York. p. 140.

[2] Huxley, Julian 1942 *Evolution: the modern synthesis.* Harper & Brothers, New York. p. 395.

[3] *Ibid.*, p. 514.

[4] *Ibid.*, p. 395.

[5] *Ibid.*, p. 514.

[6] Dobzhansky, *op. cit.*, p. 85; also cf. Goldschmidt, R. 1940 *The material basis of evolution.* Yale Univ. Press, New Haven. p. 22.

[7] Shull, A. F. 1936 *Evolution.* McGraw-Hill Book Co., New York. p. 28.

[8] Simpson, G. G. 1944 *Tempo and mode in evolution.* Columbia Univ. Press, New York. p. 190.

[9] Huxley, *op. cit.*, p. 391.

[10] Dobzhansky, *op. cit.*, p. 364.

[11] Huxley, *op. cit.*, p. 394.

[12] Dobzhansky, *op. cit.*, p. 364.

[13] Shull, *op. cit.*, p. 18.

[14] Arey, L. B. 1947 *Developmental anatomy.* Fifth edition. W. B. Saunders Company, Philadelphia. p. 177.

[15] *Ibid.*, p. 153.

[16] Dewar, D. 1931 *Difficulties of the evolution theory.* Edward Arnold & Company, London. pp. 47-51.

Dewar, *More difficulties of the evolution theory,* pp. 28-33.

[17] Davies, A. M. 1937 *Evolution and its modern critics.* Thomas Murby & Co., London. pp. 144-146; 150-151.

[18] Hamilton, F. E. 1931 *The basis of evolutionary faith.* James Clarke & Co., London. p. 175.

[19] Dewar, *Difficulties of the evolution theory.* p. 8.

[20] Shumway, W. 1932 "The Recapitulation Theory," *Quarterly Review of Biology,* v. 7, pp. 93-99.

[21] Shumway, *loc. cit.*, p. 98.

[22] Straus, Wm. L., Jr. Book Review. *Quarterly Review of Biology,* 22 (2): p. 149.

[23] Shull, *op. cit.*, p. 25.

[24] *Ibid.*, p. 25.

[25] Arey, *op. cit.*, p. 184.

[26] A medical scientist 1945 *Evolution.* International Christian Crusade, 366 Bay St., Toronto, Ontario, Canada.

[27] Shull, *op. cit.*, p. 26.

[28] O'Toole, G. B. 1925 *The case against evolution.* The Macmillan Co., New York. pp. 296-7.

[29] Hamilton, *op. cit.*, p. 175.

[30] Snyder, L. H. 1946 *The principles of heredity.* Third edition. D. C. Heath and Co., Boston. p. 393.

[31] Shull, *op. cit.*, p. 26.

[32] Dewar, *More difficulties of the evolution theory.* p. 54.

# THE DISTRIBUTION OF ANIMALS

From the book of Genesis we learn that God created animals and man. This man gave names to "all cattle, and to birds of the heavens and to every beast of the field." Animals undoubtely spread from the place or places of their origin and also from Eden, located somewhere in the region of the Tigris and Euphrates rivers. Were all land animals and birds brought to Eden, or did Adam name only the ones which had migrated to Eden from the regions of their origin? Later, "all that was on the dry land died" from the waters which covered the mountains. The portion of the earth covered by the flood was repopulated by the survivors from the ark as they spread from the mountains of Ararat.

The comments which follow are of a highly speculative nature but are presented so that readers may test them by any knowledge they possess which applies to the problem. By making guesses and testing them, we may be able to discern the most sensible ideas to be held concerning the bearing of animal distribution on the origin of living forms. Because one may make a wrong guess, he should not hesitate to guess for his assumption may aid another to find the truth.

Was the whole earth covered by the deluge? If it was, to harmonize science and Scripture there must be an accounting for the present distribution of all air breathers as though they all had been at one place at one time. If the flood was world wide, we need to account for such peculiar features of distribution as the kangaroo in Australia and the anteater in South America but neither of them in Asia Minor. We need either to trace them back to Ararat or derive them from some animals that were in the ark. However, if the flood was only a local occurrence, we may assume that kangaroos arose in Australia, were not destroyed by the flood, and have not migrated to other continents.

The flood obviously occurred after men appeared. Fossil men are first found in the Pleistocene, the most recent of the geological strata, so the flood may be considered a late Pleistocene event. The Pleistocene epoch witnessed relatively little change in the structure of animals. "In any region where the fauna is adequately known, we find, in the Pleistocene, representatives of all types now present. But, in addition, there are invariably numerous animals now extinct and most of these are of large size."[1] Therefore we cannot assume that very much change has occurred in animals after the Pleistocene and after the flood.

Let us presume for a moment that the flood was world wide and it caused a destruction of the types of animals found in the Pleistocene. The ark preserved pairs of these also, but because there are now no living representatives of such forms as giant kangaroos and mastodons, these perished after the flood. There are, however, smaller kangaroos now living. If kangaroos were in the ark and first touched land in Asia, one would expect fossils of them in Asia. According to Romer,[2] the only place there are either fossil or living kangaroos is in Australia. What shall we conclude? If the fossil evidence means that there never have been kangaroos in Asia, then kangaroos were not in the ark or if they were, they hurried from Australia to meet Noah, and as rapidly returned to their native land. Is it not easier to believe that they never were in the ark, and hence were in an area untouched by the flood, and that the flood occurred only in the area inhabited by man?

There are other well known examples of animals which have never lived in Asia. The edentates (sloths, armadillos, anteaters) are confined

(15)

to tropical America or Texas although in the past ground sloths were found in Pennsylvania.[3] Many animals have had a more extensive distribution than at present but a large number of these cannot be traced back to the supposed place of the landing of the ark. To me, this indicates that they never were in the ark. Two alternatives remain: either these animals developed after the flood or they were living in countries not affected by the flood. Because the edentates were in existence long before the Pleistocene, we are convinced that they did not originate during or after the epoch of the flood. Therefore they were not in the region covered by the waters. This leads to the observation that the flood was not world wide and may be disregarded in a study of animal distribution. The flood had no distinctive effect upon the geography of animals. Those in the ark repopulated the devastated regions; other areas already had their own fauna, which may have helped in the repopulation of the flooded regions.

Let us turn to the ideas regarding the origin of species that are presented by students of geographic distribution. "Most species are limited to definite regions"[4] An illustration is found in the honeysuckers (Drepanididae), a family of birds found only in the Hawaiian Islands. Some of them have bills for eating seeds, others for hard nuts, and one genus can loosen bark to get insects. Long bills for obtaining nectar from tubular flowers as well as insect-eating bills are pictured by Gulick.[5] All these birds, belonging to one family of eighteen genera, are descendants from some migrant to these Islands. One scarcely can believe that the Creator put them there and nowhere else, for that is not in harmony with the probable source of the Hawaiian Islands from volcanic or coral action in a not-too-distant geological past. Furthermore, only four other families of the order to which the honeysuckers are assigned can be found on the Islands in contrast with the twenty-one families in Eastern North America. The scarcity of types of Hawaiian birds indicates that the Hawaiian forms are migrants. One sees, therefore, in the Drepanididae how varied the descendants of a single kind of bird may be. All of us have believed that the domestic varieties of poultry, dogs, and cattle have descended from single ancestral stocks; why should we hesitate to admit that even more variation from a common ancestor may have occurred as in the honeysuckers? Is it possible that each species of honeysucker was separately created? It is improbable.

By noting the range of a species in relation to the range of a similar one, we are led to the idea that one gave rise to the other. The ranges of the Flicker and the Red-shafted Flicker overlap in the Great Plains, where hybrids occur, but in the Western United States the Red-shafted is common and in the Eastern part of the country occurs the Flicker, easily distinguished from the other by yellow instead of salmon color under the wings and tail. Since they differ in colors probably determined by mutations, one may be ancestral to the other. The kind in one region is conceivably the parent of the species slightly different from it, which inhabits an adjacent area. It is easier to assume that one species was derived from the other than to believe that both are special creations. By similar reasoning we may conclude that all woodpeckers may have been derived from one type of woodpecker. Caution should be used, however, for only as similar types are found to replace each other in adjacent areas can the conclusion be safely reached.

Although the distribution of animals indicates that some species have been derived from others, is there any limit to this derivation of types from other types? Do paleontological findings indicate that simple protozoa have been ancestral to other groups of invertebrates and also to vertebrates? Recall that in the Cambrian period, where first appear fossils in any number, "as if suddenly, all the principal phyla of animals are represented."[6] These Cambrian forms are "about as complex as present forms." Because three-fourths of geological time was passed before the first extensive series of fossils was preserved, "we know practically nothing about the origin of phyla." Simpson[7] reminds us that there is almost a universal absence of connecting links between orders of vertebrates. Because the missing links are inferred to be soft-bodied creatures lacking parts capable of being fossilized, there is no fossil evidence to tell us from what group the vertebrates came. We are at a loss to know the origin of the phyla, classes and orders of animals. It is reasonable to believe the earliest members were created.

The ancestors of Edentates apparently first lived in South America, spreading northward. The kangaroos and some other types of marsupials have given no evidence of residence elsewhere than Australia. Such facts indicate that there have been appearances of distinct groups of animals on the continental nuclei,[8] those bodies of land which have never been under the ocean, e.g. Australia. Where did the kangaroos come from? Either from creative activity on Australia or by modification of some other marsupial. "Cretaceous marsupials have been found in North and South America, and they are present in Eocene beds in Europe and South America. They probably reached Europe from North America, but whether they originated in the Northern or the Southern Hemisphere, either in Australia or South America is a matter for guesswork in view of the small amount of evidence."[9] A land connection is supposed to have existed between South America and Australia, hence it is believed that the ancestors of marsupials migrated from one continent to the other before the land masses became disconnected. The first stratum in Australia bearing fossil marsupials is the Miocene, two epochs later than the Cretaceous. In the Miocene only one genus is found (Wynyardia).[10] Is Wynyardia ancestral to the kangaroo? If we are correct in assuming that the members of an order may have a common ancestor, then we may conclude that Wynyardia is ancestral to the kangaroo.

Where shall we draw the boundaries of the groups descending from created animals? Draw it where paleontologists indicate there are gaps—between orders in practically all cases, also between man and the apes or their relatives. We may need to revise our conceptions as new fossil finds come to light but the boundaries of most orders are fairly well known. "During the present century a very great amount of paleontological work has been done, and many strange forms have been brought to light. These, however, have been almost always members of groups already known or forms tending to connect such groups."[11] Because marsupials are separated by structural gaps unbridged by intermediate forms from other orders of mammals, a creationist may conclude that the first marsupials were specially created. If there are gaps between families or genera within the order of marsupials, he may believe that such families also had separately created beginnings. Whenever a gap is filled then a revision of the idea of what are the special creations should be made.

Orthodox Christians have assented to this position. We find Hamilton's remark, "Ordinary variation and mutation within the genus, or perhaps in some cases within the family, starting from a fixed point as a centre and spreading in all directions, varying as it goes, would account for all these intergradients as well as the theory of evolution!"[12] My suggestion is to carry the logic one step further and say "within the order." Dewar says "the flora and fauna of oceanic islands, therefore, seem to indicate that the new species and genera, and possibly two subfamilies, have arisen by evolution."[13]

Students of Shull's text will note that "non-evolutionary hypotheses regarding the inception of species have included the assumption that species all arose within a comparatively brief time."[14] This statement does not apply to the creation viewpoint carried in these comments. The origin of the marsupials is found abruptly in the Cretaceous, along with other mammals. The earliest insects are believed to have been created in Carboniferous days, several epochs earlier. An honest creationist will ask the paleontologist what he knows of the time of origin of animals, and draw his conclusions from the data. It is obvious then that a creationist may believe in the origin of species at different times, separated by millions of years, and in places continents apart.

## BIBLIOGRAPHY


1 Romer, A. S. 1945 *Vertebrate paleontology.* Second edition. Univ. of Chicago Press, Chicago. p. 569.

2 *Ibid.,* p. 611.

3 Hesse, Allee, Schmidt 1937 *Ecological animal geography.* J. Wiley & Sons, Inc., New York. p. 105.

4 Shull, A. F. 1936 *Evolution.* McGraw-Hill Book Co., New York. p. 31.

5 Goldschmidt, R. 1940 *The material basis of evolution.* Yale Univ. Press, New Haven. p. 215.

6 Shull, *op. cit.,* p. 46.

7 Simpson, G. G. 1944 *Tempo and mode in evolution.* Columbia Univ. Press, New York. p. 107.

8 Shull, *op. cit.,* p. 30.

9 Davies, A. M. 1937 *Evolution and its modern critics.* Thomas Murby & Co., London. p. 219.

10 *Ibid.,* p. 219.

11 Romer, *op. cit.,* p. 510.

12 Hamilton, F. E. 1931 *Basis of evolutionary faith.* J. Clarke & Co., Ltd., London. pp. 188-189.

13 Dewar, D. 1931 *Difficulties of the evolution theory.* Edward Arnold & Co., London. p. 23.

14 Shull, *op. cit.,* p. 42.

# FOSSILS SUGGEST CREATION

The opinion of the majority of biologists is that "fossils constitute one of the most convincing indications of the origin of species at different periods, and of the general course which evolution has taken if the assumption be made that all forms, or at least large groups of them, are genetically connected with one another."[1] The previous sections of this work have indicated that there is a certain amount of relationship between animals in the minor systematic groups. It is probable that all the species within a genus and perhaps the genera within a family, and perhaps even all the families of an order may have arisen by a subdivision of an original ancestral species. All the thrushes may have descended from one thrush. Possibly thrushes, robins and bluebirds were all the off-spring of an ancient thrush or robin. What information is obtainable from a study of fossils that bears upon the extent of changes which animals have undergone?

Writers of the orthodox Christian faith have held varied interpretations of the significance of paleontological findings. In my opinion, the best position is that of Dewar, who writes, "Let me here clearly state my views regarding the units of creation. I do not assert that every species, or every genus, or, even, every family has been specially created."[2] Hamilton agrees with Dewar. He believes ". . . . even the most conservative Christians could find the view of a progressive creation of various forms of life over successive geological ages, in harmony with the Bible."[3]

The fossils are genuine evidences of animals which were formerly in existence. This study accepts the conclusions of geologists regarding the arrangement of the strata and the estimated age of them. One critic has questioned the geologist's ability to date the strata, however, and believes all of them resulted from the flood.[4] It appears to me that the logic of such work is based upon the exceptions to the rule rather than the rule. At any rate, a belief in the findings of geologists regarding the strata does not automatically make a person an evolutionist. There are orthodox Christians who believe the days mentioned in the first chapter of Genesis are long periods of time. Perhaps long periods occurred between the days which are described. Some hold that the strata were laid down between Genesis 1:1 and Genesis 1:2. A belief in an earth millions of years old and in successive creations separated by long time gaps is in harmony with the viewpoint of a number of theologians.

Let us see what information regarding the origin of organisms is gained by a study of fossils. Living beings are extremely ancient. The Archaeozoic strata contain graphite, which may be the result of the carbon cycle carried on by simple plants; therefore there must have been plants then.[5] "Enormous masses of limestone . . . . . also probably owe their origin to secretion by plants."[6] In the Grand Canyon there are sponges composed of silica . There is even some possibility that some worms and the protoza of the order of Radiolaria also existed in the Proterozoic.

The major groups of animals appeared suddenly in the Cambrian. "All the principal phyla of animals are represented."[7] There is no definite evidence of vertebrate animals but "even the vertebrates may not have been absent, since fishes appear in the very next period. . . .", and ". . . . only a small portion of existent types were fossilized."[8] Romer states that ". . . . it is possible that highly developed vertebrate types may have been in existence in the Cambrian. . . . "[9]

The evolutionary view is that this cannot be the sudden eruption or creation that it appears to be but it is rather a ". . . . long period of unrecorded evolution."[10] A first appearance in a stratum of a type of fossil does not convince a modern biologist that it had its origin there but it may have existed in strata previous to the time of its first appearance. For example, ". . . . the true flowering plants appeared suddenly in such abundance and variety in late Cretaceous that it is generally assumed they originated much earlier, though little fossil evidence of them in earlier periods has been obtained."[11] In the Eocene mammals are so prominent that Shull believes "the rapidity with which all this change occurred suggests immigration from some region not yet explored, perhaps Asia."[12] If all the major groups of animals are found in the Cambrian strata, one can believe either that they originated at the time the strata were deposited or that they originated in some previous age. If they arose before a stratum was laid down it appears that their time of origin and their method of origin cannot be determined except by inferences from structural similarities. If we may consider the first recorded fossils to be in the layers made at the time the organisms first appeared, then it is possible to believe in a direct creation of animals, for the major groups appeared with the major structural characteristics of their groups in complete form.

Furthermore, there is a considerable variety of forms at the time of their first appearance. Shull states that there were 300 species of trilobites in the Cambrian. There are hundreds of species of brachipods. Even vertebrates may not have been absent. There must have been abundant plant life to furnish food for the animals, none of which can make their own food but must depend upon plants, directly or indirectly. If all species of trilobites developed from one species, there must have been a time of development unrecorded in the Cambrian during which the 300 species were being formed.

I have assumed that the sudden appearance of fossils in a stratum is evidence that they were not derived from pre-existing types but had a direct origin from creative activity. One believes either this or he believes that the fossils for which he has no ancestral types were descended from animals existing in areas where fossils were not formed, in areas not yet examined for fossils, or in regions whose fossils have been destroyed.

To test these opposing beliefs, consider the extent of the changes which have occurred in the descendants of the earliest animals. Some animals flourish and become extinct. There are over 300 species of trilobites in the Cambrian, but in the next stratum are found over a thousand species grouped into 77 genera. Only 7 of these genera are the same as those of the Cambrian. This means an appearance of 70 new genera. If these are descendants of the Cambrian trilobites, then there has been a large amount of variation following their first appearance. The more recent forms do not possess strikingly different structures, but have modifications of organs already present in the earliest types. Because these modifications are comparable to the changes produced in a species by hereditary changes, called mutations, one may believe that the later trilobites descended from the earliest trilobites.

Were the trilobites changed so much in successive ages that they became new kinds of creatures? The descendants of the early species became so much altered that they were grouped into 13 families in the Ordovician, only three of which existed in the Cambrian.[13] A creationist may maintain that the original trilobites have changed enough since their creation to have become 13 different families of trilobites. "No new family arose after the Ordovician period. . . ."[14] A creationist may hold that this is the extent to which animals have changed. He can set limits to the possibility of transformation, especially because he does not find fossils which link the trilobites to animals belonging to other orders or classes of animals.

As a rule, the ancient types of animals are already divided into their classes and orders. Molluscs as different as bivalves, snails and cephalopods occur in the Cambrian. Straight cephalopods of earlier ages turn into coiled cephalopods, one of which presumably became the Chambered Nautilus. Each order of the molluscs is distinguishable from other orders, although connections between them are assumed.

Some animals have survived from their origin to the present with very little change. The horseshoe crabs occur first in the Silurian, ". . . . but by Jurassic time only the genus Limulus existed, and this genus survives today. . . ."[15] Lingula, a brachiopod, has not changed since the Ordovician. Many of the vertebrates have changed very little. The lungfish, for example, has remained much the same from the Devonian. Some groups appear, vary in later ages, and die out, or become more varied. Others survive only a few geological ages. Still others persist from their origin to the present with little change.

A survey of backboned animals also shows a limited amount of alteration. Bone, a complex tissue, is the earliest kind of skeletal material. Cartilage occurs as a subsequent tissue. Two subclasses of bony fishes were "already quite distinct at their first appearance in the fossil record."[16] An ancient order of fishes, the ostracoderms, is represented by "several distinct groups."[17] Amphibia (frogs, salamanders, etc.) are supposed to have come from the crossopterygian fishes, but ". . . . . the better known genera are probably too specialized or too late in time to have been the actual ancestors. . . ."[18]

The earliest reptiles are in the Pennsylvanian. But our knowledge of them is gained from the coal swamps ". . . we know nothing of the life of higher and dryer regions where reptiles might well have already been numerous."[19] The increase in the number of orders and families mentioned by Shull may represent merely the finding in later strata of animals which existed in dryer areas at the time the primeval reptiles were in the coal swamps but were not preserved because no strata from the dry areas have been found. The dinosaurs ". . . . were already divided at their first appearance into two distinct stocks. . . . ."[20]

Birds began as birds. Two genera, Archaeopteryx and Archaeornis, left their dead bodies in the Jurassic. "Archaeopteryx and Archaeornis were already definitely birds. . . ."[21] Mayr comments that if birds became extinct at the Archaeopteryx stage, ". . . . Archaeopteryx would be listed merely as an aberrant order of feathered reptiles."[22] He believes this genus is ". . . . as perfect a missing link between reptiles and birds as one could possibly hope for."[23] The paleontologist, Simpson, writes that these earliest birds subdivide but do not fill the gap between birds and reptiles.[24]

Mammals are the animals which have hair and milk glands. Their fossils are recognizable from their teeth. Early records consist mostly of jaws. The mammals of Triassic times probably disappeared; ". . . . we know extremely little about their history during almost the entire span of the Mesozoic."[25] Also "we know little except the dental anatomy of the forms which have been found, and the sparseness of the record suggests that many groups may well have existed but have escaped discovery."[26] Shull says that the rapid rise of mammals in the Eocene hints at ". . . . immigration from some region not yet explored, perhaps Asia."[27] The two major groups of mammals, the marsupials (opposumlike) and the placentals, occur together in the late Cretaceous beds,[28] in the time preceding the Eocene with its ". . . rapid rise of mammals. . ."[29] It is possible that some of the mammals supposed to have arisen in the Eocene were living in the age before it, but their fossils have not been found to prove it. "In the Jurassic, mammals are definitely present."[30] Note that birds arose in the Jurassic and by "Eocene time most of the present orders of birds were in existence."[31] These findings parallel the Genesis account which places the origin of birds and "beasts of the earth" near together.

The history of the horse and other members of its order, the tapirs and rhinoceroses, is fairly complete. It begins with an animal of the Eocene called Eohippus in America and Hyracotherium in Europe. It had four toes on its front feet and three on its hind feet plus ". . . . traces of the missing digits in the form of metatarsal splint bones."[32] Its molar teeth had a rather simple crown pattern. Between this animal and the modern horse there is a series of extinct animals, which show gradual changes until the modern horse, Equus, is found with a single toe on each foot and with a complicated molar tooth pattern. Simpson says, "if the change in any one character from Hyracotherium to Equus is divided into 300 steps, these steps are imperceptibly small and are incomparably less than the amount of intragroup variation at any one time."[33] Therefore the modern horse appears to have ascended from Eohippus by gradual changes no greater than those accounting for genetic differences within a species.

Bear in mind that Hyracotherium is a genuine member of the horse family, Equidae. "Despite the great difference between Hyracotherium and Equus, most of the characters of the Equidae did not change appreciably throughout their history."[34] Simpson states that Hyracotherium was a backboned animal, nourished its newborn young by milk and its unborn young by a placenta, had hoofs in uneven number, had the shape of a horse and was a horse. This is ". . . . a classificatory way of saying that the vast majority of its multitude of morphological characters were already the same as those preserved in Equus and in all equids as well as in many other more or less related animals."[35]

Note that there is a gap between Hyracotherium and its supposed ancestor, a condylarth, a member of a different order. "But nowhere in the world has any recognizable trace been found of an animal that would close the considerable structural gap between Hyracotherium and the most likely ancestral order, the Condylarthra."[36] The differences between a condylarth and Hyracotherium are less than between Hyracotherium and Equus. To an evolutionist, this means that a condylarth evolved very rapidly into Hyracotherium;[37] to a creationist the absence of bridging fossils suggests the need for a creative act to produce the ancestor of the modern horse Equus.

It is reasonable to believe in the creation of ". . . . cattle and creeping thing, and beast of the earth after his kind. . ."[38] occurring in some of the fairly recent geological periods in view of these statements which indicate the indefiniteness of the time of origin of the mammals. We know little of the early history of mammals; therefore more kinds may have existed than the record demonstrates. The appearance of many kinds in the Eocene can be accounted for by assuming that they were being evolved from reptiles in regions where no strata were being formed in previous periods. From an evolutionary point of view there could not be a sudden appearance of clearly recognized horses, camels, pigs, rodents, carnivores and monkeys. To be sure, these may not be identical with the species of today but they do belong to the same groups (orders) as our recent species. Although there were reptiles that are called "mammal-like" which lived in the ages preceding the rise of mammals and which may be considered transitional forms between reptiles and mammals, one may also believe that there were reptiles which were similar to mammals but not ancestral to them. It is reasonable to suppose that because there are many kinds of reptiles and many kinds of mammals, therefore some reptiles and some mammals would be similar to one another. A creationist believes this and does not believe that the "mammal-like" reptiles actually became mammals.

An inspection of the charts which picture the relationships of mammalian groups arranged by similarities of structure reveals the gaps between the orders of mammals. [39] [40] Simpson calls them "systematic deficiencies of record" and insists that "this regular absence of transitional forms is not confined to mammals, but is an almost universal phenomenon, as had long been noted by paleontologists. It is true of almost all orders of all classes of animals both vertebrate and inverterbrate . . . . and it is apparently also true of analogous categories of plants."[41] It appears to me that when animals are traceable to other animals by a closely graded series of transitional forms, then I may believe the early forms have given rise to the later ones. But when a group of animals is separated from another group by an unbridged gap, then, until bridges are found, one may hold that the groups so separated have arisen from independently created kinds.

Those who do not believe the gaps really existed believe animals and plants did live which bridged the gaps. Both Simpson[42] and Mayr[43] have given the following explanations of the absence of missing links. New fossil finds indicate that there were some animals and plants structurally intermediate between distinct types. Because so few fossils have been found, there is an inadequate picture of all the kinds of organisms that have lived. Furthermore, many animals leave no fossils because they have no hard parts, or their habits or habitats prevent them from leaving remains. If animals lived in areas of erosion, or in a deep sea whose sediments have never been exposed, or left their bodies in strata subsequently metamorphosed by heat or pressure, it is unlikely that we would have any trace of them. Probably the major reason why transitional animals are not found is that during the periods of change of organisms from one kind to another, there was rapid evolution in only a few individuals. If only a small number of animals were evolving during the time when the other conditions which also prevent fossil formation were at work, there is additional probability that the missing links will remain missing. Some of these reasons listed were also used by Darwin to account for the imperfection of the geological record.

Shall we believe that the gaps will be filled? If the trend of the time continues, new fossils will be found. However, until a major group of organisms such as an order, is connected to another order by a closely graded series of forms, one need not hold the interpretation which derives the members of one order from some other order.

## SUMMARY

The earliest representatives of the major groups of animals and plants are complex organisms. They are separated by structural gaps from the members of other groups. As long as the gaps remain unbridged by a series of fossils grading between one group and another, one may conclude that the ancestral types of the groups are specially created and not descended from other living types. Within the orders and families, there appears to have been descent with modification as in the series of horses.

## BIBLIOGRAPHY

[1] Shull, A. F. 1936 *Evolution.* McGraw-Hill Book Co., New York. p. 44.

[2] Dewar, D. 1937 *More difficulties of the evolution theory.* Edward Arnold & Co., London. p. 10. See also pp. 137, 138, 170.

[3] Hamilton, F. E. 1931 *The basis of evolutionary faith.* James Clark & Co., London. p. 195.

[4] Price, G. M. 1923 *The new geology.* Pacific Press, Mountain View, Calif. p. 36.

[5] Personal communication from Roy M. Allen, Sc. D. "The evolutionist likes to make the most of this line of reasoning, but the fact is that graphite does not necessarily result from organic life. It occurs quite commonly in minute flakes in many primary igneous rocks (e.g., granite, syenite, gneiss, etc.) where it could not possibly have had an organic history, excepting, of course, its occasional reduction from the di-oxide. It also occurs as fine microscopic dust in igneous rocks. I have a fine example of such in a petrographic section I made years ago of an intrusive dyke in basalt from Ft. Lee, New Jersey.

"In this paragraph you also mention the possibility that some worms existed in the Proterozoic. This is another of evolution's claims, originating in what was called Eozoon Canadense, found in the Laurentian Limestone, and hailed as fossil worm forms, but now known to be minute serpentine nodules. (I have made petrographic studies of a rock sample of this in my collection and verified the true nature of the inclusions.) I do not believe there is any evidence of worms in the Proterozoic. As a matter of fact, the placing of imaginary life forms in the Archeozoic and Proterozoic is nothing but "wishful thinking" on the part of evolutionists, in order to place the origin of life as far back as possible. This is a point we creationists do not have to cede to evolution."

[6] Shull, *op. cit.,* p. 45.

[7] Shull, *op. cit.,* p. 46.

[8] *Ibid.,* p. 48.

[9] Romer, A. S. 1945 *Vertebrate paleontology.* Second edition. Univ. of Chicago Press, Chicago. p. 36.

[10] Shull, *op. cit.,* p. 46.

[11] *Ibid.,* p. 57.

[12] *Ibid.,* p. 63.

[13] *Ibid.,* p. 49.

[14] *Ibid.,* p. 49.

[15] *Ibid.,* p. 49.

[16] Romer, *op. cit.*, p. 75.
[17] *Ibid*,, p. 25.
[18] *Ibid*., p. 140.
[19] *Ibid*., p. 171.
[20] *Ibid*., p. 229.
[21] *Ibid*., p. 171.
[22] Mayr, E. 1942 *Systematics and the origin of species.* Columbia Univ. Press, New York. p. 298.
[23] *Ibid*., p. 296.
[24] Simpson, G. G. 1944 *Tempo and mode in evolution.* Columbia Univ. Press, New York. p. 111.
[25] Romer, *op. cit.*, p. 309.
[26] *Ibid*., p. 309.
[27] Shull, *op. cit.*, p. 63.
[28] Romer, *op. cit.*, p. 316.
[29] Shull, *op. cit.*, p. 62.
[30] Romer, *op. cit.*, p. 309.
[31] Shull, *op. cit.*, p. 61.
[32] Romer, *op. cit.*, p. 426.
[33] Simpson, *op. cit.*, p. 46.
[34] *Ibid*., p. 158.
[35] *Ibid*., p. 159.
[36] *Ibid*., p. 106.
[37] *Ibid*., p. 121.
[38] Genesis 1:24.
[39] Romer, *op. cit.*, pp. 310, 325.
[40] Simpson, *op. cit.*, p. 107.
[41] *Ibid*, p. 107.
[42] *Ibid*, pp. 115-124.
[43] Mayr, *op. cit.*, p. 296.

# THE SCIENCE OF HEREDITY AND THE SOURCE OF SPECIES*

A great geneticist has stated that "No individual can claim such a mastery of all facts pertaining to evolution to enable him to present . . . ."[1] a full discussion of the facts, laws and theories of evolution. This appears to me to be especially true of the bearing of the science of heredity on evolution. Therefore this discussion is limited largely to a consideration of one question, "To what extent have species changed into other species?"

Some species appear to have descended from others. If two species can be crossed, and living offspring produced, there is proof of compatible germ cells from the two species. The obvious source of their hereditary material is descent from a common ancestor. An example of such crossing is found in two birds of the eastern United States, the Golden-winged Warbler and the Blue-winged Warbler. Their hybrids are called Lawrence's Warbler if they have black throats, and Brewster's Warbler if they do not. The hybrids show the characteristics of the parents variously distributed to the offspring. A typical Lawrence's Warbler has a black throat from the Golden-winged parent but has yellow underparts like the Blue-winged ancestor. Similarly, upper parts, wing bars, and ear patches are typical of one parent or the other. Because the hybrids do combine traits of both parents, it is probable that the two parental types were derived from one source. To be sure, this transition was not observed; surely each species could have been created separately, even though they now interbreed. However, evidence from fruit flies which indicates that a species is now in the process of subdivision into two species suggests that the idea of the derivation of the two warblers from common ancestry is the correct explanation.

A species now undergoing subdivision is Drosophila pseudoobscura. There are two races, A and B which differ from each other, not in any observable external appearance but in the arrangement of genes on the chromosomes. (Genes are the units of heredity in cells arranged serially on rod like bodies called chromosomes). The genes in one race are in a certain order; in the other race, some of the genes are inverted.[2] The races are told apart by breeding tests. Race A produces fertile offspring when mated with Race A, but when mated with Race B, sterile male offspring are the result. Here is the indication of the subdivision of the species. If the two races thus formed were to have visible character differences, one could conclude that they had become two distinct species. Already within each race there are several variations of the chromosomal arrangements. For example, in the third chromosome ". . . . twenty-one different sequences have been recorded."[3] Any subsequent rearrangement may produce a difference in external appearance, or mutations (changes in the genes) may do so; then race A will be obviously distinct from race B and two species will have been observed to have arisen from one species.

In another species of Drosophila, melanogaster, character differences have appeared before there has been any reduction of fertility between members of the species. There are hundreds of variations affecting eye colors, wing sizes, body colors, etc. Yet each kind crosses with any other to produce fertile offspring. When a chromosomal change arises which will make those individuals possessing some traits sterile with those displaying other traits, then there will be two species. If this occurs, character differences will have preceded reproductive differences in species alteration. On the other hand, in Drosophila pseudoobscura, reproductive segregation has already preceded character differentiation.

In nature, a number of species of Drosophila actually differ in the ways seen in the flies already discussed. Chromosomal rearrangements have occurred within the members of one species; ". . . . the chromosomal differences between species are identical in kind, if not in degree, with those found among races and individuals."[4] We readily accept the idea that the members of one species have a common ancestry. We should believe that flies now admittedly separate species have a common ancestry if the differences between them are the kinds that occur within one species.

If any species gives rise to a new species, may the conclusion be drawn that the first living protoplasm could have been the ancestor of all kinds of life of the past and present? Reasoning from the probability that one species may become two, biologists conclude that one form of life may have given rise to all forms. As a creationist, one may accept the origin of species from other species, called microevolution, but need not believe that a protozoon was transformed into a coelenterate, or a coelenterate into a round or segmented worm and hold that similar macroevolutionary transformations have occurred leading to man. That there is a large assumption in evolutionary belief which is not forced by genetic knowledge is evident in the following quotation, "Experience seems to show, however, that there is no way toward an understanding of the mechanisms of macroevolutionary changes, which require time on a geological scale, other than through a full comprehension of the microevolutionary processes observable within the span of a human lifetime and often controlled by man's will. For this reason we are compelled at the present level of knowledge reluctantly to put a sign of equality between the mechanisms of macro-and microevolution, and, proceeding on this assumption, to push our investigations as far ahead as this working hypothesis will permit."[5]

Hereditary studies are incapable of establishing the assumed connection between all types of life. A university geneticist in a recent lecture remarked, "The higher categories and the problems there presented are a good ways from the field where genetics can make a direct contribution." Crosses can be made between species and between genera, and only infrequently between members of different families and orders. Because species in different classes have not been crossed, there is no information about the similarities and differences in the genes of the members. We cannot put genes from two classes into one hybrid; birds and reptiles will not cross; we cannot therefore demonstrate by genetic means that birds and reptiles have come from one common ancestor. The evolutionist interprets the Scriptures to mean that the original kinds, brought forth by the command of God, such as fowls and creeping things, did not spring from one another.

Notice some of the difficulties there would be in producing the species of one phylum from a preexisting phylum. If a jellyfish came from a protozoon, a single celled animal would have to become many celled. These cells must be arranged into two layers, perhaps by the inpushing of one side of a hollow ball of cells. The inner cells must acquire contractile fibers at their bases. Certain outer cells need to produce stinging capsules and the whole mass be arranged into body and tentacles. Only in imagination can there be found hereditary changes capable of producing all of these advances which must have occurred if members of one phylum were to change into members of another phylum, but the evolutionist believes that in the past there were mutations which could accumulate to do this. The mutations which have been noted have altered organisms within a limited sphere. One of my professors said, "It would be a miracle that a mutation causing diversity would also cause convergence for an adaptive end."

## A NEW THEORY

Neo-Darwinian geneticists conceive that a species changes slowly over many generations until it forms reproductively isolated populations which can be considered two species. But Goldschmidt has questioned the ability of mutations to produce specific differences. "Microevolution by accumulation of micromutations—we may also say neo-Darwinian evolution—is a process which leads to diversification strictly within the species, usually, if not exclusively, for the sake of adaptation of the species to specific conditions within the area which it is able to occupy."[6] Consequently, he has postulated a "new" kind of evolutionary mechanism, "Species and the higher categories originate in single macroevolutionary steps as completely new genetic systems. The genetical process which is involved consists of a repatterning of the chromosomes, which results in a new genetic system."[7] No mutations need to appear; all that is necessary is translocation[8] or inversion[9] of genes already present.

This process, he feels, will produce new species more rapidly than by the slow accumulation of small mutations. Drosophila miranda and Drosophila pseudoobscura are only slightly different in external appearance, but hybrids between the species are sterile. Studies of chromosomes reveal that the species are distinguished by different arrangements of similar genes. But Goldschmidt believes that a relatively small repatterning of chromosomes will also produce considerable changes in appearance.[10] This would avoid the difficulties of conceiving a gradual change from one species to another.[11] Peculiar structures like the performed exit in a plant for the insect living in it could have been formed at a single step. The imitation of a distasteful species of butterfly by an inoffensive one need not be by a gradual change over many generations but could have occurred between parents and offspring merely by rearrangement of the positions of the parts of chromosomes.

Goldschmidt emphasizes the large amount of change which may be produced in an offspring if a pattern rearrangement occurs in the germ cells of the parents. The development of the offspring is considerably altered. The wing pattern in a butterfly (Papilio dardanus) varies markedly depending on a few differences in genes. It is supposed that a rearrangement of these genes would produce as great or a greater difference. There would be "macroevolution by single large steps."[12] However, some of the new characters in fruit flies showing greatest change from normal characters, such as the appearance of four wings instead of two, are the result of simple mutations and not the result of pattern changes. Goldschmidt presumed that alterations occur in the germ cells and survive or perish there. They need not wait until they have produced visible characters and these characters have been selected for survival or death. The altered chromosomal pattern finally selected affects a developing individual so strikingly that a decidedly different form is produced. "The first bird hatched from a reptilian egg."[13] Goldschmidt admits unfortunately no experimental attack upon this problem is at present apparent. . . ."[14]

In summary, Goldschmidt's view is this: Slow accumulation of mutations over many generations will not change one species into another but the rearrangement of the genes will produce in the germ cells a pattern resulting in strikingly altered offspring which have become different species from the parents in a single step and are separated from them by a "bridgeless gap."

This idea has not been accepted by other geneticists. Dobzhansky notes that Goldschmidt's ". . . . systemic mutations . . . . have never been observed. It is possible to imagine a mutation so drastic that its product becomes a monster hurling itself beyond the confines of a species, genus, family or class. . . . The assumption that such a product may, however rarely, walk the earth, overtaxes one's credulity, . . . ."[15] Again, ". . . . the simplicity of Goldschmidt's theory is that of a belief in miracles."[16] Goldschmidt's chromosomal mutations, which are supposed to cause new species, are no different than those which occur within races of the same species. This is to say that Goldschmidt is offering nothing new; he is merely claiming more for a phenomenon than others do.

Single genetic changes may produce striking effects but the result is not likely to be a structure working harmoniously with other parts of the body. Sumner holds that such "jumps" as Goldschmidt believes in are not integrated with the rest of the animal. To get a bird from a reptile by any other means than the slow accumulation of small changes, which is typical Darwinian evolution, would need "the direct intervention of the Creator Himself." "Only the wave of the magician's wand could have transformed the scales of a reptile forthright into the plumage of a bird."[17]

Another criticism of Goldschmidt's work is that he maintains species to be discrete groups, separated by the bridgeless gap from other species. Bridgeless, that is, by any other mechanism than the rapid transformations which he thinks his systemic mutations will produce. But species do not fit into rigid compartments. "The lack of universality of such rigid subdivisions is what has ever since Darwin been one of the main arguments in favor of evolution."[18] Certain species can easily be told apart. However, there are some species that blend into others; intermediates are placed arbitrarily. Among flies of the genus Drosophila, some types seem to belong to distinct species, other variants are classified according to the opinion of the investigator into different species or merely into different races of the same species.[19] Further, Dobzhansky and others point to the lack of definition of species by Goldschmidt so that a reader has difficulty in detecting where these "bridgeless gaps" between species exist. Consequently ". . . . the main premise of his theory is wholly unacceptable."[20]

It should be mentioned to the credit of both Goldschmidt and neo-Darwinians, like Dobzhansky, that the latter admits, "It must, nevertheless, be recognized that Goldschmidt's keenly critical knowledge has emphasized the weaknesses and deficiencies of the neo-Darwinian conception of evolution, which are numerous, as even partisans ought to have the courage to admit. It would seem that this fact along obliges anyone interested in the modern evolutionary thought to read Goldschmidt's book."[21] One also should read Dobzhansky's excellent volume and compare his ideas with Goldschmidt's theory.

## COMPLEX ORGANS

The source of structures as complex as the eye has always been difficult to explain. "Here is one of the puzzles of evolution which appears to be still far from solution."[22] Goldschmidt has listed nineteen different features which he challenges Darwinian evolutionists to explain "by accumulation and selection of small mutants."[23] Mayr has admitted that "it is a considerable strain on ones credulity to assume that finely balanced systems such as the eye or a feather could be improved by random mutations."[24]

The eye appears as if at a single jump from no definite previous form, and "transitional stages are wholly conjectural."[25] There are eight or more theories recounted in a volume by Walls. None of these is corroborated by comparative anatomy, because no animals exist possessing the stages which should have preceded the eyes of the cyclostomes (the first vertebrates), which in all essentials are like mammalian eyes. What caused the lens to be formed from the outer germ layer at the same time the retina was developing from the nerve tube is a "tantalizing mystery."[26]

"The study of heredity provides practically no evidence for a type of genetic change capable of producing the transformation of any kind of simple light-sensitive structure into a complex organ of vision."[27] A duplication of genes at one location on a chromosome of a fly will reduce the number of facets in a compound eye to a bar eye. Several eye color alterations have appeared but these are shades of red; . . . . "There has been no blue or green. In view of the frequency of mutation of eye color, one is led to suspect that blues and greens are absent because Drosophila is incapable of mutating in that way."[28] A certain mutation produces an eyeless condition. Another gives an antennalike structure "instead of an eye or combined with a rudimentary eye."[29] Not only is it difficult to see how mutations could improve the eye as it now exists, but also how they could begin to form an eye in view of the fact that the known mutations of the eye are of no benefit to it. Froriep compared the origin of the eye to the birth of Athena full grown from the brow of Zeus. Because creation is the making of something without the necessity of a prototype, this abrupt appearance of the eye in vertebrates is in harmony with the Creation doctrine.

One modern scholar has seen the reasonableness of creation. "Some people assume, entirely as a matter of faith, a Divine Creation of living substance. The only alternative seems to be the assumption that at some time in the dim past, the chance association of the requisite chemicals in the presence of favorable temperature, moisture, etc., produced living protoplasm. In other words, if one subscribes to this theory, he admits that the first protoplasm to appear on our earth was a product of spontaneous generation. Then, if he accepts the evidence of Pasteur and others against spontaneous generation, he must reverse his explanation of the origin of the first protoplasm to explain the origin of all subsequent living protoplasm from the first protoplast. In other words, spontaneous generation, according to these opponents of the idea of Divine Creation, worked when the first living substance was formed, but probably hasn't worked since. Actually, biologists are still as far away as they ever were in their attempts to explain how the first protoplasm originated. The evidence of those who would explain life's origin on the basis of the accidental combination of suitable chemical elements is no more tangible than that of those people who place their faith in Divine Creation as the explanation of the development of life. Obviously, the latter have as much justification for their belief as do the former. It is possible that the problem of life's beginning on our planet will always remain insoluble, a philosophical question rather than a subject capable of experimental investigation and solution."[30]

# BIBLIOGRAPHY

*A paper presented at the third annual Convention of the American Scientific Affiliation at Calvin College, Grand Rapids, Michigan, September 2, 1948.

[1] Goldschmidt, R. 1940 *The material basis of evolution.* Yale Univ. Press, New Haven. p. 3.

[2] Dobzhansky, Th. 1941 *Genetics and the origin of species.* Second edition. Columbia Univ. Press, New York. p. 120.

[3] *Ibid.*, p. 120.

[4] *Ibid.*, p. 148.

[5] *Ibid.*, p. 12.

[6] Goldschmidt, *op. cit.*, p. 183.

[7] *Ibid.*, p. 396.

[8] A segment of a chromosome broken away and joined to another chromosome.

[9] A segment of a chromosome broken out, turned around, and reattached to the same chromosome.

[10] *Ibid.*, p. 209.

[11] *Ibid.*, p. 244.

[12] *Ibid.*, p. 331.

[13] *Ibid.*, p. 395. The statement was made by Geoffrey St. Hilaire.

[14] *Ibid.*, p. 334.

[15] Dobzhansky, *op. cit.*, p. 53.

[16] Dobzhansky, Th. "Catastrophism versus evolutionism," *Science,* 92 (2390): 358.

[17] Sumner, F. B. "Is evolution inscrutable," *Science,* 93 (2422): 522.

[18] Dobzhansky, Th. "Catastrophism versus evolutionism," *Science,* 92 (2390): 357.

[19] Dobzhansky, *op. cit.*, p. 371.

[20] *Ibid.*, p. 371.

[21] Dobzhansky, Th. "Catastrophism versus evolutionism," *Science,* 92 (2390): 358.

[22] Shull, A. F. 1936 *Evolution.* McGraw-Hill Book Co., New York. p. 161.

[23] Goldschmidt, *op. cit.*, pp. 6, 7.

[24] Mayr, E. 1942 *Systematics and the origin of species.* Columbia Univ. Press, New York. p. 296.

[25] Neal, A. V. and Rand, H. W. 1939 *Chordate anatomy.* The Blakiston Co., Philadelphia. pp. 395, 407.

[26] Walls, Gordon L. 1942 *The vertebrate eye and its adaptive radiation.* Cranbrook Institute of Science, Bloomfield Hills, Mich. p. 108.

[27] Mixter, Russell 1947 "A Christian view of the origin of the vertebrate eye," *Christian Medical Society News.* (49):6.

[28] Shull, *op. cit.*, p. 123.

[29] Goldschmidt, *op. cit.*, p. 327.

[30] Fuller, Harry J. 1941 *The plant world.* Henry Holt and Co., Inc., New York. pp. 19, 20.

# THE EYE AS AN OPTICAL INSTRUMENT

Frank Allen, M.A., Ph. D., LL. D., F.R.S.C.

Professor Emeritus in Physics

University of Manitoba, Winnipeg, Canada

"Sturmius, says Paley, held that the examination of the eye was a cure for atheism'. Yet Helmholtz, who knew incomparably more about the eye than half a dozen Sturms, describes it as an instrument that a scientific optician would be ashamed to make; and Helmholtz was no atheist."

This quotation from Professor Ward's Gifford Lectures[1] is a reference to the celebrated popular lecture[2] of Helmholtz on vision. In view of the commanding influence on visual science which he has exerted for over ninety years since the beginning of the publication of his Handbuch der Physiologischen Optik (1856-66), his views on the characteristics of the eye as an optical instrument carry great weight. The importance of this subject from several standpoints has led the writer to assemble the latest measurements on the defects of the eye, as well as all the statements of scientific men of unquestioned authority in this field of enquiry that he could find.

First of all, the words of Helmholtz himself are given so that there may be no doubt as to what he did say.

"Now it is not too much to say that if an optician wanted to sell me an instrument which had all these defects, I should think myself quite justified in blaming his carelessness in the strongest terms, and giving him back his instrument. Of course, I shall not do this with my eyes, and shall be only too glad to keep them as long as I can--defects and all. Still, the fact that, however bad they may be, I can get no others, does not at all diminish their defects, so long as I can maintain the narrow but indisputable position of a critic on purely optical grounds....All these imperfections would be exceedingly troublesome in an artificial camera and in the photographic picture it produced. But they are not so in the eye - so little, indeed, that it was difficult to discover some of these....The chief reason (for not observing the defects) is that we are continually moving the eye, and also that the imperfections almost always affect those parts of the field to which we are not at the moment directing our attention....For the eye has every possible defect that can be found in an optical instrument, and even some which are peculiar to itself; but they are all so counteracted, that the inexactness of the image which results from their presence very little exceeds, under ordinary circumstances of illumination, the limits which are set to the delicacy of sensation by the dimensions of the retinal cones....The adaptation of the eye to its function is, therefore, most complete, and is seen in the very limits which are set to its defects....The defects which result from the inexactness of vision and the smaller number of cones in the greater part of the retina are compensated by the rapidity with which we can turn the eye to one point after another of the field of vision, and it is this rapidity of movement which really constitutes the chief advantage of the eye over other optical instruments."

The defects of the eye, which are set forth so prominently in this discussion, should be reviewed in the light of the most recent knowledge of visual optics which

1. Naturalism and Agnosticism. Black. London. 3rd Ed. 1906. 2 vols, Vol. 1, p. 6.

2. Popular Scientific Lectures. Trans. by Atkinson. London, 1873. The Recent Progress of the Theory of Vision, p. 219, et. seq.

has been assembled both by Duke-Elder[1] in his exhaustive treatise on ophthalmology, and also by Hartridge[2] in his recent investigation of the visual perception of fine detail. For no optical instrument has been examined so minutely, critically and accurately as the eye, by men of the highest genius such as Helmholtz, Gullstrand and Tscherning.

In the optical system of an instrument, such as the microscope or a telescope, the surfaces of the lenses should be perfectly spherical and they should be centered upon a common optical axis. The mathematical theory of such a dioptric system, as originally developed by Gauss, demands, in addition, that it should be of small aperture, not more than 10°, so that the rays of light are limited to the axial regions, of homogeneous media, and that monochromatic light should meet the refracting surfaces nearly at right angles. It is with such an ideal system that all practical optical instruments are compared.

Because of the nature of light and its mode of propagation, as well as the nature of the refracting media which form the image of an object, all optical systems are subject to many aberrations. These defects must in some way be overcome or minimized or the image will be more or less imperfect. The method of optical correction is to use combinations of several lenses of different refractive powers and curvatures. In the microscope, an achromatic object glass, which forms the image, is composed of as many as six lenses, and an apochromatic object glass may have ten. These lenses must be made of carefully selected glass with accurately figured surfaces, and adjusted to each other with extreme precision. The focal length is quite small. So perfect can the image be made, that magnifications of more than 1500 diameters are possible. They operate at a fixed distance from the object under observation, and the field of view is extremely small. In addition to the object glass there is the eyepiece with which the image is magnified and viewed. All the lenses are immersed in air so that the numerous surfaces have abrupt discontinuity with that medium. The telescope has fewer lenses, but the focal length is relatively large which minimizes or eliminates the most troublesome defects.

Compared with such complicated optical instruction, the human eye is rudimentary. The refracting system consists of the cornea, the aqueous humor, the lens and the vitreous humor or body. The lens, therefore, is immersed in the humors and not in air, which greatly minimizes the optical discontinuities. The indices of refraction of the two humors are practically the same (1.336), which differs but little from that of water (1.334). The lens is extra-ordinary in constitution since it consists of many layers or zones which vary in refractive index from 1.386 at the surfaces to 1.406 at the center. The cornea and aqueous humor taken together, form the major refracting medium of the eye, and the lens, by its power of accomodation, operates as the fine adjusting mechanism. The cornea-aqueous combination is about 2.5 times greater in refracting power than the lens, which is due to the relatively great difference in refractive index between the air (1.00) and the cornea (1.376), as compared with (refractive index) between the humors (1.336) and the surface of the lens (1.386). The rays of light are first refracted at the cornea and aqueous humor, second by the anterior zones of the lens, then by the core, and further by the posterior zones. In consequence of its structure, the lens has a greater refracting power if it were homogeneous with a refractive index equal to that of the core. Besides increasing the refractive power considerably, this arrangement diminishes the spherical aberration.

The fact of its decentration constitutes the main feature of the optical system of the eye. Strictly the cornea has no axis of symmetry, though the deviation is extremely small. The centering of the cornea and the two surfaces of the lens is never exact, nor are the various zones of the lens concentric with one another, and the lens itself is not concentrically placed upon the optic axis. The visual axis, which touches the fovea, makes an angle of about 5° with the optic axis upon which

1. Text-Book of Phythalmology. Kimpton, London, 1932. Vol. 1. Chaps. XVI and XVII. pp. 662-774.
2. Phil. Trans.Rou. Soc. London. Ser. B. Vol. 232, 1947, pp. 519-671

all refracting surfaces should ideally be centored. Vision therefore occurs obliquely through the system. "It is," remarks Duke-Elder, "as if one of the lenses of an optical instrument had slipped a little out of place and then when we looked through the instrument we tilted it very slightly. The deviations, however, are usually so small as to be functionally negligible, and they tend to some extent to minimize the aberrations of the eye."

It would seem, therefore, that the construction of the eye slightly violates all the conventions that apply to optical instruments. But the final result, nevertheless, is the formation of an improved image, as free from perceptible defects as those produced by the finest artificial systems.

The defects of dioptric systems, whether of glass or of the eye, comprise two main classes: first, aberrations originating from the composite nature of white light, termed chromatic aberration, and those arising from the structure of light and the manner of its propagation, called diffraction; second, monochromatic aberrations, or defects occuring with rays of light of a single wave-length: spherical aberration, the sine condition, refraction of eccentric rays, curvature and distortion of the image, and depth of focus, which is not a defect but a property of a lens.

The defects of the first group depend on the nature of light; those of the second group on the structure of the optical instrument.

The discovery of the spectrum by Newton (1664) established the composite nature of white light and the different refrangibilities of the colors. From these discoveries he erroneously concluded that an achromatic combination of lenses was impossible to devise. Apparently he gave no thought to the construction of the eye. As no chromatic aberration had ever been observed in it, Euler (1750), convinced that it was achromatic, stimulated the eminent optician Dollend (1758) to construct achromatic systems of lenses, such as telescopes, in which he was highly successful. He also discovered the chromatic aberration of the eye.

Chromatic aberration. Since white light is a mixture of many different wave-lengths and as a lens has the nature of a series of prisms, the real image of a disc of white light formed by a convex lens consists of a superimposed series of images of all colors of the spectrum focused at slightly different distances from the lens. The violet image, formed by the most refrangible rays, is nearest the lens, and the red image the most remote. The chromatic difference of focus is very small, being from red to violet, about 0.47 mm. (Hartridge). Since the eye accommodates itself to the brightest yellow-green rays, the difference of focus is normally about half the amount for violet in front of the retina, and half for red, as it were, behind it. The separate color images or aberration discs are also of different sizes, termed the chromatic difference in magnification, varying, according to Hartridge, for the usual entrance pupil of 4 mm. diameter for the eye, from 0.0216 mm. in diameter for orange light to 0.0588 mm. for blue, the letter being 2.7 times the former. In ordinary vision these defects are extremely small, and since the blue image completely overlaps the red, it neutralizes that color into white leaving a narrow blue border which, because of its low luminosity, is practically invisible. The whole effect of chromatic difference in magnification tends to counteract the chromatic difference of focus. In the eye the effects of chromatic aberration are small; and with a 2 mm. diameter pupil, as Duke-Elder remarks, 70 per cent of the light falls on a retinal area of 0.005 mm. diameter. The remaining light is more widely diffused and normally is therefore unnoticed, so that images are free from colored borders, as common observation shows.

Diffraction. Diffraction is due to the spreading of the waves of light at the edges of the wave front. The effect is the opposite of chromatic aberration since it is greatest with the longest waves, the red, and least with the shortest, the violet. It follows from the nature of light that a point of white light is always brought to a focus as a small blurred disc (Airy's disc) of light and dark bands with a bright spot in the center. No point image can therefore be formed but only a diffraction pattern, which varies in dimensions directly as the focal length of the system and the wave-length of light, and inversely as the aperture, or area of the pupil, through which the light passes. Diffraction is inherent in the nature of light and its propagation and cannot be overcome. In optical instruments, such as the microscope, its effects are not troublesome except at very high magnifications about 2000 diameters, when the image itself becomes a complicated set of diffraction patterns.

The central bright spot of the diffraction pattern of a point source receives about 84 per cent of the incident light; the first ring 1/57th and the second 1/240th of the intensity of the central area. At a distance of only 0.02 mm. from the center, the light is of too low an intensity to be seen. Since the eye has a very short focal length of 22.78 mm., the diffraction effects with a pupil 4 mm. in diameter are correspondingly small.

Comparison of the figures for chromatic aberration and diffraction shows that they have the same order of magnitude, and are much too slight to be observed in the ordinary usage of the eyes. It is to be noted that the effects of chromatic aberration increase as the pupil widens; but as this condition changes act together to leave the actual definition of the image practically unchanged with alterations of the pupil. As pupillary areas are governed by the illumination, the definition of the image is therefore the same with all intensities of light.

These are the characteristics of the images of point sources. But "the extended source", according to Hartridge, "differs from the point source in an important respect, namely, that its image is largely composed of white light, and that color fringes are only found near its margins. Each of the fringes for a 4 mm. pupil is about 0.01 mm. wide. Only parts near the edge are strongly colored, and that as the distance from the edge increased the yellow fringes become progressively whiter and the blue fringe progressively blacker, until the yellow fringe merges imperceptibly with the white interior of the image, and the blue fringe merges similarly with the black background."

The mathematical theory of optical systems of lenses demands a very small aperture so that the rays are limited to the axial regions. In optical instruments an aperture of 10° is considered to be the maximum compatible with efficiency. It is with this aperture that such instruments are designed to eliminate chromatic aberration and other defects. But the pupil of the eye is rarely less than 4 mm. in diameter which corresponds to an aperture at the cornea of 20°. With a double size of aperture, therefore, the eye is equally free from aberrations, while optical instruments with such wide apertures would probably be useless.

Spherical aberration. In a biconvex spherical lens the peripheral or marginal rays are refracted more than the axial, so that the former come to a focus nearer the lens than the latter. This effect is called spherical aberration. It can be eliminated by grinding the lens so that its curvature decreases from the center to the periphery. Such lenses are termed aplanatic. Spherical aberration is diminished by making the curvature of the anterior surface of the lens, where the light enters, greater than that of the posterior surface. In the eye the lens has the opposite orientation since the anterior surface has a smaller curvature than the posterior.

With a glass lens of refractive index 1.5, immersed in air, the ratio of curvatures to give maximum spherical aberration is 1:6. With the lens of the eye, assuming an equivalent refractive index of 1.43, the ratio of curvatures would be 1:4, if surrounded by air. Actually the ratio is 1:1.7. The small degree of spherical aberration which the eye exhibits therefore indicates that the theory of glass lenses does not apply, in this regard to a lens of variable refractive index like the eye, when immersed in media of nearly the same index of refraction.

The chief corrective of spherical aberration is the pupil which shuts out the peripheral rays of light and admits the axial and those near it. The brighter the light, the more noticeable would be the aberration; but with greater illumination the smaller becomes the pupil, and the more axial become the rays admitted. "Although the eye does not form an aplanatic system", remarks Duke-Elder, "the effects of spherical aberration are small - much smaller than the effects of either diffraction or chromatic aberration. This is due partly to the fact that the cornea is flatter in the periphery than in the center, but more largely to the fact that the lens core is more highly refractive than the periphery. Both these circumstances cause the axial rays to be refracted more strongly than the peripheral ones, a tendency which counteracts the effect of spherical aberration....Regarding the eye as a whole, the axial area is usually under-corrected and shows a positive aberration, while the periphery is usually negative."

The Sine Condition. It is clear, therefore, that, by means of the automatic adjustable aperture of the pupil, as well as by other devices peculiar to the eye itself which are not available in any optical instrument, spherical aberration in the eye is reduced to such a small amount that no perceptible blurring occurs in the visual image.

Even though a lens system is designed to produce an image of a point free from spherical aberration, a wide pencil of light does not necessarily give a clear image of the area surrounding the point. Different zones of the lens may bring light from various parts of the object to different positions, so that the image is drawn out like the punctuation mark, the comma, the tail of which points to the optic axis. To prevent this from occurring, a certain mathematical condition must exist relative to the object in air and its image at the back of the vitreous body on the retina;namely, the product of the refractive index of air, the size of the object and the sine of the angle of divergence of the rays of light from it to the eye, must be equal to the product of the refractive index of the vitreous body, the size of the image and the sine of the angle of convergence of the rays forming it. This relationship is called the sine condition. "The eye," states Duke-Elder, "appears to obey the sine condition almost exactly, so that as far as comma is concerned the displacement of the fovea to the side of the optic axis is no disadvantage."

Refraction of Eccentric Rays. Since the eye is a decentred optical system, the pupillary line and the optic axis do not coincide, but differ in position by an angle of $5^{o}$. The so-called determining ray to the foveais therefore somewhat peripheral or eccentric, and consequently suffers a little radial astigmatism. The determining rays are brought to a focus sooner than the equatorial rays which tend to interfere somewhat with visual acuity. The amount of the defect is only about one-eighteenth of the chromatic aberration, and still less of the spherical aberration; and is therefore of no significance in vision.

Curvature of the Field. It is a characteristic of a spherical lens to form a curved image on an object. On a plane surface the whole of the image cannot be in focus at once. The curvature of the field, as it is called, is corrected by having a curved image-plane. In the eye the retina is such a curved screen. Since its radius of about 10 mm. is shorter than the posterior focal distance (22.78 mm.) of the optical system, "it satisfies very closely the theoretical curvature for complete correction."

The image of an object formed by a lens is also subject to two kinds of distortion; one, in which straight lines crossing one another at right angles are either curved outwards, barrel-shaped, or inwards in the opposite direction in the image. At the same time the image is respectively diminished or enlarged in size compared with its normal area. The magnification of the ends of an object is also different from that of the middle. "Peripheral distortion of the image is almost theoretically corrected in the same way as curvature of the field," that is, but suitable curvature of the retina (Duke-Elder).

Depth of Focus. If several objects are at different distances from an optical system, their images will be formed at different distances also, so that if one is in facus the others will be slightly out of focus. The greatest distance of the objects from one to another, still having their images satisfactorily focused, is called the depth of focus. In other words, the greatest distance through which an object can be moved and still produce a satisfactory image without change of accommodation, is the depth of focus.

In the case of the eye, Hartridge has shown that with a pupil 3 mm. in diameter and with the eye focused for 24 meters, objects both at infinity and 12 meters will still be in focus at the same time. But if the eye is working at 25 cm., the normal reading distance, the depth of focus is reduced to 1.1 cm. As the pupil contracts during accommodation, the depth of focus increases. With bright light, a pupillary diameter of 2 mm. the depth becomes 3.2 cm. This property, therefore, relieves the strain of accommodation.

Loss of Light. In traversing the optical structures of the eye, light is reflected at all surfaces of sufficient discontinuity to do so. By this process a considerable quantity of light is deflected from the main dioptric path. There are six principal surfaces in the eye at which light may be reflected: the anterior and posterior surfaces of the cornea, the anterior and posterior surfaces of the lens and the surfaces between the zones of optical discontinuity within the lens. At each of them an image formed by reflected light can, by suitable means, be observed. But for four of the chief six images, the light forming them travels out of the eye and is lost. In the case of two, the light reaches the retina and tends to diminish the optical efficiency of the eye. In the case of microscope objectives there may be from 12 to 20 surfaces of such discontinuity where loss of light occurs. "The eye, however", says Duke-Elder, "sets a higher standard in this respect than most optical instruments, for the whole of the light thus lost does not equal 2 per cent." The loss of light is negligible in quantity since the refractive indices of the ocular media are so closely related, the lens being immersed in a medium of approximately the same index as itself. That of the humors of the eye is 1.336, and the refractive index of the outer layers of the lens is 1.386, so that the difference between them is extremely small; while in optical instruments the difference in refractive index between glass, 1.5, and air 1.0, is ten times larger.

Scattered Light. There is considerable scattered light reflected from the retina by which the fundus can be ophthalmologically examined. But very little of it confuses the retinal image. The shape of the fundus causes the reflected light to pass out at the pupil or to strike the insensitive anterior portion of the retina. The eye is also protected from oblique illumination by the nose and eyebrows, and the scattered light is effectively absorbed by the retinal pigment.

Halation. "Halation would appear to be of negligible amount. The term applies to the reflection of light back to the sensitive surface from other surfaces immediately above it; but the reflecting layer of the retina is so close to the rods and cones that this disturbance can be of little moment."

The defects which are recognized by ophthalmologists can all be detected by careful and expert examination. But by the vast majority of humanity they are never noticed, and they have no influence on the normal visual properties of the eye. Because they exist they can be referred to as defects, as the quotation from Helmholtz shows. But to exaggerate them grossly into an instrument that an optician would be ashamed to make, is to magnify them out of all proportion to the minute and unnoticed degrees in which they are to be found.

The mechanism of accommodation of the eye is the means by which accurate focusing of the image on the retina is automatically brought about. It is a muscular device which is optically and neurally controlled, and peculiar to the eye. It cannot be imitated in an optical instrument. The lens of the eye is quite plastic and enclosed within a transparent and highly elastic capsule which, under tension, is capable of moulding the lens into a more spherical form, but when relaxed permits the eye to resume its original shape.

In accommodation three elastic forces are involved which operate on the principle of double antagonism, that is beautifully adapted for protection of the lens from sudden and dangerous deformations. It has been fully described by Gullstrand[1]. The form of the lens is controlled by two antagonistic elastic forces, those of the choroid and the capsule; and at the same time the force of contraction of the ciliary muscle and the stronger of the two elastic forces, that of the choroid, act antagonistically. This arrangement protects the lens from the action of too strong external forces and from sudden variation of these forces. The elastic force of the capsule, that protects the change of form of the lens, is the weakest of the three, and, like all elastic forces, constantly diminishes in strength during the development of its effect, so that the movement terminates without any jerk. In the relaxation of accomodation, the greatest force producing the change of form is the elasticity of the choroid, and this force diminishes steadily during the movement, and at the same time the resistance of the lens capsule is continually increased by dilation. However, this means of protection would fail with decrease of power of the lens to change its form. Without that power every strong tendency to accommodation would result in a dangerous jerk on its structure. When it is realized that opacities, which impair or destroy vision, can occur in the transparent lens of the older person from its sudden and violent deformation, the great importance of the protective arrangement provided by the double antagonism of the forces acting in accommodation is obvious.

Listing's Law. A conspicuous and useful characteristic of the eyes is their extraordinary mobility which enables them to be voluntarily turned in perfect coordination in any direction within the obvious limits of possibility. No rolling motion of the eyes occurs as that would at once cause disorientation of the relative positions of corresponding retinal areas with a serious disturbance of vision. But when the eyes are moved in any direction, the movement is carried out with the greatest dispatch and the least expenditure of effort (Listing's Law). Operation according to this law causes the least degree of fatigue to the extra-ocular muscles which enables them to function for many hours a day. Listing's Law is a special case of the universal law of nature, known as Least Action, according to which all operations are performed with the least expenditure of time and energy. The law of economy of nature therefore operates with the eyes.

After reviewing the most recent estimates of the visual defects of the eye, it will be interesting to quote several other opinions of competent investigation.

Gullstrand[2] states: "Helmholtz's famous dictum that the monochromatic aberrations of the eye are such as would not be tolerated in any good optical instrument, is sometimes construed to mean that the eye is a very badly constructed optical affair - which Helmholtz never said and certainly did not mean. But another question that this statement raises is whether these aberrations are not serviceable and what is their purpose. First of all, it should be noted, as Helmholtz pointed out, that a limit is imposed by diffraction to the physical sharpness of the image."

In accommodation there is an elevation of the total index of refraction of the lens due to its structure which denotes a change of focusing out of proportion to the change of form that is unattainable with a lens of glass and is of the highest advantage. "The monochromatic aberrations," continues Gullstrand, "are the necessary evil

---

[1]In Physiological Optics. Helmholtz. Eng. trans. Southall. Vol. I. p. 408.

[2]Ibid. pp. 440,443.

for obtaining this advantage; and even if the convergence of the rays is not as good as it might be, the clearness of the image in good illumination is still above the limit of the capacity of the eye as imposed by the laws of diffraction. Hence, the monochromatic aberrations are a witness for the perfection of the eye, if what is meant by the perfection of an optical instrument is good convergence of rays to the degree that is needed to obtain the greatest useful sharpness of image; anything in excess of this being sacrificed in order to gain some other end."

In regard to the statement of Helmholtz, Duke-Elder[1] remarks: "As in all optical systems, many aberrations are met with in the eye. It is frequently said that these are gross, and the statement is attributed to Helmholtz that the construction of the eye is such as would not be tolerated in any good optical instrument. If Helmholtz ever said this he certainly never meant it, for, as Helmholtz fully realized, the eye is constructed with extreme attention to detail in keeping the image as physiologically useful as possible, while at the same time retaining an immense range of adaptability. It is true that the peripheral field is sacrificed to some extent for the central field; but it is the latter which is of importance for the purposes of accurate function, and the functional capacity of this region is up to the limits imposed by the diffraction of light. For yellow light and a pupillary diameter of 0.6 mm. the limit of capacity would be a visual angle of 0.3 minute of angle; for a pupil of 3 mm. it would be 0.82 minute, and only for a pupil of 4 mm. is the angle for yellow light 1.22 minutes, while for green light it is 1.05 minutes. When these figures are compared with those of the minimum visual angle (0.40 minute) it is evident that the limit of possible visual capacity as imposed by diffraction is attained by the visual acuity. It would appear, therefore, that to sacrifice adaptability in the attempt to make the central acuity greater by optical means would merely be a task of supererogation."

It has also been remarked by Southall[2] : "The process whereby the normal eye is enabled to focus on the retina in succession sharp images of objects at different distances is called accommodation, and it is this marvelous adaptability of the human eye together with its mobility, which perhaps more than any other quality entitles it to superiority over the most perfectly constructed artificial optical instrument."

In the introduction to his extensive investigation of the visual perception of fine detail, Hartridge[3] states: "The resolution of a grating test object by a microscope requires that instrument to possess lenses of fine performace. These contain as a rule achromatic combinations of carefully figured lenses which have been adjusted in relative position with the utmost precision. The design of the eye appears rudimentary in comparison with such lenses, yet practical tests with grating and other test objects show that this elementary lens system, possessing apparently no chromatic correction of any kind, and certainly no flourite-like medium which could impart an apochromatic correction, behaves in an almost flawless manner. So good, in fact, is the apparent definition of the foveal image, that if the lens system of the eye could be removed and an apochromatic lens, selected for its good performance, could be substituted, it is doubtful if the possessor of this eye would detect any improvement in his perception of fine detail. What he would certainly notice would be a very serious deterioration at the periphery of his visual fields. His original eye lens gave him an angle of view exceeding a right angle. His new microscopic lens gives him a field of view almost insignificant in comparison. He now knows for the first time what it feels like to have 'tubular vision', that is, the type of vision produced by looking down a narrow tube. But other more subtle differences would soon be noticed. His original eye lens adjusted its focus automatically, and was almost equally good for observing near and distant objects. His new microscopic lens possesses no such valuable features.

---

[1]Text-Book of Ophthalmology. Vol. I, p. 756.

[2]Mirrors, Prisms and Lenses. MacMillan. 3rd Ed. 1936, p. 434.

[3]Phil. Trans. Roy. Soc. Lond. B. 232, 1947, 523.

It has been designed to work at one fixed distance between object and eyepiece only, and a small error in this adjustment produces either over- or under-correction for spherical aberration. The performance of the eye lens is superior to that of the microscope lens, for the purposes for which it was designed."

The investigation of ophthalmologists which have been revised amply show that the optical defects of the eyes are so minute and unobtrusive in character that they are without detrimental effect on normal vision, as the testimony of the leading recent authorities indicate. It is therefore clear that the strictures on the eye in the first part of the remarks of Helmholtz are decidedly unjustified. But when the practical perfection of visual optical imagery, accommodation, rapidity of accurate focusing, breadth of visual field and mobility of the eyes, are added the remarkable acuity of vision and color sensations with which images are perceived in detail and adorned, as well as direction of vision, binocular coordination, perception of depth and adjust- of distance, all of which are automatic in action, the eyes obviously far surpass all possible optical instruments in performance.

Finally it should be remembered that the eye differs from all other optical instruments in being a living system, and is preeminently the organ of the chief intellectual sense. By itself its own minute defects are assessed. With its exquisite discriminating power it pronounces ultimate judgment on the degrees of perfection of performance attained by artificial instruments. No inferior instrument could give a valid judgment on a superior.

The Visual Apparatus As A Telescope. It is often stated that the eye resembles a camera in its construction, and in some ways the likeness is quite complete. But the eye itself comprises only half the visual apparatus, for the area striate in the cortex cannot be separated from the visual process for it is there that vision actually takes place. The striate area, however, exercises a much wider function than only to provide sensations of vision.

The area striata, or visual cortex, consists of a number of folds in the calcarine areas of both cerebral hemispheres. If spread out flat it would be oval in shape with a total surface of about 3000 sq. mm., about evenly divided between the macular and peripheral areas of the retina.[1] The area of the functional retina is about 905 sq. mm., of which the macula comprises only about 3 sq. mm. The optic nerve fibres, about 500,000 for each eye proceed from the retina to the lateral geniculate body, whence they continue in increased numbers, as the optic radiations, to the area striata. The macular fibres number about half the total; that is, there are as many nerve fibres reaching the striate area from the 3 sq. mm. of the macula, as there are from the remaining 900 sq. mm. of the rest of the retina. The macular fibres therefore spread out over the striate area of 1500 sq. mm., in the ration of 3:1500 or 1:500. There is clear evidence that both retinas are represented in the same striate area. Possibly the two macular images of an object differ sufficiently in the striate area so as to excite the perception of depth.

From these figures it follows that an image covering the macula is spread out over a striate area 500 times the size, or the image is magnified at least 500 times. Since there is an increasing concentration of cones, which function in form vision, from the boundary of the macular towards the central fovea, it is probable that in the striate area the part corresponding to the fovea is expanded in like proportion.

Marshall and Talbot[2] state that a pattern 1 minute of arc wide at the fovea expands in the striate area 100 times linearly, or 10,000 times in area. The area striata is also stated to be from 30 to 600 times as fine grained as the retina, which

---

[1]G. Elliot Smith. New Light on Vision. Nature. 125, 1930, p. 820

[2]Biological Symposia. Cattell Press. 1942. VII, 117 - 165.

would counteract its relatively coarse mosaic structure.

The fovea, which is the retinal area of distinct vision, has an area of about 0.05 sq. mm. Its cortical expansion in the striate area will not be less than 500 times this amount, or 25 sq. mm., nor more than 10,000 times, or about 500 sq. mm. (1sq. in -- 645 sq. mm.). In other words the foveal image is magnified in the cortex by an undetermined amount between 500 and 10,000 times, possibly about their mean value.

These facts and computations signify that the visual apparatus as a whole represents a telescope, of which the eye is the objective, the area striata the magnifying device, and consciousness the observer. The macular image of a man's face at a distance of about 3 meters is approximately 0.75 sq. mm. in area. To detect in this minute image all the details of form and color would be little short of miraculous. But in an image enlarged several thousand times, the discrimination of fine detail and color would be easy, as experience shows to be the case.

The optical magnification of retinal images in the cortex is obviously out of the question. But the neuro-mechanical magnification is effectively designed to take its place. In a telescope the eye-piece and the objective must always be kept rigidly in alignment. In the visual apparatus this is unnecessary since the vital connection between the two parts is neutral. Sufficient slackness in the optic nerve is allowed to permit all the delicately coordinated motions of the eyes in their orbits without interfering in the least degree with the conduction of the nerve impulses to the magnifying and perceiving area in the cortex. A flexible connection between the two parts in which optical alignment has no significance is impossible in artificial optical instruments.

To obtain a similarly enlarged image on the retina itself would mean redesigning the dioptric system of the eye in a form impracticable with living tissues. Such an eye would be much enlarged, sluggish in movement, slow and inaccurate in focusing and subject to considerable muscular fatigue as well as greatly, probably grotesquely, disproportioned to the size of the face and depth of the orbit. The increased retinal area would mostly be wasted in any case, as attention could be directed only to a very small portion at a time.

In their present form, due to their small volume (6.5 cc.) and weight (7 gr.), the eyes have a wide and easy mobility, swift and accurate accommodation, and perfect coordination, which enable rapid judgments of the form, direction and distance of objects under observation to be made. The magnifying cortex can unobtrusively and effectually supply the necessary enlargement of the images, the perception of detail and the appreciation of color. While magnifications as high as those in a microscope cannot be obtained, experience shows that they are unnecessary in the normal exercise of vision. If the retinal image were grossly defective, as sometimes erroneously asserted, the magnified image in the cortex would be useless for accurate visual purposes, and optical instruments designed for visual observation would be without value.

The magnificence of the achievement of the theoretical and practical designers of microscopes and telescopes in overcoming the natural defects of images due to the nature of light and lenses, and in obtaining the high magnifications which they have thereby accomplished, can scarcely be exaggerated. Infinitely greater is the glory of the Divine Designer not only in having contrived the far simpler and more efficient optical system of the eye, but also in establishing biological laws by which the minute details of ocular construction have been transmitted with precision to countless individuals of all generations of mankind.

Diffraction pattern produced by light passing through small circular aperature. Bands like those, but insignificantly small and practically invisible, are produced on the retina by the pupil.

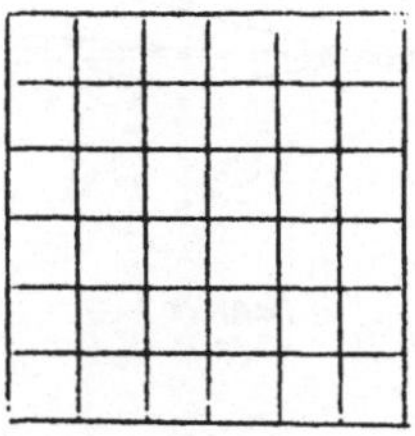

Object

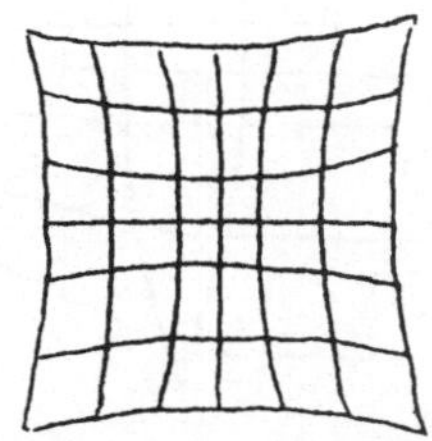

Image distorted by convex lens

Distortion of Image

Diagram of the Eye

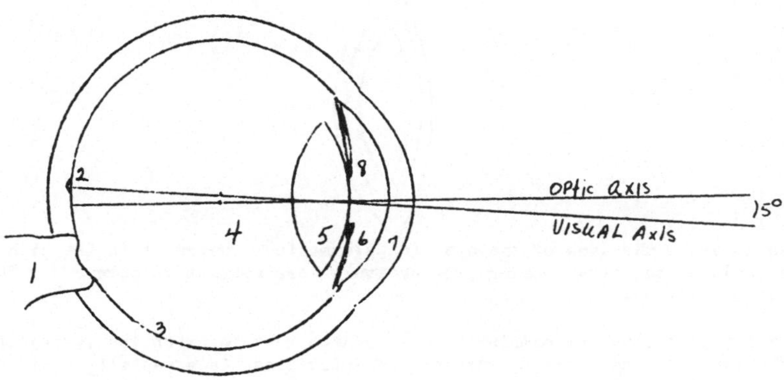


1. optic nerve, 2. Fovea, 3. Retina, 4. Vitreous body, 5. Lens, 6. Aqueous humor, 7. Cornea, 8. Iris.

Chromatic Aberration

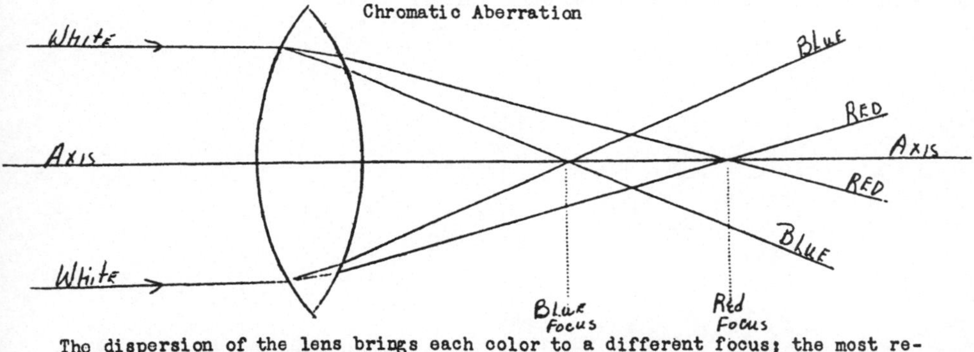


The dispersion of the lens brings each color to a different focus; the most refrangible, blue, nearest the lens, the least refrangible, red, farthest, away, the other colors between them.

Spherical Aberration

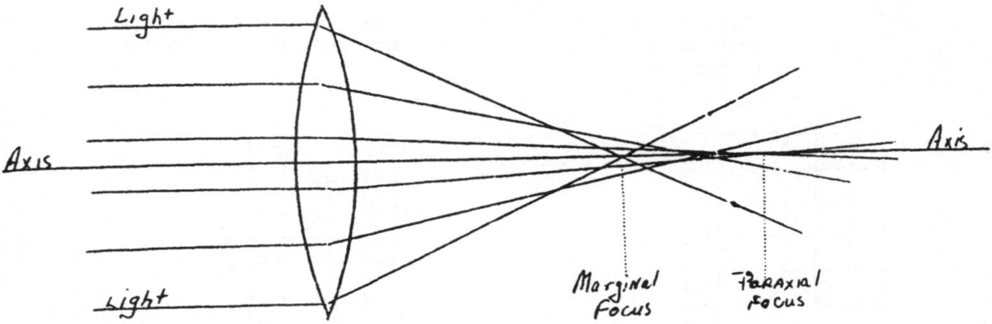


Diagram to show that rays of light passing through a lens at different places have different foci. Rays travel from left to right.

The lens of the Eye

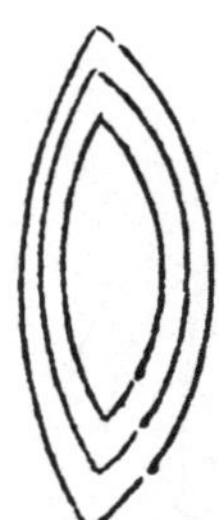

Three zones of the lens of the eye. (diagrammatic) Innermost is the high refractive nucleus, the others being progressively less refractive outwards. The zones are very numerous.

"In all probability the complexity of its architecture, with its layers of graduated curvatures and zones of varying refractivities, is especially designed to counteract and neutralize the effects of spherical and chromatic aberration. It also increases the converging power of the lens, which is thus enabled to cover a greater range of accommodation and, in addition, tends to minimize the formation of secondary images in the eye."

(Duke-Elder: Text Book of Ophthalmology, Vol. I. p. 126).

# THE SCIENCE OF HEREDITY AND THE SOURCE OF SPECIES*

A great geneticist has admitted that "No individual can claim such a mastery of all facts pertaining to evolution to enable him to present...."[1] a full discussion of the facts, laws and theories of evolution. This appears to me to be especially true of the bearing of the science of heredity on evolution. Therefore an attempt will be made to answer only one question, "To what extent have species changed into other species?"

Some species appear to have descended from others. If two species can be crossed, and living offspring be produced, there is proof of campatible germ cells from the two species. The obvious source of this hereditary material is descent from a common ancestor.

The birds of the eastern United States which hybridize readily are the Golden-winged Warbler and the Blue-winged Warbler. Their hybrids are called Lawrence's Warbler if there is a black throat, and Brewster's Warbler if there is not. The hybrids show the characteristics of the parents segregated in the offspring. A typical Lawrence's Warbler has a black throat from the Golden-winged parent but has yellow underparts like the Blue-winged ancestor. Similarly, upper parts, wing bars, and ear patches are typical of one parent or the other. Because the hybrids do combine traits of both parents, it is probable that at some past date the two parental types came from one type. To be sure, this transition was not observed; surely each species could have been created separately, even though they now interbreed. However, evidence from fruit flies which indicates that a species is now in the process of subdivision into two major groups indicates that the idea of the derivation of the two warblers from common ancestry is the correct explanation.

A species now undergoing subdivision is Drosophila pseudoobscura. There are two races, A and B which differ from each other, not in any observable external appearance but in the arrangement of genes on the chromosomes. (Genes are the units of heredity in cells arranged serially on rod like bodies called chromosomes.) The genes in one race are in a certain order; in the other race, some of the genes are inverted.[2] The races are told apart by breeding tests. Race A produces fertile offspring when mated with Race A, but when mated with Race B, sterile male offspring are produced. Here is the indication of the subdivision of the species. If the two races thus formed were to have visible character differences, a classifier would be justified in concluding variations of the chromosomal arrangements. In the third chromosome"....twenty-one different sequences have been recorded."[3] Any subsequent rearrangement may produce a character difference, or mutations (changes in the genes) may do so; then race A will be obviously distinct from race B and two species will have been observed to have arisen from one species.

In another species of Drosophila, melanogaster, character differences have appeared in large numbers. There are hundreds of kinds of flies possessing varied eye colors or wing sizes or body colors. Yet each kind crosses with any other one to produce fertile offspring. To produce two species from this single species, there

---

* A paper presented at the third annual Convention of the American Scientific Affiliation at Calvin College, Grand Rapids, Michigan, September 2, 1948.

1. Goldschmidt, Material Basis of Evolution, 1940, p. 3.
2. Dobzhansky, Genetics and the Origin of Species, 1941, p. 120.
3. Ibid.

needs to appear a chromosomal change which will make those individuals possessing one trait sterile with those displaying a contrasting trait. If this occurs, character differences will have preceded reproductive differences in species alteration. On the other hand in Drosophila pseudoobscura, reproductive segregation has already preceded character differentiation.

In nature, the species of Drosophila actually differ in the ways seen in the flies already discussed. Chromosomal rearrangements occur within the members of one species; "....the chromosomal differences between species are identical in kind, if not in degree, with those found among races and individuals."[4] We readily accept the idea that the members of one species have a common ancestry even though they now have various chromosomal patterns; why should we hesitate to believe that flies now admittedly separate species have a common ancestry if the differences between them are the kind that occur within one species?

Current biological thought holds that if any species gives rise to new species, then the conclusion may be drawn that the first living protoplasm could have been the ancestor of all kinds of life of the past and present. Reasoning from the probability that one species may become two, biologists conclude that one form of life may have given rise to all forms. As a creationist, I am willing to accept the origin of species from other species, called microevolution, but do not see the necessity for believing that a protozoon was transformed into a coelenterate, or a coelenterate into a round or segmented worm and similar macroevolutionary transformations up to man. "Experience seems to show, however, that there is no way toward an understanding of the mechanisms of macroevolutionary changes, which require time on a geological scale, other than through a full comprehension of the microevolutionary processes observable within the span of a human lifetime and often controlled by man's will. For this reason we are compelled at the present level of knowledge reluctantly to put a sign of equality between the mechanisms of macro- and microevolution, and, proceeding on this assumption, to push our investigations as far ahead as this working hypothesis will permit."[5]

Hereditary studies are incapable of establishing the assumed connection between all types of life. A university geneticist in a recent lecture remarked, "The higher categories and the problems there presented are a good ways from the field where genetics can make a direct contribution." Crosses can be made between species and between genera, and only infrequently between animals of greater distinctivenesss. Because members of categories as widely separated as classes have not been crossed, there is no information about the similarities and differences in the genes of the members. We cannot put genes from two classes into one hybrid; birds and reptiles will not cross; we cannot therefore prove that the two could have come from one. The evolutionist speculates that somehow it could have happened; the creationist follows the statements of the Scriptures which say that the original kinds were brought forth by the command of God.

Notice some of the difficulties there would be in producing coelenterates from protozoa. A single celled animal would have to become many celled. These cells must be arranged into two layers, perhaps by the inpushing of one side of a hollow ball of cells. The inner cells acquire contractile fibers at their bases. Certain outer cells produce stinging capsules. The whole mass becomes arranged into body and tentacles. Only in imagination can there be found hereditary changes capable of producing all of these advances which must have occurred if members of one phylum (major group) were to change into members of another phylum. The evolutionist believes that in the

---

4. <u>Ibid</u>., p. 148.
5. <u>Ibid</u>., p. 12.

past there were mutations which could do this although they are not observed at the present time. The mutations which have been noted have altered organisms within a limited sphere. One of my professors said, "It would be miracle that a mutation causing diversity would also cause convergence for an adaptive end."

## A NEW THEORY

Neo-Darwinian geneticists conceive that a species changes slowly over many generations until it forms reproductively isolated populations which can be considered two species. But Goldschmidt has questioned the ability of mutations to produce specific differences." Microevolution by accumulation of micromutations - we may also say neo-Darwinian evolution - is a process which leads to diversification strictly within the species, usually, if not exclusively, for the sake of adaptation of the species to specific conditions within the area which it is able to occupy."[6] Consequently, he has postulated a "new" kind of evolutionary mechanism, "Species and the higher categories originate in single macroevolutionary steps as completely new genetic systems. The genetical process which is involved consists of a repatterning of the chromosomes, which results in a new genetic system."[7] No mutations need to appear; all that is necessary is translocation[8] or inversion [9] of genes already present.

This process, he feels, will produce new species more rapidly than will happen by the slow accumulation of small mutations. Drosophila miranda and Drosophila pseudo-obscura are only slightly different in external appearance, but hybrids between the species are sterile. Studies of chromosomes reveal that the species are distinguished by different arrangements of similar genes. But Goldschmidt believes that a relatively small repatterning of chromosome will also produce considerable changes in appearance.[10] This would avoid the difficulties of conceiving a gradual change from one species to another.[11] Peculiar structures like the preformed exit in a plant for the insect living in it could have been formed at a single step. The imitation of a distasteful species of butterfly by an inoffensive one need not be by a gradual change over many generations but could have occurred between parents and offspring merely by rearrangement of the positions of the parts of chromosomes.

Goldschmidt emphasizes the large amount of change which may be produced in an offspring if a pattern rearrangement occurs in the germ cells of the parents. The development of the offspring is considerably altered. The wingpattern in a butterfly (Papilio dardanus) varies markedly depending on a few differences in genes. It is supposed that a rearrangement of these genes would produce as great or a greater difference. There would be "macroevolution by single large steps."[12] However, some of the new characters in fruit flies showing greatest change from normal characters, such as the appearance of four wings instead of two, are the result of simple mutations and not the result of pattern changes. Goldschmidt presumed that alterations occur in the germ cells and survive or perish there. They need not wait until they have produced visible characters and these characters have been selected for survival or

---

6 Goldschmidt, op. cit., p. 183.
7. Ibid., p. 396.
8. A segment of a chromosome broken away and joined to another chromosome.
9. A segment of a chromosome broken out, turned around, and reattached to the same chromosome.
10. Ibid., p. 209.
11. Ibid., p. 244.
12. Ibid., p. 331.

death. The altered chromosomal pattern finally selected affects a developing individual so strikingly that a decidedly different form is produced. "The first bird hatched from a reptilian egg."[13] Goldschmidt admits "unfortunately no experimental attack upon this problem is at present apparent...."[14]

In a sentence summary, Goldschmidt's view is this: Slow accumulation of mutations over many generations will not change one species into another but the rearrangement of the genes will produce in the germ cells a pattern resulting in strikingly altered offspring which have become different species from the parents in a single step and are separated from them by a "bridgeless gap."

How has this idea been received by other geneticists? Dobzhansky notes that Goldschmidt's "....systemic mutations....have never been observed. It is possible to imagine a mutation so drastic that its product becomes a monster hurling itself beyond the confines of a species, genus, family or class....The assumption that such a product may, however rarely, walk the earth, overtaxes one's credulity, ...."[15] Again "...the simplicity of Goldschmidt's theory is that of a belief in miracles."[16] Goldschmidts chromosomal mutations, which are supposed to cause new species, are no different than those which occur within races of the same species. This is to say that Goldschmidt is offering nothing new; he is merely claiming more for a phenomenon than others do.

Single genetic changes may produce striking effects but the result is not likely to be a structure working harmoniously with other parts of the body. Sumner holds that such "jumps" as Goldschmidt believes in are not integrated with the rest of the animal. To get a bird from a reptileby any other means than the slow accumulation of small changes, which is typical Darwinian evolution, would need "the direct intervention of the Creator Himself." "Only the wave of the magician's wand could have transformed the scales of a reptile forthright into the plumage of a bird."[17]

Another criticism of Goldschmidt's work is that he maintains species to be discrete groups, separated by the bridgeless gap from other species. Bridgeless, that is, by any other mechanism than the rapid transformations which he thinks his systemic mutations will produce. But species do not fit into rigid compartments. "The lack of universality of such rigid subdivisions is what has ever since Darwin been one of the main arguments in favor of evolution."[18] Certain species can easily be told apart. However, there are some species that blend into others; intermediates are placed arbitrarily. Among flies of the genus Drosophila, some types seem to belong to distinct species, other variants are classified according to the opinion of the investigator into different species or merely into different races of the same species.[19] Further, Dobzhansky and others point to the lack of definition of species by Goldschmidt so that a reader has difficulty in detecting where the "bridgeless gaps" between species exist. Consequently "...the main premise of his theory is wholly unacceptable."[20]

It should be mentioned to the credit of both Goldschmidt and neo-Darwinians, like Dobzhansky, that the latter admits, "It must, nevertheless, be recognized that Goldschmidt's keenly critical knowledge has emphasized the weaknesses and deficiencies of the neo-Darwinian conception ofevolution, which are numerous, as even

13. Ibid., p. 395. The statement was made by Geoffrey St. Hilaire.
14. Ibid., p. 334.
15. Dobzhansky, op. cit., p. 53.
16. Dobzhansky, Science, Vol. 52, 1940, p. 358.
17. Sumner, Science, Vol. 93, 1941, p. 522.
18. Dobzhansky, op. cit., p. 357.
19. Dobzhansky, Genetics and the Origin of Species, 1941, p. 371.

partisans ought to have the courage to admit. It would seem that this fact alone obliges anyone interested in the modern evolutionary thought to read Goldschmidt's book."[21] In my opinion, one is obliged to read Dobzhansky's excellent volume also and compare his handling of data with that by Goldschmidt.

## COMPLEX ORGANS

The source of structures as complex as the eye has always been difficult to explain. "Here is one of the puzzles of evolution which appears to be still far from solution."[22] Goldschmidt has listed nineteen different features which he challenges Darwinian evolutionists to explain "by accumulation and selection of small mutants."[23] Mayr has admitted that "it is a considerable strain on one's credulity to assume that finely balanced systems such as the eye or a feather could be improved by random mutations."[24]

"The eye appears as if at a single jump from no definite previous form, and 'transitional stages are wholly conjectural'!"[25] There are eight or more theories recounted in Walls' volume. None of these is corroborated by comparative anatomy, because no animals exist possessing the stages which should have preceded the eyes of the cyclostomes (the first vertebrates), which in all essentials are like mammalian eyes. What caused the lens to be formed from ectoderm at the same time the retina was developing from the nerve tube is a 'tantalizing mystery'.[26]

"The study of heredity provides practically no evidence for a type of genetic change capable of producing the transformation of any kind of simple light-sensitive structure into a complex organ of vision."[27] A duplication of genes at one location on a chromosome of a fly will reduce the number of facets in a compound eye to a bar eye. Several eye color alterations have appeared but these are shades of red; .... "there has been no blue or green. In view of the frequency of mutation of eye color, one is led to suspect that blues and greens are absent because Drosophila is incapable of mutating in that way."[28] A certain mutation produces an eyeless condition. Another gives an antennalike structure "instead of an eye or combined with a rudimentary eye."[29] Not only is it difficult to see how mutations could improve the eye as it now exists, but also how they could even begin to form an eye in view of the fact that the known mutations of the eye are harmful to it. According to Walls, Froriep compared the origin of the eye to the birth of Athena full grown from the brow of Zeus. Because creation is the making of something without the necessity of a prototype, this abrupt appearance of the eye in vertebrates is in harmony with the Creation doctrine.

One modern scholar has seen the reasonableness of creation. "Some people assume, entirely as a matter of faith, a Divine Creation of living substance. The only alternative seems to be the assumption that at some time in the dim past, the chance association of the requisite chemicals in the presence of favorable temperatures, moisture, etc., produced living protoplasm. In other words, if one subscribes

---

20. Ibid.
21. Dobzhansky, Science, Vol. 92, 1940, p. 358.
22. Shull, Evolution, 1936, p. 161.
23. Goldschmidt, op. cit., pp. 6,7.
24. Mayr, Systematics and the Origin of Species, 1942, p. 296.
25. Neal and Raud, Chordate Anatomy, 1939, pp. 395, 407.
26. Walls, The Vertebrate Eye, 1942, p. 108.
27. Mixter, A Christian View of the Origin of the Vertebrate Eye, Christian Medical Society News, May, 1947.
28. Shull, op. cit., p. 123.
29. Goldschmidt, op. cit., p. 327.

to this theory, he admits that the first protoplasm to appear on our earth was a product of spontaneous generation. Then, if he accepts the evidence of Pasteur and others against spontaneous generation, he must reverse his explanation of the origin of the first protoplasm to explain the origin of all subsequent living protoplasm from the first protoplast. In other words, spontaneous generation, according to these opponents of the idea of Divine Creation, worked when the first living substance was formed, but probably hasn't worked since. Actually, biologists are still as far away as they ever were in their attempts to explain how the first protoplasm originated. The evidence of those who would explain life's origin on the basis of the accidental combination of suitable chemical elements is no more tangible than that of those people who place their faith in Divine Creation as the explanation of the development of life. Obviously, the latter have as much justification for their belief as do the former. It is possible that the problem of life's beginning on our planet will always remain insoluble, a philosophical question rather than a subject capable of experimental investigation and solution."[30]

R. Mixter 1/1/49

DISCUSSION OF DR. MIXTER'S PAPER:

DR. VOSKUYL: Thank you very much, Dr. Mixter. Are there any questions or any comments that you (the audience) would like to make?

QUESTION: I would like to ask Dr. Mixter whether the number of genes increase. I notice you have one with nine and another with nine and you get eighteen and so you have a combination, but the originals were always there to begin with. Now do you ever get eighteen from nine without drawing in some other source of genes?

DR. MIXTER: Yes, sometimes in one species the number will just be duplicated so you have twice as many at the same time as before.

QUESTION: And then the combination results--

DR. MIXTER: Often in that case in bigger species or bigger specimens of the same species, but not structurally much different.

PROF. DE KONING: Dr. Mixter mentioned something about the interchange of the genes and the particular chromosomes as brought about by X-rays. Since his paper was entitled "The Mechanisms of Evolution", do you have any suggestion about how that accomplishment is made?

DR. MIXTER: Apparently it has not been produced by natural radiation because that is insufficient to produce the changes which are obvious by artificial radiation. No, they are still at a loss as to what has produced these changes.

MISS ALTA SCHROCK: Are there areas in the earth where these changes are more rapid than others?

DR. MIXTER: Yes, apparently down in some mines where there is more radiation down there. They took flies down there to see if the mutations increased and it did not do so significantly.

30. Fuller, The Plant World, 1941, pp. 19, 20.

DR. KULP: I think we might well look to Brookhaven and similar laboratories that are doing extensive evolution-biology research using all types of radiations that are now available in attempting to see just what can be done along that line. Certainly results, if they are going to come, will come from that source. I would also like to ask Dr. Mixter if he has considered the new synthetic theory for the mechanism of evolution which Simpson claims is the final answer. I have read his article in the Scientific Monthly and I have tried to get around to write an appraisal of it, and it doesn't seem to me that he has said anything new. I wondered what you would say about that, Dr. Mixter.

DR. MIXTER: Simpson's synthetic creation is essentially that natural selection is the Creator, that nature has somehow watched over the whole process and selected--not in a sense that a Divine Mind would plan ahead and get things done--but in a sense that nature would somehow control its processes. I agree with you that I don't know exactly what he means by this synethetic creation. It is definitely an attempt to avoid thinking of a Divine Creator, somebody different from a member of the natural order of things. It seems to me that it was very intangible, and just a way of saying, I believe, "something has controlled all that, so we have gotten where we are."

DR. VOSKUYL: Are there any further questions? I am sure we wish to express our appreciation to Dr. Mixter for this excellent address.

DR. MAC RAE: I wonder if this would be a fair summary, to say that if a person sets out to imagine how this house was built and says they built that wing first and then this one next, or vice versa--you have a right to advance either hypothesis you want to so long as there is no definite evidence, but if you show no evidence either way, then it is purely a guess. If one man says, "I believe that God created by a single effort, and in accordance with his plan created the earth," and another man says, "No, I believe that through a natural process this came into effect,"--so long as there is no evidence either way, one has a right to choose, and no one has a right to call the other a fool for choosing the other way.

The teachings of most of our colleges is that there is sufficient evidence of the way things came into existence; that the way our species came into existence is through an evolutionary process and that one is rather foolish not to accept it or to hold the other view.

Are we safe today in saying that there is no evidence that is at all conclusive as to any particular way in which such an evolution has occurred and therefore it remains a matter of individual choice as to which guess you prefer to make, rather than something which at present is not provable one way or the other? Is that right?

DR. MIXTER: I think all you can say is that this indicates that some sub-species have come from other sub-species and probably some species from other species. From that point on, the evolutionist assumes the process continues to work so you can get all the differentiation that appears in the animal and plant worlds. This says that there has been some change within birds since creation but that is quite different from saying that birds came from reptiles and reptiles from something else, so I think you have made a very good statement of it except this is the kind of evidence that the evolutionist works on at a minor level and that it must have worked that way in every case has been imagined.

DR. MAC RAE: You wouldn't say it was conclusive evidence that it has worked on any minor level?

DR. MIXTER: No, I think it is conclusive that on the minor level it has worked, but that is much different from saying that all has occurred that the evolutionist says has occurred.

DR. MAC RAE: A man said to me--a student said to me, "Of course, we have to believe in evolution, because you can always see the evolution of a boy into a man." To my mind, a change which takes place in the natural world is something entirely different from evolution. The development of a caterpillar into a butterfly is in no sense an evolution. Now an evolution, if you have your species-A here as definitely existing and you see it under the appropriate circumstances develop into a species -B which is now distinct and cannot cross with A, and there you have before you clear evidence of the development of a new species. If you merely say you have A and B which you can't tell the difference except that they do not cross, it is purely conjectural, and I don't see how it is provable.

DR. MIXTER: I would agree with you.

DR. MAC RAE: Then you wouldn't say it is proven, would you, but merely conjecture.

DR. MIXTER: Well, all I would say is that it would be easier to believe that some one of those has given rise to all the others than to believe that the Creator bothered to make every special kind.

DR. MAC RAE: You mean you could believe that he created 1000 but not that he created 5000?

DR. MIXTER: Well, that is on too small a level. I could believe that he created 100,000 major types and that those 100,000 have given rise to the two million species that we now have.

DR. MAC RAE: But it is merely a matter of degree?

DR. MIXTER: That is right.

DR. MAC RAE: I mean if we can believe that the creator created 100,000, then it is just as easy to believe that he could create a million. It is simply a matter of proof one way or the other and the proof is lacking. It is merely a question of which seems easier to believe.

DR. MIXTER: That is right, definitely.

DR. MONSMA: I don't think that statement is entirely correct. There is proof that there have been changes and we do know that within certain limits changes such as have been illustrated in this paper have taken place, so that we do know that there are more, well, what-do-you-call-it, species, or what else do you want to call it, than there were at the time of creation. I think that is quite definitely established.

DR. MAC RAE: Do you have any actual instances of that?

DR. MONSMA: I think so. It makes a difference as to what you want to call a species. Evidence has been found among the flies and the birds, evidence that one could cite that different forms have arisen from these similar types of ancestors.

DR. MAC RAE: You have a man and woman that are tall and maybe they have a dwarf child or a midget; and a couple of midgets have a child and it is tall. That is a striking change but it is entirely within our human species, and even if from a tall group, small dwarfs are born and keep on being dwarfs as long as they interbreed with the tall group, you wouldn't say it is a new species, would you?

DR. MONSMA: No, I wouldn't, but who is going to tell what somebody else's point of view is. Taking an example of two groups, one called 'clumpers' and the other that they call 'splitters.' If you were a 'splitter,' you would call everything a new species; if you were a 'clumper,' you wouldn't.

DR. KULP: There is another problem. You are attempting to use the criterion of breeding. That is fine until you get into the field of paleontology. Naturally when you get into paleontology, the question arises as to what makes a species. According to the average paleontologist, it is simply a question of a distinguishable difference. When his whim decides that there are enough more rings on a certain shell to make it distinguishable for purposes of classification, that is a species. He can't say anything about breeding when he is a paleontologist. The only thing he can tell you apparently in a very broad way is that mammals and birds are not found in the earliest structures you find invertebrate life, but in considerably later structures you find mammals and in the very uppermost sheet of paper on this whole big stack of volumes, you find hominoid fossils, and in a very broad way, that is about all he can say.

MISS ERDMANN: I disagree with that. I think you are talking about the greater gaps of evolution when you talk about man and the invertebrates and if you talk about just down to the level of the invertebrate paleontology there are many instances where we have indisputable changes through a succession of strata in a given number of centimeters. In other words, say you have 100 centimeters in a certain bank some place exposed, and in the bottom layer of those however many layers are there, you find one type of an animal; then in the successive ascending layers which represent successively later periods of time, you have changes in this one type of animal, and then by the time you get to the top layer, you have something entirely different from what you started with at the bottom layer, obviously something has happened there, and subjectively or not they call it two different species; but we have this direct proof on a small scale that we do have indisputable changes.

DR. KULP: I would disagree that it was indisputable, on the basis that a slight change in temperature would gradually make any species migrate along the coast so it would maintain precisely the same condition from top to bottom, so you might merely be reporting changing conditions at that particular spot.

MISS ERDMANN: You can change the strata across the country; you can check on that.

DR. KULP: If you can trace it in two dimensions it becomes of more significance but the paleontologist cannot hurdle the ecological barriers up to the present time. Until we can get physico-chemical methods to determine solubility, temperature, and pressure of the deposition, we cannot determine whether the fossil is the same or whether it has changed. I think that may come in ten years, but at the present time it isn't indisputable. It is merely evidence that there has been that change.

DR. VOSKUYL: Shouldn't we wait for the next ten years? There is a number of you that have been sitting here for at least an hour and a half, and the next two papers do not look like bed-time stories, so let's have a five-minute recess at this time.

# THE SCIENTIFICO-LOGICAL STRUCTURE OF THE THEORY OF EVOLUTION

by
Bernard Ramm, A. B., B. D., M. A.
Professor of Apologetics, The Bible Institute of Los Angeles

Many scientists do not understand the actual structure and nature of scientific knowledge. In the laboratory they are of a very practical bent of mind, and they take things just about as they find them, i.e. they do not subject their methods nor materials nor scientific heritage to any kind of critical scrutiny. The results of this are usually bad because they lead to a very biased, dogmatic scientism.

This naivete is apparent in reference to the usual conception of scientific law. We speak of "proving" a scientific law beyond the shadow of a doubt. To the contrary today logicians admit that certainty of proof is possible only in formal logic, i.e. in the logic of classes in extension which neatly clips off the perplexing problem of intension.[1] Hence logicians restrict the word "prove" to the realm of formal logic where all statements are tautologies. Cohen and Nagel put it this way: "A hypothesis becomes verified, but of course not proved beyond every doubt, through the successful predictions it makes."[2]

In the realm of material[3] implication are to be found the problems of science. In material implication we prove nothing with certainty. We can only verify with degrees of probability. The naive scientist makes his big mistake right at this point: he uses the categories of formal logic in the realm of material implication and comes up with the application of formal certainty to the material non-formal world. And so it is that many writers speak of evolution as having been proved as if as a scientific statement it were endowed with the certainty of conclusions in symbolic logic.

Broadly speaking a law of nature is generalization. It may be a simple, empirical statement such as "lead has the properties of a, b, c...n." It may be a broader statement, i. e. more general, e.g. chemical reactions occur in units or multiples thereof. This would not only include lead but all the elements. Or we may even generalize further and say that every particle in the universe is attracted to every other particle in the universe in direct proportion to their masses, and in inverse proportion to the square of the distance.

---

[1]The difference between intension and extension is somewhat as follows: To give the intention of a term is to indicate its qualities or characteristics, which is a very difficult job in many cases, e.g. the varying interpretations of man himself. To give the extension of a term is to give a list of the objects in the class or to simply indicate the class. Hence it is far easier to say X is a man (extension) than to say man is characterized by x, y, x,..n (intension). For further details on this cf. Blanshard, The Nature of Thought.

[2]Cohen and Nagel, An Introduction to Logic and Scientific Knowledge, p. 208.

[3]Formal implication is the deduction of theorems from axioms, or primitive sentences. Geometry is the classical and historical example of a formal system with formal implication. In reference to the regular connections in the natural world the problem is far more complicated. Here connection is described as "material implication." Such implication is probable.

Thus generalization arises from the observation of particulars. We do not see laws; we see only instances, or cases of the law. In the falling of a rain drop we do not see the law of gravity but what we consider to be an instance of the law. When we have seen a great number of instances that appear to be similar we endeavor to form a generalization, i.e. a hypothesis that will explain the instances by showing that they are all members of a class.

The next question is this: how do we know if we have chosen the correct hypothesis? The truth or falsity of a hypothesis is in terms of the logical and experimental entailment of the hypothesis. In a logical system no hypothesis can be both true and false. If our hypothesis gets us into logical trouble we may find out that our system is wrong (i.e. the hypothesis is right) or that the hypothesis is wrong (the system is right).[4]

From the standpoint of experimentation an hypothesis must survive rigid tests. As soon as the hypothesis involves us into a contradiction in material implication then we must revise our hypothesis, or revise our entire science of that subject, or drop out the hypothesis. If, on the other hand, the hypothesis meets with success it gains prestige. If it meets with more success it is "converted" into a law.

But whether it be a simple empirical law of the nature of lead or a very broad generalization like the second law of thermodynamics, the law is still a probability statement. It is not a logical statement possessing universality and necessity. Its validity hangs upon further experimentation.[5]

So far we have endeavored to establish two propositions: (1). We must not think of physical laws in terms of logical certainty, and (2). We must conceive them as probability statements. Our (3) next statement is this: the more general a law is the more difficult it is to establish it with a high degree of probability[6] and the more susceptible it is to suplantation by another generalization. This can be illustrated by contrasting the problem of determining the specific heat of iron with the most general laws of astronomy. Perhaps a hundred more years of experimentation will not change our present figure for the specific heat of iron but cosmological generalizations involve such a complexity of data that we cannot have such simple assurances as we have about the specific heat of iron.

Now let us take our three propositions and endeavor to apply them to the theory of evolution.

---

4 The genius of Einstein was his reconstruction of physics rather than trying to patch up Newton. Relativity was born out of logical trouble, Einstein's so-called "thought experiments."

5 For example, in reference to lead itself besides 204 they have discovered 206, 207, 208. Thus our law about lead must be further defined.

6 However "crucial experiments" that meet the requirements for such have a high evidential value such as Eddington's verification of Einstein's theory of light being bent when passing a large body and thus confirming relativity theory (1919).

(1). If our first proposition is true, then we cannot think of evolution being established as certain, universal, and necessary. Scientists who use such terms are not speaking or writing accurately or clearly. They evidence lack of understanding in failing to discriminate between formal and material implication, between verification and confirmation.[7]

Bold pronouncements then as to the finality of evolution as the ultimate theory of biology are not in keeping with the nature of scientific knowledge.

(2). According to our second proposition, evolution is a probability statement.[8] It is, to begin with, based on a great number of observations, geological, biological, embryological, and psychological. From these observations are noted similarities and from these similarities a class is extracted, e.g. "all ostriches have long necks," "this animal reproduces by internal gestation," "event X follows event Y in the embryo." Such generalizations may be quite particular such as "ostriches have long necks" and others would be quite general such as "ontogeny recapitulates phylogeny."

Finally, the great bulk of scientific data is summarized by the most universal generalization possible such as "all animals are the result of previous gradual transformations and will in turn change into other forms by the continuance of this process of transformation." This much is clear, nobody actually saw evolution at work. What are observed are specific incidents and these incidents are classed into generalizations of different compass and are in turn probability statements. Evolution is a generalization of the widest biological compass and it cannot rise in certainty higher than the simple, empirical generalizations that are foundational to it, and it is therefore a probability statement.

As a hypothesis it can only survive when it can be shown that no logical inconsistency is involved in the systematic organization of the biological and geological sciences, and no material inconsistency develops. If a logical consistency is discovered, or if a material inconsistency develops, then evolution must either be modified or discarded depending on the seriousness of the trouble.

For example, if the theory of evolution be discovered to contain a series of contradictory statements the theory must be revised or discarded. It cannot categorically state at the same time the following: (a) Life comes only from life; (b) Life arose spontaneously ages ago. Nor can it state categorically: (a) offspring tend to keep within the normal spread of variation. This task of detecting internal inconsistency in the theory of evolution is a difficult and a tremendous one, and part of the current prestige of evolution is that this is not a major note in biological studies.

Again, evolution must be revised or altered if it can be shown that it runs into difficulty with material implication. If no mechanism for evolution can be demonstrated then the theory stands in grave danger. Typical of the evolutionists is Howells who admits that there is no known mechanism for evolution yet he accepts

---

7 A classic article on such terms and the problems here mentioned is that of Rudolph Carnap, "Testability and Meaning," Philosophy of Science, 3:419-71, 1946, and 4:2-40, January, 1947.

8 The philosopher of science who has most developed the probability character of scientific statements is Hans Reichenbach, Experience and Prediction, Chapter I.

the theory without facing the logical implications of the lack of a mechanism. He writes: "And there is also the mystery of how and why evolution takes place at all . . . Evolution is a fact, like digestion. . . Nor is it known just why evolution occurs, or exactly what guides its steps."[9]

Furthermore, the geologic record might get evolution into trouble. If in the course of the next hundred years it be clearly recognized that no real intermediary forms exist in the geological record, or that they exist in such meagre number as to forbid any real prediction, then evolution must again reckon with serious change or bankruptcy. Granted, all would agree not everything was preserved in the rocks, but at least we must reckon that things were representatively preserved.[10]

To summarize, all of the statements in evolution are probability statements whose further life hangs on further evidence that may confirm or deny. To assert finality is not in keeping with what we know to be the logico-scientific structure of scientific theory.

(3) The generalizations of generalizations are the most difficult statements to make because they are at the long end of a series of probability statements and because the verification of such a statement is fraught with so many problems such as the strong element of subjectivity in the scientist, the welter of details, the number of possibilities, the various less-generalized statements that need to be harmoniously worked in.

If the history of science is anything it is the history of ruined and wrecked generalizations. Although this is a sign of the health and self-correctiveness of science, it is also a sign of its undertainty and tentativeness on the theoretical level. Two great generalizations have been accepted by western culture only to be subsequently discredited by the later developments. The first is Ptolemaic astronomy which seems to be so obvious to the naked eye. However we today pride ourselves in our emancipation from it. What was then commonplace and adequate is today listed among the mistaken theories.

Secondly, the Newtonian system acquired even more prestige than the Ptolemaic system for Newton had apparently written the very laws of God. For over two hundred years his system reigned supreme. But now after the work of such men as Einstein, Bohr, Planck, and Heisenberg, it stands as a special case of a now larger theory. Its absolutes (time, mass, and space) have been rejected. Its application to subatomic particles is an impossibility.

Besides these two great instances hundreds of others could be cited where the obvious and accepted has given way to the new such as in the theory of phlogiston, the generation of germs from slime and muck, the abhorrence of vacuum by nature. It would be well if the evolutionists would seriously face the significance of the history of science with its thousands of defunct hypotheses before it takes the theory of evolution as unquestionably established.

---

9 William Howells, Mankind So Far, p. 5.

10 In a most interesting series of charts prepared by Dewar, the latter sets forth the statistics concerning fossil remains in terms of the number of general, the percentage, the relationship to those new living, etc. Dewar claims that this is the first time this has ever been done in the history of geology. Dewar and Shelton, Is Evolution Proved, p. 61 ff.

If the evolution theory be a generalization of a host of lesser and greater generalizations then it is not possible to show that--at the present state of knowledge--all possibilities have been exhausted. Like Ptolemaic astronomy and Newtonian physics it may be subsequently replaced. We must therefore insist that evolutionists use the word "probable" and not "true" when speaking of evolution. At best it is an effort at reconstruction as to how things might have happened, not how they must have happened. It is the best guess to date that an unregenerate mind can make that rejects Biblical revelation, and without revelation might never shift its position.

Finally, what is the interpretative principle that the Christian brings to the myriad of facts, biological, geological, and the like?

The general interpretative principle that this writer adopts is that Genesis 1 records in broad outline the successive creative acts of God in bringing the universe and world to the state when it could be inhabited by man. Being a very general and broad sketch, Genesis 1 leaves considerable room for the empirical determination of various and diverse facts. Hence a multitude of facts now generally accepted by scientists would remain unchanged according to this view we are advocating.

Secondly, there is no advance upward apart from the creative activity of God. There may be horizontal radiation of life but no vertical. This is precisely the point where this view differs from theistic evolution. Evolution, theistic and naturalistic, believes in the radiation of life from lower to higher forms, from the simple to the complex. According to our view radiation can only be horizontal. That is, the "root-specie" of, shall we say the "dogs" may radiate outward into wolves, coyotes, and dogs and all the varieties of each. But there is only unraveling of gene potentialities--no upward evolution. And this seems to be in keeping with the fact that we do have in geology no demonstrable vertical radiation but plenty of horizontal radiation.

Thirdly, Genesis gives us the general movement of the origin of geological strata and life forms. The six work-days are geologico-biological work days. We expect then the basic rocks to be azoic. We expect the simpler forms on the bottom layers. We expect the higher forms on the higher levels, and man the highest form on the highest level.

This presentation is, of course, limited by space, and thus somewhat sketchy but if fully worked out we believe that it would constitute a general interpretive concept that would replace the evolutionary one because it can account for all that evolution tries to account for, and then can go on and account for the things that evolution cannot. The reason for this is, we believe, that the basis of it, namely Genesis 1, is a divine revelation.

In closing we wish to point out that to indicate weaknesses and inconsistencies in evolution theory is all right but it does not go far enough. For example, in modern psychology there are serious objections to all the major schools of psychology yet the adherents to these various schools do not give up their convictions. Convictions are surrendered when a more unifying and integrating hypothesis is suggested and demonstrated. So, we as Christians must not think we have done our job by indicating the difficulties with evolutionary theory. We must go on and present clearly and factually the Christian interpretative principle in geology and biology. If we can show that this view has the maximum of internal consistency, and has a high degree of accuracy in material prediction, then we have really done something to the evolutionary theory. Until then we fight pretty much of a guerilla warfare that may sting but does not force a retreat.

DISCUSSION OF DR. RAMM'S PAPER:

DR. VOSKUYL: Thank you, Dr. Cowperthwaite, for reading this paper, Since the gentleman who wrote this paper is not present today, and it is not the best of Christian ethics to discuss someone else's paper behind his back, we perhaps should not discuss it. However, the following paper is so closely similar to it, that I think we shall postpone all discussion until that time.

I am sure that the Secretary of our group will extend to Prof. Ramm our appreciation for his splendid paper.

## SOME PRESUPPOSITIONS IN EVOLUTIONARY THINKING

E. Y. Monsma, Prof. Organic Science, Calvin College

Some time last winter President Everest presented to the members of our affiliation a proposal that we publish our estimate of the evolution theory after we have thoroughly studied and discussed the matter and have come to certain definite conclusions. The proposed publication is to appear about ten or eleven years hence, at a time coinciding with the centennial of the publication of Darwin's "Origin of Species." This would give the members of our affiliation ample time to prepare a thoroughly scientific and scholarly work on the Christian approach to this vital and important subject. The suggestion is no doubt an excellent one. There is perhaps no greater need among us than a scholarly Biblical statement of views in regard to evolution.

Since the publication of the "Origin of Species" the theory of evolution has become established in the various fields of science to such an extent that one must admit its well nigh universal acceptance. Hundreds and perhaps thousands of books and papers have been written against it but to no avail so far as the prestige of the theory among the rank and file of the scientists is concerned. Religious leaders have attempted to stop the onrush on this theory because it seemed to conflict with the account of creation in Scripture but their attempts have gone largely unheeded. There is no use in duplicating work already done by others. If our organization is to do anything in this line it should be something different, something more fundamental than anything that has thus far been attempted. We shall have to be positive rather than negative. We shall have to construct rather than break down. If we do that, we shall have to start at the bottom, at the foundation. And it is just here that we, who believe the Scripture as God's inspired truth, differ from most evolutionary scientists of today. Faith in God and in His word is not a refuge for a bewildered mind whose native capacities have failed to comprehend the intricacies of a complex universe. It is rather a starting point and an accompaniment to the normal functioning of the mind. It is a guide to truth. We must be willing to assert that we start with certain presuppositions, test their validity, and proceed from them to the logical conclusions.

This may sound very unscientific in this age of inductive reasoning, and yet, in following this method we need be no more deductive than the modern scientist who interprets all he sees in the light of his evolutionary conception. I have yet to see a consistent evolutionary interpretation that was derived by a purely inductive process.

It is the purpose of this paper to point out briefly some fundamental presuppositions that are basic to evolutionary thinking, to evaluate them, and to give their phenomena. I mention just four. (There may be others):

The evolutionist believes:

1. That our knowledge of natural phenomena comes from nature alone.
2. That the fundamental similarities among living organisms can be explained only on a basis of a relationship of descent.
3. That the variations or changes that are observed in living organisms are unlimited in their scope, and
4. That the causes of such changes are operative today in the same way they have always been in the past.

The first mentioned presupposition, namely, that our knowledge of natural phenomena comes from nature alone, is in a sense basic to all the rest, for it is our counterpart to this proposition which enables us to see the limitations and shortcomings of the other three. The counterpart to the first proposition would read something like this:

> Natural phenomena are not known from nature alone but from nature and from the Scriptures, for God has revealed himself not only in nature but in the Scriptures as well. And it is precisely with the aid of this latter revelation that we get a complete picture of the facts.

Now it is evident that right here at the beginning of our considerations our faith plays its important part. It is here also that we can expect the most severe opposition. For what, after all, have we done with the so-called scientific method? Have we discarded it altogether? If so, we cannot expect to receive a listening ear from those who have so proudly and to a large extent so justifiably lauded its achievements during the past century. It is for us to point out that we have not dropped the scientific method altogether but that we have amplified it. We have included with our sources of knowledge the inscripturated word, because we believe it to be a part of the whole. Our unbelieving and liberal associates will not accept this inclusion as valid. We can only point out that without it we shall not be able to get a complete picture, for such questions about nature as whence? whither? and wherefore? can not be answered without it. We must be willing to take this position from the outset or lose the cornerstone of our entire scientific structure.

The second presupposition (that the fundamental similarities among living organisms can be explained only on the basis of a relationship of descent) is an assumption which seems logical when one considers that ordinarily things which are similar have a common origin. Lindsey in his "Textbook of Evolution and Genetics," for instance, devoted several paragraphs to this point.* He states among other things that, "When relationship is mentioned, the immediate thought aroused is of similarity. Further analysis shows that we cannot have similarity, i.e., relationship, without some degree of community of origin." To Lindsey and other evolutionists any similarity indicates a relationship of descent, a genetic relationship, and therefore, since living organisms have at least some structures in common, even though they be only protoplasm and cell structure, they must all have come from a common protoplasmic and cellular ancestor. This conclusion is unwarranted. Though we may be willing to admit that similarities usually indicate a common origin, we cannot conclude that all living organisms are genetically related. Sometimes similarity does not indicate such a relationship but is evidence of a common creative idea. Our counterpart for this second evolutionary presupposition would be: "Similarities in living organisms do not always indicate a genetic relationship, but they may and do in this case point to a common cause, a common author, a common creator who has made all things according to certain fundamental plans." This conclusion is just as logical as the evolutionary conclusion and we accept it because it is in accordance with the revelation in the Scriptures.

---

*Lindsey--Chapter VII--1937

A third presupposition deals with variations observed in living organisms. It is an observed fact that living organisms are not wholly static. That is, offspring are not identical with the parents. They differ from them in varying degrees of dissimilarity. The reasons for such dissimilarity can, at least to a large extent, be explained by Mendelian heredity, gene mutations, and polyploidy. It must be said to the credit of Charles Darwin that he took note of the importance of variations although the above-mentioned phenomena were not known at his time. We can not, however, proceed to Darwin's conclusions that such variations have given rise to the various groups of animals and plants that exist today. Although modern scientists recognize the inadequacy of Darwin's theory, they nevertheless conclude that variations somehow or other have in the past and do now give rise to new forms of organisms. The whole quest of the modern evolutionists is to find some explanation for changes sufficiently significant to account for the appearance of these different forms. It is admitted that our present knowledge has not yet supplied us with an adequate explanation. Nevertheless, it is the firm conviction of evolutionists that there is an explanation and that through continued study and experimentation it will some day be discovered. It is this conviction that we can not share, for it is based not on facts in the first place, but on a previous acceptance of the evolutionary process.

On the other hand, both Scripture and our present knowledge of scientific facts indicate that variations are limited in their scope. All known causes for change, such as ordinary Mendelian inheritance, hybridization, mutations, polyploidy, appear to operate within certain set boundaries, within which we observe the variation to which evolutionists attach such great significance. They argue from the specific to the general. Since they see variations which give rise to different forms within a species, they conclude that similar variations have given rise to different species, and genera, and families, et cetera. Such a conclusion is not logically warranted.

In Scripture we have the significant statement that God created living organisms "after their kind." The word "kind" is not necessarily synonymous with the word "species" as used by the modern scientist. Because of the confusion presented by the use of the word "species" for the Genesis "kind," Marsh suggests that the word "baramin" be used. Says he, "If this word were used it could present but one idea in the mind of the reader; not the broad Linnaean species, nor his narrow one, nor the modern 'species', but only the 'Genesis Kind,'"* I do not know whether the use of such a new word is necessary. If it were used, however, it should be defined as a sort of kaleidoscopic entity within which the pattern and appearance may change and vary a thousand times, but whose various appearances are always limited by the number, shape, and color of the pieces that make up the pattern. This, I believe, is a true picture of the created Kinds of organisms. They vary and change, but their pattern is always limited by the genetic composition of the germ cells. So far as I can determine there is nothing in Scripture or in science that conflicts with such a conception. All the known facts seem to conform with it.

We come now to the fourth and last mentioned presupposition, namely, that the causes for change are operative today in the same way they have always been in the past. This has become one of the most fundamental assumptions of evolutionary thought. Dobzhawsky in his "Genetics and the Origin of Species" mentions it as one of the three main assertions of evolution when he says, "All these changes have arisen from causes which now continue in operation and which therefore can be studied experimentally."** Following in the wake of Charles Lyell and his actualistic

---

* Marsh, F. L. --"Evolution, Creation, and Science," p. 162, 1944.
** Dobzhavsky, T. "Genetics and the Origin of Species," 2nd edition, p. 8, N.Y. 1941.

geology, all evolutionists have made this principle their leadstar in determining the events of the past. For, as Nordenskield puts it, "if past natural phenomena in general are to be calculated or at least reconstructed with fair probability, it is necessary to start from the present, whose course of events it is possible to survey."*

It is no doubt due to the appeal of such statements that this principle has been so universally accepted not only by evolutionists but also, credulously, by some who profess not to believe in evolution. A consistent application of this principle, however, spells evolution if not in fact then certainly in a way of thinking. A consistent application of this principle, it seems to me, does away with creation in the orthodox interpretation of this term. It identifies present phenomena of development and change with the developmental processes which we call creation. The latter has always been interpreted by conservative Bible scholars as a unique process, which was completed at the beginning of the seventh day. It should not be confused with God's continued sustaining care of his creation, and his continued operation in the universe. These we call his providence. The distinctive character of God's creative work is clearly indicated in Genesis where we read: "And the heavens and the earth were finished, and all the host of them. And on the seventh day God finished his work which he had made; and he rested on the seventh day from all his work which he had made."**

A belief in the uniqueness of the creative process brings us face to face with certain difficult questions. These questions should be squarely faced before we attempt to publish our views of the origin of this universe from a Christian scientific point of view.

Why is it, for instance, that we so readily accept the period theory of the days of creation? Why is it that we so readily accept the two billion years estimate of the age of this earth and include in that age the entire creative period? Is it not in part due to the fact that we erase the line of demarcation between creation and providence? Is it not because consciously or unconsciously we accept the principle that present natural phenomena are a measure of past events, including the events of creation? Have we not with credulity accepted the interpretations based upon this fourth presupposition?

If so, should we not first of reconsider the implications of some of our conclusions lest we find ourselves torn loose from our moorings?

Scripture tells us that God created this universe with all that it contains. It does not tell us how God created it. Does his other revelation, the revelation in nature, tell us that? Are his operations in the care and sustenance of this universe perhaps the same as those which governed his creation? This appears to be the assumption of some of us. But is this assumption in accord with the Scriptures? If it is not, we err when we come to certain conclusions which are based upon it. Paul under the guidance of the Holy Spirit tells us that, "the invisible things of him since the creation of the world are clearly seen, being perceived through the things that are made, even his everlasting power and divinity."*** Do these "things that are made" also show us the processes God used in making them? And the writer to the Hebrews tells us that it is: "By faith we understand that the worlds

* Nordenskield: "The History of Biology," p. 456, 1928

** Genesis 2:1 and 2

*** Romans 1:20

have been framed by the word of God, so that what is seen hath not been made out of things which appear."* Does the Holy Spirit here mean to tell us that present appearances do not lead us into the mysteries of God's creative work?

It appears to me that we stand committed to an evolutionary philosophy if we do not distinguish clearly between creation and providence. But how then can we explain certain natural phenomena? How can a Christian geologist for instance, explain the earth's strata without Lyell's presupposition? Will he have to return to the old cataclysmic theories of Cuvier and others? Hardly. But upon the basis of Scripture, cataclysms can not be ruled out altogether. In this connection, we may ask: Have Christian biologists and geologists (outside of Price and his associates) taken adequate account of such Biblical and historical facts as the effects of sin and the fall, the peculiar characteristics of the antediluvian period, and the tremendous effects of the deluge?

In connection with this fourth presupposition it seems to me we shall first of all have to assert that we cannot hope to explain God's creative work by means of empirical procedure. Furthermore, we can accept the assertion of Marsh** that: "any........changes which have appeared in organisms since creation have arisen through natural causes which now continue to be in operation and which therefore can be studied experimentally."

Before we as an organization are ready to express ourselves publicly and officially on the larger aspects of the evolution problem we should study prayerfully and scientifically such fundamental issues as I have tried to present in this paper. Then, with the help of God, we may at least be able to establish more firmly the faith of those who by God's grace are led to the truth and significance of that majestic pronouncement: "And God said let there be........and there was."

---

DISCUSSION OF DR. MONSMA'S PAPER:

DR. VOSKUYL: Thank you, Dr. Monsma. I do not know but what you have precipatated what will be a very lengthy discussion.

MISS ERDMANN: I have a question for information. You seem to deplore the acceptance by Christians of the idea of geologic ages. Are you implying that evidence for geologic ages is insufficient for us to believe in them or if you believe in them, where would you place them in our current interpretations.

DR. MONSMA: I do not believe that we as Christians can--at least with the information that we have in hand now--say these geological ages represent the same days of creation as pushed back into the creation period. I would not deny, of course, the fact that we have stratification and all of that, but if we are going to interpret it upon the basis of what we see now, we can go back to the end of the creation period and no further.

I do believe that personally we as Christians have not taken into account enough of what has happened since creation. There is a series of events that took place there, and I don't believe that Christian geologists have taken into consideration at all as far as the series that I mention there, the sin, the effects of sin,

---

* Hebrews 11:3

** Marsh, F. L.: "Evolution, Creation, Science," p. 24. Washington, 1944

the curse, and what the effect of that has been upon this universe.

If you want to talk about cataclysm, that is out, I know, but that cataclysm is a fact and you know the effect of it on creation. Then we also know that this whole antediluvian period was a period that is different from the present time. We know that definitely. We haven't taken that into consideration. I haven't seen it at least from the Christian geologist; and then so far as the deluge is concerned, I haven't seen any of our other Christians take that into consideration at all.

Weren't there tremendous changes in the description in Genesis there, tremendous changes which should certainly be evident in the world now. I think if we as Christians take that whole series of events into consideration we may come to a certain conclusion.

MISS ERDMANN: And then the second question I wish to ask you is whether or not the time which has elapsed since the flood wouldn't be negligible in geological effects in comparison with the time which passed up until then?

DR. MONSMA: In the first place, we don't know just how long that time is.

MISS ERDMANN: Within the limits, say, of 25 thousand years.

DR. MONSMA: I would not dare to answer that question because I am not a geologist. I would have to consider that seriously. Gedney, for instance, mentions a series of woods--of course, this reveals my ignorance here--but geologically speaking, seventeen woods on top of each other which can all be seen there and if we figure all that at the rate at which these things go on now, I suppose we would come to a tremendous amount of time. However, how long does it take to build up a woods? I am speaking perhaps now ignorantly, but I know I can say that within one hundred years you can have a full grown forest.

At the University of Illinois they have the university forestry which was planted from very small trees in 1870 or thereabouts, and I drove along there a couple of weeks ago and they are now giant trees, actually a forest. That has grown up in that short length of time, so I do believe that the possibility of woods growing up within one hundred years and having a good sized woods is quite possible and I know that some volcanic destruction of such a forest could take place in three or four days and perhaps even less time than that. I am thinking of that Mexican event down there, and no doubt conditions in early creation were different from what they are now, and what things have taken place in the past there, I am just wondering; I don't know whether the time that the geologist sets is correct, nor am I certain that the time that we have for the Biblical chronology is correct. They tell us that this early chronology is not correct, so I think we need more study on that question.

DR. KULP: I feel compelled to say something if I could have the board and some chalk.

One of the most probable facts in geology, I believe, is that the earth is close to two billion years old, and I think this can be demonstrated, at least by way of order of magnitude with as much validity as we can demonstrate many of the laws such as conservation of momentum that we meet with as physicists.

Unfortunately, historically this is true: Over the last fifty years there have been practically no Christians in the field of geology. I was trained as a chemist before I felt that the Lord wanted me to go into geology. I went into it very critically, and I am still overly critical of all information that I receive. However, most of us do not understand enough geology to appreciate the geologist's method of securing geological data. He is not one millionth the philosopher that he is usually given credit for being,

I would like to give you about four examples indicating the antiquity of the earth; and the only requirement that I would want Professor Monsma to agree with me on is that the rules haven't been changed. If he will say that the fundamental interactions between atoms, if the binding energies have changed, if God has completely changed the rules since whatever time you wish to specify, I can say nothing. However, if the rules have remained the same, if the hydrogen atom today is the same as the hydrogen atom of creation, if the combining energy of two atoms and if the binding energy of the uranium nucleus is the same, I think my case will be evident.

Really one doesn't have to go any farther than radio-activity to demonstrate the antiquity of the earth. First let's consider radioactivity. We know that uranium is something that disintegrates through a series to the end product, lead. We know that in doing so, it gives off helium atoms. We know that the rate at which this takes place is not affected by any temperature or pressure that we can put on that system at the earth's crust. That has been attempted. We have put pressure on uranium up to the equivalent of twenty miles of depth in the earth's crust which is deeper than most rocks with that are now found, since they are found near the surface; and furthermore, we have subjected it to temperatures up to three or four thousand degrees without any change.

Now if the rules remain the same, the rate of disintegration of that specified uranium nucleus must have been the same whenever this uranium atom was put somewhere in the earth's crust as it is today. What happens geologically speaking is something like this: We have some simple horizontal strata of, say, sandstone and limestone. In the course of time there may be a fracture in the strata and some molten rock or some electrolyte solutions come up of high temperature and possibly high pressure, and in that fissure they may precipitate or crystalize crystals of uranium minerals and all other minerals.

Suppose now that a small crystal of uranium mineral forms in that belt at that time. Prior to the formation of that crystal, the uranium atoms were not directly associated with lead atoms. How do we know that? We know it in two ways; first, we know it because geo-chemically or just chemically uranium and lead are quite different in their reactions, and therefore, in a belt deep in the earth's crust they tend to be differentiated from each other because they are fundamentally different chemically.

But more than that we have analyzed the species of lead that one finds everywhere in the earth's crust. Now most of you here today have heard of the term isotope where the biolgist would use the term species just as well. There are four isotopes of lead that are found everywhere you find ordinary lead, that is, lead that is not in close proximity to uranium, and the ratio of those four to each other is exactly the same. However or whenever you obtain or measure the lead that is found in a uranium crystal, you find that it is essentially composed of just one of these four species, and that one specie as we know by experiment is derived from the decomposition process. Well now, if there are some of the other prominent species in that crystal also, we know they were there because there was some primary lead present in the neighborhood and we can subtract that out before we make our computation of the ratio of lead to uranium. The ratio will increase with a known rate. If we measure the uranium:lead ratio and the species of lead that are present, then we have a very good estimate of the time since that crystal was formed.

Now this isn't the only story, because fortunately uranium decomposes into lead and while decomposing, gives off eight helium atoms and that is all trapped there. If we have another small crystal which is a magnetite crystal, that crystal will contain just a small amount of uranium. We can't measure the helium in this uranium crystal because there is just so much of it produced and most of that leaks out, but you take a tight crystal such as magnetite and you have $10^{-6}$ grams of uranium and in geological time that does not produce enough helium to produce enough pressure to explode and leak out, and we see no evidence of fracture or leakage. When we take the crystal and measure it, measure the ratio of helium to uranium and find that the age from that crystal compares favorably with the age as determined from uranium-lead ratio, we have very, very strong confirmatory evidence--but that isn't the end of it.

It also happens that the element rubidium, isotope 87 decomposes into strontium 87. Anywhere that you can find rubidium which is close to potassium, a certain fraction of it (about twenty per cent) is radioactive and decomposes into strontium. They are very different chemically, and when we get a pegmatite to freeze in such a position, we find rubidium that is independent of strontium except for this peculiar strontium No. 87, which is the result of the decomposition of rubidium 87, and again we know that the rate of decomposition is not dependent upon nor affected by temperature and pressure, and again these values in a rough way check. We have values for all of these, and they show that they are up towards two billion years old.

There are still other methods, since this process occurs and since strontium is found in ocean water and all over the world, and since the process itself is taking place in our geologic time, the ratio of strontium in the ocean water of this species is going to increase over ordinary strontium during geological time, and again in a gypsum bed, the ratio of strontium 87 to ordinary strontium No. 88 should constantly increase, and once we can calibrate that system, that becomes a way of measuring the time when that strontium was deposited out of the ocean; and that too gives confirmatory evidence.

Maybe that is all philosophy, but it is just physics and chemistry really, so let's go on to something a little more picturesque.

First I would like to go back to the petrified forests. We have out in the western edge of Yellowstone Park a formation showing seventeen successive petrified forests which is exposed in the gorge of the Yellowstone River. Now there are several things to note here. First of all, the seventeen successive petrified forests are not the geological count of that area. There are quite a number of other strata on top. There are some folded sedimentary rocks below, and of course, the gorge is not cut to the bottom of the sequence. Let's reconstruct what happened here.

First you have to lay down under water these successive sedimentary rocks and with this sandstone, that means that there were just certain conditions of velocity of water that could have washed out that particular material, because if the water is moving more slowly, fine clay would have been deposited. We find a shale or a silt bed and then we know that the water is carried more slowly. Furthermore, we know that if there is a limestone, the limestone was precipitated under quite different conditions. Therefore, these had to be laid down each under different physico-chemical conditions. After that sequence was laid down, just from what you see on the board here, you know that this area had to be folded.

Now from some of the comments I received in the discussion yesterday, it is evident that many of us do not have much appreciation perhaps for what happens when you fold rocks. The rocks that are folded are just as hard as any other rocks that you have ever seen, but they are not fractured, and they still show the individual unformed shells, deformed shells or fossils, and the undeformedsand and silt grains.

Very elementary knowledge of physics will certainly tell anyone that if you attempt to squeeze solid rock and fold it into a pattern a tremendous length of time will be required otherwise the internal friction would be so tremendous that you would melt the rock, and then of course, it would look like granite. You would no longer have the undeformed shells or nice little rounded sand grains. It would be quite a different looking thing, and in fact, it wouldn't be too difficult a thing to compute how slowly such a process would have to be for the adequate dissipation of the heat so that it wouldn't melt the rock, knowing the internal coefficient of friction. After all, that took place while this whole thing was growing. It is true, as was pointed out to me this noon that you can weather away soil if it isn't properly protected. You can take six or ten or fifty feet of soil away and loose dirt without much trouble but to weather away solid rock and bring it down to that low a level, takes a tremendous length of time.

After that took place, the area had successive petrified forests, seventeen of them. Let's concern ourselves with that for a moment. I don't claim that you can make several hundred thousand or million years for that, but I think you have to allow a little more time than Prof. Monsma wanted to admit.

In the first place, many of these forests have very considerable trees, trees in which the rings can essentially be counted because they are petrified so well, and it has been demonstrated that the rings are annular and they can be counted, and therefore you can demonstrate that some of these forests are up to one thousand years old, but suppose that a volcanic ash wipes it out in three days which it could have done and probably did; have you considered how long it takes to make fertile soil out of that volcanic soil? Before the next set of trees can even start to grow in an area -- assume that it was a wide area, and then the second one starts and so on for seventeen times, and after that is all over with, you still have some other layers laid down, and then when you finish laying all those down, the top of this formation had to be the level of ocean or an inland sea, and once all that was done, this whole area had to be lifted possibly a mile or more in the air in the Grand Canyon region, and then in that lifting process, the river had to cut down through. This is qualitative, but I think that is an impressive example.

My last example brings us to Grand Rapids, Michigan, because it is always interesting to know a little bit about what we are standing on. We are standing on something like fifteen miles of sedimentary rock. In this board illustration, I can only sketch in a very rough way the sedimentary sequence uunder Michigan, the famous Michigan basin, those rocks go down to tremendous depths. Again, each of these layers is of a different type laid down under different conditions. Very deep oil wells across the state have penetrated these.

What is the significance of the sequence here? Is it not the fact that all of these have been laid down on a very old basin which had to be provided before you even started? All of these have been laid down in a depositional trough, and that trough was constantly sinking.

The important thing is that there is a very thick band of salt which provides great wealth for the state here.

Let's consider how those bands of salt were formed. They are salt, anhydrite and gypsum, and if you were to take ocean water of about thirty miles in depth which, of course, is about six times deeper than most places in the ocean, and evaporate it, you will get something on the order of one hundred feet of gypsum. This, of course, can be computed accurately. It is on this order of magnitude because sea water is so diluted. Of course, you don't have thirty miles of sea water over Michigan, but what you must have is a lagoon type of surface under arid conditions where water is washed over the inland sea or lagoon and where you have constant evaporation and concentrating in the trough.

Over in Africa and in Mexico, the oil geologists have driven over this type of sequence. You have an arid condition and an inland sea with constant washing into a sinking trough, because some of these go up to about three hundred feet in thickness in West Texas, and if you take all of the salt, you would come to probably over a half a mile thick.

Well now, to me, that sort of thing means that the Creator either put all this in together to deceive us, which obviously isn't true to my way of thinking, or the rules have been the same and great antiquity is required for this in every one of these strata from top to bottom, and in all of them you can find fossils or fossil life of one type or another.

DR. VOSKUYL: Thank you very much. We didn't promise you this extra lecture, but we appreciate it. Are there any further questions, any more discussions?

DR. BENDER: I appreciate what we have had, and I appreciate very much the approach that Dr. Kulp has given us in the study of geology. I think, however, that the arguments that Dr. Monsma has presented in the latter part of his paper essentially still stand, and that is that we must give recognition to the viewpoints that he was mentioning. We must give recognition to the possibility of some cataclysms of various kinds, not that we will blindly close our eyes to the obvious physical facts, certainly not. That is part of our condition. I am not a geologist, but I have been interested in geology for some time and there is a number of problems that I would like to have answers for. One of them is simply the large extent of fossils and fossil content in these strata and stratified layers.

So far as one can determine now, it is difficult to find anything on the surface of the earth that would be fossilizing, plants or animals, at the present time, and so one will need to postulate some condition other than what is existing at the present time in order to have produced fossils at all, at least to the large extent in which they are found; also the very large extent of the stratified layers must have some explanation other than present conditions because nowhere do we find stratification occurring over any extended area such as these things would require. Those are merely problems. I don't know what the answers are. I hope that somebody will find the answers to them, but it seems to me that we will have to recognize them as problems and also recognize that Dr. Monsma has presented the fact that the Bible does postulate some cataclysms, and I don't see why we can't piece these things together. I think that is the tremendous task before us now.

May I say one more thing. I think that Dr. Monsma has done a very excellent job in pointing out to us that we need to be positive in our actual approach to the whole problem, and I think it certainly connects with our purpose, as will probably be discussed tomorrow, that our task as an Affiliation is not so much to add to the existing argument against evolution as it is to present a consistent treatment of the whole problem that is positive and that is Biblical.

DR. MAC RAE: I wonder if Dr. Kulp would say a word on the matter of fossils. What does make fossils and is it true that they are not being made now? What is that situation? I don't know anything about it whatsoever.

DR. KULP: I hesitate to disagree so flatly with Dr. Bender but I have seen large areas of fossils and fossil formation at the present time and I can point to half a dozen areas in the earth's crust where there are tremendous crusts undergoing stratification at the present time. One example that is very close is the present Gulf of Mexico.

They are drilling for oil in the Gulf of Mexico at the present time, and that is a very vast area taking in more than all of New England and New York put together. The area is a depositional trough at the present time as big as some of these great trough-like areas such as the Michigan basin in times past. The basin of Southern California is another one. There is a great one in the Caspian Sea and another one in the Black Sea.

Regarding the fossils, you can go down on the Florida coast and if you can get out and get with some of those well drilling companies where they are drilling through recent lime muds, you can see everything, and you can see limestone while it is still soft. You can see all sorts of fragments of fossils. Maybe you can get Miss Erdmann to tell you more about that. I was also thinking of an example of Jones Beach in New York. You can go down to the water and come back about thirty feet from the water and there is just a little rise of about six feet of sand and you can go there and look at that bank, and it is the most perfect example of a fossiliferous sand I can think of that is not stone yet. You can pick out every little shell, and every one of them is perfectly in place in a layer, and if that region were to subside or the ocean were to move out and a stream were to come down over there and deposit more material, in time that would go over into sandstone, but to me that is the best textbook example that I can think of.

I think that the geologists who have been out in the field and who have followed a lot of these oil drilling companies will recognize many places where the same type of rock and the same form of fossilization is going on today.

Now as far as the catastrophies are concerned, that is a very good point and something that we should look for. However, regarding the flood, insofar as geology is concerned, one would not expect much of a record of the flood of Noah, even if it had covered, as apparently it did, the entire earth. The reason for that is that if you had certain displacement which would allow a great tidal wave to go over the entire earth, the amount of sediment would be negligible. That doesn't come from tidal waves. It comes from the gradual erosion of mountain slopes and uplands, and therefore, that certainly may have covered the entire face of the earth. A thousand years later, subsequent erosion may have removed all traces of such an event. However, there should have been those major displacements, and the geo-physicist particularly with his seismological equipment should pick up something like that, and I think that is where we ought to look, but not to look for these shells on the top of the peak.

DR. MONSMA: May I just say this? I know I am sticking my neck out in starting the geology here, but I am very glad that Dr. Kulp has given us his viewpoints on that. I think that we should have it. The statement that I want to call attention to is the fact that my emphasis was this, that there is a difference between Creation and Providence. I believe that upon the basis of Scripture we must adhere to that, and I am afraid that at the present time geologists are not doing that. My solution is perhaps not the correct one; the solution at least, that I answered Miss Erdmann is perhaps not the correct one but I would like to consider this: If things do occur the same way now that they always have, is God operating now the same way as he did in the past? I think we should seriously consider that, and that is the point that I wanted to make.

DR. GATHERCOAL: I would like to say that Dr. Monsma's presentation is the most sound and fundamental one that I have ever seen presented in this organization, because it is based upon the word of God. Now we all recognize innumerable difficulties this vast amount of study involves, but I really believe that if it is undertaken that the Lord himself will help us most wonderfully in it. Now, in the first place, there are about fifteen thousand statements in the Word bearing upon this subject, and I really believe that if those are carefully looked up and analyzed, put into tables under the different headings and studied by scientists who know how to interpret a good many of them, that you will have a foundation that will not be overturned. It is a great amount of work but we are laying aside for this work perhaps a period of five or six years. If the work is started at once, and if it is divided up and assigned and is really undertaken in the spirit in which it has been offered here this afternoon, I think it will be an absolute success.

Now I am not going to offer to list by positions those fifteen thousand references, but I am going to make a start on them.

DR. MAC RAE: How many of them bearing on evolution or geology did you say there was?

DR. GATHERCOAL: The subject was on evolution. Geology doesn't hurt us any. That is all right, and as far as that period of two billion years is concerned, we don't need to worry about that. That is all provided for in the Word.

DR. MAXWELL: May I ask a question in regard to uranium? They seem to be using the disintegration of these materials as a clock more or less, and I wanted to ask, is there any uranium in the world that has not disintegrated? And if so, where does it come from and who started it?

When does the clock start, in other words? I think that ought to be answered. Then I have this further remark in regard to Dr. Gathercoal's remarks. I think that is in line and along the point that Dr. Monsma is suggesting, that we can work along on that as a positive line, and that is what we have all been wanting.

Just this last year or so I have been working on that line Dr. Gathercoal mentioned, and I find a tremendous lot of material. I think he is very conservative in saying fifteen thousand statements in the Bible, and therefore to my mind it is fundamental science, basic science, the truth, and from that we can really produce something which will stand the test of a true science in time, and being out of the study-method, but there are other methods, and whatever method we use we will get the same truth. So then let's start using our Bibles and our knowledge of science and see if we can't build up something constructive and positive.

DR. VOSKUYL: I will stand corrected by anyone that wants to correct me, but as far as I know there has been no such thing as pure uranium found. The isotopes are all found in a definite specific ratio and that indicates that at one time there perhaps was pure uranium, but since that time there has been that constant disintegration going on.

DR. KULP: I think part of the problem has been half-life. It doesn't all go off at once.

DR. COWPERTHWAITE: I wonder if that really answers the question which has been raised by Dr. Maxwell. As I understand Dr. Maxwell's question, it is that if we find some uranium in a pegmatite, how do we know that when the pegmatite was formed, the uranium was pure and not already partially decomposed. We get that from our theories of solution that the pegmatite is a liquid and the different minerals and crystals formed from the liquid state according to their individual solubility. Uranium mineral would have a different solubility from the lead mineral. Therefore, when the liquid solidified, the uranium would crystalize out free from disintegration products so that at the time of solidification, you would have pure uranium mineral which could start this disintegration from that time on.

DR. BARNES: I wanted to ask one or two questions purely for information. We have all been urging a positive approach to this problem. It seems to me that we have been presented with a positive scientific evaluation of some of the existing phenomena and if we accept that as positive it is necessary to provide for a great deal of time.

Now, I would like to ask this question: could the proposed or supposed lapse in time usually placed between the first and second verses of Genesis be considered useful for this purpose? I suppose that perhaps Prof. MacRae could straighten me out on that or perhaps Dr. Kulp.

DR. KULP: I would refer you to that article that you have had in your possession for some time by Prof. Ramm who very beautifully treats that subject. I think it shows that particularly from a Biblical point of view there isn't too much to commend the view, but from the geological point of view there is even less.

DR. MAC RAE: Will you explain that last statement?

DR. KULP: I think that if we are going to have a great catastrophe such as Satan and his hosts, you are going to do awful things to the neatly placed geological strata and there isn't any evidence of such a thing in the geological record.

DR. MIXTER: Are these strata in Dr. Monsma's suggestion part of God's providence or what or where are providence and creation in relation to that?

DR. MONSMA: I am not ready to answer that. That is one of the questions that I think we should face. If that is entirely determined upon the basis of present-day changes, I would say we must be careful about how we go into that. You use the same yardstick that we are using for what happens today, and I should just like to make that at the present time a warning lest we are cut loose from our moorings. I think we have to be very careful there to make certain presuppositions which perhaps we do not have to take. I know that there is a great deal of possibility in that creative period. The earth was waste and void. In the beginning God created the heavens and the earth. What does that mean, a separation of water and land? That must have been some tremendous things too, but can we on the basis of present-day phenomena determine just what it was and when it was? That is my question.

DR. MIXTER: How much of Genesis One is Creation and how much is Providence?

DR. MONSMA: The way I indicated it in my paper, you can't distinguish the whole thing, but the way I would distinguish it is when God said that the heavens and the earth were finished, we have the end of the creative period. and from then on, we can't speak of Creation any more. At the end of the sixth day it was completed, and then we get Providence. I believe that is the Biblical view of it, and I think we must realize that the evolutionist's view of it, of course, doesn't distinguish definitely. Creation is going on for him at the present time and there is no difference at all, and that is the great difference between us and the evolutionist, I think.

DR. MAC RAE: Are you inclined strongly towards a twenty-four hour day rather than a longer period of time?

DR. MONSMA: I don't believe that is the basis upon which we conclude long periods of time. I am not committed to that basis yet, and so temporarily I think possibly it was at least very short periods of time.

DR. MIXTER: The final third of my question here concerns the man that was talked about in the Bible on the sixth day, and here are the first plants and simple animals which are talked about on the first day, so it looks as though that is all part of the creative activity. Geologically speaking, that took time, didn't it?

DR. MONSMA: If you count upon the basis of present-day changes, then I think you are applying that particular yardstick again. I don't think--of course, we find plants down below there. We find them all the way along. We find other fossils too. I don't know that I get the point of your question exactly.

DR. BENDER: I would like to talk some more. Again I am not a geologist, and I know that I stuck my neck out once before here, but here is the supposition implied by what Dr. Mixter has put on the blackboard there, which I think is entirely justified. Are we correct in stating that man appears only at the top?

DR. MONSMA: May I say to him that if that is what Dr. Mixter has in mind, we find plants here (lower) and man in this layer (upper). These are fossils, dead men, dead animals, but here is the time at the end of the Creation let us say, and then of course, the picture would be that all of the death and destruction had taken place during that time of creation, during that six days. I can't quite see that that is compatible with Scripture. That is one objection that I have. We have to have further explanation.

DR. MAC RAE: I don't want to ask a question, but I would like to say a word here. It impresses me as I look at this evolutionary matter that in the field of biology and these other fields there are some problems, but nothing that is extremely difficult. As far as our attitude and relations are concerned, the thing has been thought through and worked through a great deal, and I think it is pretty much what you choose to select as far as our scientific knowlege is concerned.

There is no strong evidence that is very difficult for us to assimilate into our Biblical interpretation as we understand it today.

In the field of geology, as Dr. Kulp pointed out, for the last fifty years the field has been entirely taken over by the unbeliever. We have had very few, practically no Christians working on this material, studying it and trying to interpret it in relation to Scriptures for the last fifty years. The result is that a great volume of conclusions has been worked out which hasn't been studied. Now, Dr. Kulp is one man, and I think a very bright man, and he has been working in that field for a comparatively short period of time as yet, and naturally we are all tremendously interested in what he has found, but I think we will have to reserve judgment on what appears even quite obvious until he has worked on it and until many more have worked on it.

I think there are very serious problems in that matter of geology which will take a good deal of time for a good many of us, in order to decide what the interpretation is that is most compatible with the statements as found in the Scriptures, and so my feeling as far as the geology is concerned, it is wise for us to reserve judgment on a great many things.

There are certain points that appear to Dr. Kulp to be practical and answerable at present. I don't think that any one of us should take his word for it. We need to go further in that ourselves, and he also needs to work on it. I think if we find that the evidence is just about absolutely definite on these points, I think there is a way of interpreting the Scriptural evidence in such a way that these matters can be worked out. I can think of five or six hypotheses in the Scripture, on the basis of Scriptural questions where there is no difficulty in the matter of the earth's being two billion years old, but I don't want to adopt any one of these five or six hypotheses and stand on that; I simply don't know, and I would rather not adopt any one of them unless it is necessary, but I don't see any difficulty from Scriptural standpoint in seizing one of the five or six on the matter of the age of the earth.

The age of man is more difficult, but not insuperable.

There is one point on which we should be cautious, and that concerns the statement that Dr. Kulp made, "if the rules were not changed; if the rules remained the same." I think it is entirely possible and quite probable that the Lord has kept the rules the same for a very long period but I think what we know of the rules is a very small proportion of what there is to know. What we have been told is a very small part. In any field of past history, social or human or physical, there is always the possibility of other forces or factors that we don't know about at all, which could make tremendous changes over a large period of time and then be quiescent for a period.

I feel that when we get into the dogmatic statements, particularly about the time before writng came into existence, I think there is a tremendous possibility of other forces that we can't think of but which we can discover at some time. A man of a few years ago would say it was utterly impossible for an airplane to fly without a propeller. Today jet planes are common; there is a new factor that we didn't think of then, and we can find that over and over in all phases of science, and when we come to past history, I think we must reserve a large amount of the decision on the question, that there may be some other factor that we don't know about. Even of this matter of the decomposition of uranium, it would seem on the basis of present knowledge that the rate might not have been different, and yet I think it is far from possible to say that there might not be some other factor that was operating over a large part of the earth's surface at one time, perhaps before the flood and removed at the present time, and perhaps some other thing that might be discovered at some future time.

I feel that when we get into the past history, we should be very very cautious. You will find any book in our archeology of thirty years ago telling about the great rulers of 5500 B.C. and that 4241 B.C. is the earliest fixed date in history of the Egyptians. Today no archeologist believes that there is any date that can be given previous to 3000 B.C. because it is now regularly understood or recognized that all those dates they put of five or six thousand B.C., or all higher than 3000 B.C., are incorrect because there was no writing. The whole attitude has changed. As far as any probable date in archeology is concerned today, Ussher's Chronology could be correct.

Now ten or twenty years ago people on the basis or archeology would just laugh You can't do that today, so I would like to recommend extreme caution on past history on chronology, and I think we should have a more open mind on these matters of geology than we have as yet and I hope that there will be others like Dr Kulp who will go into the archeological work quite thoroughly in the next few years.

DR. VOSKUYL: Thank you. Well, that seems to conclude the speeches for the afternoon. We wish to express our appreciation for the kind attention which our visitors have given us. We are glad that you could be with us.

---

# BIOLOGY AND CHRISTIAN FUNDAMENTALS *

R. L. Mixter, Ph.D.
Professor of Zoology
Wheaton College, Wheaton, Ill.

Scientists are not forced by their science into any particular religion. Goldschmidt[1] has a man from Mars descend to inspect human beings and decide what significance there is to the differences in human structure. If this Martian should look at the religious beliefs of scientists he would find them as varied as in one of the graduate schools I attended. The head of the department was a Unitarian. The nature study teacher embraced the Catholic faith and the specialist on conservation appeared to have no religious beliefs at all. Across the campus a professor of engineering was a fundamentalist. Each of these men was acquainted with the same scientific method of induction and had the college library at his disposal. Science neither makes one religious nor keeps him from religion.

Therefore, one may examine the controversies which have occurred between science and religion unhindered by a "weight of opinion" held by scientists. The scientific method is the Scriptural method. "Prove all things; hold fast to that which is good."[2] The scientific method is that of observing the data, experimenting with phenomena, correlating the findings and drawing conclusions from them.

In spite of their use of the same method scientists may draw different conclusions. In the department of anatomy of a university in the city of Chicago, a famous anatomist and his pupil believed that the lining of the lung is a very thin membrane formed by cells touching one another, derived from the inner germ layer. In the same building were other anatomists who held that the blood capillaries were only partially covered by cells of the middle germ layer. Both of these groups of men had the same technique at their disposal but held to contrasting beliefs.

Outside their fields of research, scientists have opinions on many matters. Consider the variants in political ideas among university teachers. Not all of them adopt the same forms of recreation. In worship they bow before the God of the Christian or reverence eternal forces and matter. As scientists they try to leave the spiritual forces out of reckoning and yet in drawing their conclusions they speculate at length. A recent book on processes in evolution illustrates this. Simpson[3] holds that only physical hypotheses should be used in scientific work but is willing to construct these guesses on data which he admits are inadequate. He does not rule out the spiritual in fields where the physical has so far failed to explain natural phenomena. Another famous biologist believes that "we are the intellectual and spiritual children of .... that most influential character in all history, Jesus Christ."[4] He does not believe that the world was formed by chance. He cannot believe that the world and man are the products of pure accidents.

Because scientists have such differing beliefs in their own specialties and in every sphere of life we should not hesitate in our study of problems to follow the advice of one of the greatest of modern scientists whose brogue has him put a profound truth in this expression, "Vat iss de evidence?"

1. Goldschmidt, <u>Material Basis of Evolution</u>, 1940. p. 121
2. I Thessalonians 5:21
3. Simpson, <u>Tempo and Mode in Evolution</u>, 1944, p. 76
4. Conklin, <u>Man, Real and Ideal</u>, 1943, pp. 181, 190

In science a theory is satisfactory if it is consistent within itself and explains a group of phenomena. To me one of the chief Christian evidences is the ability of Christianity to meet the needs of man. If he needs comfort there is the 23rd Psalm. He is challenged by Hebrews 11. Some of his curiosity is satisfied by God's answers to Job. He receives conviction from the apostle Paul and derives his character from Christ Himself. The Christian Scriptures consistently meet the needs in all phases of man's personality.

Consider some of the facts of Biology which are in harmony with great doctrines of Scripture. First, organisms are machines. Carlson and Johnson call their textbook of physiology THE MACHINERY OF THE BODY. They say, "the working hypothesis of the biologist is that eventually the phenomenon of life will be explained in terms of physics and chemistry."[5] Whether their ultimate goal will be realized is questionable, but many features of body activity are so explained.

In 1801 Paley argued that if one observes mechanical features in the body a machinist is needed to explain them. Where there is a contrivance like a watch there must have been a contriver. The more we find the body to be a machine, the more we are indicating there must have been a machinist greater than the body in order for the body to be produced. One of the best sellers in 1947 shows us that in order to get by chance combination one of the simple proteins of which the body is composed it would be necessary to have 1 followed by 243 zeros years. This is the beginning.[6] Not only must the proteins be formed but in addition there must be enzymes to control them and cellular organization to give them stability. It may safely be stated that by the activities of pure chance there is no chance that life could have been spontaneously generated. Morrison has similarly argued in his book, MAN DOES NOT STAND ALONE, which was reviewed in the Reader's Digest of December 1946 under the title, "Seven Reasons Why a Scientist Believes in God."

We see then that the facts of biology are consistent with our first fundamental: there is a purposeful, intelligent Creator.

A second important fact in Biology is that organisms vary. Creatures are not confined to fixed types. The earliest human family has produced "all races that dwell under Heaven." These races are of one blood. The blood groups so important in blood transfusions are found in all races. If the phrase "of one blood" refers to germinal potencies it is very apt for any race may breed with any other and produce fertile offspring.

Paleontology also indicates that human beings have always been human beings. Franz Weidenreich stated that all fossil men belong to the genus Homo.[7] Present day human beings are not all alike nor were all fossil men like present day human beings in all respects, but all are human in the essential characteristics of structure and behavior.

A similar situation exists in animals. Within a group of animals such as the family of horses there is indication that there has been some change from one form to another. It appears reasonable that the first horse may have had four toes on a front foot; a modern horse has only one. The first horse appeared to have teeth adapted to a browsing diet; today's horses are definitely of the grazing type.

5. Carlson and Johnson, The Machinery of the Body, 1941, p. 4
6. du Nouy, Human Destiny, Chap. 3.
7. Science, Vol. 104, No. 2709, 1946, p. 516

But all this change is within a limited group of animals. There is no fossil to close the gap between supposed ancestors of horses and the first horses. It is also true that there are "systematic deficiencies of record" between all comparable groups of animals and plants.[8]

The science of heredity adds the same type of conclusion. It is possible to make crosses within a species as in human crosses; between species as the mating of horse and ass; between genera like the crossing of the cabbage and radish and even between members of different families and orders as seen in the cross of killifish by the mackerel. But not one has ever seen the result of an attempted cross between such widely separated animals as the members of different classes such as reptiles and birds. As one of my university professors put it, "the answer to the origin of different classes of animals is outside of the field where genetics can make a contribution." We see, therefore, that although there has been some variation within the limits of the minor groupings of animals there has been no change demonstrated from one major group to another. This is in harmony with the Scriptural idea that God created the great sea monsters, the winged birds and the swarms of living creatures of the water after their kind and created man in His own image.

There is a third phenomenon of Biology which reflects on a miracle. It is called parthenogenesis, which means the production of an animal from only one parent. Gregory Pincus receives credit for much of the research in this field. His work was popularized a few years ago in Collier's by the title of an article on rabbit production, "No Father to Guide Them." Pincus operated upon female rabbits, by cooling stimulated the tubes down which the eggs were descending and let the animals come to term. A rabbit was born without a male parent.

One hesitates to compare this with the Virgin Birth which is the greatest of the miracles and yet the possibility of its occurrence may not seem so remote because of the demonstration that even mammals may have offspring without fathers.

Even miracles need not be inconceivable. A speaker in our chapel remarked "if the Bible said a tadpole swallowed Jonah, I would believe it." This appears to me to be asking one to believe something inconceivable, to accept something that cannot be imagined. By contrast a miracle is not an impossibility; it is the result of an Unusual Cause who does not distort nature to produce His results. Animal parthenogenesis suggests that the miracle of the Virgin Birth is within the realm of the possible. Thus a third great fundamental of our faith can no longer be questioned by biological laws. God has stepped into men's shoes in the person of Christ to reveal Himself to us and bring us back to Himself.

A final observation from Biology is this: all complex organisms die. The material of which they are composed disintegrates but we believe that the body of man encompasses a mind. The body came from an extremely minute speck of living material formed by a union of cells from each parent. Through this tiny bridge the hereditary characters of the parents are passed on to the offspring. Morrison said a thimble would hold all of the hereditary determiners necessary to produce all humanity now living. If the physical contribution to man's person demands so little material, it is possible that the entrance of mind into man requires no physical bridge. This was Paley's reasoning in his famous work on Natural Theology.

If so little substance is used in the early formation of a person, may the spirit not leave the body without the aid of any material? The body dies and decays,

8. Simpson, op. cit., p. 107

we describe the changes as we can observe them. What happens to the spirit we cannot see. Vernon Kellog remarked that we describe the before change conditions but the after change conditions may be the most important. To some biologists the only immortality is the passing on of our hereditary characteristics to succeeding generations, plus any contributions we have made to society.[9] But if so small a speck of matter is necessary to maintain biological continuity, the Creator can provide spiritual continuity without any matter. Thus we conclude

"Dust thou art, to dust returnest
Was not spoken of the soul."

Students of nature should remain confident that the facts of their research will fit into the pattern the Bible outlined for the formation and preservation of living beings.

9. Conklin, op. cit., 1943, p. 179

# DELUGE GEOLOGY

J. Laurence Kulp, Ph.D., Columbia University
New York, N.Y.

## INTRODUCTION

### General Statement

The theory of deluge geology is summarized by H. W. Clark, The New Diluvialism, Science Publications, 1946, p. 205 as follows, "The pristine earth was destroyed by one great overwhelming catastrophe, the Flood, the Deluge, of Genesis 6, 7, and 8 and that this Flood was the direct or indirect cause of most of the major geological features of the earth." This statement is in complete disagreement with the conclusions of trained geologists the world over who believe various strata have been deposited over tremendous periods of time according to known physical and chemical laws. Since geology as a science is concerned with the history of the physical and biological earth, it deals with a subject important to a Christian view of the world. Although a large proportion of the geologists of the 19th century were at least nominal Christians and suggested means of harmonizing the science of geology with the Scriptures, the field was soon lost to the rising tide of secularism. In the early 1900's George McCready Price wrote the first of a stream of books and papers on "Flood Geology". Because this view gave apparent support to certain popular interpretations of Genesis, Price's writings received wide distribution in evangelical circles. Numerous would-be apologists who wrote for the Christian public such as Rimmer, Nelson, O'Toole, Hamilton and others assumed Price's Flood Geology to be sound science and a weapon against evolution. This theory has grown and infiltrated the greater portion of fundamental Christianity in America primarily due to the absence of trained Christian geologists.

It is the purpose of this paper to present selected fragments of the geological evidence in terms of the physical and chemical laws which apply to geological processes and materials. The greatest difficulty in accomplishing such a purpose is the adequate visualization of geologic features. To understand geology rightly one must visit the fields, the mountains and valleys, and the seasides armed with a reasonable preparation in physics and chemistry.

### Literature

The most complete account of flood geology by G. M. Price is to be found in his "The New Geology" 1923. (For fifty years the thesis has hardly been modified as in one of his latest, "Common-sense Geology" 1946.) In 1941 the Bulletin of Deluge Geology was started, edited by G. M. Price and C. B. Courville for further discussion and propagation of the thesis. This publication has apparently been replaced in 1946 by "The Forum" of which Volumes I and II comprising 130 pages over 1946-1948 have appeared.

Recently another point of view in flood geology has appeared in a book by Clark referred to previously. Clark differs considerably with Price in that he admits many of the accepted geological facts which Price denies. These admissions make flood geology even less tenable.

### Training of the Exponents of Flood Geology

G. M. Price according to his brief autobiography in Vol. I. No. 4 of the Bull. of Flood Geology had two years at Battle Creek College in classics and theology.

Later he took a teachers training course at the Normal School in New Brunswick which included some elementary study in natural science. This ended his formal education.

The primary proponents of flood geology have been of 7th Day Adventist persuasion and have assumed from their theological tenets that a 6-10,000 year span is required for all life. Price writes in the above Bulletin, "From childhood I had always believed in the Flood and ... I had become convinced of what that book (Patriarchs and Prophets, Mrs. E. G. White 1890) teaches about the Flood as the cause of the chief geological changes." It is evident that he did not approach the subject without a preconceived pattern to which the data must conform.

The authors of the articles in the flood geology bulletins include two M.D.'s, one Ph.D. chemist who makes a number of incorrect statements on chemical principles, and the others who list no degrees. The other prominent writer on Flood geology is H. W. Clark who has a master's degree in biology. None of these men have done any geologic field work such as mapping or studying paleontological or structural sequences.

## Basic Errors

There are four basic errors in the flood geology writings which appear persistently:

(1) There is confusion that geology and evolution are synonomous. The evolutionist does claim support from geology, and geology generally predicates at least physical evolution but geology itself is the independent science of the rocks of the earth's crust, their properties, origin, and history. The basic principles which geologists have discovered about the earth's crust can be deveolped without recourse to fossils.

(2) It is assumed that life has been on the earth only a few thousand years, therefore the flood must account for geological strata. A good example of this is the constant reference to the present as a "zoologically impoverished world." There are certainly many more animal types represented in the whole fossil record than are living. The present world is only "zoologically impoverished" if all these fossils lived at one time and were all destroyed by the flood. However if these lived in various times over 500 million years as seems to be the case, then the biological statistics of this era are little different from others. It is a case of assuming in a major premise that which is to be proved.

(3) The physical and chemical conditions under which rocks are formed and, later, under stress, folded are not understood. This accounts in part for the illusionary confidence of the sincere flood geologists.

(4) The historical perspective accounts for the continuation of basic error. Price first wrote before radio activity was applied to age measurement, before oil wells had penetrated over three miles of sedimentary strata and perforated certain parts of the sedimentary column like a sieve, before Hedberg's classic work on the conditions under which sedimentary rock will be formed, before Grigg's experiments on the plastic flow of rock under sufficient pressure and temperature, before M. King Hubbert's calculations defining the state of rocks while undergoing folding, and before extensive sampling and age measurement of the finely laminated, continuous, but very ancient deep sea sediments.

## THE SCIENCE OF GEOLOGY

### Uniformitarianism

Like other fields of physical science, geology assumes that the world about us operates according to law and order. Much has been written by Christians against "uniformitarianism" while actually we must believe in it. We believe God created the universe according to a blueprint and that it operates according to predetermined laws. While it is true that God may supersede any of these laws at any time, it is certainly not the normal course of events, nor is it perpetuated for any great length of time. Therefore in geology the present operation of erosion, transportation, sedimentation and compaction is the key to these actions in the past. The geologist does not believe everything that has happened to the earth's crust in the past can be identically duplicated today. He does believe that the present is the key to the past and that the same physical chemical laws have applied to all phenomena of geology. This is accepted generally by Christians for other sciences and should be extended in their thinking to the science of geology. Another viewpoint is represented by Prof. Monsma of Calvin College who distinguishes "creation" from "providence". He thinks we can accept the extrapolation into the past of most physical quantities as measured at present except time. Thus he would infer that phenomena such as radioactivity operated by different rules in the "creation" and have the present rules due to "providence". This view is considered unreasonable because it would require God to have changed hundreds or thousands of physical laws or situations all in different ways by different amounts to give a consistent appearance of great antiquity but actually all were created a short period of time ago. Thus radioactivity, the accumulation of the salt in the sea, the heat radiation from the earth, or the speed of light and the distance of stars, all indicate millions of years of earth history yet are all independent, depending for their observation on entirely different laws and events. So it will be assumed the rules haven't changed and that they can be applied.

### Physical Chemistry of Rocks

#### Rock Types

There are three major rock types, igneous, sedimentary, and metamorphic. The igneous rocks which are those formed by a cooling silicate melt, intrude and cut the other two types of rock. The most common igneous rocks are granite and basalt. Sedimentary rocks are made up of particles derived either from the erosion of previously existing rocks or by chemical or biological precipitation. The major types of sedimentary rocks are shale (mudstone), sandstone, conglomerate, and limestone. There are many minor varieties and combinations of these four. From shale through sandstone to conglomerate the difference is primarily in particle size. Either igneous or sedimentary rocks may, under sufficient temperature and pressure become changed into metamorphic rocks. If this change is carried far enough, fossils are destroyed and still further many of the characteristic sedimentary features are obscured.

#### Formation of Sedimentary rocks

The first stage in the formation of a sedimentary rock is either erosion by chemical or mechanical means of previously existing rock, or in the case of chemical sediments, of precipitation by some chemical agency. First consider the detrital rocks-shale, sandstone and conglomerate. Whether the material is derived from chemical decomposition or mechanical abrasion, the process is extremely slow. Consider granite tombstones. How much has been chemically decomposed in historic time?

After material is broken up by mechanical or chemical means it must be transported by wind, ice or water. It has been determined that for sand and gravel the size of the particle that a stream can carry varies with the sixth power of the velocity of the stream. Thus the larger particles are quickly segregated on the basis of speed of the carrying medium. Fine particles of colloidal dimensions are carried in suspension by sluggish rivers and are precipitated only upon entry to the ocean water rich in electrolyte. Far out from shore, over the continental shelves there is very little sediment. Over these areas rates of sedimentation are so slow that a few feet may represent several hundred thousand years of continuous deposition. (See papers by W. D. Urry in Bull. of the Geol. Society of Am. and Am. Journ. of Science). The sedimentary rocks of the earth's crust were not produced by such deep water sedimentation but rather along continental shelves and in great inland seas such as the present Gulf of Mexico and Hudson's Bay. Offshore, where detrital material is lacking, limestone may form if the sea is sufficiently shallow and warm. Some limestones consist largely of shells of organisms but the bulk of the limestone in the sedimentary strata is in large part chemical precipitate. It must be emphasized, contrary to statements by flood geologists, that practically every sedimentary rock type known can be shown in the process of formation today with or without fossils.

After obtaining the detrital material and transferring it to a place of deposition such as a river delta, how does this loose mud become rock? Hedberg's great work in Venezuela elucidated this process under typical conditions. (Am. Assoc. Pet. Geol. Bull. 10, p. 1035 and later Am. Jour. Science (1936) p. 241.) He was able to study the gradual compaction of river mud into hard rock shale with oil well cores. The area he chose is well known geologically from the intensive drilling. No essential change in sedimentation type is observed for two miles vertically in this area. Hedberg demonstrated that loose sediment must have at least a mile of additional sediment piled on top of it to lithify it. A mile of water will not suffice because the rock forming process is one of squeezing out water. Additional work by other scientists has confirmed these findings and further shown that essentially similar thicknesses must be applied to limestone and sandstone. Here then is a quantitative measure of the rock forming process. When we see a sedimentary rock on the surface, it must of necessity have been at least a mile deep in the earth's crust at some time in its history. Therefore if we observe a great thickness of sedimentary rock such as along the Rocky Mt. Front which has been folded and eroded to yield a much younger conglomerate, it is known that rocks were already lithified before this erosional process for the second cycle and had been covered by at least a mile of additional material which has been eroded off. The time dimension is large for such a process.

In Price's writings, the statement is frequently made that the fossiliferous strata are relatively thin. For example in Vol. I No. 4 of the Deluge Bulletin, "Everyone knows that the fossiliferous strata are seldom more than a mile or so in total thickness - often only a few hundred feet and below them is granite or primitive rocks which contain no fossils." The "everyone" in this statement must not include geologists for it is well known that in the various geosynclines and basins of sedimentation there can be found up to 20,000 to 30,000 feet for each geologic period. (For the relations of the various geological periods to each other as they are mentioned in the discussion, see Table 1.) Strata in the northern coast of

Table 1
Geological Time Table
(as generally accepted)

| Era | Period | Epoch | Date |
|---|---|---|---|
| Cenozoic | Quaternary | Recent | |
| | | Pleistocene | 1,000,000 |
| | Tertiary | Pliocene | |
| | | Miocene | |
| | | Oligocene | |
| | | Eocene | |
| | | Paleocene | |
| Mesozoic | Cretaceous | | 60,000,000 |
| | Jurassic | | |
| | Triassic | | |
| Paleozoic | Permian | | 200,000,000 |
| | Pennsylvanian | | |
| | Mississippian | | |
| | Devonian | | 300,000,000 |
| | Silurian | | |
| | Ordovician | | |
| | Cambrian | | 500,000,000 |
| Cryptozoic (Pre-Cambrian) | Keweenawan | | |
| | Huronian | | |
| | Timiskamian | | 1,000,000,000 |
| | Keewatin | | |

Gulf of Mexico have been demonstrated by oil wells and seismic methods. A simple calculation will show that were all of the soil and loose rock in the U. S. A. bulldozed into a basin the size of this Gulf Coast geosyncline, it would hardly be full. But this basin is only one of dozens throughout geologic time and this represents only the time from Cretaceous to the present. Table 2 shows the total thicknesses (maximum) of sedimentary rocks observed in the Frazer Geosynclinal belt, which trends from So. California to Alaska, for each geologic period. It should be obvious that no flood of short term duration could possibly account for these strata.

These great accumulations of sediments occur due to the filling of subsiding basins called geosynclines. One of these geosynclines is commonly truncated by another at a later time. In any particular period the sequence of sediments in the geosyncline is normally (1) conglomerate, (2) sandstone, (3) shale, (4) limestone.

The origin of coal, when the evidence is understood, also precludes flood geology from being correct. Coal is found in every period of geologic time but is most abundant in Central North America and England during the Carboniferous. That coal results from the accumulation, compaction, and alteration of plant debris has been demonstrated without question. There are two theories of origin, one that the material accumulated in situ where the vegetation grew and fell, the other that the material was transported. Most all deposits can be shown to fall into the first type - certainly all those of large areal extent. Since it takes a hundred feet of loose vegetable matter to form one foot of coal it can be readily calculated that if all the vegetation were scraped from ten times the areaof the midwest coal fields and compressed in this sedimentary basin it would only supply enough plant matter to make a few of the thirty significant coal horizons.

The evidence for the in situ origin of coal beds is voluminous but a few items will be presented to show the nature of the data. Further details can be obtained by examining the scientific literature on coal or referring to some standard work on the subject such as Coal, by E. S. Moore published by Wiley, 1940.

1. There are large accumulations of vegetal matter forming in swamps at the present time, some of which, as the Great Dismal Swamp of Virginia and North Carolina, or the many large swamps of Sumatra, are on a scale approaching those which gave rise to coal seams of considerable extent.

2. The purity of the coal, its freedom from mineral matter suggests the collection of the vegetation in swamps rather than in deposits where it has been transported with other sediments.

3. Numerous tree trunks with their roots firmly embedded in the underlying clays occur in the coal seams and in some cases the rootlets pierce fragments of buried wood in the clays.

4. Old soils on which the trees grew lie beneath the seams in some places.

5. The arrangements of various portions of plants with respect to one another is not, as a rule, that of transported material.

6. The lenses of cannel in bituminous coal indicate patches of open water in swamps where spores would collect in great quantities rather than deposits forming part of an ordinary sedimentary formation.

Table 2

Maximum thickness for various periods in the Frazer Geosynclinal Belt (So. California to Alaska)

| Period | | Maximum Thickness |
|---|---|---|
| Cambrian | | 12,000 feet |
| Ordovician | | 10,000 |
| Silurian | | 14,000 |
| Devonian | | 8,000 |
| Mississippian | | 15,000 |
| Pennsylvanian | | 5,000 |
| Permian | | 15,000 |
| Triassic | | 25,000 |
| Jurassic | | 28,000 |
| Cretaceous | | 13,000 |
| Tertiary | | 15,000 |
| | Total | 160,000 feet or about 30 miles |

Many minor features give the geologist important information on the origin of certain strata. Example (1) Gravels deposited by streams are generally clearly graded in texture and the pebbles are rounded in proportion to the distance from their source. Glacial gravels on the other hand are poorly sorted and commonly show parallel striations caused by being dragged over abrasive surfaces. These markings are never found on water carried pebbles. (2) The peculiar rare and heavy mineral aggregates may distinguish one source area from another for the detrital material used in making up the rock. (3) Glacial varved clay could occur only under the unique physiochemical conditions of low electrolyte glacial melt water which rushes into glacial lakes during the summer and is very quiet with an ice cover during the winter which permits slow sedimentation of the colloidal fraction. This has been demonstrated both in the laboratory and by sampling over a period of years of present day glacial lakes. There is no question any longer that glacial varves are annual. (4) Conglomerates are most interesting in that they contain pebbles of sedimentary rock which has already become lithified and which due to warping or folding has been exposed to erosional agencies. Conglomerates which rest on an angular unconformity generally show pebbles of all of the underlying tilted beds.

Fossils

Fossils occur sporadically in sedimentary rocks. Many thick formations contain no observable fossils while others are richly fossiliferous over large areas. This is due to the type of deposition and the nature of the life entombed. Along present shores the same variation may be observed. A recent three foot high wave cut bench cut in unconsolidated sand at Jones Beach, N. Y. gives a beautiful exposure of undisturbed shells in the sand awaiting further burial. The appearance is that of a fossiliferous sandstone. Off the Florida coast and around the Gulf, away from large deltas, fossiliferous lime muds are undergoing compaction into limestone. The end product will duplicate any fossiliferous limestone of the Ordovician system except for the fauna.

Much has been made of the geologist "reasoning in a circle". It is said that the geologists assume that evolution is true and thus dates the rocks primarily by the fossils that are in them. He simply makes the simpler fossils and the less complex, younger. This is a misrepresentation of the situation. While it is true that the geologist will tentatively label a rock as Cambrian because it contains certain distinctive fossils, he does not do it because he accepts evolution. He does it because hundreds of thousands of careful observations have shown that for large periods of time, or if you please, for thick sequences of rock, there has been a definite fossil sequence which is the same the world over. While gaps in the record occur locally, and these are expected of an unstable crust which with its buckling and doming allows sedimentation at one time and erosion at another, the sequence observed is always the same so long as undisturbed sedimentary strata are dealt with.

Orogeny

The geologist has excellent evidence that during the past the earth's crust was not static but constantly changing. Such changes are still going on but at a very slow rate. The Scandanavian peninsula for example, is emerging from the water at a measurable rate. Similar observations have been made on many parts of the earth's crust. In the rock strata these broad warps or narrower intense welts are shown by basins or troughs of deposition where the sediments are much thicker than in immediately adjacent areas. Invariably after a certain period of deposition either a broad warp or compression folding brings these rocks to the surface where erosional agencies can act on them. Subsequently they are again depressed and the new sediments truncate the old at some angle. Such is called an angular unconformity (or nonconformity) and since it commonly involves first the planing off by erosion of the tilted strata prior to deposition of the next cycle (which in each case recall must be superseded by at least a mile of sediment) a considerable time, far more than that available to historians, must be admitted.

The Flood geologists attribute such folding to soft muds and sands produced by the deluge. Physically this is impossible. M. King Hubbert, Director of Geophysical Research for the Shell Oil Company contributed an excellent theoretical investigation on the matter of scale models. He showed that in order for any laboratory attempt to reproduce earth conditions, it must keep all the dimensions of the model to scale. By using the known viscosities of solid rock at the temperatures and pressures of 5 to 10 miles in the earth's crust, it can be shown that the comparable scale model material for a laboratory compression experiment would have to be modeling clay or shoemakers wax. Experiments with these materials under controlled conditions have produced Appalachian type folding, and in general the type of deformation observed in rocks. If the sedimentary rocks were once unconsolidated debris at the same time and the entire muddy mass was subjected to compressional stress the result would be a chaotic mixture of material. The laboratory model to scale would be observed by squeezing a rubber flask which contained a layered mixture of alcohol and water. It would simply mix and spill out as a more or less homogeneous mass. Nothing of the type of thing observed in folded rocks could be obtained.

Other objections to the compressed mud suggestion is that the shear and tension fracture patterns can only be interpreted in terms of essentially solid rock. Further if the deformation has gone very far, high temperature and pressure minerals take the place of those of sediments. Pressure and temperature gradients are sufficiently well known for the earth to know that these minerals can only be produced at considerable depth on a regional scale. Also most of these intense metamorphic minerals are anhydrous.

Griggs, in his various articles in the Bull. of the Geol. Soc. of Am. and elsewhere, further showed by laboratory experiment that if rocks such as granite and limestone were subjected to confining pressure and temperature equivalent to great depths, the rocks would deform plastically to give the type of material found in nature.

Thus, strata is first lithified at considerable depth. Subsequently, subject to greater depth plus compressional and tensional stresses, it is deformed to yield folded rock strata so commonly observed by the geologists in the areas of orogeny (folding).

Price has claimed repeatedly that recumbent folds or thrust faults have been invented by paleontologists simply to save face when the fossil sequence was reversed. He says "In all examples of this sort (reversal) which are now known from all parts of the world, the physical evidence of the strata gives no indication of anything abnormal." By "anything abnormal" he means folding or thrust faulting on a large scale. To such an unwarranted assertion the person who wishes to evaluate the question must go map the rocks for himself. The fact is that such faults occur only in belts of obvious intense deformation which make up only a very small fraction of the sedimentary column. Commonly described areas are the Alps, Scottish Highlands, Appalachian-Taconic region, and the Canadian Rockies. Price always refers to a statement by McConnel of the Canadian Survey Report of 1886 on the Structure of the Canadian Rockies which emphasizes the conformability of the pre-Cambrian limestone on top of the Cretaceous shale. Price then claims on this basis that thrust faulting is a fiction. The difficulty is that he has listed a small part of the quotation which gave clear evidence for such thrusting. A more complete quotation from the report follows:

"The fault plane here (in the Bow Valley) is nearly horizontal and that two formations, viewed from the valley, appear to succeed one another conformably. The Cretaceous shales are bent sharply toward the east in a number of places, but with this exception have suffered little by the sliding of the limestone over them, and their comparatively undistrubed condition seems hardly compatible with the extreme faulting which was necessary to bring them into their present position. They are, however, very soft, and doubtless owe their immunity to this fact. It is otherwise with the overlying limestones, which have been strongly corrugated in many places, and are often whitened and cracked in the vicinity of the fault plane, the cracks have been subsequently filled with calcite. Enclosed argillaceous beds have been turned into schists and the banded appearance of much of the limestone is, no doubt, due to the shearing caused by the thrust. The limestones west of the fault are often bent by their pressure against the beds on the eastern side into a succession of sharp folds and are occasionally completely overturned. They also show disturbances in the altered and cracked appearance of the strata in the immediate vicinity of the faulted line."

The mobility of such shale has been demonstrated in the laboratory. 70 years of additional detailed mapping in the Canadian Rockies has thoroughly confirmed the thrust faults - on physical not paleontologic grounds.

## The Time Dimension

The time required for these processes to yield their results is obviously the most critical in the whole discussion. If the various rocks labelled Cambrian, Ordovician, Silurian and so on can be shown to have been formed at 500 million, 400 million, 350 million years ago respectively there could no longer be any argument about the flood doing the work 6000 years ago.

In geology there are numerous evidences of great antiquity. Some of these have already been mentioned. Relative ages are agreed on by all, flood geology writers included. The clear superposition of strata in undeformed areas is sufficient. For absolute time, recourse must be made to other features. Qualitative indications can be made from rates of weathering, rates of mechanical erosion and sedimentation, the number of unconformities in a sequence indicating uplift and erosion, rates of deformation, rate loss of heat from the earth's surface, accumulation of salt in the sea, etc. In Yellowstone Park there is a stratigraphic section of 2000' exposed which shows 18 successive petrified forests. Each forest grew to maturity before it was wiped out with a lava flow. The lava had to be weathered into soil before the next forest could even start. Further, this is only a small section of the stratigraphic column in this area. It would be most difficult for flood geology to account for these facts. Price suggests that vulcanism is due to coal burning underground which is being formed from vegetable matter trapped during the flood. This would mean that hundreds of times the total world coal reserve must have been used in melting rock at depth to account for the prodigious thickness of volcanic rocks through the stratigraphic section. A chemist might inquire where the oxygen came from for the combustion.

There is an interesting unconformity bordering the San Joaquin Valley in California. The Pliocene Tulare formation was deposited on older sedimentary beds, later folded back on itself, then was eroded level so that Pleistocene sediments could be superimposed. After more than a mile of this material was laid down the whole area was tilted and eroded to give the present relationship. This is one more of the excellent qualitative evidences for the great antiquity of these formations.

Evaporites

One of the most convincing arguments of a qualitative nature of the antiquity of sedimentary strata is derived from a study of the gypsum and salt content of the strata. Great thicknesses of gypsum and salt occur in several places in North America such as the Silurian strata of Michigan or the Permian of West Texas and New Mexico. In the latter area the gypsum alone totals around 1500 feet in thickness. It is not continuous vertically but some of the pure gypsum beds may be several tens of feet thick. The beds have an areal extent of thousands of square miles and are underlain and overlain by fossiliferous sedimentary strata. Gypsum is formed by precipiation of the salt out of ocean water under intensely arid conditions. The presence of large bodies of salt therefore means long time evaporation in restricted bodies of salt water with continual subsidence of the land. Since the evaporation of 1000 vertical feet of sea water yields only 0.7 foot of gypsum, it would take a body of water some 450 miles high to evaporate out the amount of gypsum found in the West Texas Permian basin. This can only mean continual influx of ocean water plus sustained evaporation. Such natural salt pans are numerous in coastal areas of arid regions. The Gulf of Karabugaz, on the eastern side of the Caspian Sea, is a well-known example of concentration of salt water behind a bar with a narrow shallow inlet. Here amid the surrounding deserts, evaporation is so rapid that a current flows continuously from the Caspian Sea into the Gulf augmenting the Gulf waters with some 350,000 tons of salt daily.

An idea of the rate of evaporation can be obtained from the Dead Sea where careful measurements show an evaporation of 10' of water per year. Since this is the fastest evaporation measured it will be applied to the West Texas case. Here about 5,000,000 feet of water must have been evaporated. This would require 500, 000 years as an absolute minimum. Recalling that these gypsum layers are intercalated, underlain and overlain with fossiliferous sedimentary rocks measuring up to four or five miles in total thickness, great spans of time are essential for any rational interpretation.

Quantitative

The only quantitative methods of getting absolute time are based on radioactive isotopes in the earth's crust. While much remains to be done in the refinement and development of technique, it is thoroughly established that approximate absolute dates may be put on rocks with as much theoretical justification as the weight of an object by a balance.

The method is very simple in principle. All radioactive atoms of a particular specie of an element disintegrate at a constant rate characteristic of that particular specie or isotope. This rate is unaffected by any temperature, pressure, electrical or magnetic fields, mass, impurity, or indeed anything that could be done to the atom under normal surface conditions. It can be calculated that these rates do change as stellar temperatures of the order of millions of degrees are reached. But such temperatures the earth has never known since it was a separate body in the universe. The various geochemical processes in the earth's crust tend to separate one element from another, thus it is that uranium is found in the mineral uraninite or pitchblende while lead is found in galena. The uranium in the uraninite crystal so formed from ore solutions emanating from the depths of the earth's crust was disintegrating throughout its history at its constant rate but the lead being produced was constantly separated from it by these chemical processes at depth. The result is that when the uraninite crystal forms it is essentially free from lead and hence as it continues its constant disintegration lead accumulates. Thus when a geologist finds such a crystal and brings it back to the laboratory an analysis for uranium and lead coupled with a knowledge of the rate of production of lead gives him the age since the uranium crystal was formed. Since such crystals are found in ore bodies which crosscut other rocks, the other rocks must at least be older than this date. But what if the uranium crystal was in some way contaminated with some common lead from somewhere. This simply means a little more work. Ordinary lead has a different isotopic composition from uranium lead and hence by mass spectrometer analysis the presence and percentage of ordinary lead can be measured. Actually on most uraninite specimens, four independent age measurements can be made. There are present the two uranium isotopes, U238 and U235 as well as thorium isotope 232. Each of these isotopes has a known long half life (the time for half of the material to disintegrate) and each yields its own specie of lead. Further, since the ratio of U238 to U235 is constant, but the rates of decay are different, the ratio of the lead produced from U235 to that produced from U238 changes in a known way with time. This is the fourth independent age measurement that can be made on the same specimen.

Hundreds of measurements of uranium-lead ratios have been made. In all cases the ages have been in agreement with the order already worked out by geologists on the basis of stratigraphy and structure. An idea of the precision of the method is shown in Table 3 which gives the measured ages of the Marealabian pegmatites of the Baltic shield in N. E. Karelia. The specimens come from four different pegmatites and were analyzed by three different chemists.

The uranium lead method has yielded ages of rocks from early pre-Cambrian to Pliocene.

Radioactive age determinations are not restricted to uranium-lead ratios. The heavy radioactive elements give off helium in the course of their series, decay to lead, and this is trapped in the crystal lattice. Thus a helium-uranium ratio should also give the age of the mineral. With helium, however, there is the possibility of leakage throughout geologic time. It has been shown that certain minerals such as magnetite retain their helium almost completely while others have known leakage rates. In general the helium age would be a minimum age. Two important conclusions come out of the helium work to date. (1) For a given rock type the relative ages of rock strata from all over the earth agree with the accepted geologic labels and show Paleozoic rocks to be of the order of hundreds of millions of years old. (2) The smaller number of very carefully measured specimens on minerals which retain their helium 100% are in complete agreement with uranium-lead ages. Again much work of refinement and of technique and the amassing of more data is going on but the method has been demonstrated to be valid by the hundreds of consistent results obtained.

In addition to these methods there are a number of others which are entirely independent but which give identical results. One of the most promising of the newer methods is the measurement of the rubidium 87 - strontium 87 ratio. Rubidium 87 decays to strontium 87 with a half life of about 5 x $10^{10}$ years hence can be used to date the older rocks. Ordinary strontium is dominantly strontium 88 and hence by mass spectrometer analysis the radiogenic (i.e. from rubidium) can be separated from common strontium. The amount of radiogenic strontium, of course, is alone used for the age ratio. Further, methods are being developed for the direct dating of sedimentary rocks.

A recent symposium on these methods published by the ASA gives details on the various methods and contains a bibliography of the more important scientific papers dealing with the subject. Anyone unconvinced of the validity of the radioactive method of dating should study these papers following the mastery of a book on the elements of radioactivity or nuclear physics.

Within five years most of the important stratigraphic sections will have been dated in absolute time to an accuracy of 25% of the true age. The dates will range from the present back to earliest pre-Cambrian (ca. 2,500,000,000)

## Mechanisms of Flood Geology

G. M. Price suggests a shock from outside the earth (astronomical) which might cause great tidal waves sweeping twice daily around the earth from east to west at speeds up to 1000 m.p.h. at the equator. After the waters started receding climatic conditions would be so altered that for centuries there would be increased rainfall causing faster erosion than today. Further he assumes the clouded conditions produced cooling and glaciation. The alternative that Clark suggests is that the initial geography of the earth was that of a low broad flat area with low narrow seaways which are supposed to correspond to presently observed geosynclines of deposition. He appears to want less exciting conditions. He offers strong rising currents along narrow waterways to account for the stratified rocks. Since Clark admits of the universal fossil sequence he can use quieter water. He says the sequence is simply an ecological one with the animals that lived at lower altitudes being engulfed first.

All of the potent objections to flood geology presented previously apply to both of these views. There are a few additional facts which remain to be considered. Since the important sedimentary rock types are being formed today under observable conditions it is reasonable to assume that 1000 m.p.h. tidal waves would cause a very different result. Such a process would, it seems, continually mix and remix in the most chaotic fashion the soft materials of the earth's outer surface. It would do relatively little to solid rock and certainly could not be called on to produce miles of thinly laminated colloidal sediments, nor is it reasonable to assume that the fossils would be unbroken lying in neat horizontal layers as is the usual case. The fact that deep sea caves show extremely slow unbroken sedimentation of a most delicate sort back hundreds of thousands of years by radioactivity measurements also precludes wildly rushing water. But then, if quieter water is assumed where is the detrital material to fill even a few of the great sedimentary basins? Why should most geosynclines show the sediments to coarsen to the uplifted side or sides showing clearly that the materials which fill the basin were eroded from the adjacent land mass and that solid rock was being eroded largely by chemical weathering which would play no part in a flood since it involves oxidation and the slow action of carbonated water on the insoluble silicates. If the fossiliferous strata are simply a reflection of ecological zones why are ocean and continental deposits constantly interlayered after the other throughout the geological column. All Tertiary rocks are certainly not continental.

Table 3
Ages from Uraninite Specimens from Marealbian Pegmatites
Baltic Shield (Report of Comm. on Geol. Time 1946-1947 p. 43)

| Sample No. | | Apparent Age in Million Years |
|---|---|---|
| 1 | | 1760 |
| 2 | | 1760 |
| 3 | | 1795 |
| 4 | | 1790 |
| 5 | | 1795 |
| 6 | | 1780 |
| 7 | | 1710 |
| 8 | | 1700 |
| 9 | | 1715 |
| 10 | | 1750 |
| 11 | | 1800 |
| 12 | | 1805 |
| 13 | | 1730 |
| 14 | | 1790 |
| 15 | | 1805 |
| 16 | | 1800 |
| 17 | | 1775 |
| | Average | 1770 |

Those that contain the mammals are continental--because mammals generally live on land but there are deep water Tertiary deposits as well as deep water and shallow water Cambrian deposits. The difference is that in the continental Tertiary deposits mammal bones are found whereas in continental Cambrian rocks no such are found. Whether a rock is continental or marine in origin can be determined by the mineralogy independent of the fossils it contains.

Finally it might be of interest to observe the effect of large tidal waves, the tsunamis of the Pacific. During a tsunami the water often rises very much like a tide so that erosion which accompanies the wave may come largely from the rushing back of the water down the slope or from the ordinary wind waves which ride in on top of the tsunamis and which are able to attack an area that has not been adjusted to wave attack and may, therefore, yield very rapidly. Sand dunes and earth embankments inside the beach may be greatly truncated but the beach outside this eroded area may show little sign of change in profile. Many such cases were found in the Hawaian Islands after the tsunami of April 1, 1946. Two features claimed by flood geology are emphatically not observed even in miniature in tsunamis action. (1) Solid rock is not appreciably eroded. (2) Sedimentary deposits similar to that which could form rocks like those observed in the geological sequence are not formed.

## Conclusions

1. The theory that a relatively recent universal flood can account for the sedimentary strata of the earth is entirely inadequate to explain the observed data in geology. The major propositions of the theory are contradicted by established physical and chemical laws.

2. The science of geology precludes certain interpretations of Genesis but does not make impossible acceptance of plenary inspiration of the scriptures.

3. This paper has been negative in character because it is believed that this unscientific theory of flood geology has done and will do considerable harm to the strong propagation of the gospel among educated people. On the positive side it is important for the Christian geologist to carefully study the truly superficial erosional and depositional effects on the earth to obtain evidence on the extent and nature of the flood of Noah.

Comment on the "Deluge Geology"
Paper of J.L. Kulp

By an A.S.A. Member.

In an earlier letter I had stated that I wished to say more about the discussions on geology at our Los Angeles meetings. I feel that this is a very critical area for our consideration both because it is a very critical area for our consideration both because it is an important area and because very little is being done in this area by those who have the Biblical viewpoint. I am very happy that Dr. Kulp is working in this area and I appreciate very much his Christian spirit and his scientific approach. I am inclined to think that he may still be too much influenced in his own thinking by the orthodox geological viewpoint with which he has been associated and which, of course, is the only scientific viewpoint which has had systematic development. I feel also that Miss Erdman is perhaps too much committed to the orthodox viewpoint.

To be more specific regarding Dr. Kulp's paper on the deluge geology, I should like to make a few comments. In my mind the paper seemed to lack something of the scientific approach in that it began with the viewpoint that Dr. Price's flood geology is all wrong and that his ideas are not worth considering. The rest of the paper was then a presentation of arguments against the flood geology concept. Even though one would grant that this conclusion, that the flood geology concept is all wrong, is correct, the method of the paper seemed to be inadvisable inasmuch as this was definitely considered to be a disputed issue on which we wished to have some light. At least for my own thinking I would have appreciated much more if he had presented facts and arguments on their own merit and drawn his conclusions at the end of the paper on the basis of these arguments rather than to have used the paper merely to bolster his pre-determined conclusions. Personally I hold no brief for the flood geology concept as presented by Dr. Price because I do not feel that I am informed enough to pass judgment on many of these issues. I do feel strongly, however, that the members of the ASA are competent to understand the facts and the issues involved and that they deserve a clear presentation of these facts from which they can then themselves draw conclusions. I do not think that the issue is too clear in the minds of many of the members of the ASA. I would welcome continued presentation of facts and discussion of the problem. Let me say again that I appreciate very much the efforts of Dr. Kulp and I believe that the affiliation should continue to foster his efforts in this direction. I hope that some other good scientists may be inspired to contribute also to this problem.

I do not know whether my presentation of my feelings on this point in this way can effect any good, but I am hoping that by airing my views there may be some influence brought to bear on the future program and efforts along this line.

Incidentally, I am wondering whether you had noticed on the inside front cover on a recent Scientific Digest a statement regarding the age of the fossil remains of one of the prehistoric human forms. This is very interesting in that it places the remains definitely into the inter-glacial period, if I remember correctly, rather than at a much earlier time, as has usually been considered. I believe there are new techniques being developed for the dating of these things which should be pursued as vigor ously as possible in trying to dissolve some of the discrepancies that now exist between the orthodox geological view and a Biblical view.

# THE PLAN IN THE FIRST CHAPTER OF GENESIS

By R. T. Fulwood

General Outline

I. Gen. 1:1: ORIGINAL CREATION, or beginning of the earth's career.
The duration of this first condition not given.

II. Gen. 1:2: A RUIN (tohu): Dissolution, or chaos.
The duration of this second condition not given.

III. Gen. 1:3:
to 2:3: RE-CREATION, or resurrection from ruin in six days.

* * * *

THE SIX DAYS WORK AND THE REST DAY

(Beginning at verse 3)

| First three days: MATTER, or Physical Divisions: | | Second three days: MOTION, or Organized Powers: | |
|---|---|---|---|
| 1st Day: | LIGHT | LIGHTS (organized) | 4th Day. |
| 2nd Day: | AIR divides WATERS | AIR & WATER LIFE | 5th Day. |
| 3rd Day: | (a) EARTH APPEARS<br>(b) VEGETATION | (a) EARTH LIFE<br>(b) MAN | 6th Day. |

7th Day: COMPLETE HARMONY, or REST

* * * *

Probably the majority of those who are more or less familiar with the account of creation in the first chapter of Genesis do not suspect the existence of the precise, clearly defined plan which it contains. A long period of study of this chapter has revealed some very interesting features of this plan which it is the writer's purpose to present as clearly as possible, and having thus condensed as much as he was able, he would suggest that the article be read over the second time in order to arrive at a clear understanding of the deep significance of the plan.

The Hebrew equivalent of the word "create", meaning to bring into existence a new thing, is used on three occasions only in the entire chapter, viz:--in the first verse, in verse 21 (Fifth day) and in verse 27 (Sixth day), or at the introduction respectively of Matter, the animate Soul and the human Spirit. On other occasions the word "made" is used, signifying the working with material previously existent.

There is a definite distinction in character between the subjects of the first three days, which are masses, stationary in relation to their surroundings, representing Matter, and those of the second days which are multitudes of separate units, either organisms or comprising organizations but all moving, and representing Motion, or Force. This grouping is a further evidence of method or plan in the chapter.

Another marked feature is the order in which the subjects are presented and the correspondence between the first and fourth, second and fifth, and third and sixth days regarding the nature of the subject treated ("Light" and "Lights", etc.). By referring to the outline above and reading across the page this will become clear. This is carried out so completely that even the form of a supplementary addition on the third day of the first group is repeated on the third day of the second, the words "And God said," introducing a new subject, occurring twice on these two days, only, appearing but once on each of the other days.

The above features are merely external evidences of careful arrangement but they are doubtless sufficient to show that this account is no haphazard collection of assorted legends as has been claimed.

Seven Hebrew words, in the first verse, state the original creation.

"In (the) beginning God created (the) heaven and (the) earth." By omitting the word "the" which does not appear in the Hebrew language we have the same number of words in English. This is an isolated statement, denoting an aoristic point in time: possible ages upon ages prior to all the succeeding portion of the account. "Seven" among the Hebrews, and in its usage throughout the Bible, indicates the completion or "perfection" of the matter in hand and the inference is therefore perhaps fairly warrented that a finished or completed condition of the earth at that time is signified by these seven words. This view is not only warranted by the text but is in perfect analogy with the spiritual doctrine developed later, there is no contradiction of it by any other scriptures and it is in complete accord with the discoveries of science. God is light, and his work perfect from its beginning. He does not create ruins, nor work in darkness, which condition we find the earth in in the second verse. This condition and method are the work of the Adversary.

The second verse reads: "And the earth was without form and void;
"and darkness was upon the face of the deep. And the Spirit of
"God moved upon the face of the waters." This would indicate a lapse from its previous complete state, if the above inference is correct. In any event there is no indication as to the duration of this desolate condition. It may have lain thus for long ages. Geology indicates such lapses in its history.

The six days work in restoring the earth to its present condition really begins with the third verse of the chapter. These days are divided, by the nature of the subjects, into two groups of three days each, followed by the one day of rest which should have been included in this chapter.

Reference to the summary above will show that the entire range of Matter, viz:- the Ether, Gases, Liquids and Solids, in their ordinary, typical representatives which are Light, Air, Water and Earth, respectively, is comprehended in this first series of days. The supplement at the conclusion of this group is Vegetation, and it is a noteworthy fact that Vegetation is the combination in one organism, or unit of life, of all the physical divisions that have preceded it in these three days, in its chlorophyl, or coloring and assimilating force (Light), its breathing leaves (Air), its sap (Water), and its fibre, or mineral matter (Earth). It thus sums up and completes the subject of Substance, or Matter, and it would even appear that this elementary and purely physical, or material, form of LIFE is the ultimate object for which the various physical formswhich have preceded it exist, and their culmination; for the humble, fragile blade of grass, while almost infinitely inferior in appearance to these mighty masses of matter, is superior to them in its nature, having the power of growth and reproduction. And while it lays hold of and uses them all in its internal economy, they combine to serve it externally, also, as the Light, or "radiant energy" as we now have it in the sun, by means of the Air causes evaporation of the Water which is carried over the land in clouds, falls in rain and moistens the Earth in order to produce food for the plant. Thus these various divisions converge into the plant as a focus, and having been combined, and converted into food for the animate creation through the wonderful transforming power of its minute cells they are expanded again by the process of growth. This is nothing less than the finger of God. "Adaptation to environment," "survival of the fittest," the theory of the transmission of life to this plant from some other by the agency of light, and other such hypotheses which have been offered to account for the appearance or development of life on the earth, fail to recognize to give any head to the marvelous arrangement of the mighty system which we see in practical operation every day for the perpetuation of life.

The plant being the focus through which inert material is converted into life fittingly occupies the central place in this six days work, at the termination of the third day; and there is a gradual expansion of development in the following three days which confirms the fact of the termination of the third day being a focal point. This will become clearer as the subject is developed.

It is plain that Vegetation completes this first series, or group of days, and this view is further confirmed by the fact that on the succeeding day - the fourth - the account returns to the subject of the first day, - "LIGHT," beginning another series; but it is now in the form of an organized system of "LIGHTS" in their relation to the earth.

"MATTER" having been summed up in this first series of days, and combined in the mystery of life in the plant organism, this subject may be considered as closed, and we are led to expect something beyond Matter in the second series. We find it in

MOTION or FORCE

Force, or Power, per se, is always invisible, but the effect of Power is motion in some form. The inert earth is dependent upon the sun for its enercy and all the power expressed in motion on the earth is derived from the sun. Therefore, on this fourth day, beginning a new series with "POWER," made visible through motion, as its theme, the scene of operations is transferred from the earth to the heavens, where the source of the earth's energy is located. Hitherto in the account the physical earth has been the theme, but now a new field is entered upon where invisible powers as exhibited by their effects claim our attention. These powers, it is true, are expressed through widely varying visible, physical bodies, but the hidden powers themselves are the true subjects of the second three-day group.

On the Fifth Day the moving soul is created with the power to Love and Hate (higher in quality than the mightiest mechanical power) and with the new power of automatic, or self-directed movement. The Hebrew equivalent for the word "create" (Bara) meaning to bring into being that which did not previously exist (as distinguished from "making" or adjusting, in different relations, previoulsy existent things) is used on three occasions, only, in the entire chapter, as has been stated above, and it is striking that these occasions should be the introduction of the Physical (BODY), the Emotional (SOUL) and the Intellectual (SPIRIT) factors, which comprise everything that is contained in the universe as we know it.

On the Sixth Day the embodied Spirit, Man, is created. Man is the epitome of the universe. As Vegetation, in a supplement to the first three-day series, sums up the several divisions that have preceded it, so Man in a corresponding supplement to the second three-day series sums up and combines in his triune being all that has preceded him in the entire six days. And as the plant has the power to lay hold of the incorporate within its purely physical system the various material, visible elements surrounding it, so Man, the invisible although embodied spirit, is able to lay hold of invisible forces, surrounding him, to adapt them to his uses according to invisible laws, chemical and mechanical, and to conceive and create an entirely new combination of matter and force, such as the steam engine, to serve his uses.

The third division, as shown in the summary, is the seventh day which is set apart from the other six.

The first three-day group is static in character: its units or divisions, as a whole, maintain unvarying relations towards each other.

The second group of three-days is dynamic: its myriad variety of forms is in constant motion - and unrest, the animal world pursuing and being pursued: a condition shared by mankind morally.

Following this is the seventh day, characterized as a day of Rest: not extinction. Rest implies life, but life without disturbance, in complete harmony with its environment.

## NATURE OF PHENOMENA PRESENTED ON THE SEVERAL DAYS

A brief review of the days will show a logical progression in the order of their subjects, and just the order that we might expect in a material world projected by a Spirit-God, i.e. from spiritual, or unseen, to material forms, as in the first three-days; and from invisible and unrecognized powers to conscious and recognized capacities and faculties as in the second group. The account is extremely brief and condensed, considering the vastness of the subject and apparently merely serves as an introduction to the Career of Man on the earth. But while the phenomena are presented very simply, as they would appear to the untutored human being in all ages, the successive developments and adjustments are entirely logical in their order.

1st Day-LIGHT: On this day light is brought in on the face of the deep, where, as stated in the second verse, darkness had been. Contrary to the popular notion it is not stated that light was created on this day; after its use in the first verse the word "create" does not appear again until the fifth day is reached. The light appears here without a visible source, and it is light, per se, apart from its source or its effects, that is the subject here.

Light is the most ethereal or spirutual of all the phenomena recognizable by our senses. We are cognizant of its presence or absence but its nature has not yet been satisfactorily analyzed. Light and Heat are classed together under the general term of "Radiant Energy," and the chief characteristic of light is radiation in all directions. It is thus pre-eminently a centre of influence; it reveals true conditions wherever it appears, and it is the visible indication of the presence of energy.

2nd Day-AIR: The "firmament," or expanse, is the elastic sphere of air, some forty miles in thickness, which envelopes the earth. (See the fifth day, where fowls fly in the "open firmament of heaven.")

The air is the great means, or medium, (a) of communication: while creating no sound itself, it is the means of communicating all sounds, particularly speech, by which we communicate our thoughts. (b) It is also the means, or method, of life of every organism, plant or animal, on the earth. (c) It is the means of producing vegetation by distributing moisture over the earth in the clouds.

Thus the Air is characteristically the great agent, means or servant.

3rd Day- (a) EARTH: On the first part of the third day no new feature is introduced, but there is a new adjustment of the earth and waters, the dry land being raised above the waters, which had theretofore buried it beneath them.

The solid earth is inert and stable. Acted upon, but not active and thus characteristically an "Effect" as distinguished from the former features.

The principles of Cause, Means and Effect are thus represented in their due order in these first three days.

3rd Day- (b) VEGETATION: Vegetation is the elementary form of organism, or life, wherein different parts contribute harmoniously to form a living whole or unit; and as noted above it is either composed of, or represents in its economy, the three forms of matter, viz: gases, liquids, and solids, and the hypothetical ether.

Reviewing the series briefly we find a

UNIFORM PROGRESSION FROM SPIRITUAL TO MATERIAL FORMS

Before the six days work was begun the watery waste, shrouded in darkness, lay before the Creator.

Then light, the most ethereal, spiritual element recognizable by the senses, is brought in upon it.

Then the Air, or typical gas, invisible but having density, divides the waters above from those beneath, the latter, of course, being those enclosed within the air sphere.

Then the Waters, or typical liquid, with still greater density, are gathered together and given their place in the scheme.

Then the Earth, proper, or dry land, the most solid, stable and dense of all these forms of matter, is resurrected, as it were, from beneath the waters.

Finally the Plant, the combination of these forms in an organism, springs up out of the earth.

Note the orderly progression from the non-material, yet visible, to the most solid material. Beginning with the visible representation of power, which is nearest in character to the Creator ("God is Light") who is a Spirit, and descending through gradually denser forms to a solid body.

The ingressive development is also noteworthy: The light is seen upon, or outside of the watery waste; then the air sphere enters and divides the waters; then the waters beneath the air sphere in their turn are parted and the solid ground is brought up from beneath them. Finally, out of the inert land itself, thus resurrected from burial, the Plant, or living organism appears: Life springs out of that which was dead and buried. The Plant continues as the type of resurrection life in nature, drawing its nourishment from that which is dead and decayed, giving it a new, living body and transforming it into wholesome food for higher types of life. Again: it flourishes through the summer, dies, apparently, in the winter, and rises again in new life in the spring. It appears here on the third day, which is the day or resurrection throughout the Bible (Hosea 6:2; Luke 24:7; Rev. 11:11, etc.) That it is a symbol of resurrection life is confirmed by the Great Teacher himself in John 12:24: "Except a corn of wheat fall into the ground and die it abideth alone; but if it die it bringeth forth much fruit." He applies this to his own death in the 32nd and 33d verses of the same chapter. And a very beautiful symbol it is of resurrection life: leaving the earth behind, as it grows, it rises toward the light from heaven which it reflects, partially, in the varied color of its leaves and flowers and fruits, as the Christian partially and imperfectly reflects some features of the moral glories which together give us some idea of the great white light of the character of God.

4th Day-LIGHTS: On this day the present existing relations between the earth and the heavenly bodies are established. The earth bears its own witness of a glacial period, showing that these relations have not always continued just as we find them today. And if the six days work of the chapter be a resurrection of life, and of conditions which permit of life which have become extinct in a desolate, dead and buried earth, and we believe the record shows it is, then the restoration of light and heat to it by the sun at this time is in complete accord with the remainder of the account. It is not stated that either the sun, moon, or stars were created at this time. On the other hand, given a Creator, in the beginning, he must have absolute power to effect changes in that which he has made when and how he will, and cannot even be fettered by the "laws of nature" without making him inferior to those laws. Such a Creator is unthinkable. Laws are incapable of changing their operation, are without volition or consciousness and are thus inferior even to mankind, and a God who is incapable of suspending or changing his laws at will is no God.

"Light" as a subject was introduced on the first day, where it appears without a physical source, indicating that Light, per se, apart from its source or effects, was the subject there. In the description of this Fourth Day the effects or relations of the "Lights" toward the earth are dwelt upon at length, and the heavenly bodies themselves are passed over with slight mention. The "Lights" rule over the earth in that it is dependent upon them for light, heat, power and guidance, and its changes of season and "times" are determinded by its constantly varying attitude towards them. The relations between these heavenly bodies and the earth, the components of a vast complex organization, more accurate than the finest mechanism, are not maintained, as are the relations of matter in the plant organism of the preceding day, by visible, physical contact, but by great, invisible, counteracting Forces. And from this day forward, in the account, while differing physical forms are presented, it is their invisible forces, powers or being which are the subject, these being far higher and more important than the mere material through which these powers find expression.

On the Fourth Day Motion, or change of location, which is visible manifestation of power in action, is introduced. Tremendous in its sweep through space but on the other hand entirely involuntary, and impelled and controlled by blind, external force. Hence it is inferior in character to the almost infinitely more limited but self-impelled and directed motion of the animals which appear on the succeeding day.

5th Day--AIR AND WATER LIFE: On this day fish and fowl are created. Here we find an organism - not a mere mass of matter - in motion which is voluntary and self-controlled, the forces that move the body being lodged within it, and the controlling force denominated the "living soul." The method of motion of the birds and fishes, which swim in the air and water as the planets in space, is a complete contrast to the circular, orbital mode, in one direction only, without resistance, of the preceding day, being produced by a reciprocal, or forward and reverse movement of wings or fins, and progress being made by opposing force from within to resistance from without in the air and water. In its physical structure the animal is usually bilateral or "paired" in almost all its parts or organs, many species mate or live in pairs, and birds and fishes characteristically among animals migrate or move to and fro in large masses at certain seasons, as impelled by a common instinct. In its physical structure and movements the number "2" appears to be stamped on the animal, and this appears as well in its soul, or emotional qualities.

The invisible soul of the animal is in communication with the visible creation by five channels of perception, or sensation: with the light by the eyes; with the air by the ears and nose, and with the earth and its vegetation by touch and the mouth. The animal soul manifests through these channels, two great conflicting qualities, broadly speaking, which comprehend all lesser ones, viz: - Love and Hate, or Attraction and Repulsion. And as love is constructive and uniting in all its degrees of manifestation and hate is destructive and disintegrating, conversely, these opposed qualities show the animal to be a concrete representation of Force and Resistance, or Positive and Negative influence, or in other words a visible expression in an organism of duality, conflict and opposition.

6th Day--(a) EARTH LIFE: As remarked before, this Sixth Day, or third of the second group, is divided into two parts like the third day of the first group. On the first part of this day the land animals or four-footed beasts are made. There is, of course, no radical difference in nature between these and those previously created. They are a new adjustment, adapted to land travel with their four feet as the birds and fishes with their wings and fins are to the air and water, respectively. The method of progress on the surface of the ground is more limited than that of the birds and fishes, and it is among the land animals that man has found these which are readily subject to him and which he keeps in flocks and herds for his service.

Note the remarkable parallel between this new adjustment and nothing more, on the first half of this day and the same feature on the first half of the third day, regarding the earth and water. (See "3rd Day - (a) EARTH").

6th Day--(b) MAN: The creation of Man, apart from the animals and as superior to them, is the prominent feature of this day, and is emphasized by the two parts into which it is divided. The word "Create" occurs in this connection for the third time, as noted above, showing that this being is a new and higher order. There is no radical difference between the body of man and that of the animal. The animals also exhibit soul qualities, or emotions, very similar to those of man; but the animal soul is incapable of conceiving, or being influenced by, an object or purpose outside of the range of its five organs of sensation, or consciousness of visible surroundings. We can conceive of no mere animal voluntarily going to the stake to be burned for the sake of allegiance to an invisible being who has never been visible to it. Such things have been done by the embodied spirit, Man. Spiritual realities are invisible, and the invisible spirit alone can apprehend them. The man was created independent; not bound by the law of instinct as the animals, although he shares its operation to some extent with them. He is able to make new laws for himself, governing his own conduct and that of others, and even to modify the operation of physical laws by opposing one to another. The invisible spirit of man manifests itself thus in its perception of and active relation to the invisible realm of force and law.

Vegetation at the close of the first series of days is the organizer of physical forms, but mankind at the close of the second series dominates and organizes invisible forces, according to invisible laws, producing visible effects according to his will. The human spirit thus dominates its surroundings and by ruling and controlling them shows itself to be "first" or supreme, as the animal soul, by its continuous conflict, manifests its position to be "second."

Reviewing this second series we find that it sets forth the visible effects, in action, of successively higher

POWERS: FROM BLIND ACTIVITY TO CONSCIOUS PERCEPTION.

1. On the first day of this series (4th) day the source of all mechanical power on the earth is shown and we see the effects of its action, in combination

with the invisible power of Gravitation, in the regularly recurring changes of seasons, day and night and months and years on the earth, resulting from the motion communicated to it. Unconscious, unresisting obedience to external power may be said to characterize this action, or mode of motion of the earth.

2. On the second day (5th) an emotional power, the "living soul" is presented: a being conscious of tangible, visible objects, by the senses, but of nothing intangible or invisible; expressing opposed qualities of Love and Hate towards other beings; having the faculty of self-directed movement by the exertion of force from within upon outward resistance, motion and progress being thus produced by conflict; a method far higher than that exhibited on the preceding day, i.e. by blind force from without overcoming inertia.

3. On the third of these days (6th, second part) a spiritual power, Man, is created in the image of the creator ("God is a Spirit"). Here is a being conscious of himself as an invisible, individual "Ego"; perceiving invisible forces and laws surrounding him, and even apprehending the Creator himself; capable of controlling and guiding his own and other's actions, and of manipulating material and subjecting it to the action of invisible forces and laws to produce such effects as he chooses. This dominant, conscious force is the opposite extreme from the mechanical motion seen on the first of these days.

Finally it will be noted that the gradation is perfect: Automatic, unconscious motion in subjection to external force, followed by internal and external conflict, and this succeeded by conscious control or supremacy over forces and material. As in the first series of days the development was from what makes visible, or the light itself, toward material density and visibility, so in this series of successive powers the development is from unconscious activity, through consciousness of visible surroundings, to consciousness of invisible powers, and thus of the invisible Power - the Creator - God. The first series is objective development; the second, subjective development, and the purpose is the same in both series, viz: - The revelation of the invisible Creator in his material universe and to his created being, Man, in whom the material is united with the spiritual.

In mankind the highest power on earth - Intellect - is united with the highest type of physical organism.

These six days of work are followed by a Seventh, designated as a day of Rest, which is a day without an evening. It has a beginning but no ending. The temporal is thus united with the eternal, and on this day, which extends out into infinity, God rests. It is numbered as an earthly day, but it is "sanctified" or set apart from all the others and nothing is created or made in it. But why should the Creator, the source of all power, rest? And why does the infinite God mark out an earthly day on which to rest? It is a "7th" and thus be inference a "perfect" one and it goes beyond his six days creative work which culminates in Man. If we turn to later Scriptures they tell us of one who answers to this descritpion, in whom the temporal is united with the eternal in a perfect being who is called "Emmanuel" or "God with us," and we are told in Colossians 2:16,17, that Christ is the body of which this seventh, or sabbath, day is the prefiguring shadow; who was not created (as nothing was created on this seventh day) but of whom it was said "in the fullness of time God sent forth his son, made of a woman" (Gal. 4:4); and again "Wherefore when he cometh into the world, he saith, Sacrifice and offering thou wouldest not, but a body hast thou prepared me" (Heb. 10:5); and again "therefore also that holy thing which shall be born of thee shall be called the Son of God" (Luke 1:35); in whom the Creator found rest, as stated in John 1:32-34, "I saw the Spirit descending from heaven like a dove and it abode upon him;" and "Upon whom thou shalt see the Spirit descending and remaining -- this is the Son of God;" who was separated from other men, - "holy, harmless, undefiled and separate from sinners" (Heb. 7:26), as this seventh day is set apart from the other days. In whom we find rest, as well as the Creator, as in Matt. 11:28, "I will give you rest." Concerning whom it is said "For it pleased the Father that in him should all fullness dwell; and having made peace by the blood of his cross by him to reconcile all things unto himself; by him I say, whether they be things in earth or things in heaven." (Col. 1:20) So that complete harmony and rest in God's creation are to be found eventually in Him.

The foregoing exposition of the phenomena in this chapter does not however, set forth the principal underlying theme, which, as elsewhere throughout the Bible, is spiritual truth. If we fail to consider this we shall miss entirely the Creator's thought in giving us this record.

## SPIRITUAL SIGNIFICANCE OF GENESIS I IN THE LIGHT OF LATER SCRIPTURES

As in the microcosm - Man, the material portion of his being serves to give outward expression to his invisible, spiritual nature, so in the great universe, or microcosm, the material but serves to reflect the spiritual of which it is the visible shadow. The Spirit God is infinitely superior to his material, created universe, and to Him the material, vast as it may be, is but as shadow compared with the spiritual reality. Hence the prominence given to spiritual themes as compared with material in the scriptures. The creation of the material world occupies but one chapter: the spiritual history of mankind requires volumes. These volumes are divided into two sets of books known as the "Old Testament" and "New Testament." The Old Testament is bounded by an earthly horizon and is devoted to the Creator's dealings with his earthly people, the Hebrews, or natural descendants of Abraham, as the children of the first Adam. But the New Testament begins with the "Kingdom of Heaven" and its theme is the Second Adam, the "Lord from heaven" (I Cor. 15:45-47) and his heavenly people, the church, which is a new, spiritual body not mentioned at all in the Old Testament. The spiritual realities and powers set forth in the New Testament are not visible to the natural eye except by their effects, but are revealed only to the spirit by faith.

The analogy between the foregoing contrasted dispensations and the contrasted groups of days, on the first set of which different visible, physical divisions are set forth, while in the second three days hidden powers only as manifested by their effects are seen is obvious. And this is no mere arbitrary arrangement but is due to the inherent and inevitable unity and harmony of Truth, whether expressed in the physical or spiritual realm, so that we should expect to find the same order of procedure in both.

In God's revelation of himself we find Him distinguished as three persons: Father, Son and Holy Spirit. The Father, or first person of the trinity is revealed as the originator, source of life, "first cause" or power. In 1st John we read "God is light." Light in the material world is the manifestation of power in action and is a radiant source of influence whose presence reveals the truth of conditions, and the correspondence of the properties of light, as shown in the first day of our chapter, with the revelation of God as Father is obvious. Furthermore, the effect of light on this first day is to dispel the darkness which surrounded the chaotic earth, and the revelation of God has the same effect on man, morally, as shown by 2 Cor. 4:3-6 "God who commanded the light to shine out of darkness hath shined in our hearts, to give the light of the knowledge of the glory of God in the face of Jesus Christ."

The Son, or second person, is the dependent, obedient one, limited and controlled by the will of the Father; the one who does not act from himself, but serves, carrying out the Father's will, and "took upon him the form of a servant." He is the Mediator or means of communication between God and man (I Tim. 2:15); He is called the "Word" (or means of communication - message); "the Word made flesh" (John 1:1,2); the "obedient servant" (Phil. 2:6-8) and the sustainer of all things (Heb. 1:3 and Col. 1:17). The analogy between the natural characteristics of the Air, on the second day, and these spiritual ones of the Second person is apparent. The air is a sphere, thus expressing limitation; it is that of which words are formed, which are the means of communication; it is the mighty servant, carrying the life-giving moisture from the oceans over the land; and finally it is the means of sustaining life in every breathing, living thing.

The air sphere being introduced into the waters has the effect of dividing them, sustaining the portion of them which is raised up and purifying it of its salt, so that it descends on the earth as rain, giving life to the vegetation and running water to mankind, for his uses; the remaining portion being left in its original condition. So Christ, introduced into this moral scene, divides mankind at once into two parties, well illustrated by the two thieves on the crosses beside him. Both were equally guilty, but one acknowledged him as Lord and is with him today in Paradise, raised up and purified, according to his promise, the other reviles and despises him, and is - where? So, also, in each individual soul the entrance of Christ by faith at once produces conflict of desires, the old, dead nature to continue in sin and the new, or Christ nature desiring to be free from it, as shown in Romans 7.

The Holy Spirit, the third person of the Trinity, pervades the material and works through it but remains ever invisible and his presence and activity are manifested whether in the material or spiritual realm, only by the transformations he effects in the subjects of his work. The third person of the Trinity is always spoken of as effecting resurrection, as in Romans 8:11; I Pet. 3:18. He is the great Organizer (I Cor. 12:11-13); the producer of fruit (Gal. 5:22). The resurrection of the dry land from beneath the waters on the third day and plant or living organism springing up from it "whose seed is in itself," producing fruit, manifesting by its growth the mysterious, hidden life force within it and transforming dead and decayed material into beautiful, living forms, together present a perfect analogy with the work of the Spirit in the Spiritual realm.

Thus the Father is the source of Truth and Life, the Son or "Word" declares the Father, or is the means whereby His character is made known to man, objectively, and the Holy Spirit works in the hearts of men subjectively to make the message effective and productive.

Returning, for a moment, to the second verse of the chapter, the earth is there seen in a state of complete ruin and desolation: darkness covers the face of the deep and the waters have submerged the earth, so that there is no place for life to exist. This is also a graphic picture of the spiritual condition of man away from God. How two new elements from above are introduced: the light from heaven dispels the darkness and the air, or earth-heavens, divides the waters, raising up a part of them and turning that water which had been death and a burial shroud to the earth into the means of life upon it by the rain-bearing clouds. On the third day the land is raised from beneath the waters and then the two life-giving elements from above are united by the agency of the third, the invisible life force with the two resurrected and readjusted ones from beneath (i.e. the purified waters and the earth) in one living organism - the Plant. This is the doctrine of resurrection life in symbol: The Holy Spirit, the great Organizer, unites the spiritually dead mortal with God himself in a new and living spiritual creation, "and this life is in His Son." Christ came as a Hebrew, at the close of their dispensation and in fulfilment of the promise to Abraham that in his "Seed" should all nations be blessed (Gal. 3:16). He grew up as a tender plant (Isa. 53:2) the humble herb of the field, and as the corn of wheat must fall into the ground and die in order to bring forth fruit, so His death is the source of all spiritual life for men, through His resurrection. Death is the food of life throughout nature, and every morsel of animal flesh we eat symbolizes that death is necessary in order that we may have life. God teaches in His word that death was necessary to His Son that we might have spiritual life with Him. So nature's book is teaching us also, every day, a direct contradiction of the theory, built on hypotheses and not facts, of the "Ascent of Man." For we do not see the evolution, or transmutation, of one species into another going on about us in nature, but on the contrary in its four great kingdoms, the Mineral, Vegetable, Animal and Human we see a wonderful process of DEATH AND RESURRECTION going on constantly, wherein the higher form of life reaches down, as it were, and takes the lower up into itself, the lower dying out of its existence as a lower form and being incorporated with the higher in a new life, with a new identity. Thus--

The Vegetable takes up the Mineral;
The Animal takes up the Vegetable;
The Human takes up the Animal and Vegetable--

and there is one step higher, which we find in 2 Cor. 5:17: "Therefore if any man be in Christ Jesus he is a new creature (or 'creation')." In other words--

God takes up the Human in new, or resurrection, life.

The doctrine of the Book and the doctrine of nature agree. Who made both?

Plant life from the dead and buried earth, or the meeting point of life and death on the earth, at the close of the third day, is immediately succeeded by the shining forth of the Sun in the heavens on the fourth. So Christ - "being found in fashion as a man, humbled himself, and became obedient unto death, even the death of the cross. Wherefore God also hath highly exalted him, and given him a name which is above every name" Phil. 2:8,9. Having ascended to heaven, after His resurrection, He shines forth from there as the world's spiritual light. On earth it was the lowliest of the low, like the wheat which is not a lofty tree, but a lowly herb, yet produces the principal sustenance of mankind, but now in heaven He has all power in heaven and on earth, is the "Prince of the kings of the earth" and "his countenance is as the sun shineth in his strength". Rev. 1:5,16. So on this fourth day the earth's light and source of power is seen as a physical body from which the light emanates, and the earth's relationship with this body is defined as a subject one. The earth revolves around the sun as its centre and is bound to it by an invisible force. Here the Christian's relationship to the risen Christ is set forth. Having received new, resurrection life he is bound to Christ, the "Sun of righteousness" (Mal. 4:2) by a spiritual bond which cannot be broken, although he may turn away at times and become cold; yet the life has not departed as that in the earth is not lost during its winter's sleep. This relationship involves change, or vicissitudes, on the earth - cold and heat, day and night, etc., just as obedience to Christ involves trial to the Christian in this world. But as the vast storehouse of physical energy derived from the sun is available for man, and only needs the application by him of the proper means to render it useful, so spiritual energy is made just as free to the Christian by obedience and prayer.

As the animals on the fifth day make progress by opposing the force within to the resistance from without, and each has its own peculiar way of getting its food and manifesting activity, so the Christian rejoices in the power of this new life but finds resistance to it in all the world around him, including his own old nature, and that the only way to make progress, spiritually, is in the exercise of his new power against the resistance. He finds food for his new life in the Word of God, but this is not laid out for him in a convenient set of formulas which will call for no effort for their understanding. It requires the exercise of his new powers to get this food (just as all animals must exercise their powers to get theirs) and he becomes vigorous, spiritually, in proportion as he digests this food and exercises his "own proper gift" spiritually. He is called upon to live a life of love and service, building up himself and his fellows, or in other words live a Positive life of faith, hope, and love, but on the other hand he may give way to hate and selfishness, destroying or tearing down himself and his fellows in a Negative course of doubt, synicism and egotism and his course will determine the measure of his eternal benefit or loss.

As the land animals, presented on the sixth day, are divided into "Cattle, the beast of the earth and creeping things," so men may submit themselves to their Lord for His discipline and training in His service, like the "cattle," or domestic animals (and are proportionately useful in the general scheme of things as they do so); or they may run wild, taking their own imperious way, like the "beast of the earth," leaving destruction and misery in their train, or they may simply grovel and wallow in earthiness like the "creeping thing."

On the second part of this day the Lord of the whole creation is seen as a man, but associated with him is a companion who was taken from his side during his deep sleep (Gen. 2:21-25), just as "the church of God which was purchased by his own blood" (Acts 20:28) is the result of His death, and is referred to as His bride, sharing in all he has. (Eph. 5:23-33; Rev. 21:9-17).

Finally there is Rest in all the work fully and well accomplished on the one perfect day, the seventh, referred to already as the one in which the temporal and eternal are united, as they are united in Christ in whom all God's purposes are to be consummated (Eph. 1:10; Col. 1:20).

There is also an historical outline contained in the chapter - ("Known unto God are all his works from the beginning of the world" Acts 15:18) - which has not been touched upon, as it would prolong this paper unduly to take it up now, but doubtless sufficient internal evidence has been developed of a logical, orderly progression of idea, corresponding with the development of spiritual life as set forth in later scriptures, to show that the Holy Spirit of God is the author of the whole book, according to its claim.

It has long since been remarked that the order of development shown by the researches of geology corresponds with the order shown in this chapter, viz: - that the earth was once a watery waste, later solidified, and that vegetable life appeared first, followed by fish and bird life (in the order named), then the quadruped and finally man. Also the simple processes of nature which we see in operation around us every day teach us the principal of new, or resurrection, life, on a higher plan, after death, which is in perfect harmony with the teaching of the Bible throughout. Therefore, this view has the advantage over the theory of evolution that it rests on present-day facts "which may be observed and verified as often as we like," and is not based on hypotheses and deductions, as the theory of evolution notoriously is. Furthermore, those who hold the view which agrees with the Bible teaching are not obliged to retire behind a curtain of many millions of years, like the spiritualist into his dark cabinet, "their wonders to perform," but can point to present-day processes of nature in confirmation of their faith. Finally, if nature set out on her career by a gradual process of developing lower into higher forms of life, at what point did she reverse the process, so that what we observe taking place every day, at present, is the exact contrary of this method: i.e. we see lower becoming higher forms by being lifted up by the higher, losing their former identity in the process, and being translated, as it were, into a new form of life. In other words they are raised by power from above, and all power on the earth is from above. The rain from heaven which sustains life on the earth was raised by power from above in the sun. The heat value in coal, oil, and wood was all derived originally from the sun, or in other words, from above. Furthermore, man is able to control, within certain limits, and very larely to modify the forms and habits of the animal and vegetable kingdoms beneath him, although if he withdraws his controlling guardianship they very quickly revert to their original type. But their showing themselves to be plastic to a limited extent in the hands of a higher power, in accordance with the law which we see in operation everywhere, proves exactly the contrary of their being able to raise themselves. Yet Mr. Darwin, in his "Origin of Species" devotes the first half of his book to citing what everyone knows, and has known from very ancient times, of the possibilities of breeding to modify species "only so long, however, as man, the higher power, retains his control" as a basis for his argument in the latter half of his book that the species have been able to modify themselves along the same lines, or in other words, to raise themselves to a higher plane by power from beneath, or within themselves. Nature's laws do not contradict themselves after this fashion.

"Now of the things which we have spoken this is the sum:"

If man has come up from an indefinite past of millions of years, slowly advancing and developing, then he is bound for an equally indefinite future and has no positive knowledge of any authority to which he is responsible for his actions beyond the light of his own reason. As a matter of fact, this logical result of that position is now being taught in the principal universities as the only binding code of morals. If this be true, then the Bible is a collection of the most colossal falsehoods that could be imagined, for it makes the most definite and precise statements regarding man's past and his future, as well as his responsibility to his Creator-God, and claims the authority of the eternal God for these statements. It is time for the learned gentlemen who have been patronizing the Bible as a good book, and at the same time flatly denying its statements, to take their stand either with it or against it. One cannot consistently hold to evolution and the Bible at the same time, for there is nothing equivocal about its statements.

The teaching of evolution is shown to produce lawlessness, so far as any fixed standard of righteousness is concerned, and there is no doubt that the rising tide of lawlessness and crime witnessed in recent years is largely due to the discrediting of the Bible by its false friends, and the consequent loss of respect for its authority as the revelation of God, which in turn removes the restraint which it formerly exercised over the actions of men.

* * * * * * * *

# WHY GOD CALLED HIS CREATION GOOD

William J. Tinkle
Ball State Teachers College
Indiana

At the close of the third, the fourth, and the fifth days of creation God stated that what he had created was good. At the end of the sixth day he looked over all that he had made and pronounced it all very good. He did not say that it was primitive with potentiality for development, but good. These statements have been a point of attack for the critics of the Bible. The Christian, on the other hand, recognizes that God's evaluations are true and his decisions are final, so that the created world must be good even if it does not fit our convenience. But the discoveries of science help us to appreciate how God's statements were true.

The beneficence of water, oxygen and sunshine can not be doubted, but man does find foes and competitors among the creatures and organisms that make life difficult for him, and in his rage he hurls at them every epithet except good. It is true that there is a school of philosophers who say that we should acquiesce meekly and accept the world as we find it. Their spokesman was Alexander Pope, who stated in the "Essay on Man," "Whatever is, is right." Now Pope's plan was to place in polished form the ideas of the leading men of his day. These leading people in Western Europe during the eighteenth century were enjoying immense prosperity and so appropriating to themselves much more than they had earned. Little wonder that they could say, "Whatever is, is right."

God has the power to force all of his creation, including man, to do absolutely right. To explain why he does not do so would take us too far into theology. But the problem of this paper is to dissolve the discrepancy between God's evaluation and our evaluation of the universe. When we fail to agree with his statement it may be that we do not have the right definition of the word good, or it may be that the organic universe is not so good at present as it was when it was created.

Among created things that seem bad we find weeds, insect pests, predatory animals, pathogenic parasites, and defective strain of people, animals, and plants. As for weeds, many of them have uses. Dandelion furnishes a wholesome salad, medicine from the root, and honey and pollen from the flowers. These products of the flower are valuable to bees, coming at a time when workers need to be raised to gather later harvests of honey, notably that from clover. Weeds cover the soil, tending to prevent erosion, and add their decaying bodies to improve the soil. Man in his haste forgets that soil should be covered to protect it, and nature, looking to the centuries ahead, beneficently covers it with plants. But some of these plants conflict with man's plans to raise crops, and he calls them weeds. Many weeds are controlled with little difficulty at their original home but in a new region are more troublesome. The prickly pear cactus, *Opuntia*, is a useful feed plant, and I recall a farmer in Texas holding it over a little fire with a pitchfork, burning off the thorns so that he could feed it to his cow. It was taken from the United States to Australia to serve as pasture for the cochineal bug, *Coccus cacti*, from which dye is made. In the new country it spread and covered an area larger than England and Wales by 1922.* In 1925, 3,000,000,000 caterpillars of the moth *Cactoblastis* were distributed and they have practically eradicated the cactus.

This principle of increased reproduction following migration applies also to mammals, birds, and insects. At its home a species fitted well into the economy of nature but in a distant part of the world it thrived too well and become a pest. An

---

* Enc Britannica

example is the mongoose, Herpestes mungo, which in its native home of India is adroit at killing snakes. It was introduced into Jamaica to rid the island of rats; which it did rather promptly, increasing in numbers all the while, then started to devour birds. By this time the people wished they had not brought it in. Other examples are the rabbit in Australia, the English sparrow and Japanese bettle in America. God said they were good, but not necessarily good for all parts of the world.

Another seeming discrepancy in the Creator's planning is the existence of predatory animals. Some have conjectured that this was not the original plan, in other words that the lion did not eat the lamb in the Garden of Eden. This view is made plausible by the prophesy in Isaiah 11:7 that "the lion shall eat straw like the ox." If this is to be his food in the future perhaps it was his original diet. But on the other hand it is implausible that the lion was once an herbivorous animal in that his teeth are made for tearing flesh, not for grinding plants. If animals did not kill any living things, their food would be very limited. It they eat plants they take the life or at least decrease the vigor of the plant. Remembering that every seed contains an embryo plant, what multitude of lives are snuffed out by a grain-eating animal! Consequently if no life were taken the only food would be fruits, not including the seed, plants and animals that had died a nautral death, and milk, but not in an amount that would rob the calf. Consider also how general this habit of predation is; the wren is as much a bird of prey as the eagle.

But if God planned that one organism should be the food of another it need not make a problem for us. "A day less or more at sea or ashore, we die--does it matter when?"* Death is preferable to a prolonged, decrepit old age. Wild animals and birds seem to be happy and there is no evidence that they worry about impending death. They may not even suffer much when overcome by a predator. When David Livingstone was pounced upon by a lion that crushed his shoulder he testified that he felt no pain. His biographer** states, "Now everything became dreamlike to the lion's victim. He felt neither pain nor fear." Later, however, he felt excruciating pain from the eleven tooth wounds and the splintered scapula but not while he was under the power of the predator.

That this was not an exceptional case is set forth in the book, "Is Nature Cruel," written by an Englishman, J. Crowther Hirst. Writing to big game hunters and missionary doctors he secured the record of sixty-six men who were pounced upon by bears, lions, tigers, leopards, and panthers. Sixty-four of them felt no pain nor terror. If this be true of man with his intense emotions it should be more true of animals. God is merciful to them in the hour of death.

When we consider the bacteria that cause disease, however, who would have the complacency to say that they are good? We must recognize that most bacteria do not cause disease, but decay, thus returning to the soil the plants and animals that have died. It is certain that God created bacteria of decay for soil could not be fertile without them; and it may be that the pathogenic ones arose from these beneficent ones by mutation.

While this statement may seem like speculation, it is certain that comparable mutations have occurred. A mutation usually results in the loss of something, and if the bacteria of disease arose from bacteria of decay they lost the ability to grow by using compounds in the soil; instead of this they live within an animal or person and get food compounds already formed.

Mutations and other chromosomal changes have made many organisms inferior to the created type. Albino rats lack vigor as evidenced by the fact that they have

---

* Tennyson, A, "The Revenge"
**Eaton, Jeanette; D. Livingstone Foe of Darkness, p. 68

been turned loose but failed to become established in the wild. The wild Drosophila is a better type of fly than any of its numerous mutant types, having more vigor and reproductive ability. In the beginning there were no children born deaf, blind, feeble-minded or albino for these mutant types had not yet occurred. The chromosome number, size, and gene content were still as the Creator had planned them. Had we been in the reviewer's stand looking over the parade of creatures we would have agreed that they were all very good.

By his patient and careful breeding for many centuries man has concentrated certain genes in his domestic animals and plants, developing the Jersey cow for butter-fat and the Angus breed for beef, the sugar beet for sugar and the mangel wurzel for cow feed, etc. Man has the ability thus to control and improve by intelligent planning, being made in the likeness of God. In this respect also the organic world is good, being favorable raw material for man's breeding art.

The problem is not entirely solved in this paper for it is a big one, and bound up with the bigger problem of the origin of evil. Our failure to recognize the goodness of the universe comes mostly from two causes: either we do not know the meaning of the word good, or there have been changes for the worse since the creation. Here, as in many other studies, the discoveries of science help us to appreciate that God's Word is true.

I still can't see the good of a rattlesnake! - Ed.

* * * * * * * * *

DISCUSSION:

Dr. Monsma: I noted that Dr. Tinkle distinguished between animal life and bacteria so far as their food habit was concerned. When he speaks of a herbivorous and carnivorous habit of animals he believes that animals before the fall, at the time of the creation, were carnivorous. When it comes to bacteria he wishes to assume that there were no pathogenic forms, but that by mutations pathogenic forms have come about later, after the fall, I assume, and it seems to me there is a little inconsistency there. I am just wondering why we can't assume that before the fall the animals were herbivorous and not carnivorous, and that they have become carnivorous due to mutations; that these teeth changes that he mentions and other changes, no doubt more radical changes too, have taken place by mutations since the fall. The reason why I suggest that is because in Scripture we do have an indication that man and animals ought to be herbivorous. We have no indication that God created carnivorous organisms until after the fall and after the flood. When we read Genesis 1:29 and 30, we read, "And God said, Behold I have given you every herb yielding seed which is upon the face of the earth and every tree in which is the fruit, and tree yielding seed, to you it shall be for food, and to every beast of the earth, and to every bird of the heaven, and to everything that creepeth upon the earth wherein there is life I have given every green herb, (notice the green) for food and it was so." That is immediately followed by this statement, "And God saw everything that he had made, and behold it was very good."

Mr. Turekian: Some dinosaurs were carnivorous, and were here before any indication of a fall.

Dr. Bullock: In that connection there is one verse that was brought to my attention some time ago and that is the sanction that God puts upon the carnivorous habit in the Psalms. In the 104th Psalm, for example, in the 21st verse we are told that "the young lions roar after their prey and they seek their meat from God." In other words they are getting sanction put on their carnivorous habit.

Miss Erdman: I appreciate Dr. Bullock's position because this is a matter that has been bothering me for an awful long time. I think that instinctively we turn from the idea of carnivorous animals as being good, at least it doesn't seem as though that is the kind of thing our God would approve. And yet as I have been thinking about it over this past year, the problem that Mr. Turekian has brought up of course is very obvious. Either the animals were carnivorous before the fall, or there must have been some sin on the earth before the fall. But there is another problem involved here too, and that is the balance of nature. If it were not for the carnivorous habit of animals, I wonder if you realize that man, within the space of probably about two days, would be run off the face of the earth simply because of the rapidity with which certain insects and bacteria, etc., would accumulate. I am sorry I don't have the figures on that with me now, but it has been demonstrated in some classes in which I have been that if it were not for the carnivorous habit in animals, man simply would not exist any longer. And this to me is at least another facet in the situation that ought to be considered, possibly pointing out that the carnivorous habit was ordained of God in order to keep things -- well, to make the earth a fit habitat for man.

Dr. Bullock: I'd like to give a first hand example of that. In the state of New Hampshire, for a good many years, there has been a bounty on bears and an overprotection of deer, so that at the present time there is a very serious danger of over population of deer, and by the way the overpopulation of deer hasn't served to the benefit of the hunters. The bear are actually being faced with extinction. So it is now being considered to remove the bounty of bears and actually protect the bear.... That is just one example that we have at the present time of the effect of the carnivorous habit.

Dr. Monsma: Like Miss Erdman, I am very much interested in the question because I think it is very important. I believe that what Miss Erdman has said, of course, applies to today, and I also believe what Dr. Bullock has said about conditions the way they are today. But, I believe that we should distinguish between today's conditions and the conditions before the fall. We are, after all, interpreting things upon the basis of Scripture and nature as has been said, and that is why I believe that we should take Scripture into consideration. Now, one of the greatest and most tremendous things that is related to us in the Scriptures is this matter of the fall. And we must take that into consideration in our interpretation. The effects of the fall are tremendous, and I'm afraid that sometimes we forget about that. As far as the cycle of nutrition in nature is concerned, well, I think that plants form a very important part, and bacteria too. In the main part of that cycle animals and man are on the subsidiary part depending upon that main cycle. Present conditions, I will admit, would not allow of all the organisms, animals and man being our basis, but we do not have the ideal condition at the present time. I think we should always take the tremendous effects of the fall into consideration; and we don't know exactly what the conditions were there but there is no indication in Scripture that there was anything of a carnivorous habit before the fall so far as I know it. The quotation that was read here from the Psalms, of course, in the first place, meat refers, I think, to food in general. In the second place, I am willing to accept that it is meat but that of course is a condition after the fall when the eating of meat has been approved and was put into that particular situation.

In this matter I am not trying to be dogmatic.... I believe you got from my paper. It is a big question in which we are just feeling our way, but there are some more things that need to be said. In the first place, the bacteria of decay are not ...They are sacrificed: they live upon dead material. So, that hadn't ought to make a discrepancy in my paper, it seems to me. It would be very hard for plants and animals in general to be sacrificed. The amount of food would be very, very much limited, you see. Of course, there are some animals that are sacrificed. Little water slaters, for instance. A little animal about the size and the build of the sow bug which is perhaps more familiar. It lives in water. It lives on dead plants in the water. So there are some animals, of course, that live on dead material but most animals would find it hard to do that. And then about the lion and other carnivorous animals getting the teeth that they have through mutation. Mutation doesn't usually give us that sort of thing. Those teeth are too well formed. There is nothing defective about them as we usually find in a mutation. We must guard against the habit that if we think a certain change would fit our theory to say, "Well, a mutation gave that sort of change." Mutations will give us certain types of things but they don't give us just everything. And then, I do not believe in making such a big distinction between animal life and plant life. You go down to the smaller types of animals and plants and you finally get to the place where there is no distinction at all, and of course every biologist knows you find some organisms that are claimed both by the botanist and the zoologist. And, if animals do not bear excruciating pain and if they do not live in terror because of impending death, it seems to me that their becoming the food of another organism isn't so bad as we are inclined to think. We are inclined to mash a finger and, oh, how that hurts, and then we think if we break an arm or a leg or if we should have a worse accident the pain would be in proportion. In many cases it isn't at all. When I had my arm broken the doctor gave me a handful of pills to take when the pain became too great. I didn't take a one of those pills. But I have mashed a finger a number of times and it always hurts. We have to beware lest we make these predations in nature really worse than they are. I wouldn't insist too much on this point but it seems to me it's the general plan of God that certain organisms should be the food of other organisms.

Mr. Saarnivaara: We read in the Bible that after Adam and Eve had fallen into sin God said to Adam that he had to eat his food with the sweat of his eyebrow and that the earth would produce for him thorns and thistles. It seems that these words of the Bible indicate that some definite changes took place after the fall in nature. Possibly thorns and thistles existed before the fall, for we read that God created every kind of plant, but it is probable that when God cursed the earth after the fall such a change took place that the thorns and thistles and other weeds and also harmful bacteria and other things belonging to that category began to increase and disturb the happy life of men and perhaps even animals, so that some changes in the nature took place after the fall. For we read that God cursed the earth because of man. And because God cursed the earth we may conclude that after this curse the earth became different from what it was immediately after creation, so that the words of God--everything that He created was good--do not directly apply to the conditions after the fall. They apply to the conditions before the fall. And the second thing is that in the flood or deluge great changes took place again. We read in the Bible that God destroyed the face of the earth. He did not destroy the whole earth but the face--the surface part of the earth was destroyed and changed. And only after the fall we read of the seasons of the year and of cold and heat. It is quite obvious that this earth became less fertile and its climate became more rigorous after the flood. Therefore, two changes have taken place after God created this earth. The first change through the curse of God after the fall into sin, and the second change through the judgment of God through the flood. Therefore, calling good the works of God and everything that He created does not apply to the earth as it is now. It applies only to conditions immediately after creation. This earth is in a cursed condition in a two-fold sense and because of that there are so many troubles and things for us until...we know that God is the standard of what is good and what is bad. And if God created carnivorous animals and said that they were good for the sake of the balance in nature and for other reasons, of course we have nothing to add to the statement because God determines what is good and what is bad.

Mr. Turekian: There is nothing original, that I can see, about that change of the face of the earth. I think Dr. Kulp said it for us, that a flood of that dimension would have hardly any effect on the face of the earth. And possibly there was weather before the flood too because it seems that we had at least several ice ages with varying weather. I think most geologists will agree that there was a variation of weather of some sort over the face of the earth before the flood.

Dr. Mixter: Regarding mutations being good or bad. Most of them are bad, but a small percentage apparently are good for this reason: natural species which can be crossed are different in mutational differences of a greater or lesser amount and those natural species existing in nature are apparently very well adapted to it. The differences between different species of cotton and between different species of fruit flies are really mutational differences. Now we have not seen enough mutations over enough periods of years to actually see one species change into another as the result of mutations, but we can find that they do differ in the mutations which they possess. The differences that they possess are differences of good qualities, so I think it is unsafe to argue because most mutations found in the laboratory are not valuable that none of them have been or could be.

Dr. Tinkle: We have to guard, however, against one possible error there. The differences that we find between varieties and species that have come from the same cource are not all mutations. Some of them are segregations, and if these original forms were created heterozygous then that gave an opportunity for several differences to arise--quite a number of them in fact--limited in extent but not all of these differences then would be mutations, but some of them segregations just as we find differences in the progeny of a cross. Cross two different kinds of peas--the different shape of the seed, the color of the seed, the color of the flower, and so on; then in the second generation you will find diversities that are segregations.

* * * * * * * *

# THE PALEONTOLOGY OF THE HORSE

Cordelia Erdman
Instructor in Geology
Wheaton College

One of our contemporary historical geology texts contains this statement: "The horse was a native of North America from early Eocene to late Pleistocene time and underwent most of its development here. Skeletons assembled from successive horizons reveal a gradual evolution in teeth, limbs, feet and size hardly equaled for any other stock of animals." (Dunbar, 1949, p. 468). Horse evolution has long been cited by evolutionists as a classic case of directional progression. On the other hand, non-evolutionists have been inclined to question the validity of the fossil sequence upon which the case is based; or at any rate they have wondered how unconsciously subjective the arrangement has been. Therefore it is worthwhile to make a brief survey of the following situations:

1) The postulated ancestry of the horse.

2) The relative ages of the strata containing equid fossils

3) The trustworthiness of this postulation.

The Postulated ancestry of the horse. Due to the capricious nature of fossilization, especially the fossilization of vertebrates, very often a group which is most interesting proves to be the one which is represented only by scanty and scattered remains.

Witness the known occurrence of only three specimens of birds in all the rocks of Jurassic age--one of them represented merely by a feather! One hesitates to draw conclusions from such evidence because there are few boundaries to check the free play of imagination. It is with relief, then, that a paleontologist turns to consideration of the perissodactyls, the mammalian order to which the horse belongs, for there is no dearth of fossil horse material. Romer declares, "Among all vertebrates there is perhaps no group whose fossil history is better known than that of the order Perissodactyla, the 'odd-toed' ungulates" (Romer, 1936, p. 319). Several of our great museums are store-houses of skeletal material which is sufficient for much research, although the specialist constantly meets problems which make him wish that these collections might be augmented. It is true that complete skeletons are rare, but there are enough skulls and long bones and partially complete skeletons to afford evidence of the many types of horses which have existed.

Not only is there an unusual abundance of horse remains but also a widespread distribution of those sites where such remains have been found. Thus our knowledge of the history of this group is not limited to one locality but encompasses a considerable span of space and time.

Based on this cumulative evidence, the following horse series with its concomitant changes is in good standing at present: The first true horse of North America is named Eohippus because of its occurrence in Eocene strata, although it now appears that the European Hyracotherium is the same genus and that that name should have priority. This little animal was about a foot high, the size of a fox terrier, with a short, slender face and eye orbits in the center of the skull so that the face length was one-half of cranial length. Its limbs were relatively long and slender and were unique in that the front feet bore four toes each, whereas the hind feet bore only three. Vestigial toes brought the number to a total of five for each foot. The teeth, always of significance in vertebrates, were low crowned and in most species were bunodont, a condition in which the surface is smoothly undulating rather than ridged or folded. From this it may be inferred that Eohippus was omnivorous in diet, just as from his feet we infer that he was a forest dweller rather than a plains animal. A short gap or diastema existed between premolars 1 and 2, and all the premolars were simple and triangular. Romer comments, "[Eohippus] seems surely the beginning of the horse line but is probably close to the stock from which came the other odd-toed ungulates as well" (Romer, 1941, p. 143).

In Middle and Upper Eocene strata a slightly different horse is found, namely, that which is called Orohippus. This animal was much like Eohippus, but larger. The central digit of the foot is relatively and absolutely larger than that of Eohippus, and the splints of the non-functional toes of the hind feet have disappeared. The gap between premolars 1 and 2 persists, but the latter premolars (3 and 4) have lost their simple shape and have become molariform.

In Epihippus of still later Eocene time the premolars 3 and 4 have become morphologically identical with the molars, and there is no gap between 1 and 2.

The ensuing Oligocene epoch witnessed the appearance of Mesohippus. This horse was about the size of a full grown collie dog or a sheep. Distinctive features are the presence of only three toes both front and back, all of which touch the ground; the reappearance of the diastema; the molariform appearance of the second premolar, and the beginning of ridges on the crown of the cheek teeth (a condition known as lophodont dentition).

In later Oligocene times (Upper Oligocene strata) Miohippus exemplifies these same characteristics but in a larger body (about three feet high at the shoulder).

Miocene time is notable for a great and important change in the dentition of the new forms which appear. This change is the introduction of hypsodonty, or a high-crowned condition. The height of crown in Oligocene forms was only approximately half that height attained by Miocene forms. It is believed that this transition accompanied a change of habitat, and that from this time on the horses became dwellers on the plains where the rough grass demands high crowned teeth.

Parahippus is the horse in which hypsodonty is first suggested. Also in Parahippus teeth for the first time have some cement covering the crown. The diastema is well developed. This horse, not quite the size of a small pony, had the central of its three toes very pronounced, but its side toes were barely touching the ground. Merychippus of the Miocene attained pony size. Its dentition was definitely hypsodont (high-crowned) with cement playing an important part and with the pattern of the cusps essentially as in horses of today. The side toes were even more reduced than in Parahippus and failed to touch the ground, so that weight was now being borne by the central toe. Accompanying the new dentition was an enlargement of the jaw which lengthened the face relative to the cranium, thus displacing the orbits to the posterior. (Merychippus is thought to have given rise to another pony-sized form, Hipparion, which migrated to Europe in the next epoch).

It is postulated that the main line of equine evolution continued in North America during Pliocene time. Pliohippus continued the trend toward larger size, and according to Edinger there is no sharp demarcation between this and the Merychippus group. However, the average member of the genus was the size of a large pony. The upper teeth are notable for a strong curvature anteriorly. The length of its jaws approximated the modern status. Some species became truly mondactyl (one-toed) but retained splints of the side degits. By the beginning of the Ice Ages of the Pleistocene epoch true one-toed horses were almost universally present.

Equus, the one-toed horse, had and has members that are larger than any of the preceeding forms, though our familiar Shetland pony illustrates that not all members of the genus were outstanding for size. The teeth of Equus are larger, higher and more complicated than those of his fore-runners. For some reason yet unfathomed all horses became extinct in North America at the close of the Ice Ages and were re-introduced by man.

This series of horses is considered to be a phylogenetic series, that is, one in which each type is ancestral to the next. Of course this ancestry cannot be proved, but if the sequence is a valid one, it seems overwhelmingly probable that it is a phyletic sequence. And if it is a phyletic sequence, then one must admit that a certain amount of spontaneous change has taken place within this group. Tilly Edinger, who has done extensive research on the brain cases of these various forms, says, "Ours is not a series arranged by human judgment, but a phyletic series from consecutive geological periods; it represents the progressive evolution of one brain" (Edinger, 1948, p. 158). Thus she demonstrates her faith in the validity of the sequence as a phylogenty.

Relative ages of the strata containing equid fossils. In order to ascertain whether various horses did appear in the fossil record in the sequence which has been indicated, we must determine the relative ages of the strata containing them. This problem of equating units of time as they are represented by rock layers belongs to the realm of stratigraphy. Stratigraphic studies are, therefore, basic to the construction of any portion of a geologic timetable. Rule 14 of Nomenclature and Classification for the Geologic Atlas of the United States, reads, "The fundamental data of geologic history are: 1) local sequence of formations; and 2) the chronologic equivalences of formations in different provinces. Through correlation all formations are referred to a general time scale . . ."

Geologists admittedly have made many errors in correlation. Grabau tells us that "before fossils were extensively studied, similarity of superposition and lithic identity were taken as the guides to synchroneity..which..led to many erroneous correlations" (Grabau, 1913, p. 1121). It is implicit in this statement that the key to correlation lies in the fossils themselves, but there are methods of correlation which are valid and which are independent of fossil evidence. Schuchert explains the underlying principle of stratigraphy which he calls the Law of Periodicity: that there are cycles of earth movement, of sea invasion and land emergence, of erosion and sedimentation and cycles of organic evolution. Thus at no one place on the earth is there a continuous record of sedimentation of life since the beginning of time. Rather, the record in any one place is somewhat piece-meal and interrupted by gaps. A widespread emergence with ensuing erosion thus gives convenient demarcation for a unit of time and serves as a basic for correlating the strata deposited immediately before or after the emergence.

The non-paleontologic methods of correlating strata may be summarized thus, following Dunbar: 1) lithologic similarity; 2) similar thickness; 3) similar position in a sequence of formations some of which are known to be equivalent; 4) continuous tracing between outcrops; 5) lateral gradation and intertonguing of one into the other when the formations are lithologically dissimilar.(Dunbar, 1949, p. 12). All of these are locally useful and fully utilized by the geologist. Nevertheless there are many occasions when none of these criteria are appropriate and in such cases the paleontologic methods must be used. These are 1) occurrence of identical faunas, and 2) fossils that are dissimilar but are known to occur together elsewhere and therefore indicate the same age (Dunbar, loc. cit.).

Dunbar defends the paleontologic method of correlation by saying, "We now know, of course, that different kinds of animals and plants have succeeded one another in time because life has continuously evolved; and inasmuch as organic evolution is world wide in its operation, only rocks formed during the same age could bear identical faunas" (Dunbar, 1949, p. 8). Thus, since paleontologic correlation rests upon an evolutionary concept, we may not use such correlation to determine authentic horse phylogeny.

Actually, although the paleontologic method is most widely used for dating rocks, most geologists agree that it must be used with caution. Schuchert points out that very dissimilar faunas of the same general types of animals may be equivalent in age (Schuchert, 1923, p. 23). Nearly forty years ago Grabau cautioned that although "..most modern stratigraphers believe in the possibility of essential chronological equivalency of formations characterized by the same faunas..such equivalency is not necessarily indicated by the similarity of faunas, and..a given fauna may appear earlier and continue longer in some sections than others" (Grabau, 1913, p. 1125). He further declared that chronologic equivalence "has been repeatedly assailed by emminent geologists," (p. 1124), and cites Huxley, Geikie and H. S. Williams. More recently James Gilluly, past president of the Geological Society of America, made the following significant comments:

> If several species that arose in different centers eventually come to live in a particular area, it is not only conceivable, but has been shown by careful collecting, that there order of superposition may be different in different parts of the area. In other words, 50,000 year..units (are) far too short for distant correlations. When we take into account the entire fauna we never find such reversals, but the fauna as a whole does not, of course, change so quickly as the individual species constituting it. These reversals, therefore prove that even in theory it will never be possible to be certain of the precise equivalence of beds in widely separated regions... The finer the sub-divisions the paleontologist recognizes,

the more likely he is to find such reversals to superposition. This point is philosophically crucial, for all our geologic histories are based on paleontology and it can therefore never offer us a chronology capable of indefinite refinement" (Gilluly, 1949).

What, then, of the stratigraphy that is pertinent to the horse? There is evidence that early in the Cenozoic Era there was considerable disturbance of the earth's crust along the belt of the present Rocky Mts. which caused the structure of the Rockies to be defined and at the same time brought many of the great basins of the West into being. Streams flowing down from these mountains locally deposited layers of sediment on the plains to the east, but to the west, sedimentation was mostly confined to filling of the basins. Hence the Cenozoic layers of sediment west of the Rockies occur chiefly in structural basins such as the Uinta basin of Utah and Colorado, and the Bighorn, Wind River, Bridger and Wasatch basins of Wyoming. No layer designated as Eocene is found east of the mountains.

In mid-Cenozoic time there came a gentle warping along the Rocky Mountain axis which eventually caused the streams to cut down through their own deposits so that these ancient layers are now exposed to view. It is in these Cenozoic deposits that the fossil horse is found. Correlations from east to west of the Rockies can be made only by use of paleontologic methods, because no strata are continuous between the two areas. Furthermore, for the same reason, even inter-basin correlation west of the Rockies must be made by comparison of faunas. Thus in order to achieve an unpredjudiced picture we must lay aside the time-designations given to these various layers. We turn, then, to an examination of the geographic locations from which the fossil horses are known.

Little Eohippus is found at Greybull, Wyoming, in the Bighorn basin. Orohippus is known from the Bridger Basin, and Epihippus from the Uinta basin.

Mesohippus has a wider geographic distribution. Species are known from Montana, but are relatively abundant in the South Dekota Badlands. Here Mesohippus occurs in four successive horizons (Chadron; lower, Middle and Upper Oreodon zone of the Brule). In the immediately overlying horizon (Leptauchenia zone of the Brule) Miohippus has been discovered. Miohippus, in turn, is known also from the Lower Rosebud formation of South Dakota which is stratigraphically above the Brule formation. Above the Rosebud is the Harrison formation of Nebraska in which the same species of Miohippus occur. (Other Miohippus fossils lie in the John Day basin of central Oregon). In both the Lower Harrison and the Lower Rosebud primitive species of Parahippus make their appearance.

Members of the genus Parahippus become very abundant in the next highest layer (Upper Harrison), and it is further found in the Sheep Creek beds which overlie the Harrison, and finally in the stratigraphically higher White River beds. The same species are known from Nevada, Oregon and California. The Sheep Creek individuals are associated with primitive species of Merychippus, whose more representative species are found in the Pawnee Creek beds of northeast Colorado which appear to be stratigraphically above the Sheep Creek beds of Nebraska and below the Little White River beds. Merychippus species are also known from California and New Mexico.

Pliohippus has a geographic range which surpasses those of the proceeding forms. It is known from the Upper Snake Creek beds of Nebraska where it is just above a layer that contains a species of Merychippus. It is found also in association with Parahippus in the Little White River beds of South Dakota, and occurs in the Fort Niobrara formation of Nebraska which succeeds the Little White River. Additional species were fossilized in California, Texas, Oregon, Nevada, and Colorado.

Equus, the genus to which our modern horse belongs, left his early remains in Osborn's Loup River formation, which comprises glacial deposits of the Great Plains.

In summary, Eohippus, Orohippus, Epihippus, and Equus are stratigraphically isolated either in basins or otherwise, and therefore their relative ages cannot be determined without resorting to paleontologic methods of correlation, a procedure which we have ruled out a priori in this case. The remaining horses which have been mentioned all have representatives in South Dakota-Nebraska strata whose order of deposition can be determined fairly easily. Therefore at least this portion of our originally cited sequence appears valid:

Mesohippus is succeeded by Miohippus; Miohippus is partly contemporaneous with Parahippus but is eventually replaced by it; Parahippus subsequently is joined by Merychippus which then takes the ascendency while Parahippus continues. Pliohippus is found in beds above Merychippus and above those in which the final forms of Parahippus occur.

Obviously it cannot be proved definitely that any one of these forms was not contemporaneous with any or all of the others. This is a negative argument, albeit an important one. If, however, all these forms were contemporaneous, it is an almost unbelievable coincidence that they just happened to become fossilized in a deceptive sequence, for as Scott says, "Between Hyracotherium and Equus there is an immense difference in all parts of the dentition and skeleton, so great, in fact, that without the intermediate steps of modification, hardly anyone would be so bold as to assert that Equus had descended from Hyracotherium" (Scott, 1941, p. 911).

The trends represented by these successive forms are unmistakable: the increase in size, the development of a monodactyl condition, the appearance of hypsodont, curved teeth, the lengthening of the face. Surely there is reason to fit the Eocene horses into the picture, since they readily fall into line with these trends, and it appears equally valid to regard Equus as the culmination of the processes of change.

Our discussion thus far has dealt with the postulated main line of equid evolution, showing that certain genera which probably descended from one another have manifested trends of development. This same phenomenon can be observed on the level of the species. The Sheep Creek beds of western Nebraska have already been mentioned. Here in different horizons of a layer of sandy shale Osborn found fragments of five individuals of the species Merychippus isonesus. Each of these had distinguishing characteristics of sufficient magnitude that Osborn considered them five separate varieties. The bottom-most individual he assigned to the variety M.isonesus primus, and used secundus, tertius, quartus, and quintus to designate the others in the order of their superposition. Each one was larger than the one which occurred just below it, although the limb bones in M. isonesus quartus were the same weight as those of M. isonesus tertius. The limbs of the second, third and fifth varieties form a series of increasingly robust limb shafts. In M. isonesus secundus the ulna and radius are compressed but not fused. In the overlying variety these bones are actually fused for a short distance. These and other successive changes in this series vividly demonstrate that horses were constantly in a state of flux and that their changes exhibited a significant amount of pattern even in minor trends, or minor portions of major trends.

The whole question of pattern in the development of successive genera is fascinating. Granted that our horse series is a phylogeny, we have then admitted that a certain amount of evolution has taken place, and further, that this evolution has not been haphazard but has proceeded along well-defined paths, including increase of body size, development of one-toed feet, and so forth. Such "straight-line" development has been called orthogenesis, and according to Romer "has played a prominent part in paleontological thought," being "usually coupled with the supposition that some mysterious principle lies behind the observed phenomena"-- "a deity, or 'Nature'" (Romer, 1949, p. 105, 107). Romer continues by pointing out that the history of the horse was long thought to be undeviatingly orthogenetic, but that now the family tree is believed to have many branches and that the succession which we have presented is only the unbranched portion of that family tree. Then, since the definite trends which we have cited in this succession cannot be ignored, he offers this summary of current thought upon the subject, following closely after Simpson (Simpson, 1949).

> Orthogenetic phenomena, then, are probably much less common than they were thought to be. But even in cases where straight-line phenomena are present, there is no need to postulate any teleological principle to explain them. Phyletic lines of this sort are reasonably to be considered as due to orthoselection, process of increasingly improved adaptation to a relatively stable environment. Under such conditions any deviation from the 'normal' line would be negative as to survival value, and would tend to be eliminated; the potential branches of the 'tree' would tend to be pruned by selection before they became marked enough to become apparent in the fossil record" (Romer, 1949, p. 107).

Thus the horses have given us strong indication that directional change has occurred. Incidentally, to call this change "progress" is only valid insofar as it can be demonstrated that Equus is better off in his environment than Eohippus was in his, and this, of course, cannot be demonstrated. Beyond this point the paleontologist cannot go, and he must turn to genetics for an explanation of how such changes may have arisen initially.

## BIBLIOGRAPHY

Dunbar, Carl O. 1949. Historical Geology. John Wiley & Sons, Inc.

Edinger, Tilly. 1948. Evolution of the Horse Brain. Memoirs of the Geological Society of America, Memoir 25.

Gilluly, James. 1949. Distribution of Mountain Making in Geologic Time. Bull. of the Geological Society of America, vol. 60.

Grabau, Amadeus. 1913. Principles of Stratigraphy. A. G. Seiler & Co., New York.

O'Harra, Cleophas. 1920. The White River Badlands. South Dakota School of Mines, Depart. Geol., Bull. 13.

Osborn, Henry F. 1890. Equidae of the Oligocene. Memoirs of the American Museum of Natural History, n. s., vol, 2, p. 87-100.

Romer, Alfred S. 1936. Vertebrate Paleontology. University of Chicago Press.

_______. 1941. Man and the Vertebrates. University of Chicago Press. 3rd ed.

_______. 1949. Time Series and Trends in Animal Evolution, in Genetics, Paleontology, and Evolution. Princeton University Press.

Scott, William Berryman. 1941. Part V. Perissodactyla, in the Mammalian Fauna of the White River Oligocene by Scott and Jepsen. Transactions of the American Philosophical Society, n.s., vol. 28.

Schuchert, Charles. 1923. Stratigraphy of the Eastern and Central United States. John Wiley and Sons, Inc.

DISCUSSION:

Dr. Mixter: Eohippus is North American. There was an equivalent specimen in Europe called Hyracotheriam, which Miss Erdman referred to. That apparently became extinct rather early. This sequence is all North American, and this portion of it, as she emphasized, is found in superimposed strata so we're pretty sure that these specimens were descendents of one another.

Dr. Saarnivaara: Professor Frank L. Morse, says in his book, Creation, Evolution, and Science, which was published in Washington, D. C., in 1944, that the hoof of a modern horse has been found in Colorado deeper in conformable strata than Eohippus. Do you know anything about that? And how to explain it correctly?

Miss Erdman: I have heard rumors of Equus, (that is the modern horse), occuring in strata earlier or deeper in the earth than Eohippus, but in none of the literature with which I was dealing did I come across any reference to this. And I have never had any specific documentation for that. I would be very glad to know about it. I'd like to know who found if, if he was a competent geologist, if he knew his stratigraphy, and what the stratigraphic conditions were at the site of the discovery. And before we have those facts I don't believe we can make any comment; at least we can't make any definite commitment as to the occurence of Equus in much older strata. If we did find Equus in that much older strata it would certainly indicate that this horse sequence was merely a coincidental sequence, which as I tried to point out in my paper I think would be a very large coincidence. I am afraid that I can't give you any positive answer on that because of not ever having found the documentation for it.

Dr. Monsma, Chm.: Probably you have the documentation for it Dr. Saarnivaara.

Dr. Saarnivaara: I have read it in the book I mentioned.

Miss Erdman: You have read it in a book but, you see, if the book was quoting someone else, you would really have to know who he was quoting.

Dr. Kulp: This is quite trivial, but I wonder if Dr. Hartzler or Dr. Bender or someone from Goshen couldn't go in the library and get that book of Morse's and see if there is a reference. I certainly would say that if this is true, it is a very significant thing. But we have also read in some of the literature labelled "apologetic" and written by people who think that all geology can be interpreted in six thousand years, the statement that there was a shoe found in Mesozoic sandstone out in Nevada. There was no reference given there either, and that sort of thing you have to treat like flying saucers, I'm afraid. We should be very diligent to find any kind of available data like that, but the chances are that the specimen was taken out of a cave or some loose slump material. We must be very careful, but we certainly should go after it. If someone thinks they have some information, let's tract it all the way down and find out where it is.

Dr. Tinkle: I came across that same statement but it was several years ago and in a non-scientific newspaper; therefore, it requires much more search. But I am very much interested in it. But what I rose to ask about is that we have more of a discussion on size. Now there are some three toed horses that have been found but have considerable size. But again I have not read this recently and I cannot give the names or just where they are found, but I should like to have more discussion on that matter.

I have some reasons for wanting to know more about size. In the first place the loss of toes can readily be explained as has been explained through mutation. But if there was a gradual increase in size at the same time that there was a gradual decrease in the number of toes through mutation, that's something that we don't find in our modern genetics. It usually doesn't work that way. It isn't often that mutation gives us added size and for a series of mutations, each one to give us an added size, would be something that is almost not found at all. And then, it may be that environment has something to do with size. Perhaps some of you saw the small horses that were taken from a canyon in California several years ago. It seemed that those horses had been trapped there a number of years ago, at least they could not get out. They could see them from the heights above. Finally they rigged up an expedition, let some men down in a cable basket and after considerable difficulty they caught three of these horses and brought them to the surface and exhibited them rather widely. How many of you saw them? I see two hands, I believe. Their backs were about the height of these tables. I saw two of them and I asked them about the third. They said the third one was taking care of a colt and, therefore, they did not care to exhibit it at that time. And then I asked them how the colt was getting along. And they said, very well, and they could see that it would be considerably larger than its parents, which indicates clearly that at least part of the small size of those horses was due to environment: they just had not gotten enough to eat. Now, it may be that the size of some of these fossil specimens was determined to a certain extent by their environment. That is something that is a little hard for us to check on at the present time. However, in cases where there are a large number of fossils we perhaps would not think that the small size was due so much to lack of food. I'd like to have a little more discussion on size.

Dr. Mixter: There is quite a discussion on size in Simpson's *Tempo and Mode in Evolution*. I'll give it to Dr. Tinkle.

Don Earl Boatman: I'd like to know on what basis they call the *Eohippus* a horse. It is my understanding that there were 300 major differences and I'm just a layman and trying to exercise some horse sense, but it seems to me that the person who would call that a horse is a jackass. Three hundred differences seems to me would disqualify the creature to be a horse. Now I'm hoping that someone can explain why they call this one a horse. That's for information.

Dr. Mixter: There were 300 minor differences as mentioned by Simpson. You see, he said that there was no appreciable difference between the two. This is the reason for thinking it's a horse. It's a backboned animal; it has hair and nursed its young by milk; it was attached to its mother by a placenta--part of the after-birth; it was a hoofed animal--it had hoofs on each of the three toes and four toes; it was an uneven-number-of-toes animal; it was a hippomorph, that is, horse-shaped in general curvature of the back, and so on; it was an equid, which means in many details of structure it was horse-like, "which is a classificatory way of saying that the vast majority of its multitude of morphological characters was already the same as those preserved in the modern horse." You can find statements in these books that if we only had *Equus*, the top one, and *Eohippus*, the bottom one, we might not connect them, that is, we would think they are entirely different genera or even different families. But it is because we have the gradual stages in between that we connect the two, and it's that sort of thing that has been found in a number of instances between species, between genera, that makes it look as though descent with modification the simplest explanation of their origin. And yet I maintain that the great gaps between the orders indicate that the first members of each of the orders was a creation.

Irvin Wills: I'd like to address a question to Miss Erdman concerning how complete fossil forms of these horses have been found. In other words, have you found complete fossilizations intact or *in situ* of say *Eohippus*, *Eppihippus*, or some of those?

Miss Erdman: I believe that I partly answered this question in my paper in referring to the abundance of horse material which is actually found, and there I stated that we do not to my knowledge have many if any complete skeletons of horses but that we have many partially complete skeletons, that is, certainly the significant parts of the horse, such as the limb bones and the skulls and the teeth which are probably the most important of all in dealing with vertebrates and particularly with horses. We have one specimen here which has a certain part of his anatomy preserved and here we'll have another specimen with a complementary part of the anatomy preserved, so that through comparing the types we are very readily able to compile a complete horse. In other words, if you have overlapping parts of two different specimens sufficiently complete to assure you that you are dealing with similar or the same animal, then it is perfectly valid to supplement one by means of the other, so that although we do not have any one complete series of intact complete skeletons we have enough material to make up such a series.

Dr. Mixter: Simpson's figures show 397 specimens of members of the family from lower Eocene, 54 from middle Eocene, 11 from upper Eocene. In the next stratum, Oligocene, 30 from the lower, 125 from the middle and 39 from the upper.

Dr. Monsma: It is always true however, isn't it, that when these specimens are studied there are two things that the investigators have in mind. One thing is the fact of actually finding the specimen and then the evolutionary idea, how does it fit in with it. I think that always plays a part and I suppose it should, in some way or other if we are studying the subject of evolution.

Robert Fischer: I'd like to ask a rather non-technical question. It's been suggested that the Christian in his own mind accepts creative acts of various types and yet in view of a quotation which was given about some of the previously considered missing links being found, an appreciable number apparently, it seems as though we are in a rather weak position and almost approaching a position of accepting creation merely to cover up our present ignorance. Is our position thus weak or are those really significant missing links definitely not being found.

Dr. Mixter: This is a very significant question. You see the evolutionist believes the gaps will be filled. We have said that the absence of specimens is in favor of the revelation we have which says that essentially there were no specimens filling the gaps. So it's a matter of believing whether they will be filled or not. Simpson said in 1944, that not only is there a gap between Eohippus and previous ancestral types, but this is true of all the thirty-two orders of mammals, and in most cases the break in the record is still more striking than in the case of the horse record for which a known earlier group does at least provide a good structural ancestory. Condylarthra look like horses in some respects so they could have been the ancestors. The earliest and most primitive members of every order already have the basic ordinal character, and in no case is an approximately continuous sequence from one order to another known." "In no case:" says this leading specialist in the subject. "In most cases the break is so sharp and the gaps so large that the origin of the order is speculative and much disputed. Of course, the orders all converge backward in time to different degrees. The earliest known members are much more alike than the latest known members, and there is little doubt, for instance, but that all the highly diverse ungulates did have a common ancestroy, but the line making actual connection with such an ancestory is not known in even one instance." He says there is a systematic deficiency of record and it looks to me as if this systematic deficiency is significant.

Dr. Monsma: It is interesting that you read this word speculation here. I recall a course I once took in the evolution of algae in which the teacher repeatedly stated, "Well, these are speculations--these are evolutionary speculations," and I think that these men will admit that too. When we are dealing with this evolution we have what we called the other day "a great deal of imputation," and it falls within the realm of speculation. Now how much speculation is that? Is there too much speculation to warrant our adherence to the theory or is there an expected amount of speculation here? That does make a great deal of difference. Of course, our viewpoint does enter into the answer to that question, it seems to me, - our fundamental outlook on the thing.

Mr. Breneman; Regarding that matter of the gaps in the record, I would just like to make an observation which is advanced by those who believe must emphatically in evolution. You can take it for what you think it's worth. Experiments have been performed on some of the lower animals at least which show that when you put the animal under severe stress, such as extreme temperature, starvation or dehydration and lack of water, sufficiently severe to produce death in a very considerable percentage of the individuals, that this greatly increases the rate of mutation of the genes. Now, as to the missing record, one of the things that we always notice that whenever there is a non-conformity of the geological record that we see a sudden occurence of a new species on the two sides of non-conformity. Now, what has produced this non-conformity? There is probably no universal answer over the world, but a very common one, of course, is a rise in the elevation of the continent which produces wide erosion, a condition which does not make it very favorable for the preservation of specimens during that period of erosion, and such specimens as are washed down in the streams are deposited out in the ocean which later, when the continent submerges, becomes deeply buried in sea deposits that we cannot inspect. I just wonder to what extent those things have taken place--that these changes have been produced under severe stress of living conditions and the severe stress occurs at the time of elevation of continents during which the changes and mutations are taking place and when the conditions are not favorable for preserving of the record.

Dr. Tinkle: Whenever we get to talking about change through mutation there tends to be a dirth of duscussion on another type of change which I think we should not overlook, and that is change through segregation and recombination of genes. It brings about considerable change. Not if the animals or plants are members of the pure line, but we do not find pure lines in nature. Pure lines have been formed by man. Our strains of inbred corn are examples of practically pure lines. They are not entirely pure lines. But where animals or plants are heterozygotes then in future generations we can expect segregation and recombination of genes and it results in considerable change. We don't have to be believers in evolution to accept that. Genetics teaches us that. We've known that for a number of years. I mustn't take more time on it. I have an article on this subject that will be coming out some of these times in the Journal. I have expressed my ideas more at length there.

Dr. Monsma: I think we can perhaps close this discussion by calling attention to this fact that Dr. Mixter has pointed out, that changes can occur, and again Dr. Tinkle has pointed it out, but that these changes, according to these men are definitely limited to certain possibilities that have been put in the organism by the Creator and that they do not go beyond certain limits.

Dr. Kulp: I'd like to have a general question answered from biology. I'd like to know if there is any modern experiment to prove definitely that stress of temperature or pressure may increase the mutation rate. It has been shown that increase in radioactivity does.

Dr. Monsma: Are mutation rates changed due to pressure and temperature stress? Will someone answer that question for us before we leave this noon?

Dr. Kulp: Not temperature and pressure. I said, pressure or temperature.

Dr. Monsma: That's a logical change, isn't it? Any pressure of any kind--do they change mutation rates?

Dr. Tinkle: That is true in some cases. It's not only x-rays and ultra-violet light but also heat will increase mutation rate in certain organisms--certain seeds--and also the age of the seed has something to do with it. In Datura or the Jimson weed in which so many mutations have occurred they find that when the seed is old the mutation rate is higher. They have gotten this seed to live for as much as eight years, and seeds that are seven or eight years old will have more mutations than fresh seed.

# METHODS OF DATING THE EARTH AND THE UNIVERSE

Delbert Eggenberger
Research Chemist, Armour and Company

The problem of the age of the earth has been a difficult one to solve. There has been an almost constant friction between the estimates of the scientific world and those of the theological world. Most of these differences were needless inasmuch as some estimates in both camps were quite tenuous. Before proceeding further with the resolution of these differences we shall examine the methods of science in setting the age of the earth and of the universe.

Suspicion that the earth was immensely aged was raised by Halley in the early eighteenth century from a study of salt deposited in the sea. Later in the same century, Hutton, basing his conclusion upon the study of geologic formations, stated that he could find "no vestige of a beginning". During the nineteenth century, estimates of the earth's age were rampant, with virtually no upper limit. Kelvin, however, basing his results on studies of temperature gradients in the earth, assuming it to have started from a molten ejection from the sun, concluded that the earth must not have been present for more than 400 million years. He later reduced this figure considerably. However, the discovery of radioactivity with its continuous generation of heat in the earth's crust vitiated Kelvin's work.

More recent work in several directions has indicated an age of several billion years. It appears that the ages of the earth and of the universe are of the same order -- at present, no difference can be detected by direct measurement. In this paper, methods will be considered under three general headings, namely, terrestrial, solar (referring to extraterrestrial bodies within the system), and cosmic.

## Terrestrial

Radioactivity measurements comprise the most important method of dating the earth. Details of the method and its limitations may be found in the American Scientific Affiliation publication entitled "The Age of the Earth" (1948).

The ages of the oldest undisturbed rocks are given as about $1.8 \times 10^9$ years for Siberian uraninite and $2.0 \times 10^9$ years for uraninite from Manitoba. The latter is particularly interesting in that the rock affords three checks, one with uraninite containing uranium, another with monazite containing thorium, and a third with mica containing rubidium. The ages range from 2.0 to 1.6 billion years. (1), (2). Undisturbed rock ages provide the lower limit to the earth's age.

It was suggested by H. N. Russell that the age of the earth's crust could be determined by taking representative values for the quantities of uranium and the corresponding lead isotope. The assumption is, of course, that lead isotope 206 all came from uranium parentage. This investigation has been carried out by several workers. Holmes (3), in studying collected results, found a concentration of values around $3.35 \times 10^9$ years. This figure, or one only slightly lower, has been verified by others. (4), (5). Therefore, it appears that the age of the earth's crust can be placed somewhere around $3\frac{1}{2}$ billion years.

One further support of the age of several billion years has been suggested by Gamow. (6). Nuclear transformation theory requires that the original abundance of $U^{235}$ be about the same as that of $U^{238}$. With a knowledge of the half-lives of each of these isotopes and the present abundance ratio of 1 to 139, it is possible to calculate the approximate time of their equal abundances. This figure is of the order of several billion years.

## Solar System

The present state of the moon's motion is that of a spiral of increasing radius. Calculations by Sir George Darwin indicated that the time required for the present state to develop, assuming the moon had its origin near the earth, would be several billion years. Darwin evolved a tidal theory of the formation of the moon, which theory has been quite generally discarded since the work of Jeffries. The latter showed that frictional energy within the combined body would be so great as to preclude such destructive resonance as Darwin postulated. However, the possibility that the lunar satellite originated at or near the earth is at present considered reasonable.

The ratio of helium to uranium or thorium in meteorites, studied by Paneth and others (8) gives ages ranging from six million to seven billion years. However, since the smaller ones have the higher helium content it appears they have been affected by alpha particle bombardment. The upper limit is concluded to be about three billion years for the age of meteorites.

## Cosmic

Star clusters are plentious throughout our galaxy. Chandrasekhar (9) has shown that the probability of a star's escape from the gravitational field of the cluster in time $t$ -- such escape velocity having been attained statistically by an "evaporation" process resulting from the buffeting action of the other stars in the cluster -- is $1 - e^{-t/t_o}$ where $t_o$ depends on parameters of the cluster. This buffeting results from gravitational forces acting through a distance. If $t = t_o$, this probability is 0.63, which may be considered the order of lifetime for the cluster. $t_o$ for Pleiades is $3 \times 10^9$ years. Since clusters such as Pleiades are quite common yet in our galaxy, that figure is taken as a reasonable average age for the clusters.

The study of binary stars affords another statistical method of estimating the age of the universe. The influence of an outside star near a binary pair is felt more by the nearer of the pair. The result of these approaches is an increased radius of rotation and ellipticity of the orbit. A sufficient accumulation of such approaches results in a dissolved partnership between the binary stars. Chandrasekhar (10) has calculated this disruption time for our galaxy as $t = 2.2 \times 10^{15} a^{-3/2}$, where $a$ is the semi-major axis in astronomical units ($1.5 \times 10^8$ km.). Studies of the distribution of separations between stars of binaries result in the conclusion that a time of the order of several billion years is correct. Bok (11) has studied both binaries and clusters and concluded that an age of 3 to 4 billion years is quite probable.

Paradoxically, while the time-scale calculated by use of the expanding universe model would be the simplest and most direct, the value found has been somewhat difficult to reconcile with the others. The figure of 1.84 billion years obtained by using a simple classical model is considerably lower than the some three billions demanded by most other methods. Russell suggests that expansion may have been slower at first because of the smallness of space then. Tolman (2) applied relativistic corrections to the simple model and found that it produced an even smaller time, 1.24 billion years. He has claimed some success in expanding the time-scale by assuming the non-homogeneity of matter in space.

The evidence of the age of stars while lending some rough support to the general picture is rather inconclusive.

Another general evidence is afforded in the temperature drop of the universe with time. Jeans demonstrated that the temperature would drop inversely as the square root of time. Assuming a beginning temperature of $15 \times 10^9$ °A -- neccesary for nuclear transformation to occur -- and a present temperature of 50 °A, the time elapsed comes out about three billion years.

## Conclusion

It appears, then, from our examination of several independent lines of evidence that the beginning of our present universe was some three billion years ago. Bishop Ussher, in the seventeenth century, concluded that the Biblical creation was 4004 B.C. The inclusion of that date in a large portion of Bibles published since that time have led many to feel that Biblical and scientific cosmogonies were irreconcilable. The result has been an unfortunate breach between two groups, both of which are in error. The one group holds that the very old date indicated by the methods of science is in conflict with the Scriptures and therefore the latter is not the inspired work of God. The other group, accepting the Bible as the revelation of God, holds to a comparatively recent creation in spite of much clear evidence to the contrary.

It has been shown by Hebraists that the words translated in Authorized and other versions as "day" could as well be translated some long period of time and "morning and evening" as "beginning and ending". Also "begat" does not necessarily imply a direct son but may mean a descendant.

The conclusion to be drawn is that the Scriptures themselves do not teach a short time-scale but its teachings are quite consistent, and without strained translations, with a very old date of creation and with long geological processes.

## References

(1) H. N. Russell, Science 92, 19 (1940)

(2) R. C. Tolman, Rev. Mod. Phys. 21, 374 (1949)

(3) A. Holmes, Nature, Lond. 157, 680 (1946)
-- , ibid, 159, 127 (1947)
-- , Endeavour 6, 99 (1947)

(4) C. Feste and M. Santangelo, Ann. Geofis. 2, 503 (1949)

(5) C. Morelli, Ann. Geofis. 2, 417 (1949)

(6) G. Gamow, Lecture before Physics Club of Chicago, 1949.

(8) W. J. Arrol, R. B. Jacobi, and F. A. Paneth, Nature, Lond. 149, 235 (1942)

(9) S. Chandrasekhar, Astrophys. J. 97, 255 (1943)
-- , ibid, 98, 54 (1943)

(10) S. Chandrasekhar, Science 99, 133 (1944)

(11) B. J. Bok, Mon. Not. Royal Astro. Soc. 106, 61 (1946)

(12) P. Rossier, Arch. des Sciences 22, 89 (1940)

* * * * * * *

# THE AMERICAN SCIENTIFIC AFFILIATION - THE FIRST DECADE

by

F. Alton Everest

Part of an address given at the Sixth Annual Convention at New York City, August 30, 1951

The ten-year period, 1941-1951, was a time of great significance to the world. During this time a second world war was fought which quite definitely silenced voices raised in optimistic hope for a peaceful world. International relations during this period were in a chaotic state. Sneak attacks, subversive activities, the lack of honor in international relations, the dishonest practices in high places within our own government, the deplorable flourishing of organized crime, juvenile delinquency--all these things have brought a feeling of insecurity and despair to our people and the people of the world.

In 1939, scarcely two years before the opening of this significant decade, there was announced the hypothesis of nuclear fission and its experimental verification. By the end of 1941 an extensive review of the whole uranium situation was completed and full impetus was given military exploitation of the fission effect. In 1942 the famous Manhattan District was organized culminating in that fateful birth of the atomic age in the gray dawn of the New Mexico desert on July 16, 1945.

These striking events in the military and political world had, perhaps, at least some effect on the changing attitudes in the scientific world. The pendulum seemed to be swinging away from the rigid materialism of the past. Heisenberg's Uncertainty Principle pointed out that strict determinism at least did not apply in the realm of the quanta. While the leap from so-called "free will" of the electron to the freedom of the human spirit is too great a jump for many to make, these new concepts were having their over-all impact on modern scientific thought.

The equivalence of mass and energy--a concept dramatically demonstrated on that morning in the New Mexico desert--also seemed to pull some of the underpinnings from beneath materialistic philosophies.

Let us not be confused--a knowledge that all our material world is but a manifestation of energy does not necessarily make a spiritual-minded religionist of the physicist. Nor should we feel that this revolution has taken place when we see cosmologists put forth their hypotheses of a practically instantaneous condensation of all the elements from a sort of hot cosmic stew and equate the event to the creation of Genesis 1:1. All these things may have their cumulative effect, but we still have to deal with scientists who, in the main, do not know our Redeemer, and yet who are somewhat more inclined toward a consideration of spiritual things. Today we find less inclination among scientists to brush off the claims of the Bible with a wave of the hand. This is illustrated by the Professor of Zoology of a large Western university who each year shows all of the Moody Science Films to his large class--gospel message and all--as he puts it, "to present the other side of the question in all fairness." Such instances may be few and far between, and the general trend away from rigid materialism may appear to be all but swallowed up in the familiar glorification of human wisdom, but it seems incontrovertible that the trend is there.

This is a brief sketch of conditions existing just prior to and immediately following the organization of the American Scientific Affiliation. There probably never was a more significant period in history--the growth of science and the opening of the atomic era--and a period that presents its own unique problems to an effective witness and defense of the faith.

It would be unfair and distinctly untrue to leave the impression that, in the midst of these great problems, the ASA was born in solitude. The ASA owes much to other similar organizations some of which rose briefly only to fade, others which are contemporarily functioning today.

The Victoria Institute of Great Britian was founded in 1865 to treat the problems arising between science and the Holy Scripture with the view of reconciling any apparent discrepancies. This organization has had a long and useful life and is actively continuing its work today. It has a total membership of around 600 and its Journal of the Transactions contains many valuable papers.

The Religion and Science Association was organized about 1928. Dr. L. Allen Higley, then of Wheaton College, was one of the founders. Other names associated in this early endeavor are Ben F. Allen, and Mr. Clarence Benson, and others. Its activities were abandoned at least ten years ago.

The Kelvin Institute had its origin in Toronto, Canada, about 1935. This group was founded by Mr. Arthur C. Custance, formerly a member of ASA. Our own Dr. John F Howitt was one of the early members. At its peak it had some 50 members in England, Scotland, Australia, USA, and Canada and produced a number of very interesting papers. However, its activities ceased during the early 1940's.

The Evolution Protest Movement, active in England for more than two decades, did not find ready acceptance in the USA. Sensing this, Dr. Arthur Pierson Kelley thought that an indigenous group patterned along the same lines would be better received and proceeded to found The Creationist Society. This proved to be entirely abortive. For some time, however, Dr. Kelley issued a series of papers under the name The Landenberg Review which carried on the anti-evolution cause. As far as it is known, this effort has been abandoned.

In addition to these there have been a number of local organizations devoted to the study of the relation of science to Christianity. Typical of these was the Nature and Scripture Study Club of Grand Rapids founded in 1935 by Dr. John P. Van Haitsma of Calvin College, one of the founder-members of the ASA. This group disbanded prior to 1942, at least partially because of Dr. Van Haitsma's failing health.

About 1938, the Society For Study of Deluge Geology and Related Sciences was formed. This group published The Bulletin of Deluge Geology 1941 through 1945. To an outsider, what then happened is somewhat obscure, but apparently reorganization resulted in the use of the names Natural History Research Group and also The Society For the Study of Natural Science. A publication called The Forum came out with one or two issues under the latter-named society. Men associated with these groups include Prof. George McCready Price, well-known advocate of deluge geology, Dr. Cyril B. Courville, Dr. Moulerus Couperus, Mr. Ben F Allen, and others. Correspondence with members of this group and a perusal of its publications leads one to conclude that it was founded principally to defend the so-called deluge interpretation of the earth's stratigraphic history and to contend for a recent formation of the earth. As many of its leaders are also leaders in the Seventh Day Adventist group which strongly promulgates these particular interpretations, it is felt by many that the society, directly or indirectly, was influenced by the teachings of this denomination.

There are no doubt many other groups not named here that were formed during the 1930's and 1940's to correlate scriptural interpretation and new scientific facts. All these groups had one thing in common. They were anxious to demonstrate the harmony of God's Word and God's Work in nature. Some may have struck too negative a note, some may have been unwise in their emphasis, some may have been downright wrong in either their exegesis or their science--but their motives were commendable and we shall leave to the great Judge of all things the ultimate value of their labors.

I believe it was in the wise providence of our God that the men that founded the ASA and who were responsible for its nurture during the early years were relatively ignorant of these other groups. For this reason their influence on the ASA was negligible and it is only in retrospect that we can see how fortunate this was in the formation of the policy of the ASA which the testimony of ten years' growth indicates as sound.

What is this policy? For what purpose does the ASA exist? The constitution states that the object of the Affiliation is to "correlate the facts of science and the Holy Scriptures." We are vitally interested in determining the proper interpretation of the creation account, the flood of Noah, Joshua's long day, etc., but we consider it distinctly improper for the ASA to become so enamoured by particular interpretations of these accounts that we shift our efforts from study to propaganda. Dr. Allan A. MacRae, prominent archaeologist and past Vice-President and member of the Executive Council of the ASA has wisely put it this way:

> "To my mind it would be unfortunate for the Affiliation to go on record strongly in favor of any one of the various views. It seems to me that its purpose should be rather to show that the Bible as correctly and carefully interpreted, and without any twisting whatever, leaves room for every scientific fact at present known, and does not contradict any scientific fact as yet discovered, however much it may be at variance with some particular theory built upon these facts."

Now this does not mean that ASA members do not have some pretty strong personal convictions concerning these interpretations--they do. Furthermore, there is a wide range of interpretations held among the membership. This is only natural and it is a very healthful situation--one that causes one to search his case thoroughly before submitting himself to one of our justly famous discussion periods! Thus in <u>The Journal of the ASA</u> you will find a paper supporting a particular interpretation and a little later another one apparently demolishing it. We consider our job well done if we can present a Bible-teacher, a pastor, or a university student an adequate survey of the various views held on a given problem and the historical and scientific data pertaining to it. The problem of the church is principally one of plain ignorance of the many and complicated factors entering into a wise interpretation of the Scriptural accounts.

Another member of the ASA, Dr. Harley Barnes, a geologist, stated the following views six or seven years ago which I believe to be a fair expression of the prevailing attitude of the majority of the present Associates and Fellows of the ASA:

> "The essence of my attitude towards evolution and the Bible as a Christian geologist is that Christians should be non-evolutionary because the Bible does not give unequivocal grounds for being anti-evolutionary. Few, if any, Christians who are scientists now would say that the Flood of Noah produced all fossils or that the earth is only 6000 years old; yet there was a time when persons holding contrary opinions were generally labelled modernists. The Bible said the same thing then as now, but our interpretation has changed because discoveries have narrowed the field of interpretation possible to rational men of faith. We are faced then, with the realization that the Bible allows numerous interpretations of the creation account, but our choice of 'the' interpretation has been limited to those which do not conflict with accumulated scientific observations. The viewpoint of scientists changes continually as additional data fills in the picture, shifting the perspective. Let us not repeat the mistake of earlier Christians by forcing the Bible to speak in the language of current or recently current theory.
>
> "The devout 'Fundamentalist' persecutors of Galileo and Copernicus helped discredit the Bible to many thinking persons because they were morally certain it specifically taught the Ptolemaic astronomy. After accumulated data overwhelmed these pious but bigoted spirits, a new crop of second guessers came forth with the 'obvious fact' that the Bible had agreed with the former heretics (Galileo and Co.) right along, so that 15 years ago I was impressed by the quotation of verses 'proving' that the earth was round. And if science should find sufficient information that the earth after all really has the shape of a dodecahedron, someone would find a verse in the Bible which had said so all the time!
>
> "The Bible does not pretend to give a technical description of these matters. What it does say is accurate but generalized; some passages are figurative, but we are not always able to determine which ones--Isaiah 40:22 or Revelation 7:1. Let us not tie the Bible to any modern (or Victorian) scientific theory. If it can be interpreted more than one way, let us admit it and when all the pertinent facts are known rejoice in the Wisdom which had been revealed but until that time not understood. Perhaps the Bible suffers less from too little literal interpretation than from too much literal interpolation."

In November, 1940, while the writer was on the faculty of Oregon State College, a message was received from Irwin A. Moon who was holding his famed Sermons From Science demonstrations in nearby Salem, Oregon. A meeting was arranged in Salem during which Mr. Moon told of the great need he had encountered among young people for an organization of reputable men of science who were also Christians. The young people flocked to him with their questions, eager for reassurance that modern scientific knowledge does not rule out faith. That afternoon Mr. Moon named several qualified men of science whom he believed would be interested in forming such a society.

Several months passed during which Mr. Moon was in constant touch with Dr. Will H. Houghton, the late President of the Moody Bible Institute of Chicago, concerning ways and means of bringing this group into existence. Early in 1941 at the request of Mr. Moon, the writer examined the constitutions of many existing scientific societies and drafted one which was considered suitable for such a society of Christian men of science as was proposed. Considerable attention was also given to a fitting name for the society.

In June, 1941, Dr. Houghton addressed a comprehensive letter to a number of men who had shown their interest in the formation of a group for the correlation of science and the Bible. In the recent biography of Dr. Houghton, A Watchman on the Wall,*. Dr. Wilbur M. Smith entitles a chapter "Through Science to the Souls of Men" in which Dr. Houghton's letter is quoted in its entirety. Dr. Smith states that this letter might "be called the birth certificate of the American Scientific Affiliation." Dr. Smith further say "The letter itself, I think, will prove in days to come a significant document--." In this letter the men to whom it was addressed were invited to meet in Chicago September 2 to 5, 1941, to "canvass the possibilities of a larger conference" and eventually, perhaps, to organize a new society. Dr. Wilbur Smith writes, "Out of this letter has come a strong, vigorous organization which today is undertaking a number of projects which, the Lord willing, in the days to come will be the means of great help to those who are troubled about the complicated relationships of contemporary science with the Word of God."

Five men journeyed to Chicago as a result of Dr. Houghton's invitation. Their expenses were borne by a donor unnamed at the time but who has since been identified as the saintly "Breakfast-Table Autocrat" and intimate friend of Dr. Houghton, Henry Parsons Crowell. Those who attended were:

Dr. Irving A. Cowperthwaite
Prof. Russell D. Sturgis
Prof. Peter W. Stoner
Prof. John P. Van Haitsma
Prof. F. Alton Everest

Dr. Houghton, Mr. Moon, and Mr. H. Coleman Crowell, the son of Henry P. Crowell and Vice-President of Moody Bible Institute, conferred with the group from time to time, but Dr. Houghton's statement in his letter of invitation was faithfully carried out when he said "Some of us are initiating this, but we haven't any desire to control it, and certainly it would be limited if it were known to be sponsored by any one educational institution--. We want you to be very sure the group will be entirely free to make its own plans."

The American Scientific Affiliation was organized at this time although its constitution and name were not officially adopted until a few months later. The precious time together was spent in considering future programs, membership qualifications and prospects, constitution, aims and purposes, spiritual standards, and many other vital topics. Dr. Cowperthwaite was elected Secretary-Treasurer and Prof. Everest Chairman and the group adjourned.

The next few months were filled with intensive correspondence and the laying of plans for a large national meeting in 1942. Three months after the Chicago meeting, on December 7, the Japanese attacked Pearl Harbor and such grandiose plans as a convention were eliminated by travel restrictions.

---

* 1951, Wm. B. Eerdmans Publishing Company, Grand Rapids, Michigan.

Looking back on it, the Lord's overruling hand in interfering with our plans by the war was probably the best thing that could have happened to the embryonic group. We were forced to work and study independently during the 1941-1945 period and were given practically no opportunity to arouse each other's ire in heated debate at close range. Rather, this time served to clarify the issues, streamline the organization, bring in new members and to begin work on the book later entitled, Modern Science and Christian Faith which is now in its second edition.

While war-time travel restrictions practically eliminated civilian travel, the chairman was not only able, but was forced to do considerable travelling over the United States in the interests of his wartime work. This provided opportunities for dozens of conferences and contacts which increased the effectiveness of the work greatly over complete reliance on correspondence.

Articles of incorporation for the ASA as a non-profit organization were filed with the Secretary of State of California in August, 1943. Also during 1943 we began voting in new members after cautious consideration of their qualifications. By the close of the war, the membership was about 50, and by the end of 1950 about 130. The adoption of the new constitution with the new membership classifications has naturally resulted in a sudden spurt over the average rate of approximately 20 new members per year. At the time of this 1951 convention in New York the membership is about 220. The growth has been steady--we have never had ambitions for a large organization or we would have had lower standards for membership. After all, there are just so many scientists, and a very small percentage are interested in Christian things, but the conservative policies adopted are rapidly gaining the confidence of these qualified persons.

It seems entirely proper that the greatest strides of the ASA be taken during the tenth anniversary year. Without doubt, the adoption of the new constitution with no dissenting voices (one man expressed disagreement with one portion) was a remarkable expression of the members' confidence in the leadership of the Council. This is especially true when it is realized that, for those members not qualified for Fellow grade, it meant virtual disenfranchisement. This new constitution's distinguishing feature is its placing in the hands of the Fellows the voting privileges and the policy responsibilities, yet opening the Associate grade for those whose interest in and enthusiasm for the work of the ASA is in no way measured by their lack of advanced scientific degrees. Thus, maximum participation is coupled with maximum protection for the high spiritual and scholastic standards of the group.

Another great stride was Dr. Mixter's selection for the Presidency after nine years of service in this capacity by the writer. Sudden upsurge in all activities of the ASA is evidence cited to uphold this statement. One of the greatest experiences in the life of the writer has been the thrill of working side by side with men of the ASA--men whose faith has been tried in the crucible of spiritually sterile scientific criticism, men who have devoted their lives to the study of God's handiwork in nature and who see there the infinite resources of the One in Whom we live and move and have our being. It is our priceless privilege to join together in proclaiming the salvation which He has offered to a lost world.

# BIBLICAL CHRONOLOGY AND GENESIS 1:1-3*

G. Douglas Young
Shelton College

As scientists, we tend to look at words as though they had "technical" point meanings only. We say, "This word means this, and that word means that." Frequently we fail to realize that each word has a more or less easily definable area of meaning. Few, if any, represent a point of meaning. Even "X" has ten definitions which cover half a column of space in Webster! Many words had a well-established usage before our particular science developed to the point it has today. Many words had, and have, an established usage from which the current usage has departed by change or modification. Some of us, and many in the rank and file of Christians, as judged by writings and conversation, it would seem, fail to recognize this simple principle of semantics. A failure to recognize the fact that words are not points of meaning, but areas of meaning, leads to misunderstanding and error. For example, the word "evolution" has a well-established general usage. We can refer to some aspect of it which happens to be true as "evolution," or we can use some narrow technical definition of it with which we can agree, and say, "I am therefore an evolutionist"--which is not the case at all, of course, except as we use the word in that very limited sense, ignoring all the other well-established parts of the meaning. We then actually speak erroneously, and others are misled.

The same is true in other fields, for example, sociology, economics, philosophy, where well-established words are redefined, and we who know the newer technical usages fail to recognize the validity of the other usages, and so cause unnecessary confusion in popular thought.

---

*Paper presented at the American Scientific Affiliation Annual Convention, New York, August 28-31, 1951.

This also is true in the Bible world. "Arbitrary" definitions are given to certain Biblical words and then the rank and file come to erroneous views upon what the Bible says about "creation," "begetting," "sonship."

Words are defined contextually. We may be able to trace the derivation, but there may be no relationship between that and the current usage. The definition of a word is made from the sum total of its several usages in actual contexts, whether written or spoken. The meaning of a word in a given environment or situation is determined by that environment or situation. Specifically, in the problem immediately before us we must do two things:

1. Negatively--we must not read English meanings and usages back into the Hebrew words, and

2. Positively--we must set meanings for the words we study which are consistent with the area of meaning that they cover <u>in the Hebrew</u> contexts.

## I--The Word "Begat"

1. In English the word means that two successive generations are spoken of. "X begat Y" means that Y was born into the house of X by his wife Z.

2. In Hebrew usage "begat" carries that meaning, but it carries more as well.

Matthew 1:8 says, "Joram begat Ozias." If we compare I Chronicles 3:10-12, we can see that Ahaziah, Joash, and Amaziah are omitted, yet the passage can and does say "begat."

This word, in Hebrew, includes in its area of meaning the idea "ancestor of."

## II--The Word "Son Of"

1. In I Chronicles 9:12, Jeroham is called "the son of Pashur." Comparing Nehemiah 11:12, we find that there are three names between Jeroham and Pashur. "Son of," therefore, means "descendant of."

2. In Ezra 7:3, Azariah is said to be the son of Meraioth. If we compare I Chronicles 6:6-9 we find six names between Meraioth and Azariah (a very interesting homoioteleuton and so a textual error, a plain mistake, or as is proposed here, a shortening).

This last case is doubly interesting because Chronicles uses the word "begat," while Ezra says "son of."

Now let us apply these more correct, even if less "exact," meanings. First, however, one further illustration is necessary.

## III--The Illustration of Terah

Genesis 11:26 says, "Terah lived seventy years, and begat Abram, Nahor, and Haran." Was Abram born, as would seem to be indicated here, when Terah was seventy years old? Compare 11:32--"And the days of Terah were two hundred and five years: and Terah died in Haran;" Acts 7:4, where we are told that Abraham left Haran after his father died, and Genesis 12:4, which says, "Abram was seventy and five years old when he departed out of Haran." It is thus seen that Terah was 205-75, or 130 years old when Abraham was born. This illustrates the Biblical usage in the dates in the chronologies.

## IV--Genesis V

(I do not say we must take it this way, I simply say it does not do injustice to the meaning of the Hebrew usage of the words if we do take it this way. I say also that only an arbitrary assignment of a meaning to the words (point, not area), makes it impossible so to take it.)

Cainan, who is mentioned in Genesis 5, lived seventy years and at that point became the ancestor of (begat) Mahalaleel, and he lived 840 years after that point. Cainan lived 910 years and died. How old was Cainan (on the analogy of the Abraham and Terah illustration) when Mahalaleel was born? Was Cainan even alive when this particular descendant, sufficiently important to be included in the list, was born? One thing is quite certain--we cannot establish a chronology by adding these numbers. Another is equally uncertain, namely, how long it was from Abraham to Noah. It could have been a few thousand years, or many.

This much also is certain--God gave us a list of the important men of that period in order, with doubtless many omissions, and He gave us also the fact that all died eventually, even though they did have long lives. The effects of sin eventually caught up with them--death came.

## V--Genesis 1:1-3

To few Biblical words has so much injustice been done as to the word "create." We believe in an ex nihilo creation, on the basis of John 1:1-3. Many "define" bara' to mean "make out of nothing," and so read the known ex nihilo creation back into Genesis 1:1, where it quite possibly is not intended at all.

The word needs to have its definition rephrased for popular consumption. Compare the use of this word in the following cases: Jeremiah 31:22, "The Lord hath created a new thing in the earth, A woman shall compass a man"; Psalm 51:10, "Create in me a clean heart, O God"; Isaiah 41:19, 20, "I will plant in the wilderness the cedar. . .I will set in the desert the fir. . .the Holy One of Israel hath created it (the new sitaation)"; Isaiah 43:1, "But now thus saith the Lord that created thee, O Jacob, and he that formed thee, O Israel (a new relationship)"; and Numbers 16:30, "And if the Dord make (create) a new thing, and the earth open her mouth, and swallow them up. . .(Dathan and Abiram)." Obviously a making of something material out of nothing is impossible in these contexts. What is made, created, is a situation that did not formerly exist. Might it be so in the case of Genesis 1? I think so.

Moses was not interested, particularly, in the original act of creation as was John. He was interested in telling us in broad outline how the earth got to be as it is now. Thus he says, "In the beginning when God was creating (not ex nihilo of the matter, but something new--the present form and order which did not formerly exist) the heavens and the earth, the circumstances were as follows: (1) the earth was without form and void, (2) darkness was over the face of it, and (3) the Spirit of God was moving over it. In such a case and time God said (verse 3 with its first was conversive, and hence its first step forward in the developing of the story), 'Let there be light,' and there was light."

If this interpretation can be justified on the basis of philology and syntax, then the so-called, popular, "gap" between verses 1 and 2 turns out to be a matter of philosophy and not a fact of revelation.

The facts are as follows:

1. The first waw conversive comes at verse 3.

2. The three preceding clauses are circumstantial ones. Compare Gesenius, paragraphs 156 and 141e.

3. The first word of the Bible is a noun in the construct state, and it is followed by a finite verb. This idiom is well attested in Semitic as a temporal clause. Compare Koran, Sura 37:144, where it is said of Jonah that he "tarried in his (the whale's) belly until the day when they were raised," and compare most especially Moses' own dialect and usage at Deuteronomy 4:15, "on the day that the Lord spake unto you in Horeb. . ."

In Conclusion:

The Genesis account does not set temporal limits in its lists of names before Noah; nor does it know, in and of itself, anything of a creation, destruction, and recreation; nor does it (while in no sense denying one) have anything to say specifically about the ex nihilo or original act of creation.

The Chairman, Dr. I. COWPERTHWAITE, asked for discussion of Dr. Young's paper.

Mr. WILLIAM J. SCHEPP: I am sorry that it wasn't quoted Isaiah 45-7 where bara is translated "create" and Ezekiel 23-47 where it is quoted "dispatch." I realize that Isaiah 45-7 is a very controversial passage of scripture. The Universalists have used it--the Jewish people have used it with the idea that God created evil and whereas if you take the word "create" out of Isaiah 45-7 and translate it on that basis, "I form the light and dispatch darkness. . .I make peace and dispatch evil. . .

To use the translation of bara in that particular sense overcomes the idea that we must water down the word "evil" there to make it affliction or sorrow.

Mr. ROY ALLEN: I was just thinking of one verse where bara is used that is, I believe, very significant as emphasizing the points of Dr. Young. I think it's the 102nd Psalm where David says, "And the people which shall be created shall praise the Lord."

He is a new creation. Old things are passed away and all things become new. We get a beautiful picture of just the people of whom David is speaking when he prophesied that the time will come when there shall be a people created that will praise His name, and were a new generation, a rival priesthood. Because we're not created as the speaker said, out of nothing, but God takes the old nature and makes something that didn't ever exist before, to a new birth. I praise the Lord for that verse, too.

Mr. DOUGLAS BLACK asked for clarification of the speaker's conclusion on Genesis 5 regarding men living before someone came to be and then living afterward.

Mr. G. DOUGLAS YOUNG: I'd like to preface, because it's going into the record, that I did not say that that must be the interpretation of Genesis 5, but on the basis of the Terah-Abram illustration, it conceivably could be that Cainan lived 70 years and then had his first child. The important person in that genealogy or in that group was Mahalaleel who may have been born 60, 70, a thousand, two thousand years later as far as we know. There's no way that we can set the time limit. In the case of Abraham and Terah it happened to be 60 years. That's 60 silent years, and therefore in these other cases, it might be 60 or 70; it might even be longer. There's no way that we can say on it. But the thing that is significant in Genesis, the 5th chapter, is that Cainan was an important man in the sight of God. For some reason he was listed there, the same as certain ones were singled out for listing in the 11th chapter of Hebrews. Here is the immediate successor. The next one significant in his line is Mahalaleel and the terrifically vital factor that no matter how long anyone may live, they all eventually die so that the problem that used to block my thinking was that after he begat so and so, he lived so many years and he died. On the basis of the Abraham-Terah illustration, it's perfectly possible that that verse means--he lived 70 years, he began his begetting. Mahalaleel was the important man that followed him. This particular man, after he began his begetting, lived 810 years and he died. He lived 910 years all told, but he died.

- - - - - - - - - -

# *Creation Days**

DR. J. OLIVER BUSWELL, Jr., Ph.D.

President of Shelton College

I wrote my paper years ago and many of you have copies of it so this afternoon I shall talk about my paper rather than read it. If any of you do not have copies of this little article on the length of the creation day, ask at my office. I think we have enough to go around.

*This paper was transcribed from a recording of a talk given at the Sixth Annual Convention of the American Scientific Affiliation in New York, N. Y., August, 1951.

While Dr. Young was giving his paper, I sat there saying, "Hallelujah." I say that not because he's my colleague but because I thoroughly agree with what he has said in regard to the interpretation of the Scripture and because I feel it so important and so necessary that we should realize the meaning of words. God in His infinite providence has chosen to reveal His Word to us in language, not in a meta-language, not in terms of mathematical logic, but in terms of the usage of words and phrases, grammar and syntax, which have historical context and whose context may change in the process of the generations.

My paper this afternoon is confined to the use of the word "day" or "days" in the Scripture, particularly with reference to the first chapter of Genesis and, as you may suppose, I believe that Moses uses the word "day" in the first chapter of Genesis in a figurative manner. When I say, "a figurative manner" immediately people say, "Well now, just how long were those days?" Well, how long is a figure of speech? What is the figurative use of the word "day?" When you say the day of Martin Luther or the day of George Washington, what are you talking about? Can you put any precise limits on the figurative use of the word "day?" Dr. Young and the Old Testament people study in 46 languages. I heard about an Old Testament man who could actually keep still in 50 languages. That was a great achievement! But any language that I ever heard about has the figurative use of words for periods of time, the figurative use of the word "day."

When you think of your Bible, the phrases "the day of the Lord" "that day" "the last days" "the last hour" come to mind. Christ said, "it is the last hour," when He was speaking to His people there in the days of His flesh. "It is the last hour"—and that hour has already been over 1900 years long. The figurative use of the words for parts of time, the word for "days" ought to be familiar, from the Scripture. We should not have any great difficulty with this concept.

A text that is frequently quoted in this connection comes from the 90th Psalm. "A thousand years in thy sight are but as yesterday when it is passed or as a watch in the night." Some people forget about the "watch in the night" which, depending on whether you are using a Semitic watch or a Roman watch is three or four hours long, or the period of time in the night when anyone stands on watch. Some would say, "A day must mean a thousand years." Why, No! A day is a thousand years or a watch in the night, either one.

Peter says in the first chapter of the Second Epistle, "This one thing must not be forgotten . . . that one day with the Lord is as a thousand years and a thousand years is as one day." This isn't arithmetic at all, but it means that God has plenty of time. God has all the time that there is. Infinite time belongs to Him.

We suppose that Moses wrote the 90th Psalm. We have every reason to believe that the tradition is correct. The 90th Psalm refers to the great eternity of God. "Thou hast been our dwelling place in all generations, before the mountains were brought forth. Or ever thou hadst formed the earth and the world, even from everlasting to everlasting. Thou art God." The great eternity of God, "before . . . or ever thou hadst formed the earth and the world."

Turning now to the first chapter of Genesis, we note what Augustine pointed out. Augustine could not be accused of interpreting the first chapter of Genesis on any geological presuppositions. He was simply giving what he thought Moses meant. Augustine said these days could not possibly be 24-hour days because the sun, moon, and stars were not given into the firmament of Heaven for times and seasons and days and years until within the fourth day. The solar day system was not set up until within the fourth day. Augustine saw that long, long ago. It was within the first "day," the day-night series was set up. It was within the first "day" that God created day and night by dividing the light from the darkness. It was not until the fourth "day" that the sun began to govern periods of time.

One of the most clear evidences of the figurative use of the word "day" in the first chapter of Genesis is in Genesis 2:4. "These are the generations (the word generation here means "stories about") the heavens and the earth, in their being created in the day of the making by Jehovah God heaven and earth." It's all one day. Now how long were the days when six days equal one day? If you're speaking figuratively then clearly six equals one, because you're simply taking a different figurative perspective. There can be no way to escape it that later Moses refers to the entire period of six days as one day.

In the fifth chapter of Genesis, there is another reference. (I don't claim to be a Semitic authority. I studied Hebrew and I still study Hebrew to some extent. Dr. Young is our authority here.) But here you have it in Genesis 5, where God created Adam . . . "And he called their names Adam." The plural pronoun ending is given here. "He called their name Adam in the day of their being created." There in the fifth chapter of Genesis also you have the creation record referred to as one day. Actually this one verse might refer only to the sixth day of creation, but it seems more probable that it is a general reference to the whole creative record.

I should like to add just this to what Dr. Young said. He may not agree with me in detail. We have academic freedom here, you know. We don't have to agree on every detail, but the creation of Adam is referred to in the plural. They are the Adamses. There is the theory, you know, that some of these names may be names of dynasties or names of households. We know that that is the Hebrew usage. It sometimes is difficult to know whether you are talking about one individual man as we would speak of an individual man in the practice of our western language or whether you are talking about a household. In the prophecies of Ezekiel, it seems very clear that the dwelling of Israel under David, their king, does not refer to David as an individual but to the house of David and to Christ as the son of David. So here the creation of Adam is referred to in the plural and there are, I think, other suggestions that possibly many of these names in the fifth chapter of Genesis may be names of households.

Furthermore, as Dr. Young so ably brought out there obviously are gaps in the story. The farthest thing from Moses' intention was to allow us to figure out a connected chronology. The scriptural attitude toward chronology as Melvin Grove Kyle shows, (see his **Moses and the Monuments**) is to present great events, great epochs, great concurrent or great successive events, and then to emphasize spiritual principles. Compare the genealogy of Christ in the first chapter of Matthew, observe the obvious and intentional gaps. This illustrates the Jewish usage in such matters. The Old Testament simply does not give us any data whatsoever with which to figure the antiquity of man upon this earth.

Summarizing then the evidence for the figurative use of the word "day", Moses, in the context, at least once in the second chapter and I think again in the fifth chapter refers to the entire period as one day.

In the 90th Psalm he refers to God's time by way of contrast with our limited measures of time. In the New Testament the word "day" is frequently used figuratively and the chronology is obviously and intentionally discontinuous.

I submit therefore that in fairness to the Scripture, not in riding a hobby or trying to work out some presupposed idea of our own but in fairness to the Scripture, we should take the usage of the word "day" as figurative in the reference to the "days" of creation. Obviously Moses points out certain great facts,—God created the Universe. He created the light and the darkness. He brought the dry land up above the waters. He brought the vegetation. He brought the various orders of life, and he created man in His own image. These great days of creation overlap, obviously. There is no measuring of them. Moses is using the word figuratively and it makes perfect sense to read it in this light.

**Discussion**

Dr. I. A. Cowperthwaite: This paper by Dr. Buswell is now open for discussion.

Dr. H. H. Hartzler: With reference to the argument that the word "day" may be a longer period of time as evidenced by the fact that the sun was not created until the third day, I would like to submit that surely if we are going to have vegetation on this earth that the sun and the present order of day and night existed before day 3. How could you envisage several million years existing, during which we had all these things, vegetation, plants growing with no sun? Certainly in the fourth day when the sun is said to be set for these funtcions it means that the sun is then visible.

Who, in this audience, would be aware that the sun exists if clouds hung over the sky all the time? You would still be aware of day and night. You would not be aware that the sun exists, nor the stars, nor the moon. It certainly seems, in all fairness to the Scriptures, that we should suggest, as has been suggested before, that this just means their appearance.

I would also like to add a point on the usage in chapters 2 and 5 of the "days" as being an extensive period of time. It does not say in my Bible, as I read it, that the word "day" in chapter 2, verse 4 refers to the total creation. It says, "the creation of the heaven and the earth," not necessarily the appearing of dry land or of plant life, cattle, and man. Likewise, in verse 5, certainly that refers to only one of the specific days.

Dr. G. D. Young: For a long while there was a psychological block in my mind, in understanding biology . . . I had some in college! I was a biology major and our school was entirely devoted to the theory of evolution. There was a psychological block in my mind in the matter of understanding that record and the record of the Bible. I was a little surprised to see that in our own Symposium (Modern Science and Christian Faith) that that which caused that block still exists. We have a chart there . . . Day 1, Day 2, Day 3, Day 4. I have been trained since I was very small to see those as necessarily chronological and quite likely of the same approximate duration. I think that when that volume is re-edited that that which caused this psychological block could be removed. It should be pointed out that these days are not necessarily successive.

Obviously the fourth day in which the sun, moon, and stars were made to appear is quite a different sort of a day from one of the other days such as, for example, the third or the fifth. Yet when it's put in the way it is in this chart we can not very well expect our boys and girls in high school to receive it in any other sense than that of successive periods of equal length. We are perpetuating that psychological block. Those days could, and doubtless did, over lap. If we are interested in harmonizing the Bible record and geology, we know that some of the things that were made on the fifth day existed also on the third day. That is, the geologic strata where the third day items appear predominantly also contain the fifth day animals. To sandwich the fourth day of equal duration in between and split up days 3 and 5 makes a psychological barrier that need not exist. I think we should be very careful in how we set up such charts in the next issue of the Symposium. That is why I was glad that Dr. Buswell got to the point before he stopped his message.

Dr. J. O. Buswell: With reference to Genesis 2:4, I think the key is not merely in the word "day" but in the pronoun with which the phrase is introduced. In English . . . "These are the generations of" . . . the antecedent of the pronoun has a clear and obvious literary meaning if we take it as referring to all that goes before. You have a structure in orderly style, an introduction . . . "In the beginning when God created the Heavens and the earth . . . (so and so)." And then you have a conclusion, "These are the generations of the heavens and the earth in their created." So it is the pronoun which must be studied.

What does he say was done in one day? It seems to be a summary of what goes before and I think I made it clear that I do not hold that the sun was created in the fourth day but it began to function for days and years. That is an explicit statement. The word is nathan "and he gave them into the firmament of Heaven for times and for seasons and for days and years." That is to say that when the sun was invisible behind the bank of clouds, you couldn't exactly have a calendar or at least it would be complicated and difficult, but on the fourth day he gave them into the expanse of Heaven to function for times and for season and for days and for years.

Dr. D. Block: It seems obivous that our discussion of Genesis 1 is to arrive at an acceptable way of harmonizing so-called geological time, be it long or short, with the Biblical chronology. Is that not so?

Dr. J. O. Buswell: No! It is to bring out what Moses said and what Moses means and then afterwards to study geology. But our only question here is whether or not these days were, in Moses mind, 24 hours or periods of time of undesignated length.

Dr. D. Block: Allowing that these days are of varying lengths of time, man was created then during the sixth period of time. However, God saw fit to bring that about. Apparently He was in the process of creation and came to completion at the end of the sixth day or period. Then we read that on the seventh day God rested from all the work that He had made and then further on that Adam lived so many years into supposedly recorded time. In all told the years of Adam were 937 years and he died. Now, I am just naive enough to feel that Adam was brought to completion within the sixth period of time. He had to live within the seventh period in which God rested and into another length of time. With this answer I know you will think that we are still in that seventh period of time and God is resting and that, for me, is impossible, because that means God provided his redemption in the Lord Jesus Christ on the cross of Calvary during the time when he was said to be resting. He was, if that be true then more busy or working harder in creation that in providing redemption and to me that doesn't mean to be the burden of the Scripture. Now you see the problem I'm facing. How old was Adam when he finally died . . . 937 years or 937 plus one more eon? Thank you.

Dr. J. O. Buswell: I think that question is adequately answered in the pamphlet. The Scripture says in Hebrews 4 that we are now offered the privilege of entering into the "rest" which is referred to in Genesis 2. Quoting Genesis 2, "God rested." The invitation is given to us to enter into the precise "rest" so that according to the author of the Epistle of the Hebrews God is now resting in a sense which is specified in Genesis 2. What is the sense in which God rested? He rested from His work of creation, "from all His work which He had created and made." Between the ending of the creation and the ending of the millennium, (Some of you may not think that there's going to be a millennium but you'll be surprised! It'll be a happy surprise too. You'll be there.) God is not creating. He will then create a new universe. He is not creating now though the work of redemption and providence goes on. The seventh "day" of Genesis 1 and 2 is the age between creation and the new heavens and the new earth. The seventh "day" is a figurative day but Adam's years are evidently literal years.

Dr. Roy Allen: The Scriptures very definitely say that God created the light and then it says that he called the darkness night and the light day, so, if that were a solar day, the period of darkness involved from Genesis 1:1 until God said, "Let there be light" must have been only of 12 hours duration. It seems rather that that was an indefinite time; the darkness at least, so that if the night were longer than 12 hours, then the solar day must be longer than that. So putting these all together, I think that the position that Dr. Buswell has taken is pre-eminently scriptural and I heartily agree with him.

Dr. Norvell Peterson: You know, like Michael Angelo said, the fundamental unity and nonessential charity, and I observe that we have both, are here. So on such a culture I would like to submit another view which I won't be at all arbitrary or dogmatic about, but it sounds very nice. I think that God is, if anything, orderly, and I think God, through the Holy Spirit, recorded something here, and what we are trying to do is to get at what it is. Now I suspect that these days were a thousand years, as Peter said, all of them, and I'll tell you why. I mean it just sounds reasonable.

In the first place God worked for six days and on the seventh day He rested. Now he (Adam) lived almost a thousand years, but with the death of Adam, with the translation of Enoch, we start another day. Then we come on down another thousand years and what happens? Abraham is born and we start another day. We come another thousand years and what happens? The dynasty of Israel was established; David becomes king. We come down another thousand years and what happens? There's Calvary. Another thousand years and we come into the time just before the Crusade which brought about the Reformation along with other things beginning the consummation of the corruption of the sixth day. God has been working very hard for the prior 6,000 years to take care of us. Then He's going to rest. But when is He going to rest? In a thousand years we'll have another day. That will be God's day of rest. And that's the way I like to think of it.

Mr. Karl Turekian: This is all very nice but where do you get any basis for two weeks . . . it's 14 days which are going on now. The first seven we can argue about because it's written down in the Book. With the consideration of the last seven you've become more than a mystic and are not on scriptural grounds. I don't think you have any right to assume that it is a thousand years for a day. I think we would have to discount the whole thing that you said as far as any scientific basis goes.

That's about all I can say because it's truly philosophical and has no place in a scientific discussion of "days" at all.

Dr. J. O. Buswell: I skip the question, "Why couldn't God do it faster?" And all I need to say is that no one has said what God could do. The question is, "What did Moses say . . . what did Moses mean?" God could make it all realistic and make it look like a thousand years or a million years but God probably didn't do that.

Mr. W. Pass: It hadn't quite crystallized in my mind but it seems from the standpoint of geology we have been forced to make some investigation as to what Moses actually said and the historical point of view on the point of time which has been stretched into an indefinite period of time. Is that correct?

Dr. J. O. Buswell: I don't think so. I don't think that's the process of thought at all.

Mr. W. Pass: What I had in mind . . . we've seen in the geological problems that arise that we are forced to a more indefinite time for the days. If I am misrepresenting your position I'd like to be clarified on that because I feel at the other end when we come to the day which man was created, we seem to lean to a point period. That is, if it were a long period of time when man was created then we are playing into the paleontological evidence of the anthropologist or paleontologist himself who also is interested in the indefinite period of time for the creation of man. Would you comment on that and clarify my thinking.

Dr. J. O. Buswell: I think we must clear ourselves from ascribing eisegetical motives to us Biblical exegetes. I very diligently avoided anything from geology. The first hundred years in the millennium, I'm going to look up Larry Kulp and I'm going to get him to teach me some geology. I'm going on a trip tomorrow all day and I'm going to learn some geology, but I don't know any geology. I do know a little bit about how to read the Scriptures. This problem as to the meaning of those days did not originate with modern geology. I pointed out the opinion of Augustine, who certainly antedates any of our modern geological data so far as scientific correlation is concerned. Augustine brought up the problem.

The early church fathers had a tendency to run into weeks. The number 7 in semitic usage is a round number and it suggests perfection or well-roundedness. You'll find in Hypolitus and Ireneus and a dozen or so of the church fathers that the history of man is going to be 7,000 years long. Because John refers to a thousand years of blessedness, well, click, that refers to a sabbath. So you've got to have 6,000 years before. Many of the fathers, however, got in 5,000 years before Christ for some reason or other and then in the year 1,000, you will remember from your church history, there was a great stir through Christendom. They said, "This is the end of the week, this is Saturday night and the Lord's return is going to come." In the year of 1,000 people all over Europe sold their property. They did just the same thing the Millerites did in about 1830, you will remember. They figured out the date the Lord was going to come . . . the millennium was right here. That tendency to put things in sevens is a very common tendency. It goes clear back to the Greeks in the fourth century, B.C. They used allegorical numbers in interpreting Homer. The tendency crops out all of the time in Christendom. People like these numbers, you know. They go on and on with the analysis of numerics.

Sober exegesis has always been opposed to mere speculation. I think we should emphasize the fact that the argument as presented was not based upon any motive other than the historical, grammatical exegesis of what Moses did say.

Dr. G. D. Young: I think the block in our thinking here is our Indo-European usage in which we link time with the word "day." My specialty, if I have one, is semitic poetry. I see it in the poetry of Ugarit pre-eminently, in the Egyptian poetry, the Babylonian poetry, the Syrian poetry, and in the Biblical poetry. It's all through here. Everything that's worth happening, if you're going to tell it in poetic languages, has to happen in sevens.

Baal had no house. He sent for the deities and one day they brought bricks and straw, and on day 2 they brought bricks and straw and on day 3 they brought bricks and straw and on the seventh day it was there. On day 1 the bricks and the straw were consumed by fire, on day 2 they were consumed in fire, and on the seventh day, lo and behold, it comes out silver and gold. Moses is painting a great picture here—this is how the lights came on, and this is how the birds that fly, and this is how the things that swim in the water came on, and the things that creep and at the end of it all, there is man. We ought to divorce in our thinking, I'm quite sure, the idea of minutes and seconds from the usage of the word "day" in the semitic context of this chapter.

Mr. J. Watson: I am persuaded in my mind that we should avoid the Greek concept of symmetry in our thinking and that is why I think of this as a Hebrew problem. To answer Mr. Peterson, I would quote a theologian, Calvin by name, that it's poor exegesis to hang any doctrine of Scripture upon one point, and I would quote Dr. Buswell as saying that a day is a thousand years with the Lord and also as a watch in the night. Now, I think, you would only find one Scripture verse—that one in Peter—to support such a doctrine. On the basis of exegesis I think it is a very poor conclusion.

Dr. J. L. Kulp: If a geology student may be permitted to come onto the platform on biblical exegesis for a minute, it is very gratifying to me to hear this dissertation this afternoon purely from a Biblical, theological point of view. For I think that Dr. Buswell is intellectually honest when he states that this interpretation is what he thinks is the best Biblical interpretation independent of any physical evidence. Now it's very nice to be able to be comfortable in a particular view of interpreting Genesis and maybe a thousand years concept may be comfortable to some. Whether it's comfortable or not has relatively no meaning. What we should be interested in is whether it is true. Therefore, since this is a scientific gathering, we might as well add the P.S. that Dr. Buswell's is possibly the best Biblical interpretation, I think that anybody who has studied geology very carefully, will be ready to admit that it is the only possbile interpretation, broadly speaking, that can be harmonized with geological science.

Dr. Coleman: I just had one verse in connection with the point that was brought up about the sabbath which I think is worth contributing in John, chapter 5, verse 16 . . . "Therefore did the Jews persecute Jesus and sought to slay him because he had done these things on the Sabbath day. But Jesus answered them, "My father worketh hitherto and I work." I think the question was raised as to what happens to the seventh day that God rested if you consider them not as days but as periods. The verse in John would clearly seem to indicate that the Sabbath was broken, as Dr. Buswell suggested, at the time of the fall of man through sin and then the greater work of the redemption began, so that you have whatever definite period would exist there and then the work of both the Lord and His Father in the great work of redemption of man began.

Dr. A. Eckert: I would like to ask one question and that is—What did the Hebrew scholars think of the interpretation of these days? Do you have any record of their exegesis in this part of Genesis?

Dr. G. D. Young: I think we have the analogic type of thing when they refer to the history of the world by thousand year periods. They had that thousand year concept and so they divided the history of the world into thousand year periods. I do not know whether they applied this method of interpretation to the creative days or not. In one method of Jewish exegesis it would not be impossible to so take them.

Dr. I. A. Cowperthwaite: Did you have a final summary, Dr. Buswell?

Dr. J. O. Buswell: Yes. I am sure Dr. Young was referring to the rabbinic tradition, not to anything in the Old Testament period, nor anything that comes to us with any New Testament authority. The rabbinists followed this kabalistic system of numerical interpretation, and you simply do not have scientific exegesis in the kabalistic literature anymore than you have in Philo, Alexandrinus, and the way in which he interpreted things, imposing a numerical philosophy upon the plain language of literature. The Bible-believing Jews in the time of Christ do not give us anything to indicate that these days were regarded by them as a thousand years.

But it is the Hebrew prophets to whom we look for a true exegesis of the books of Moses, and they are constantly pointing to the spiritual principle but there is nothing in Isaiah or any of the later prophets to indicate that they regarded these days as thousand year periods or that they regarded them as 24-hour day periods. As to "my father" working, Dr. Allen says he thinks that God rested because he had nothing more to do and then sin came in and he had to go to work again. Of course, I can appreciate that way of looking at it, but it doesn't fit my Calvinism at all. I cannot think of the Lord God almighty as idle. When we have the new heavens and the new earth, they that are holy will be holy yet more and those who are in the lake of fire will be filthy yet more. Its a progressive dynamic universe from start to finish and I would only re-emphasize the fact that the Scripture says not that he folded his hands and sat there and took a nap. No! God did not become idle. I'm not idle on the Lord's day. Any Christian is not idle on the Lord's day. God stopped his work of creating. That's all the work he stopped. The work of providence always went on. He took pains to go down and walk with Adam in the garden and talk with him and commune with him and reveal Himself to him. God's work of providence always went on and God's work of redemption began the very minute that sin came into this world. Christ, in John 5 is referring to the providential work of God which always goes on. The answer is equivalent to saying, "God works on the Sabbath day in His work of providence and in his work of redemption, and therefore, although the law of the Sabbath is clear and we need this day of rest, the work of redemption and God's work of providence has to go on."

When God gets through with this earth He will cast it aside and make another one. Then will begin a work of creation . . . "I make all things new." He will create a new heaven and a new earth. God rests from the work of creation, from the end of the creation in Genesis on until he creates a new heaven and a new earth. God never rests from all activities.

Dr. G. D. Young: (This discussion by Dr. Young was in connection with another paper not offered for publication. It is placed here because of its contribution to the subject discussed above.—Ed.)

I wouldn't bring this up except it has come into the discussion at two different levels and I have heard it also brought up in the Society of Biblical Literature where most of the folks do not accept our conclusions. It is the problem of the making of the sun on the fourth day. It was for a long time a problem to me, and apparently it still is a problem to some. It has already been pointed out that the world "made" in the statement that the sun was made on the fourth day is a word that is used a little over 2,000 times in the Hebrew Bible. Its meaning ought to be well known. 1500 of the occurences, or more, should be translated "do." It is, as a point of fact, in the standard Hebrew dictionary given with the word "do" 1500 plus times. And about 500 times it is rendered 'make." Among those occurences with the meaning "to make" there are several very interesting ones. One man came in from a long hunt and I think the king summoned him in or something like that and he hadn't had a shave for a long time so he "made" his beard. He prepared it. He trimmed it. He didn't make it "de novo."

On another occasion, a lady "made" her toe nails. She didn't make them at that moment. She prepared them in the prescribed manner for whatever social function she was about to engage in. That is the way the word rendered "to make" is used. We shall couple with this the fact, as Dr. Buswell pointed out, that it is associated with the word "natan" which means "to set" or "to put" for some purpose. On the fourth day the sun was prepared and set for a specific function. The Hebrew would not support the idea of the making of the sun on that particular day. That illustrates the simple answer to many of the alleged problems that we meet and the problems that are alleged from some points of view to exist in the creation story. I'd like to take this opportunity to put into the record something that's a very deep conviction in my heart. I know the geologists do not want me telling them anything about their science. Yet my hair stands on end as a theologian sometimes at some of the things I read in the A.S.A. Journal concerning the meaning of Hebrew words and Biblical concepts. I would like to suggest that our strength may lie in our united action. There's a very real danger that since we are among wolves if we do not hang together we will hang separately!

# The "Kind of Genesis and the Species" of Biology*

WILBUR L. BULLOCK Ph.D.
Assistant Professor of Zoology
University of New Hampshire

During the evolution-creation debates that were so popular in the first quarter of the 20th century one of the most important conflicts centered around the problem of "fixity of species." At that time it was generally assumed that "kind" of Genesis 1 and "species" of biology were synonymous. Therefore, since Genesis implied that these kinds or species were stable units, the Christian became an ardent defender of the fixity of species. Indeed, precedence for a stand on the immutability of species had already been set by such biological greats as Agassiz and Cuvier while the main exposition of the concept could be traced back to Linnaeus himself. In the past 50 years, considerable thought and study has gone into the species concept and some important changes have been made that bear heavily on this old problem of species fixity. A challenging problem thus presents itself to the Christian biologist. What is the relationship between "kind" and the newer concepts of species, and is there a Christian concept of species? Mixter (1950) made a contribution to this problem by pointing out that the word "kind" is a rather general term that may have several meanings. He further pointed out the difficulties involved in assuming synonym for "kind" and "species." Without any further consideration of the former term at this time, let us consider the problem of species as it manifests itself in that area of modern biology known as taxonomy or systematics.

One of the oldest approaches to the study of the living world around us has been in the cataloguing or classifying of plants and animals. The resultant type of biology—systematics or taxonomy—has, in recent years, been drastically revised to play a major role in outlining the course of evolution. From this has arisen the current twofold aim of modern taxonomy: (1) to name and describe animal forms and (2) to arrange these forms in an order that will indicate their evolutionary relationships. For the Christian, it now becomes necessary to analyze these objectives of modern taxonomy and to attempt to discern any correlation between the facts of this taxonomy and the Scriptures.

At first sight the naming and describing of animal forms seems to be a relatively simple task. Certainly Adam had no great difficulty in Genesis 2:19,20 where we are told: "And out of the ground the Lord God formed every beast of the field, and every fowl of the air; and brought them unto Adam to see what he would call them: and whatsoever Adam called every living creature, that was the name thereof. And Adam gave names to all cattle, and to fowls of the air, and to every beast of the field." However, our difficulties soon become apparent when we actually try to catalogue the forms we see in the world around us. This dilemma is perhaps best illustrated by the remark of the entomologist: "Where was I to put a beast or a bug when the next one that's exactly like it is entirely different the next time you look at it?" Not only does this exclamation of exasperation indicate difficulty in assigning an animal form to its taxonomic niche (or species) but it intimates that the niche itself is none too reliable.

The task of defining the "species" is one of formidable proportions. Such is the nature of the task that many have been led to deny the reality of the species. Historically, we can follow through from the fixed species of Linnaeus (1758) to the relative, "passing parade" concept of Haeckel and Darwin, to the modern polytypic species of Mayr, et al. As these attitudes varied in time, so do they vary with the particular animal or plant group involved and with the approach of the taxonomist. In this manner there is little correlation between the species of bacteria, protozoa, helminths, insects, or vertebrates. For example, witness the absurdity of applying a bacterial species concept to helminths as done by Wilhelmi (1940).

> "Species" of helminths may be defined tentatively as a group of organisms the lipid free antigen of which, when diluted to 1:4000 or more, yields a positive precipitin test within one hour with a rabbit antiserum produced by injecting 40 mg. of dry-weight, lipid-free antigenic material and withdrawn ten to twelve days after the last of four intravenous injections administered every third day.

Among the earliest concepts of species is that which considers the species to be a group of animals of similar morphology and not overlapping in this respect with any other group. This is basically the concept of Linneaus (1758). It is a much maligned concept but is still used today by many modern taxonomists. Gates (1946) relies considerably on this concept in his evaluation of primate evolution. Students in many of the invertebrate phyla use it in cataloguing new forms. Indeed, it would be impossible to begin a taxonomic study of an unworked group without a morphological introduction. However, this concept has its shortcomings. The selection of characters that are "significant" tends to become quite subjective. Besides, the significance of these characters varies from one group to another. Rensch (1934) illustrates this by noting that folds in the shell mouth of certain snails are reliable for species identification whereas in other snails individual variation makes them useless. Witness also the widely divergent views on the classification of the Trematoda depending upon whether suckers, excretory pattern, reproductive system, or life cycle is used as its basis.

Another approach to a species concept, and one that has achieved considerable popular acclaim, is on the basis of interbreeding. One of the better known manifestations of this concept is illustrated by the statement that "all forms belong to one species which can produce fertile hybrids." This is a very appealing definition but it has many limitations. (1) It cannot apply to the Protozoa or to hermaphroditic groups. (2) There are too many obvious exceptions. For example, several well known species of Drosophila are capable of producing at least some fertile hybrids. Moore (1946 a,b) showed that crosses between geogra-

---

*Paper presented at the Sixth Annual Convention of the American Scientific Affiliation, New York, August, 1951.

phically widely separated races of the frog, Rana pipiens produced defective "hybrids" whereas crosses between two **species** R. pipiens and R. palustris, from the same area produced normal hybrids capable of reproduction. Still another example of the unreliability of this definition is illustrated by the shad genus, Pomolobus. These marine fishes go up the bays and rivers to spawn in brackish water during the spring of the year. At least, in the region of Durham, New Hampshire, P. aestivalis and P. pseudoharengus overlap in their spawning seasons. Much interbreeding is now taking place so that it is quite difficult to distinguish the two species. Thus the presence or absence of fertile hybrids is not a reliable species character.

A somewhat improved version of the interbreeding species concept is that of Mayr (1942). "Species are groups of actually or potentially interbreeding natural populations, which are reproductively isolated from other such groups." This definition avoids a consideration of hybrid sterility but still retains the limitations shown by the sterility concept. It can not apply to many invertebrate groups nor does it explain the hybrids between sympatric species.

Still other species concepts have utilized genetics, physiology, or immunology as their basis. However, general application of these criteria have yielded little but confusion in the taxonomy of the greater part of the animal kingdom. It is not surprising then to find many adherents to what Mayr (1942) characterizes as the "practical species concept"; namely, that a species is a systematic unit which is considered a species by an authority on the group. This concept reduces the species to the depths of subjectivity. It creates an ephemeral, highly limited species that serves only a given group of authorities in a given group of animals or plants.

Thus it can be seen that at the present time there is no real concept of species. Indeed it is easy to become pessimistic about the very existence of the species as a valid biological unit. However, most authorities in the field of taxonomy and evolution still maintain that the species is a real and useful entity (Mayr, 1942; Goldschmidt, 1940; Huxley, 1940). This is particularly true for animal groups which are well "worked" taxonomically. In these groups the various factors in variation (geography, temperature, altitude, etc.) are most clearly understood, and it is in these groups (e.g. birds) that the "polytypic" species dominates the scene.

The polytypic species is a concept that had its beginnings with Kleinschmidt (1900) in his Formenkreis theory. This Formenkreis was a sort of super species which took into account the factors involved in the variations of populations under differing environmental conditions. Kleinschmidt further recognized the strong possibility of supernatural intervention in bridging the gaps between related Formenkreis. Such a concept—minus the supernatural intervention—is the basis of Goldschmidt's modern mutationist theory. Unfortunately, what Kleinschmidt lacked in genetics, Goldschmidt lacks in making any contribution to the species problem. In general, the polytypic species has led to the lumping of many varieties, previously described as distinct species, into only a few species. Such a concept applied to man would lead to the inclusion of all living and fossil man into one or at most two species. This seems to be in greater harmony with the Scriptural observation that "God hath made of one blood all nations of men." The reverse procedure as applied to man (Gates, 1946) leads to a great diversity of terms and serves only to support favored phylogenetic fancies. Some species, as pointed out by Huxley (1940), appear to vary little and hence are considered monotypic.

One more problem should be considered. That is the "subspecies." To Goldschmidt (as well as Kleinschmidt) the subspecies merely indicates the multiplicity of variations **within** the species. To the majority of present day evolutionists (e.g. Mayr, Dobzhansky), however, the subspecies represents an incipient new species. Should this latter concept prove correct then Christians must certainly look for the units of creation (kinds?) elsewhere than at the species level. Should mutationism triumph then the polytypic species concept might solve the problem. Much long and laborious taxonomy must be performed before anything definite can be said about the limitations of the species unit.

In summary, the species concept in modern biology is far from settled. The prevailing opinion is that there is a real entity in the species but whether this can be defined to satisfy all plant and animal groups is problematical. Consequently, the species, at present, is basically human concept. It is an interesting biological problem but probably does not bear any real relationship to the "kinds" of Genesis. It is possible that further developments in this field will shed light on this complex problem. Meanwhile, it is ill advised to champion the cause of fixity of species under the banner of Christianity.

**Selected References**

Dobzhansky, Th. 1941. *Genetics and the origin of species.* Second edition. New York.

Gates, R. R. 1948. *Human ancestry*: from a genetical point of view. Cambridge, Massachusetts.

Goldschmidt, R. 1940. *The material basis of evolution.* New Haven.

Huxley, J. S. 1940. *The new systematics.* Oxford.

Kleinschmidt, O. 1930. *The formenkreis theory and the progress of the organic world.* London.

Mayr, E. 1942. *Systematics and the origin of species.* New York.

# The Evaluation of Radioactive Evidence On The Age of the Earth*

ROY M. ALLEN, Sc.D.
368 Ridgewood Ave.
Glen Ridge, N. J.

Man is naturally an inquisitive being, ever anxious to know the answer to every question arising to vex his imagination. "How old is the Earth, and when did life begin upon it?" are two over which mankind has been pondering for many years. Ussher's chronology was but one attempt to supply the answer. It sufficed so long as believers in the inspired Scriptures interpreted the Genesis record to imply six solar days transpiring between Genesis 1:1 and the creation of Adam. Realization that the "yoms" of creation were "God's days," defined elsewhere in the Scriptures (Ps. 90:4, and II Pet. 3:8) apparently for the specific purpose of throwing light upon Genesis 1, together with unbiased analysis of the record itself resulted in a reorientation of Christian thinking. It should be emphasized in this connection that this has not been

* Paper presented at the Sixth Annual Convention of the American Scientific Affiliation at Shelton College, New York, August, 1951.

at the expense of belittling or questioning the inspired record. Rather does it serve to prove its literal accuracy when properly read.

It has not been primarily those accepting the inspiration of the Scriptures who have been responsible for pressing the birthday of the earth back into a remote past. Mechanistic evolutionists who require an enormous amount of time for the operation of any of their numerous theories have ever been the aggressors along this line.

Many are the approaches to a solution of the question tried out as yardsticks with which to measure earth time, among which might be cited: The salt concentration of the seas; the stabilization of the tides; the time required for the earth's crust to reach its present temperature; the rate of attrition of surface rocks; their re-deposition and solidification as sedimentary rocks, together with many others.

To apply any of these involves the making of certain assumptions but such procedure is common in scientific reasoning. The main objections to their testimony lie in the lack of agreement between them and their failure to indicate as great an age for the earth as mechanistic evolution requires. It is not surprising, therefore, that a "made-to-order" yardstick such as radioactive disintegration should be pressed into service. In the beginning, early estimates of the earth's age, as proposed by Kelvin, Rutherford, and others were around 400 to 500 million years although a few men of science went as low as 20 million. All figures were soon increased materially, Sir James Jeans putting it at 1450 million years. Later estimates went to 2000 million while the present tendency is to place a possible ceiling appreciably higher than this, even exceeding 3000 million years.

The Symposium on "The Age of the Earth" published by the A.S.A. in 1948 constitutes a fine presentation of the fundamentals involved in radioactive methods of dating. It makes unnecessary at this time a further review of the methods themselves, of which there are several. Briefly summing up the evidence on which radioactive dating is being accomplished, we can itemize it as follows:

1. All radioactive elements are continually decaying or decomposing (through giving off energy) into elements of lower atomic weight.

2. The rate of decay is specific for each element; some so slow their lives are measured by many billion of years; others, largely intermediate radioactive elements, may have lives measured in minutes (or less), hours, days, or years. The average life of radium, which is itself an intermediate element, is around 2500 years. To reach complete decay requires a much longer time than this. For practical purposes the rate of decay is expressed in half-life, that is, the time when one-half the original total energy would be gone. Then the average life is equal to 1.45 times this figure, a value derived by integrating the curve of total life.

3. It is possible to determine the half-life of all radioactive elements very accurately.

4. The rate of decay in every case has been found to be absolutely constant. These rates, so far as we are able to determine, are unaffected by any known external condition (e.g., such as temperature, etc.) and by such evidence as is available, identical in past ages to the present rates of decay.

5. The end products resulting from decay of natural radioactive elements have all been determined. In the case of those of higher atomic weight than lead the stable end product is some isotope of lead and the gas helium.

6. These conditions being so, we can understand the logic of assuming that an analysis of any rock or mineral originally containing a radioactive element, e.g., uranium for example, to determine the ratio existing between the element and its lead isotope present with it (and perhaps the helium ratio also) will indicate the age of the rock or mineral in question.

Numerous conditions prevail which complicate the making of accurate determinations. For instance: 1. Leaching out of some of the uranium. This will indicate an older rock than actually is the case. 2. Lead may have migrated from the mineral under test, giving too low a figure.

3. Helium may have diffused from the mineral (which appears to be a common condition) also resulting in a low figure, usually lower than the corresponding U/Pb ratio from the same sample.

4. There is always the possibility of helium being present in the rock which did not originate as a radiogenic product.

5. The same can be said for the possible presence of lead not resulting from the decomposition of the uranium. In this case this is always assumed to be indicated by the presence of lead isotope 204 which can be isolated by means of the mass spectrograph.

Other conditions are also recognized, some of which will become evident later, but all in all, radioactive methods of dating appear, at least on the face of it, to have much merit.

It is unfortunate that the majority, if not all of those engaged in this type of research are ones definitely arrayed on the side of mechanistic evolution. Such being the case we can expect a natural bias in favor of a maximum time scale. While, except for one all important issue which we shall consider later, it makes no difference in the literal accuracy of the Genesis account of creation, how old the earth may be or how long life has existed upon it, we are anxious for unbiased facts. This calls for a literal application of Paul's admonition, "Prove all things, hold fast that which is good" (2 Thes. 5:21). Where mechanistic evolution dictates figures our position might well be, **"Timeo Danaos et dona ferentes."**

With this in mind, the question I wish to raise at this time is, "Are there factors in radioactive dating which have not been taken into consideration, or given due weight?" I believe there are and that the A. S. A. should present these to the scientific world for further study.

The first issue will take us back to that time considered "the beginning." In what stage was the earth then, as indicated by radioactive evidence? That a beginning of some sort is predicated is apparent from the assumption that uranium (and other radioactive elements present at that time) had not yet evidenced any decay. This is equivalent to saying that radioactive elements did not exist previously. They therefore must have been formed at some specific stage in earth formation. This is certainly in accord with Genesis 1:1.

Regardless of what theory one elects to accept as to how the solar system came into being, the fact remains that all evidence points to a common origin for the sun and its planetary system. The formation of all the elements, probably starting with hydrogen (since this still represents 81.76% of the total volume of the sun, helium accounting for 18.17% of the remainder) and progressing to those of higher atomic weight according to some fixed law ordained by the Creator, must have been a part of the creative process.

As to the origin of radioactive elements, it is logical that radioactive decay be a reversible process. Just as the formula for decay is $U^{238}$ — energy = $Pb^{206}$ + He, so the formation of uranium in the beginning can be expressed $Pb^{206}$ + He + energy = $U^{238}$. How the Creator brought about conditions in the beginning to effect this synthesis must be considered at this time, as far as positive knowledge is concerned, one of the "secret things belonging unto God," (Deut. 29:29). There is, however, in the known existence of cosmic radiation, perhaps millions of times greater in earth's formative period, one suggestion of the means He might have employed. Also present processes for producing transuranium elements offer other explanations.

Presumably all the steps involved in the decay of $U^{238}$ to $Pb^{206}$ were reproduced, in reverse order. But notice this involves the pre-existence of both helium and all the natural isotopes of lead. Stable isotopes 206, 207 and 208 frequently are referred to as radiogenic since they represent stable end products of radioactive decay. The assumption appears common that these came into existence through decay processes. There is not only no logic in this assumption but ample evidence it is not correct. Nor is it an assumption held by research workers in age determinations.

Of all known elements about three out of every four have from one to several natural isotopes. This is true of lead also. The atomic weight of ordinary lead is 207.21. This figure results from the relative abundance of all stable isotopes of lead, usually found together in the following percentages (Nier's figures):

| | |
|---|---|
| 204 — | 1.48% |
| 206 — | 23.59% |
| 207 — | 22.64% |
| 208 — | 52.29% |

Helium and lead are both present in the sun but no radioactive elements of higher atomic weights than lead are known to exist in it. The sun apparently represents a condition where their presence is not possible, although the building blocks out of which they can be formed are at hand from which they can be produced under proper conditions.

Regardless of the state of formation the earth was in when the reaction producing the radioactive elements was completed it would appear to be the time indicated by radioactive evidence as "the beginning." But still an uncertainty exists.

In a reversible process such as under discussion there are obviously three stages to consider, all of which conceivably might be included in the **total** age of the earth, viz:—

1) The formative period during which energy was in excess and utilized in the formation of radioactive elements. If this were a slow process, as appears logical, it might even be measured in terms of billions of years. This would mean nothing to an eternal Creator. This time would not be indicated by radioactive decay.

2. An equilibrium stage during which energy was not flowing in either direction. Here again might be involved a long time period.

3) The present unbalanced condition in which energy flows out of radioactive elements. The beginning of this stage would be dated by radioactive evidence. Even here, however, another factor might affect the time figures. While all evidence indicates decay rates have always been constant, there exists the possibility that transition from a state of perfect equilibrium to one of normal decay might have been so gradual as to retard materially the decay rate at the start. Such would again increase the actual age of the earth over that indicated.

Our reasoning thus far relates to the energy factor only. Others have to be considered. As stated, both lead and helium must be present, probably in abundance, judging from the percentage of both existing in the earth today. Since a natural process of producing radioactive elements is not known to be operable in the earth under present conditions (except for the formation of $C^{14}$) the limiting factor must have been the supply of high potential energy. When this was used up, or conditions on the earth so modified radioactive element building could no longer continue, the process stopped, even though both lead and helium were present in excess. By analogy, the condition existing might be likened to the formation of silicate minerals in the presence of excess silica. The latter finally forms the free silica of acid type rocks. In a case of this kind, how are we to prove that some of the helium and lead isotope found in a given analysis are not in part residual, remaining **in situ** within the mineral from the beginning? A simple illustration may serve to make this possible condition evident.

I own a French clock, bought over forty years ago. It was guaranteed to run 400 days with one winding. Through the years I have proved this claim justified. A friend visits me and becomes interested in the clock. I tell him of the 400 day operation and by coincidence while he is at my home the clock stops. I see him take out a pencil and start figuring. Soon he announces that I must have wound the clock on or around July 1st of the previous year. He has believed me and had faith in the clock. No wonder he is surprised when I tell him he is wrong, for I remember winding it on Christmas day. Is the fault with his mental process or figures? Not at all; I explain that when I wound it I was in a hurry so did not take time to wind it all the way up! Just so, like the portion of unwound spring in the clock, the presence of untransformed helium and lead isotope (206, 207 or 208, as the case may be) associated with a radioactive element could throw all calculations as to the age of the earth entirely off; the indicated age would be too great. This condition has been recognized and all possible means to nullify its effect on accurate age determination have been employed. But still the uncertainty persists because the only indication of untransformed lead is assumed to be the presence of $Pb^{204}$ also. When mass spectrographic analyses indicate the presence of $Pb^{204}$ the presence of ordinary lead $Pb^{207.21}$ is suspected.

Were it not for the fact that radioactive decay, comprising as it does three different radioactive elements altering the proportions of lead isotopes so that ordinary lead ratios presumably no longer hold good, it should be possible to multiply the percentage of $Pb^{204}$ present by 66.5 (i.e., 98.52/1.48), then subtract this value from the total lead of all isotopes present, considering only the remainder as radiogenic lead. But this, of course, will not work. About all that can be assumed is that some of the lead present was not radiogenetically produced and that the true age of the rock is under that indicated.

It is interesting to note in Nier's classic work the common absence of $Pb^{204}$. In only a few instances does it show up to confuse the determination. He assumes its absence to indicate all the existing isotopes of lead to have resulted from disintegration of $U^{238}$, $U^{235}$ or Th. But if the synthesis (we will call it that) of radioactive elements commenced with ordinary lead ($Pb^{207.21}$) it would be selective from the start and the various isotopes would soon cease to be lead, in their progress upward toward the end of the series. Unfortunately we cannot know the conditions existing in the formation of the earth at this time, hence can only theorize. Should the temperature have been high, as appears reasonable the lead would have been in a molten state but after a time the element under formation would possess a higher melting temperature. For instance, lead melts at 327°C (620°F), radium at 960°C (1760°F), and uranium and thorium at around 1850°C (3360°F). When freezing temperature of any intermediate radioactive element was higher than that of the earth or the rock magma in which it was located it would solidify. Unaltered lead (e.g., all of $Pb^{204}$ and such portion of the other isotopes as remained) would still be fluid and might be dispersed elsewhere. In this case we would not expect original lead to be present. It is also possible, on other grounds, that radioactive element generation, even in its incipiency would affect a segregation of all lead isotopes, $Pb^{204}$ being rejected while the others, having an affinity for helium in the presence of energy (especially if catalytic action played any part in the synthesis) were retained. Varying local conditions might result in different final results, hence, as in the case of one of Nier's analyses, the presence of 0.884% of $Pb^{204}$ could be accounted for. It is the numerous uncertainties in the problem which render true interpretation of analyses, when translated into terms of earth age difficult and inconclusive.

Once it is realized that minerals like uraninite are not chemically pure combinations as we would expect, but (to employ a modern expression understood by all) contain a little of everything but the kitchen sink, the force of this line of reasoning is appreciated. Dana's analyses of 26 different specimens of uraninite emphasize this fact. They show in addition to lead,—iron, silicon, phosphorus, zirconium, manganese, magnesium, bismuth, copper, zinc, arsenic, sulphur, and elements of the rare earth series. But most suggestive of all is the universal presence of nitrogen in fairly large amounts. Of course analyzing for helium was not done at the time these analyses were made, but assuming the accuracy of the determination we can see the strong possibility that nitrogen accompanied the helium. In this connection it should be noted that next to hydrogen and helium the elements occurring in the sun in greatest volume are respectively oxygen, magnesium and nitrogen. Since nitrogen is not in combination with uraninite it must have been occluded in its atomic lattice from the beginning. Of course there is a possibility of an error in determining the gas as nitrogen, it might have been helium. It was not until several years later that Ramsey discovered helium in the mineral uraninite.

An interesting sidelight on radioactive dating comes from a comparison of age figures compiled from these analyses of uraninite (Dana's "System of Mineralogy," 6th ed., reprinting of 1922, first printing 1892) with figures of Goodman and Evens (1941) quoted in Dr. Rex's article in the A. S. A. symposium. Dana's analyses cover 26 mineral specimens, made prior to the discovery of radioactivity. In compiling figures from them it was necessary to consider the $UO^2$, $UO^3$ and $ThO^2$ as a unit, in checking against PbO. Two groups of Connecticut uraninite are given. Five from Glastonbury average 234 million years and three from Branchville 310 million. These compare with Goodman and Evan's figure for Connecticut uraninite (locality not stated) of 366-371 million years. Even though the figures themselves are a good check they suggest the presence of some of the factors I am pointing out, since all the pegmatites of Connecticut should be of substantially the same age.

Five analyses of Norwegian specimens in Dana's list average 775 million years while two more (Arendal) give a higher figure of 925 million years. These compare with Goodman and Evan's value of 1085 million years. Again a fair check but the same comments apply. It also suggests the present tendency to establish as high a figure as possible. Dana also gives an analysis for Colorado uraninite in which the PbO is only .70%, yielding an age figure of 52 million years. This figure compares favorably with one determination by Goodman and Evans of 54 (He) or 69(Pb) age, quoted by Professor Stoner in the symposium.

There yet remains another, and I believe more plausible, objection to figures derived from radioactive determinations as they are now accepted. If the premise of the origin of radioactive elements through reversible addition of energy to lead isotopes be allowed, the logic of a step-by-step build-up through intermediate elements is sound. Action and reaction should be equal but opposite in direction.

It is, of course, well known that uranium does not transform directly to lead, but through a multitude of steps, or intermediate elements, all radioactive themselves, some with long life, others extremely short. Granted a reverse process, at any stage of build-up all these intermediates would be present after sufficient time had elapsed to allow the initial process to reach the last of the series, $U^{238}$, or other, as the case may be. Stop the process at any instant and a portion of the element which started out as a lead isotope would be present in each transitional radioactive element. In other words, radioactive equilibrium should be present in a buildup process just as much as in decay. This means that should decay commence following the stopping of the process, every intermediate element would begin decay as a parent element in that their contained energy has never attained a higher level. Helium and lead would be produced which never originated in the beginning of the series. Consider the possibility that only a fraction of the additive energy had opportunity to reach the level of $U^{238}$, with a correspondingly larger portion resided in U-II ($U^{234}$). $U^{238}$ has a half-life of 4.51 x $10^9$ years while $U^{234}$ has a half-life of but 2.33 x $10^5$ years. Still other percentages of $Pb^{206}$ (and helium) would be located in Ionium (8.3 x $10^4$ years) and in Radium (1590 years) and so on all down the line. Today we could be measuring the $Pb^{206}$ ratio and assuming it all originated in $U^{238}$. This same situation could apply in principle to every radioactive method so far suggested evcept the $C^{14}$.

Let us consider the problem from still another angle. The fact that the $Pb^{207}/Pb^{206}$ ratio in a given determination is employed as a check comparison (and considered as one of the most reliable of all radioactive methods) against U/Pb ratios is an implication that the origin of $U^{238}$ and $U^{235}$ was contemporary, both starting decay at the same time. Since both are first in their corresponding cascade series this seems logical, but raises another question. The average life of $U^{235}$ is roughly only about 1/6th that of $U^{238}$. Assuming that whatever energy formed $U^{238}$ was also available for $U^{235}$, the percentage of their corresponding lead isotopes available as building blocks being almost identical, (i.e., 23.59% and 22.64% respectively) the amount of each uranium isotope in the beginning should have been about equal. Due to their different rates of decay there now exists in the earth only .72% of $U^{235}$ as against 99.276% of $U^{238}$. This ratio of 1:139 is quite constant throughout the earth, wherever uranium is found. Does it not suggest the true yardstick with which to measure time is $U^{235}$ and not $U^{238}$ and the maximum age of the earth is nearer the average life of $U^{235}$ than the higher figures set by other ratios?

Nier appears to have sensed the validity of this argument and by figuring backwards on the basis of relative decay rates comes to the conclusion that $U^{235}$ originally was not present in so large a proportion as $U^{238}$. However, granted a difference in the degree of complete synthesis to end products as I have suggested and also a tendency of one to commence decay at an earlier date than the other, his estimated original percentages may be entirely off. This is a line of research that should be followed up in every one of its possible ramifications.

An approach to the age problem, not primarily of the earth but of the universe as a whole has been via the constitution of meteorites. Apparently the original idea responsible for this line of research was the establishing of a firm basis for the helium method, since there should be no loss of helium from a meteorite. While this constitutes a valuable line of research and should yield important information of a cosmological nature, actually it is subject to the same factors entering into the origin of radioactive elements as earth studies.

Helium ages, as determined from some meteorites have exceeded those from terrestrial determinations, being appreciably higher than those derived by other means. Even here, however, the picture is not clear since recent evidence indicates that considerable of the helium of meteorites may have been derived from cosmic radiation. Such would nullify age figures derived from meteorites by the helium method. If age figures determined from meteorites can be considered as evidence of the time of radioactive element formation throughout the entire solar system (or the galactic universe) it raises questions of vital importance relative to age dating in the earth. For instance:

1. Was the general formation of radioactive elements simultaneous throughout the solar universe?
2. If so, what was the spread of time required for the operation of the process and was it completed universally at substantially the same time?
3. Did disintegration begin universally at approximately the same time? In other words, did age time clocks start ticking all together?
4. Can we agree that the time when radioactive decay commenced constituted the true birthday of the earth, that is, the greatest age indicated by any radioactive age determination either on the earth or in a meteorite?
5. If so, is the age of minerals showing a great divergence from maximum age of two or three billion years to possibly as low as 20 million to be predicated on a recrystallization of the radioactive mineral with partial or complete elimination of previously formed end products during a reforming of the rock masses in which they occur? Is this a logical assumption?
6. Is the alternate assumption of radioactive element formation simultaneously with the origin of the rock masses logical?

These should be thought-provoking to everyone interested in this age determination problem. It is not my intention to analyze them in detail. I shall merely mention some implications which must follow as a result of applying them to the dating of geological ages. Granting the premise of uniform radioactive element formation at some distant period in the past and subsequent initial decay, let us say roughly three billion years ago, then follows, **pari passu,** that any radioactive mineral showing an age less than this has undergone modification and re-crystallization at a later date. Such processes can be demonstrated petrographically with practically every other known mineral, hence appear logical with radioactive minerals also. The problem then becomes one of demonstrating the degree of elimination of earlier formed end-products. Some of these must have been eliminated else all age determinations would be identical. But it does not appear possible by any known means to solve this problem. In view of this every age figure except the highest, determined from truly primeval rocks unaltered from the beginning, and the lowest, where elimination of previous end products has obviously been complete, can be viewed with suspicion.

If we take the other horn of the dilemma and assume original formation of radioactive elements to have occurred at the time of formation of a given rock we must admit that so long as radioactive element formation was taking place the earth was still in an embryonic state of development and therefore its true age is that of the youngest rock, possibly only 20 to 50 million years old. In this case the spread between maximum and minimum figures (three billion to twenty million) represents the time the formation of the earth was in a state of flux, a period of gestation, as it were, of an earth about to be born. Then figures for events transpiring on a substantially completed earth should be predicated on the minimum age figure.

On the whole, the former premise appears more logical since it is difficult to conceive of radioactive decay going on for some billions of years while at the same time reverse build-up reactions are occurring at other places possibly not far removed. Nevertheless this provides no accurate dating for various so-called geologic ages. On the basis, however, that substantially consistent conditions prevailed in the formation of all rock masses, relative ages of various rocks can be correct even though the figures themselves be on the high side. On such assumption a complete new scale of rock ages could be worked out independent of those established from fossil dating. Much has been made of the fair degree of correspondence in figures derived from various radioactive methods of dating (though $Th/Pb^{208}$ and $U^{238}/Pb^{206}$ often fail to agree) but if the same factors enter into all methods alike this should be expected. Errors should be consistent all along the line.

This analysis of radioactive age determinations has not been made with the idea of condemning the methods **per se**, but rather for the purpose of putting them on a more firm foundation. Properly interpreted they can be of great value in establishing geophysical facts. It is these facts we want and proper methods of approach will ultimately yield them.

It would seem the past tendency to establish the age of the earth from the beginning, dividing this time into various ages, is the wrong approach. As pointed out, this beginning means little, if anything, from a practical standpoint. A far better method would be to establish recent dates accurately, then work backward into the past, little by little.

From the standpoint of the fundamental A. S. A. position there exists what might well be accepted as the basis on which to calibrate all types of pre-human dating. Reference has been made to the one exception to long-time dating on which we, and every Bible believing Christian must take an absolute stand. It is in regard to the time when the man created in the image and likeness of God came on the scene. Mechanistic evolution is more concerned in placing the origin of Homo sapiens in the far-distant past than in achieving any other single objective in its program. The reason is not far to seek. The great gap existing between man and the highest of the anthropoids (not on physical or anatomical grounds but on the cultural side) on the basis of an evolutionary development demands almost an eternity of time. Those of us who have followed evolutionary estimates in this respect can recall how the figure, starting years ago at around 25,000 years has been continually increased, to 50,000, 100,000, and now to many times this latter figure. And it has always been predicated on the gradual development of an entire race of beings.

Contrasted with this concept is the Scripture pronouncement that man did not evolve but was created a single individual from whom the entire human race came. This man, created in the image and likeness of God was endowed with God-like attributes from the beginning. Man, as he exists on the earth today is actually a degenerate, possessing in many respects only vestigial evidence of what Adam was before he sinned. And on top of all this he is spiritually dead, typically animalistic in nature, (See Ecc. 3:18, II Pet. 2:12, Jude 10) hence the need for the new birth to restore what man lost in Adam.

The Scriptures provide only a rough estimate or approximation of the time of Adam's creation, such being deduced from a study of the complete written Word in the light of secular human history and archeological (not paleontological) evidence. On the basis of the genealogical record in Genesis 10 the date of the flood must be moved back to at least 3000 to 3500 B.C. to meet definite secular history. If the pre-deluge genealogies are incomplete (which on the face of it appears unlikely) the time of the beginning of the human race logically can be set at 7000 to 8000 years ago, certainly not over 10,000 years as a maximum. To concede more than this is merely accepting the unproved claims and assumptions of evolution.

The question might be raised at this point,—What has this figure to do with radioactive age dating? The connection is a subtle one and possibly overlooked, even by scientifically trained fundamental believers. It can be illustrated by a simple object lesson. Suppose we take an elastic cord, say 12 inches long in its unstretched condition and on it mark off the inches with ink dots. Next we stretch it until it is three feet long. What is now the measure of an inch. It has become three inches.

Whatever the age of the earth from the beginning, evolution has divided it into time inches, or ages, ascribing to each a portion of the whole. We are not at this time interested in the relative percentages alloted to each so-called age although an obvious re-adjustment of these is possible. For instance, if pre-Cambrian time actually were 95% of the whole instead of the 75% usually credited to it, it would modify all other figures accordingly. The point to be stressed here is the effect on **all** time values, especially that credited to man's advent, when the overall time is stretched beyond its correct figure. It is this correct figure, whatever it may be, in which we are vitally interested. To obtain it the first move must be to take the tension out of the elastic cord until an inch actually measures such.

In other words, let us repudiate any dating of positive human finds on which evolution places a figure in excess of 10,000 years as a maximum. Rough stone artifacts associated with obvious human remains need not testify either to a great age or to "stone age" man in an early upward advance. With equal reason they can witness to degenerate man of the race of Adam, probably pre-deluge and of the line of Cain. The Scripture record puts Cain in the land of Nod (Wandering) from the beginning, hence he and his progeny may have reached the confines of all land areas of the earth in the Scripturally alloted time of 1650 years ante the flood. Certainly no one claims the stone artifacts in use by the American Indians less than 500 years ago to testify to stone-age man in an upward progress. The degree of culture evidenced by the Toltecs, Mayas, Aztecs and Incas a thousand or more years before, completely nullifies such conclusion. This entire concept is based solely on a mechanistic evolutionary belief.

It is not my intention to discuss the pros and cons of paleontological evidence but brief mention of it is pertinent to the complete picture. To one accepting the Scripture record of man's early history, devoid of all evolutionary bias, the travesty of mechanistic attempts to demonstrate a great age for man on the earth is pitiful in the extreme. First we have them digging up a lone tooth, a jawbone, or mayhap a portion of a skull-cap which often is not exhibited to all the world for examination and opinions as to its true identity. This is followed by some pronouncement claiming its human relationship, though often disputed by other authorities who do not agree. Sooner or later, however, contra opinions are forgotten or dismissed as of no value and it is read into the record as a definite proof of man's early origin.

On the whole the greatest argument against acceptance of such occasional finds as human links is the paucity of the finds themselves. On Scripture grounds alone the evidence should be more conclusive. From Adam on the record (Gen. 5) states "they begat sons and daughters." This implies an absolute minimum of four children for every generation, let us say an average of six. Granting a generation to be 50 years (i.e., longer than at present) there would be 33 generations in the 1650 years between Adam and the flood. In 12 generations Cain's descendants alone would number nearly one million while in the remaining 1000 years before the flood the number would go into staggering figures. Add to these the descendants of Seth and any other children of Adam and the population of the earth must have been large indeed. Where are the bones of all these? But the real argument is,— "If man did not start with Adam but existed on the earth for countless ages prior to his time, where are the literal billions of remains of all these?" Let us stand fast on this one issue, though we concede eons of earth time prior to God's creation of man.

One of the best, as well as the most recent of radioactive methods is $C^{14}$ dating. The results of late tests reported by Arnold and Libby (Science, Vol. 113, 2927, Feb. 2, 1951) are extremely valuable and enlightening. It is to be hoped this method will not be suborned by those desiring to make it prove more than direct determinations indicate. It is easy to do this by tying together organic material and assumed human remains without positive proof they are of the same age.

It would seem desirable to make radioactive determination on uranium-bearing minerals completely altered from their original form, such as gummite, coracite, carnotite, autunite, torbenite, uranosphaerite, etc., as possible follow-ups of $C^{14}$ dating. They might extend the range of the latter. Yet even with altered minerals consideration should be given to the question as to whether alteration has effected separation of earlier end products. A case in point is Nier's determinations on two samples of Bedford (N. Y.) cyrtolite giving figures of 300 and 375 million years. These are consistent with other values from the nearby Connecticut pegmatites, all of which should be of the same geological age. But cyrtolite is an altered zircon, changed during relatively recent years, so gradually as to retain all the original constituents of the parent mineral. Incidentally, in connection with uranosphaerite. because of its $Bi^{209}$ content a study of this mineral might yield data on the $4n + 1$ radioactive series and add knowledge of this group of elements, the existence of which in the natural state has been questioned.

A practice which should be adopted as standard is the blind determination of uranium-bearing rocks and minerals without knowledge of the supposed geological age of the speciment under test. This would achieve two results—nullify previous bias as to what ought to be expected, and with results freed from possible error, indicate the true age of the rock. The force of this suggestion is evidenced by a quotation from Goodman and Evans, "More specific geological dating is needed in the formulation of a useful lead time scale." No comment is necessary.

It appears more attention should be given to the mineralogical occurrence of uranium and thorium minerals. For instance, why do they occur largely with the acidic rocks rather than the basic? Ordinarily one would expect them to be associated with the basic ferromagnesium rocks (like the metals of the platinum group) because of closer agreement in specific gravity. What of the over-all metallic content of the rocks in which they occur most abundantly? How about relative ages of the acid and basic rocks themselves and their relative position in the earth's crust? I am referring, of course, to the distinctly deep-seated plutonic series and not volcanic or hypabyssal rocks. Attention to such considerations might help to throw light on the age question through the correlation of all possible data.

Another suggestion might be made at this time. Radioactive age dating research has largely been confined to minerals found in rocks the ages of which are assumed to be indicated by basic stratigraphical evidence. It is generally accepted that sedimentary rocks themselves cannot testify directly. This assumption may be incorrect. All radioactive minerals originally present in the rocks from which the comminuted material was derived must still exist, even though widely disseminated. Because of their high specific gravity it should be possible to separate out enough of these comminuted minerals to test for U/Pb ratios. This type of test could be tried out through preliminary study of unconsolidated sands and detrital material. Even in this case some errors on the high or low side might be expected. Certainly tests on relatively pure zircon and monazite sands, also magnetically separated magnetite and ilmenite should prove interesting.

As a part of the complete picture it may be of interest to consider pleochroic halos and the help which might be derived from them. These are discussed in F. Alton Everest's contribution to the symposium. So far interest in pleochroic halos in relation to radioactive studies has been confined to evidence of the constancy of radioactive emanations over long periods of time. There is a possibility, however, that further study of them might provide age data also. Presumably this would have to be done on a petrographic basis, supplemented by microchemical methods.

While halos are frequently referred to in the literature, it would seem that much of the information regarding them is derived from Joly's work on them, supplemented by that of Henderson. Inasmuch as most of their work on them was done on split biotite rather than with petrographic sections, it is not at all complete.

Pleochroic halos are mentioned as occurring in about a dozen different minerals, a few of which may be questioned due to false interpretation of the petrographic slide. They are common in pleochroic minerals and those with marked absorption in one crystallographic direction, such as biotite and other micas, tourmaline, cordierite and some of the amphiboles. The halos invariably surround a nucleus of some other mineral of which there are also about a dozen recognized. The more common ones are zircon, titanite, cassiterite, thorite and uraninite. That the cause of the halos is the presence of a radioactive element was announced by Joly in 1907. Previous to this they were a mystery. That the extent of the halo beyond the nucleus represents the length of the paths of alpha particles of the various radioactive elements in the nucleus as they penetrate the matrix mineral is also known. A few items of interest regarding them which I have observed in my own study of them within petrographic sections may be of interest.

1. They occur only in rocks which are definitely acidic. With the disappearance of free quartz the halos are absent. Examination of many hundred petrographic sections in my own collection has confirmed this.

2. As already noted, they occur only in minerals which are themselves pleochroic or dichroic. No matter how acid the rock may be, if only quartz and feldspar are present no halos are evident. Yet the radioactive element responsible for them may be there since they show up as soon as biotite occurs in the

same rock. This suggests a more common occurrence of radioactive minerals than evidenced by halos. One point I have not been able as yet to check as thoroughly as I should like is whether the feldspars in rocks containing halos always include some of the potassium series, orthoclase, microcline, or anorthoclase Should this prove to be the case it might indicate that formation of radioactive elements was brought about through catalytic action of potassium. Catalysis plays so important a part in straight chemical combinations, may it not also have been an agent in the synthesis of common elements and energy?

3. The radioactive element responsible for halos apparently can occur in any one of three forms, viz., (a) As a minute particle of a pure radioactive mineral,—uraninite, thorite, radium, or other; (b) As minute inclusions of a radioactive mineral within a secondary host mineral of larger size, evident as opaque particles in an otherwise clear nucleus; (c) As radioactive elements apparently dispersed within the atomic lattice of some host mineral.

4. The extent of the halos in every case must be measured from the outside contour of the nucleus mineral. The shape of the halos conforms to that of the nucleus. Most references to them call them spheres but this is true only when the nucleus mineral itself is an approximate sphere. They can be oval, angular, or extremely irregular.

5. While reference to halos in the literature refer to the nucleus as being extremely minute, actually they occur also around mineral grains of fairly large size.

6. The extent of the halos around the inclusions varies over a wide range, even with the same nuclear material in the same matrix, but all sizes fall into definite groups. My measurements are, in microns, 5, 7, 10, 17, 20, 23, 27. and 33. Joly's figures correspond with these except he does not include the smaller sizes and does include 39 (38-40) which I have not run across. Halos sometimes show two, or even three definite rings or zones, indicating the presence of more than one radioactive element, each with its own specific alpha ray path. Another frequent condition, probably also explained by the same mixture of elements is the amount of diffusion at the border of the halo. The halo can be extremely sharp at its outer edge or very diffuse, with all gradations between.

7. Halos are manifested in two ways,—one which shows as a definite color or absorption when viewed in non-polarized light and independent of the rotation of the stage, and a second which is only evident in plane polarized light, varying in intensity with the rotation of the stage. In this latter case the direction of greatest absorption always corresponds with that of the mineral in which the halo occurs. These phenomena are associated with the crystallographic direction in which the matrix mineral is cut and are apparently independent of the radioactive element responsible for the halo.

8. There is a great range in the intensity of the halos, even surrounding the same nuclear mineral within the same matrix. This must result from the concentration of the radioactive element in the nucleus. It can be explained either by an old condition where the emanations are dying out, or, as appears more logical, it may result from the initial percentage of radioactive element present in a given nucleus. This latter view receivs confirmation from the frequent presence of identical mineral grains (e.g., zircon) within the same matrix showing no halos whatever.

9. Joly refers to negative halos in biotite which he suggests are analogous to photographic solarization. These are areas where the maximum absorption, which in biotite is normal to the acute bisectrix, or crystallographic C axis, has been completely destroyed. The cause of it is at present unknown although several different theories have been offered to explain it. In rock sections this only shows up as a lighter area around an obvious nuclear inclusion, and is rather rare. More light should be thrown upon this when the relative ages of the biotite where they occur is known.

**Discussion**

After Dr. Allen's paper was read the session leader, **Dr. I. Cowperthwaite,** called for discussion of the paper.

Dr. J. L. Kulp: I would like to divide my comments on Dr. Allen's paper into two parts. It really was two papers. The latter paper, on pleochroic halos, I would like to commend highly. I have looked over the literature on pleochroic halos and it appears that these pictures are the best compilation I have ever seen. I think that some precision densimetry on them could lead to very interesting results and possibly to some more precise data in the range that has not been obtained before.

The first part of the paper, however, had to do with radioactivity measurements. With all kindness, I would like to state that Dr. Allen's training was that of a metallurgist and therefore he might be excused from making errors of fact and concept in atomic physics. I am sure that I would make many more.

My remarks may be organized into four areas.

First I would like to make remarks on certain propagandistic statements. Secondly, I would like to comment on the new objection which Dr. Allen has raised to the radioactivity method of age determination. Thirdly, I would comment briefly on the accepted problems which he brought up and indicate what they were. You will find the answers to those standard objections. And finally, I will briefly mention what I consider to be minor errors in the text.

It is stated tnat the evolutionists are the people who wanted to push things back. Radioactivity measurement "was pressed into service" to save the day. Another statement was made to the effect that those engaged in this work have a natural bias for producing the maximum time. Now these come in the first category of propaganda. I think they are quite unjustified. There is no conspiracy between the atomic physicist and the evolutionist in trying to discourage some poor Christian. The atomic physicist would probably delight in making paleontologists turn their fossil orders upside down. They are not in league and are quite independent scientists. They start from different premises and they get answers by entirely different means. The atomic physicist and the physical chemist uses quantitative measurement. There is no getting maximum time. This is very important because Dr. Allen referred to it a number of times, that is, that science has been pushing this date back. As a matter of fact, that is not true. Astrophysicists who are concerned about the beginning of things, Genesis I:1, have brought the accepted time for the beginning of the universe, at least so far as they can measure, from $10^{10}$ to $10^{12}$ years down to about 4 x $10^9$. That's a tremendous drop. At the present time the concurrent opinion is that the age of the universe is on the order of four billion years whereas previously—24 years ago—it was $10^{10}$ or more.

Thus the age of the universe has been brought down rather than pushed up. The age of the earth, however, has been pushed up a little bit. Twenty years ago it was still accepted that the earth was roughly two billion years old, from relatively rough data. Today, more measurements have been made and it is more probable that is is close to three billion. Nobody expects that this is going to continue to be pushed up. The nature of science is such that all of our answers are approximate but they gradually lead us to a mean which is sometimes approximate truth.

Now for the second point, Dr. Allen wants us to worry about the possibility that transformation from the nuclear equilibrium state to one of normal decay might have been so gradual as to retard the decay right at the beginning. He is perfectly right in pointing out that at one point in the history of the universe there was an equilibrium between the building up of uranium atoms and the decaying of uranium atoms. However, what he is worried about is this. He said if this building up process continued considerably into the history of the earth, then obviously if we simply measured the decay rate and used this to calculate the equilibrium time (age of the universe), we would have an anomaly because the building up of the uranium would add to decay. Now, offhand, this might seem like a reasonable objection except for two things: first, the temperature at which atoms are built up and, secondly, the temperature of the earth's surface since it was formed as a planet.

The temperature that is required for atoms to build up is on the order of one to ten million degrees. This is why, for example, you must use an atom bomb to explode a hydrogen bomb because you have to get a million degrees of temperature before fusion takes place, before particles will start sticking together to make bigger particles. Now I hope most of you are aware of the melting point of most silicates in the crust of the earth. It is on the order of a thousand degrees. Obviously therefore, once the earth was formed as a planet and geology started, there could not be any building anymore and that is the time that we are talking about in the beginning of the earth.

I do not think that we should take time to discuss the common objection to the method, since we have done it at other ASA meetings and answers are available in textbooks on nuclear physics. The method of age measurement is straightforward. Consider a series of sedimentary strata: a crack develops, some molten material comes in, and there your elements crystallize in their own particular minerals. Uranium will crystallize into uranium minerals and lead will crystallize into lead minerals. The uranium crystal which is formed at the time that the molten material crystallized is at the time that lead starts to accumulate in that uranium crystal. It is separated by chemical means prior to that. Now that this uranium crystal is in place with lead accumulating in it, the question is what can happen to that to give you an incorrect age. One obvious thing is that uranium might be leached out later in its history by ground water. This is possible. But there are ways of measuring how much has been leached out. There are ways of measuring whether the uranium in a particular crystal is in equilibrium or not by quantitative methods.

In the introduction Dr. Allen stated that one reason we should mistrust these methods is that the methods of getting at the age of the earth give different answers, such as by the salt accumulation in the sea, radioactive methods, and accumulation of sediments.

Now let's take this one simple case. Suppose we have numbers and measurements of any physical property, the correct value of which is 41.00 and suppose we have one kind of machine that can measure this within plus or minus 0.1. Suppose we have another very inaccurate machine which can measure to plus or minus 10. If one method gives 40.95 plus or minus 0.10 and the other 32 plus or minus 10, the results would be in complete agreement. So the methods for estimating the age of the earth vary greatly in precision but are consistent within their degrees of error.

Another suggestion was that uranium 238 and uranium 235 at the equilibrium state should be equal. This is not true. They should be proportionate to their stabilities and their stabilities, to the first approximation, are proportionate to their half life.

One other thing that might be mentioned is the matter of the strong evidence for man having been on earth much longer than 10,000 years, as we have presented at some of the other meetings and as written up in the ASA Journal. There is abundant evidence of a qualitative nature that man is much older than 10,000 years. But quantitatively we can measure this now by carbon 14. Unfortunately we do not have as many measurements as we need to fix this to any degree of certainty. However, a number of measurements have proved that man was in North America at least 11,000 years ago. There are many geologists, archeologists, and anthropologists who readily agree that man is very recent, relatively speaking, in North America. If this is true, then quite obviously the attempt to hold a 10,000 year age for man is impossible.

I think it is also interesting, in this connection, that Dr. Allen left carbon 14 out of this problem because of the fact that carbon 14 was not in this original isotopic buildup at the early phase of the universe. In other words, carbon 14 is not subject to this particular objection which he raised.

There are several other methods besides uranium-lead. One very important one is rubidium-strontium. Another one is potassium-argon or potassium-calcium. The Rb-Sr is certainly equally good for old rocks and the fact that by the Rb-Sr method you can get exactly the same age within the experimental error as in uranium-lead, on different metals, from different pegmatites, in the same geologic situation is a very nice independent check on both methods.

Dr. I. Cowperthwaite: Is there further discussion of Dr. Allen's paper? If not I'll ask Dr. Allen if he wishes to comment.

Dr. R. Allen: The time is almost up and I didn't even make notes of all of the comments which Dr. Kulp had but I can show you perhaps a few on which we do not agree. Now, I think that I am absolutely correct in saying that the fundamental concept of the evolutionist, the mechanistic evolutionist, is to get time back just as far as he can for the operation of his theories. Now we're not primarily concerned here with the age of the earth.

Perhaps I was not as clear in making these statements which Dr. Kulp criticized about the tendency to stretch the time as I should have been because I had in mind the fact that the longer you can stretch the earth's history back into the past, the greater the length of time that you have available for life upon the earth and that is evolution's desire, to stretch that just as far as possible ,and I think I could have proven that by quotations from many authorities.

He mentioned, for instance, or gave an illustration here as though the several different methods of evaluating time, apart from the radioactive, all agree if you allowed a sufficient variation in the degree of accuracy. Now, that is not what I had reference to at all. For instance, Lord Kelvin figured out the age of the earth on the basis of the time required for the stabilization of the tides and the announced figure, as I recall it, was 40 million years as the age of the earth.

The evolutionist raised such a holler when he announced that figure that they said they required at least 400 million years for life on the earth and therefore that his figure was wrong. They succeeded in getting him to re-evaluate it and he did boost it up to 100 million years or so but he said, "That's all I'm going to give you."

Now, you see, there is in these various types of evaluating time quite a variation. I didn't intend to go into that phase of the matter at all.

There are many other comments that he made, if I had jotted them down, that I would have liked to have answered. I'm rather surprised that he did not evaluate as well as I would have liked, the primary concept that I gave of the reversible process. It makes no difference as to the time when this occurred. He mentioned, for instance, the fact that I said that the time could be longer because there could be perhaps a greater, that is a slowing down of the reversible process.

Now all chemists, I think, will agree with me that in a reversible process you have between reaction in one direction and reation in another direction a period or a phase in which you have stability. For instance in the manufacture of sulphuric acid through a platinum catalyst, you build up $SO^2$ in hyposulphuric acid to $SO^3$ by the addition of oxygen at a certain temperature. Now, when that temperature is exceeded, there is a range there where it is stable and if you go beyond that temperature, it starts to reverse and so you do have that period of stability and that was, in my estimation, a minor point.

I had in mind one other comment that he made but it has slipped my mind now. I wish I'd taken some notes, but I didn't think I was going to have much time to reply so I didn't do it but I would like an appreciation or a denial from him of the logic of my reversible concept because if that be allowed and if there is an equilibrium in the build-up, then it certainly does upset our figures enormously when the disintegation starts and decay commences, a large percentage of the lumps of the $Pb^{206}$ would not start with $U^{238}$ but every one of the intermediate steps, of which there are in the neighborhood of 14, would be for the time-being a parent level, a parent element.

I think the time is getting so short and I don't like to keep you any longer and I don't think I shall extend my comments any further. I certainly thank you for your attention.

# *Stratigraphy and Paleontology*

CORDELIA ERDMAN
Instructor in Geology, Wheaton College
Wheaton, Illinois

The early days of geologic investigation belong to the latter half of the 18th century. At that time nobody was paying much attention to fossils, and rocks were classified on the basis of their gross characteristics. The great controversy of the day was between the Neptunists, who believed that a receding, once world-wide ocean was responsible for forming all rocks, and the Plutonists, who maintained that many rocks had cooled from a molten state. Hutton, champion for the Plutonist school, also contended that in the case of those rocks which were composed of sediments, older beds were covered by successively younger beds. This seemingly very obvious principle is basic to all subsequent stratigraphic reasoning. It is known as the Law of Superposition.

It was not long before a second law of stratigraphy became apparent. A young English surveyor's assistant (William Smith) in 1796 began collecting fossils from the rock layers he traversed in England and Wales. As he studied these, he discovered that each layer contained its own distinctive fossils. One day when he was calling upon a cleric of the Church of England, he picked out unlabeled fossils from the man's collection and told him the locality from which they had come. The amazed churchman was quick to see that Smith had stumbled on something important, and he encouraged him to study further. As a result, there was published in 1815 the first geologic map of England; and, with even more far-reaching effects, the principle of Faunal Dissimilarity was set forth, the principle that each rock unit in a given series contains a fauna distinguishable from all others in the series.

George Cuvier, the great anatomist and paleontologist, carried on Smith's idea. He, too, recognized that each major rock unit carried a distinctive fossil assemblage. Since he believed in the fixity of species, a dogma which says that every species is a separate and direct creation, he could only explain sudden appearance of new forms in successive layers on the basis of local catastrophes followed by immigration of life from distant places. Alcide d'Orbigny continued to uphold this catastrophism as the explanation for faunal dissimilarity, but he postulated re-creations in place of immigration. Deperet, according to R. S. Lull (1947, p. 6), quotes his statement of 1848:

> . . . twenty-seven times in succession, distinct creations have come to re-people the whole earth with its plants and animals after each of the geological disturbances which destroyed everything in living nature. Such is the fact, certain but incomprehensible, which we confine ourselves to stating, without endeavoring to solve the superhuman mystery which envelops it.

Meanwhile, without worrying about any "superhuman mystery," Charles Lyell was busily investigating the rocks around Paris. He adopted and adapted Smith's rock units, arranged in order of superposition, and in 1833 he presented to the geologic world a stratigraphic column which was really the first geological timetable. As his groups or systems of rocks became identified over more extensive areas of the continent, it became increasingly evident that the principle of faunal dissimilarity was indeed a law of nature. The Catastrophist philosophy thus had reason to become more firmly entrenched. It was supported by the accumulating evidence that rock systems were separated by erosion surfaces or at least by lithologic discontinuities which indicated a fairly rapid change in the environment. Since these were synchronous with gaps in the fossil record, that is, the types of life fossilized above and below the discontinuity did not grade into one another, the belief developed that they represented "recognizable and universal dates on the geologic calendar" (Krumbein and Sloss, 1951, p. 15).

Now all of this was done before the publication of Darwin's "Origin of the Species" (1859), and although Lamarck had advanced evolutionary ideas previously, these had had little influence on the early stratigraphers. This means that the geologic timetable was based on objective fact alone and was not formulated to support evolutionary concepts, nor was it even an outgrowth of evolutionary concepts.

However, the modern period of paleontology and stratigraphy did begin with Charles Darwin, because, as Neaverson states (1928, p. 4):

> In the light of evolutionary theories, the gradual increase in complexity of organization as shown by fossils passing from earlier to later rocks assumed new significance and the mutual interdependence of paleontology and stratigraphy was more emphasized.

In other words, the scientists threw over catastrophism altogether and accepted evolution **as an alternative for a phenomenon which was already** well established. According to the new theory, successive fossil faunas represented not re-creations but groups which would merge into one another in a time direction except for the interruptions in the sedimentary record.

**Determining equivalent age.**—The Christian world has long raised its eyebrows at geology. Many times one hears doubt expressed concerning the validity of the geologic timetable or concerning the validity of using fossil content to determine equivalent age of rock units. Much of this doubt grows out of a confused belief that all ideas of organic development are inherently contrary to special creation; and out of the erroneous assertion that "geologists date the fossils by the rocks and then turn around and date the rocks by the fossils." Both of these concepts have been vigorously propagated in much pseudo-scientific "Christian" literature.

We have seen that the geologic timetable is based upon the two principles of stratigraphy: 1. in a normal sequence of rock layers, the oldest must be on the bottom and the youngest at the top; 2. rock units contain their own distinctive fossil assemblages. The problem is to know how to equate or compare the ages of layers occurring in geographically separated sequences: in other words, how to fit rock strata into an over-all chronology.

Many times the stratigrapher is able to compare data from two areas which are not widely separated; or he can actually trace layers from place to place: thus he can establish their time equivalence on non-paleontologic grounds. In other cases such tracing is impossible. Then the stratigrapher depends upon similar fossil assemblages alone to make his correlation.

> In comparing the rocks of two separated districts the geologist has to rely entirely on the fossils present in those rocks. Supposing the fossil assemblages to be identical, the rocks are held to be contemporaneous (or synchronous), even if they occur in such widely separated countries as England and America. (Neaverson, 1928, p. 7.)

The reasoning is this: similar fossil assemblages could only occur at approximately the same time in geologic history: therefore, rocks containing such similar assemblages must be approximately equivalent in age. The next question is, is this **valid** reasoning?

First of all, note that a consistent relative order of appearance of varying organisms is demonstrated in every area where the Law of Superposition can be applied. Dunbar writes:

> Where the structure of the rocks is simple, the succession of faunas that occupied the region from age to age can be determined by studying the fossils of successive formations. When this succession has been confirmed by wide experience in many regions of undisturbed strata, we may be confident that different forms of life succeeded one another in this order in time on the Earth . . . The relative time of existence of a vast number of kinds of animals and plants has now been established, and their place in the geologic column has been confirmed by the cooperation of geologists the world over. This is not a theory derived *a priori*, but a discovery painfully and tediously worked out by the systematic study of the faunas of rock formations *first carefully located in the geologic column*. (Dunbar, 1949, p. 8,) (Last italics mine.)

And again: "The appearance of trilobites and dinosaurs and three-toed horses is not fortuitous and irregular" (Dunbar, 1949, p. 8). However, the fact that we can establish relative ages for the various kinds of organisms in one location, plus the fact that these same relative ages hold good in other regions, too, does not necessarily guarantee equivalent age of like kinds. Paleontologic correlation rests upon this issue, "Does any characteristic occur only once in time, or may it be repeated at successive intervals?" If a characteristic type of assemblage does occur only once, then assemblages of this type must be time equivalents. There seem to be several possible explanations for the appearance of similar fossil faunas and floras in divergent regions.

1) One may revert to catastrophism and regard the organic assemblage of each layer as a special creation of God. If there was an extensive series of world-wide catastrophes (a view not substantiated by geology), then the creation of organisms which ensued was probably (but not necessarily) a creation of like forms with world-wide distribution. This would make them of exactly the same age. If one postulates only local catastrophes, then he must say that God always re-created organisms in the same sequence (first trilobites . . . later ammonites . . . then reptiles, etc.). This would invalidate the age of equivalence of layers containing like forms.

2) One may believe that all species existed from the beginning and that recurring fossil groups merely indicate the recurrence of certain environmental conditions. In this case, fossil sequences must be due to progressive changes in environment which caused immigration of life from other areas. This, too, ,would invalidate age equivalence.[1]

3) The evolutionist says that "Different kinds of animals and plants have succeeded one another in time because life has continuously evolved; and inasmuch as organic evolution is world wide in its operation, only rocks formed during the same age could bear identical faunas" (Dunbar, 1949, p. 8).

4) One may believe that different kinds of animals and plants have succeeded one another in time because God created the great categories of organisms at successive points in time, giving to each the ability to be changed and developed within God-ordained limits. Then like assemblages represent a cross-section of time, and the degree of equivalence depends upon the rapidity of dispersal of the changing forms.

There is little in the way of geologic evidence to give credence to the grosser forms of Catastrophism. "There is no longer much general support among stratigraphers for the principle of correlation based on synchronous diastrophism" (Krumbein and Sloss, 1951, p. 308). If local catastrophes are involved (and there are many of these recorded in the rocks) there would be no need for re-creation, for the population would be reinstated by immigration. But if fossil sequences are always and only a record of successive immigration of forms which all existed from the beginning, there remains to be explained the amazing coincidence that they have always immigrated in exactly the same chronological order (considering life as a whole) or that they have immigrated in such a fashion as to build spurious sequences (considering directional change manifested within a genus or family). Also one might ask where was the source area from which all these animals came. Therefore we are left with alternatives 3) and 4): thorough-going evolution or modifiction within successively created categories. In either of these cases, closely similar faunas represent generally equivalent segments of time.

**Index Fossils.**—The Law of Faunal Dissimilarity deals with faunal assemblages. It is by means of such groupings of species that the "finer resolution of time-rock distinctions may be accomplished," for, as Krumbein and Sloss continue (1951, p. 306),

1. It might be well to mention here one ingenious suggestion of the catastrophists—namely, that during some great world-wide flood which rose to progressive heights, the first forms to die would be those living at low elevations, then those living at successively higher elevations would be engulfed. It is asserted that complexity of life increases with elevation. Thus simple forms of ocean life would be buried first, later land reptiles, and ultimately mammals. This theory fails to take into account at least two factors: 1) the new categories of life which appear in successively younger strata are often interlayered with older categories of life. For example, in the walls of Grand Canyon there are several hundred feet of rock containing marine fossils exclusively. Above this there are more than 1,000 feet of rock containing the footprints of land animals. These layers are overlain by 500 feet of layers bearing fossils of corals, brachiopods, crinoids and other marine animals. 2) many fossil sequences are known which manifest cumulative variations in a time direction. Examples of this are corals which tend to increase their number of septae in successively younger layers; or oysters which are more and more coiled in successively younger layers. It is hard to conceive of a flood manifesting such selective action.

Of course the most sweeping criticism is that the idea of increase in complexity of life with increase of elevation is only superficially tenable.

If the approximate biozone (historical span) of a number of component species in an assemblage are known,--although each biozone may be relatively long, it is probable that only in a short time-rock interval can all the species be expected in the same assemblage.

For grosser distinctions recourse is had to index fossils, those forms which have a very wide geographic distribution but to have everywhere a narrow time range. Krumbein and Sloss point out (1951, p. 305):

> Time-rock correlation by index fossils would be much more confidently applied if there were many truly cosmopolitan species . . . (but) the experienced stratigraphic paleontologist can usually place the strata involved in their proper system with some assurance. At least, the fauna can be placed as either high in one system or low in the succeeding one. Since the biozones of most species persist through several ages and commonly through more than a single epoch, application of individual species to correlation of specific stages or series is rarely effective.

Dunbar, writing for the beginning student in geology, also emphasizes the possibility of determining relative age by use of index fossils:

> During the last one hundred years, paleontologists in many parts of the world have cooperated in gathering such a mass of information that it is now as easy for a trained specialist to identify the relative . . . age of a fossiliferous rock formation as it is to determine the relative place of a sheet in a manuscript by its pagination. (Dunbar, 1949, p. 52).

The concept of index fossils is used extensively. However, the paleontologists would be the first to admit that in some cases hasty application may have led to invalid conclusions. Constant rechecking and revision eventually point up such inconsistencies.

Perhaps now we are in a position to examine the charge that geologists are guilty of circular reasoning. Their actual procedure is: 1) to observe the chronological sequence of rock units in a given area, and 2) from this to determine the chronological sequence of the contained fossils. (This is "dating the fossils by the rocks"). Then the fossils of another area may be compared to this standard sequence and assigned a relative geologic age by applying the concept of time equivalence for index fossils or like assemblages. This is not circular reasoning. It is a piecing together of information with reference to a standard.

**Present status of paleontologic correlation.**—How much reliance does the stratigrapher place on this fossil evidence in his effort to determine the age equivalence (or non-equivalence) or rock units? There are actually many means of correlation other than the paleontologic one. Krumbein- and Sloss explain that (1951, p. 288)

> Normally, two or more methods are employed in the solution of any individual correlation problem and no generalization can be made as to which methods are most useful in any individual case. Some of the methods that may be used are: 1) tracing layers laterally; 2) lithologic identity; 3) relative position in sequence; 4) structural relationships. These methods are useful only in relatively local situations.

However, Lull describes the process of correlation thus (1931, p. 19):

> Having settled the sequence of strata in a given region, fossils are sought for, especially such genera and species as are sufficiently distinctive to be considered "horizon markers." Once determined, these serve to identify the age and, therefore, correlate the rocks containing them wherever they may be found.

Neaverson says (1928, p. 7):

> It was once suggested that faunal migration is such a slow process that by the time a fauna reached the opposite quarter of the globe its original centre of dispersal would be occupied by a very different faunal development . . . Later it was realized that the time needed for migration is negligible compared with the period taken by the deposition of a rock series, so that essentially similar faunas are nowadays held to be contemporaneous . . .

Stratigraphic paleontologists recognize the limitations of their practice. James Gilully, past President of the Geological Society of America, in 1949 made an honest admission which sums up the situation. He said:

> If several species that arose in different centers come to live in a particular area . . . their order of superposition may be different in different parts of the area . . . When we take into account the entire fauna we never find such reversals, but the fauna as a whole does not, of course, change so quickly as the individual species. Even in theory it will never be possible to be certain of the precise equivalence of beds in widely separated regions. Paleontology . . . can never offer us a chronology capable of indefinite refinement.

It would be safe to say that Gilully's comments reflect a growing concern within the ranks of geologists themselves. The current use of paleontology in stratigraphy has been undergoing some rather telling analysis and criticism in the past few years. Witness this excerpt from a paper entitled "Geological Correlation and Paleoecology" by Robin S. Allan (Allan, 1948):

> Because of the sterility of its concepts, historical geology, which includes paleontology and stratigraphy, has become static and unproductive. Current methods of delimiting intervals of time, which are the fundamental units of historical geology, and of establishing chronology are of dubious validity. Worse than that, the criteria of correlation—the attempt to equate in time, or synchronize, the geological history of one area with that of another—are logically vulnerable. The findings of historical geology are suspect because the principles upon which they are based are either inadequate, in which case they should be reformulated, or false, in which case they should be discarded. Most of us refuse to discard or to reformulate, and the result is the present deplorable state of our discipline.

This undoubtedly represents an extreme position; but Allan's plea for more use of principles of sedimentation and of paleoecology in interpreting and correlating rocks is a plea which finds an echo in many other geological writings. Raymond C. Moore, writing in the same year as Dr. Allan, stated firmly that assemblages of fossils could be used as indicators of geologic time only if subsequent and antecedent assemblages were known from the same area. He writes further (Moore, 1948):

> Possibly excepting correlations based on occurrence of shortlived free-floating or swimming organisms of the sea, such as graptolites and some cephalopods, or rapidly migrating land animals and plants, most intercontinental correlations of fossil species or assemblages cannot be accepted as indicating precise contemporaneity.

However, these objections pertain to refinements within the timetable. As noted earlier in this paper, geologists have reason to be satisfied concerning the over-all sequence in which organic forms have made their appearance. But when it comes to deciding what life forms were characteristic of the smaller time units, there is far less satisfaction. The smaller the time unit, the more difficult the problem. Nevertheless, modern research in sedimentation and paleoecology will doubtless point the way toward more reliable and effective use of paleontologic methods of stratigraphic correlation.

### Bibliography


Allan, Robin S. *Geological Correlation and Paleoecology*, Bull. G.S.A., vol. 59, 1948 (Jan.).

Dunbar, Carl O. 1949. *Historical Geology*. New York, John Wiley & Sons, Inc.

Gilully, James. Presidential Address, Bulletin of the Geological Society of America, Vo. 60, No. 4, 1949.

Krumbein, W. C. and L. L. Sloss. 1951. *Stratigraphy and Sedimentation*. San Francisco, The W. H. Freeman & Co.

Lull, Richard S. 1931. *Fossils*. New York, The University Society, Inc.

................ 1947. *Organic Evolution*. New York, The Macmillan Company.

Moore, Raymond C. *Stratigraphical Paleontology*. Bull. G.S.A., vol. V9, 1948 (April).

Neaverson, Ernest. 1928. *Stratigraphical Paleontology*. London, Macmillan & Co., Ltd.


### Discussion

Mr. S. C. Sinclair: Many of you folk may be wondering why the A.S.A. is divided on the extent of mutational changes. One group believes that a sequence such as the modern horse from Eohippus is possible while the other group denies that any such change could occur. The answer to this difference lies in the perspective of the two groups. One group is looking at it Paleontologically, the other group Genetically.

Fossil sequences such as the horse fit beautifully into an evolutionary trend if such a thing can be shown to exist, but circumstantial evidence of this sort can not prove the theory in the absence of more direct evidence,, such as has not been found in Genetics. Really the sort of Evolution one believes in, whether organic, theistic or threshold is decided by where one leaves God for chance. Variety from heterozygosity is not Evolution.

A certain taxonomist I know, though he knew the disfavor of Lamarckism yet was so impressed with the ingenious manner in which animals are adapted to the environmental niche they fill, could not help but feel that some mechanism for specifically adapting them to their environment must exist. I agree with him that adaptations do not look like haphazard mutations but show the hand of almighty God; and I believe the Bible bears me out that basic species patterns are the design of God. Job 39:13-17, 5-6 and 40:15,19.

I am convinced that we shall not make any real headway in our testimony for Christ until we make an open break with the theory of Evolution.

Descent from a common ancestor can not be the only criteria of gene identity, for, fortunately for our diabetics, the genes in sheep, pigs, and horses responsible for insulin production, form its protein structure so identically that purified hormone preparations from any of these sources can be injected without fear of reaction.

I believe I have in my mind about what I would consider a hybrid difference between two individuals and a species difference. Dr. Lammerts feels he can classify plants on this basis, and that, confusion as to what is or is not a species difference is merely lack of experience. Hybrid differences usually involve the presence or absence of single genes, whereas species differences typically involve a different gene pattern as well as one or more different genes. Though most of the genes are shared in common, their integration in a successful individual in a given environment constitutes their uniqueness. For example the aquatic spider differs in several essential ways from a similar terrestrial spider. The water repellent hairs that hold the bubble of air over its abdomen must be of the right rigidity, of just the right number per square inch, growing on just the right part of the body, and of the right shape with a hook on the tip of just the right size and angle. The more complicated things such as instinct and physiological fitness must also be right.

If changes such as bunodont teeth and hind feet bearing three hooved toes to hypsodont teeth and monodactyls are possible, then changes such as a low receding forehead to a high sharp one, heavy protruding jaw to a flat vertical one, fang-like interlocking canines to even dentition, and stooped posture to an erect one are theoretically possible, fossil sequence or no fossil sequence. If changes such as these can be definitely proved then we must look further than genetics for a mechanism. Why not let us admit that God is responsible for these changes, if such can be proved, and if it's God, it's not Evolution. The lack of a fossil sequence is a pretty flimsy defense for the faith of God's precious ones. And why quote Simpson as an authority when Evolution is his alternative to faith in the God of Creation?

I don't believe we fully appreciate the effect this theory has had, not alone on biology students but on philosophy and theology and practically every area of human thought. The apostle Paul said that if meat cause my brother to stumble then I will eat no meat. So may we say that if My Evolution cause my brother to stumble then I will have nothing to do with this philosophy of chance.

An area needing further study is the distribution of organisms in time (geologically) and space (geographically). Could it be possible that a pattern of plant and animal life was established in one or several areas and subsequently spread in all directions considering ecological barriers? If so a lag in distribution might cause apparent sequences in the earth's strata. Another factor, mentioned by Dr. Kulp, A.S.A. Journal, 1-3, 23 (1949), is that necessary for soil preparation for forest growth. Certain characteristic changes in plant life would occur beginning with bramble bushes, possibly, on up to the evergreens that require shelter for proper rooting. With these floral changes would go the animal changes characteristic of them. These changes might also look like geologic sequences.

These considerations might harmonize facts of distribution which are now evidences of Evolution.

Dr. Wm. Tinkle: I wish to say that this paper by Miss Erdman is a fair statement of the position of the standard geologist. However, ,the standard geologist is an evolutionist. I appreciate this paper very much for there is a great deal of truth in it; however, in some respects I must be very critical.

Regarding index fossils, in order for a certain species of fossil to be a good index, somebody must tell us when that species started (created or evolved) and also when it became extinct, at least in relation to other species and preferably in relation to time. Suppose that we take a special species of trilobite as an index for a certain formation. Then a new formation of rocks is searched out and we find that species of trilobite there. What is the conclusion? The new formation is the same age as the old since they have the same fossils. Are there not also two other possibilities? Perhaps this species of trilobite could have lived earlier or later than previously indicated. There is still another choice. This species of trilobite may have lived in a wider region than we had formerly thought. Again, is the geologist omniscient? What about early geologists? Smith was not educated. Lyle was educated but in law. Sedgwick was trained in Latin and Greek.

Dr. J. L. Kulp: I don't believe we would expect the early geologists to have had training since it was not possible to obtain such training. But no present geologist accepts the index fossils because anyone else says so. The rocks are still there. You can still find the same sets of fossils and since the early work of the men you have mentioned, thousands of trained men have gone across the face of the earth wherever the stratigraphic section outcrops appear and have found sequences wherever the outcrop is not deformed. And I should add that only about one per cent is deformed and there is always considerable evidence of it.

Taking the total fossil picture the broad outline of life is clear and undebatable and this is exactly the sequence which is described in Scripture. This makes it distinct from any other early cosmology.

# Creation In Terms of Modern Concepts of Genetics and Physics

WALTER E. LAMMERTS

JOHN C. SINCLAIR

There has long been a need among Christian men of science for a working correlation of the facts of our environment that will be consistent with what we know of God. The more generally accepted theories today attempt this correlation apart from God. The purpose of this paper is to give a consistent correlation of certain genetical and physical data that will in addition, be true to the revealed Word of God.[1]

A recent paper by Alpher and Herman of Johns Hopkins University gives a simple account of how they believe the elements could have been formed in the beginning. That there was a beginning some 2 to 3 billion years ago is evidenced according to these authors, by the amounts of radioactive elements such as Uranium and Radium, and by the present rate of expansion of the Universe. The expanding universe is what one would expect if the universe was once densely packed and was blown out by an explosion. The uniform relative abundances of the elements throughout the universe, except for explicable exceptions, as measured chemically on the crust of the Earth and meteors, and measured spectroscopically on the stellar bodies and gas clouds, points to a common origin of the elements prior to the condensation of the stellar bodies. Known nuclear reactions at the temperature and pressure that must have existed in the initial exploding mass, can account for these abundances.

Several hundred seconds after the initial explosion, the universe was uniformly filled with a very hot gas consisting of neutrons, protons, positrons, electrons, neutrinos, and electromagnetic radiation, (principally high energy light rays). Tremendous collisions between particles took place that at first shattered both collidants, but as the gas expanded and cooled, the colliding energy fell so that more and more of the particles stuck together after hitting, resulting in larger and larger aggregates. The larger aggregates had a greater ability to capture neutrons. As the neutrons became excessive the aggregates or nuclei became radioactive to electron emission, thereby raising the number of protons and hence its atomic number. It then continued to grow by further neutron capture. After an half hour to an hour the temperature had dropped so low, and the capure of neutrons and their radioactive decay had so reduced their number that little or no further synthesis could occur.

This theory assumes the particles and their properties, and the cohesive forces that bind the nucleus; and assumes it's initial high density and temperature. Granted these assumptions, they are certain that all other properties of the elements and their compounds can be explained by the chance aggregation and random motion of the sub-elemental particles. However, the assumptions they have made logically lead to the periodic table of the chemical properties of the elements and the present form of the physical universe, an order that God designed. Where they assume, we acknowledge the Hand of God. This theory as has been mentioned,[2] fits in well with a creationist point of view and may well have been the technique used by the Creator in originating the elements. The suddenness and finality of the beginning postulated by Alpher and Herman is in sharp contrast to theories, such as the theory of Evolution, which assume that given enough time anything could happen.

At first sight there would seem to be an infinite numer of elements possible due to the infinite number of ways in which the sub-elemental units could be combined. But actually there is a limited number due to certain basic principles of exclusion. At first sight of the organic realm of living protoplasm there would seem to be an infinite number of possible basic designs or species, but actually there is a limited number of stable combinations of individual genes or genetic factors though no two individuals are identical. The genes in the organic realm seem to be the building blocks created by God just as the protons, electrons, etc., are the building blocks in the inorganic realm. The basic character of the genes is evidenced by the existence of similar enzyme systems throughout the breadth of organic life. However the total original complex of genetic factors from which all individual species were formed may have had no physical existence prior to the appearance of the individual species, but existed only in the mind of God. (All living things are able to organize lifeless elements into their own protoplasm. During the life of an individual the elements carbon, hydrogen, oxygen, etc. that once composed it, change; some of them many times. The protoplasmic patterns then are the lasting entities, not the momentary physical expression of them. Hence these patterns need not have had any physical existence prior to the moment God supervised their first physical expression. God then united and harmonized the individual genes or factors into discrete species patterns. This uniting and harmonizing act of God would explain why changes or mutations are usually deleterious. How and when God gave his species patterns a physical expression is not a problem when we realize that God was the one who did it, and that it did not and could not have occurred by any chance. The heterogeneity and the degree of variation possible in each species pattern is characteristic of each species and was characteristic at the time of Creation.

A brief presentation of a few of the authors reasons for rejecting the usual theories of gradual origin of the variation we find in the organic realm is now in order.

According to the geologic time scale as generally accepted by geologists, approximately 1,000,000 years have elapsed since Eocene times. According to orthodox paleontological thought the ancestor of the modern horse first occured in Eocene times or at least skeleton remains are first found in stratified rock generally attributed to this era of geologic history. As stated by Cordelia Erdman[3] "This little animal was about a foot high, the size of a fox terrier, with a short, slender face and eye orbits in the center of the skull so that the face length was one half of the cranial length. Its legs were relatively long and slender and were unique in that the front feet bore four toes each, whereas the hind feet bore only three. Vestigial toes brought the number to a total of five for each foot. The teeth . . . were low crowned and in most species were bunodont, a condition in which the surface is smoothly undulating rather than ridged or folded. From this it may be inferred that Eohippus was omnivorous in diet, just as from his feet we infer that he was a forest dweller rather than a plains animal."

Obviously this group of animals differed from the modern horse in so many features that only by the use of a certain amount of "scientific imagination" called deduction can one understand the reasons causing paleontologists to consider Eohippus as the ancestral type from which our modern horse evolved. By successive stages or changes during Oligocene, Miocene, Pliocene and Pleistocene larger animals having fewer toes, longer legs and teeth more and more adapted to a plains grazing life gradually evolved until finally the modern horse appeared on the scene by the beginning of the Ice Ages in Pleistocene times.

From the genetic viewpoint the number of differences, each based upon at least once allelomorphic pair of genes or factors, between the presumed ancestral Eohippus and the modern genus Equus must be at least well over 100. What are the time requirements for the accumulation of such changes, i.e., the occurrence of the various original mutations and the subsequent incorporation of them into the genetic system of Eohippus in homozygous condition? Fortunately the answer to this question has been thoroughly worked out by Patau[4] and a brief presentation of his conclusions are now in order.

Though admittedly most mutations confer no selective advantage, Patau makes the assumption that mutations do occur occasionally having a selective advantage of 1%, i.e., confer upon the animal in which they occur some superior physiological behavior or physical characteristic giving it a 1% advantage in survival and reproduction. Such a mutation according to Patau would increase in frequency from .01 to .1 percent of the population only after 900,230 generations. Another 100,511 generations are needed to increase the frequency to 100%. Certainly the time for natural selection alone to effect a change in a large population is enormous even geologically speaking. And very few mutations have been shown experimentally to confer any selective advantage. In small populations mutations are incorporated more rapidly, but one must assume very small populations indeed to appreciably shorten the time involved. Now the modern horse has a breeding cycle of about three years. Even assuming a yearling breeding cycle for the ancestral Eohippus it would take about 1,000,000 years for even one mutational change to become a constant feature of the entire species. Even assuming population 1/10 the size postulated by Patau, one would still have the slow rate of accumulation of one mutation per 100,000 years. Though several mutations could of course, be simultaneously undergoing fixation, obviously the number necessary to account for the many differences between Eohippus and the modern horse could hardly have occurred since Eocene time by natural selection of advantageous mutations.

The situation is the same no matter what groups of plants or animals is considered. As pointed out by Lammerts and Tinkle[5] the differences separating even very closely "related" species are very many. The only way changes are known to occur at present is by mutation and the overwhelming number of mutations are actually harmful to the organism in terms of survival. A particularly lucid review of this situation is presented in the American Scientist by C. B. Martin. The article entitled, "A Non-geneticist looks at Evolution," shows most conclusively that "all mutations seem to be in the nature of injuries that to some extent impair the fertility and viability of the affected organisms." He further states, "I doubt if among the many thousands of known mutants one can be found which is superior to the wild type in its normal environment."[6] Even however assuming for the sake of the argument that occasional mutations are beneficial, their chances of incorporation in the genetic system of the organism or species are small indeed as discussed above. A satisfactory explanation of the remarkable diversity of the organic realm in terms of evolution by mutation and natural selection is indeed difficult, even assuming that the long periods of geological time postulated by geologists are correct. (An assumption incidentally, not granted by the authors.)

Is it not then time to consider the bearing of Alpher and Hermans concept of the formation of the inorganic realm on our ideas of the origin of specific diversity?

As discussed above by Sinclair, the old idea of the necessity of postulating enormous periods of time for the origin of the elements and development of the material universe has been seriously questioned by these physicists. Granted the right conditions of enormous heat and a potential system of various units of mass, the whole series of elements could have been formed almost instantly even on the basis of chance alone. How much more should we as Christians recognize the omnipotence of God in place of the chance reactions postulated by Alpher and Herman?

Applying this reasoning to the organic realm it seems quite reasonable to us that God first set up the gene system as the immaterial basis of life. By this we mean that a certain total number of distinctive genes or allelomorphic factors constitute the complexity of the many kinds of plants and animals. We do not imply, however, that all organisms have the same number of genes, though some genes are, of course, common to all organisms. Thus on the basis of economy of effort a wise Creator would certainly use the same genes in all organisms wherever possible, i.e., whenever the same function was to be achieved. Thus in all probability an amoeba and man have the same genes making possible the conversion of the various food compounds into protoplasm. For though the steps involved in digestion of food by mammals are more complex than the digestion of food by the amoeba, in all probability the final conversion of the organic chemicals to protoplasm is similar. Again all organisms may well depend on basically the same genes for the phenomena of cell division. Plants in terms of this creation concept differ from animals by virtue of a gene system making possible the synthesis under the influence of light, of water and carbon dioxide into sugar giving off oxygen which in turn is used by all animals. On the basis of economy of effort a wise Creator probably used the same set of genes or genic system to accomplish photosynthesis in all plants. All birds probably have many genes possessed "in common" so to speak since in their creation certain basically identical problems of adaptation to air flight had to be met. When we as students of nature attempt to group plants and animals into classes, orders, families, genera and species we are attempting to correctly understand and classify the enormous complexity of God's organic creation in terms of which groups belong together by virtue of possessing most genes in common. The grouping then of all genera resembling the rose into the family Rosacease **if properly done,** would merely mean that more genes were possessed in common by species of this group of flowering plants than were shared by a species of the genus Rosa and let us say a species classified in the genus Geranium of the order Geraniales. In no sense of the word would this classification imply relationship due to descent from a common ancestor. The relationship would be the result of creation from the original "stockpile" or complex of genes. My various friends of the evolutionary school of thought so frequently express the thought that creationists must have the ridiculous idea that one species can in no way be related to another. Or stated in different terms, they believe that if two species are shown to have a large number of homologous genes, i.e., genes which are identical, evolution from a common ancestor is thereby proved: Surely on the basis of economy of effort a wise Creator would use the same genes wherever possible and consistent with His obvious desire to create an organic world of marvelous diversity. The remarkable similarity of the eye of the octopus to the vertebrate eye so well described by George Barry O'Toole[7] certainly cannot be due to any inheritance of genes from a common ancestor since the two classes of animals are obviously not closely related. From the evolutionary point of view the resemblance is apparent only, and marine zoologists go to considerable lengths attempting to show the importance of minor differences in eye structure in the octopus and the vertebrate eye as being proof of independent origin.

On the basis of the above creation concept, however, such similarities in the organic structures of relatively unrelated plants or animals is just what one would expect to find, i.e., similarity of structure and function because of the use of the same gene system whenever possible.

In order to allow for variation in the environment, all species, some more than others, are endowed with a number of genes capable of responsive adaptation, the resulting character expression being greatly influenced by the environment. Thus peach varieties vary greatly in the amount of winter chilling they require, some being practically evergreen and thus adapted to semi-tropical climates while others require more than 1200 hours at 40° F. or lower to break dormancy and leaf out. It is of interest that varieties having a long chilling requirement are also the ones which first go dormant in the fall, thus being better adapted to survival of sudden drops of temperature in the fall, and accordingly, having the maximum degree of winter hardiness. Fortunately, the number of factors involved in this adaptive responsive is small enough that varieties may readily be bred combining almost any degree of winter chilling requirement with desirable fruit and flower characteristics. A series of varieties adapted to the almost sub-tropical conditions of Southern California, combining large double ornamental flowers and good quality fruit have accordingly been recently developed and introduced.*

Many characters of both plants and animals are dependent on quantitative factors present in heterozygous condition. Rather rapid adaptation to changes in the environment within a few generations is thus possible. Undoubtedly this potential of individual variability is as much a characteristic of certain species as any of their physical characters such as flower shape, color or petal number. Indeed, we must realize that in many genera and species we may often have specific and varietal diversity simply for the sake of variety as such. Thus Camellia **cuspidata** is a remarkably constant species having inconspicuous small white flowers, whereas Camellia **japonica** is a highly variable one exhibiting many thousands of beautiful varieties. Yet both are diploid species having the same number of chromosomes. The limits of individual variability within the species were then also set at the time of creation along with the potential physical expressions of the gene systems.

The total number of genes necessary for us to assume for this concept is by no means infinite, or even very large. Possibly a total of less than 100,000 would suffice. Even only 12 different gene pairs give us 4,096 different combinations of true breeding organisms. Not only were all species created almost simultaneously according to our concept, but also all the potentially possible individual variations were anticipated and their limits set.

We must also realize that originally, along with the creation of the organic kingdom, a perfect world in which the gene systems making up the myriad of plants and animals could function perfectly was also created. The gene systems and the environment in which they functioned were perfectly attuned. Obviously to secure such complete balance practically instantaneous creation was necessary, since each part of the organic realm is so dependent on every other part. One of the fundamental weaknesses of the evolution theory is the failure to appreciate fully this relationship. Thus, how could the animal life of the ocean exist for millions of years without the corresponding complement of land plants? Or could the many flower feeding types of insects live without the higher forms of flowering plants?

As long as a perfect world existed, complete balance was possible. But with the entrance of sin a vicious unbalance began leading to a more and more imperfect relation of the gene system with the environment both internal and external. Even the very basic system of gene reproduction by chromosome division and transmission became subject to flaws and imperfections. Hence we witness the large number of mutations which are mostly defective in one way or another.

It is believed by the authors that a careful study of the facts of genetics, embryology, geology and paleontology will show that all the marvelous complexity of the plant and animal kingdoms was created from a common total sum of genes arranged in various integrated patterns of complexity. Their materialization as recorded in Genesis was accomplished in a very few days.

* Daily News 2, 3 and 4 Star varieties, originated by W. E. Lammerts of Descanso Gardens, La Canada, California and Altair originated by H. C. Swim, Armstrong Nurseries, Ontario, California.

### BIBLIOGRAPHY

1. Alpher, R. A., Herman, R. C., Johns Hopkins Magazine, March, 1951.

2. Eggenberger, D., Gamow's Theory of Element Building, J.A.S.A., 2 (3), 23- ('50).

3. Erdman, Cordelia, Paleontology of the Horse, J.A.S.A. 2(4), 25-36 (1950).

4. Pateau, K., Die Mathemathische Analyse de Evolutions

Vorgange, Seitschrift für Ind. Abstam. und Vererbungslehre. 76, 220-228.

5. Lammerts, W., Tinkle, W. J., Modern Science and Christian Faith, Ch. 4, Biology and Creation, 78-83, 90-93 (1950).

6. Martin, C. P., A Non-Geneticist looks at Evolution, American Scientist, Jan. 1953, Vol. 141, no. 1, pp 100-105.

7. O'Toole, George Barry, The Case Against Evolution, MacMillan co., 1926.

# A Criticism of the A.S.A. Monograph on "Creation and Evolution"

*Editor's note:* The following letter was forwarded to Dr. Mixter by another party. Correspondence has failed to disclose the writer's identity, except to reveal that he is a university scientist.

## I. Regarding logic

There is in logic what is called the *principle of parsimony*. This is exceedingly important, as it is, or should be, used in every attempt at synthesis of theories from empirical facts. It simply states that if there are two or more possible mechanisms which account for the facts, the simpler of them should be accepted until further facts make necessary a more complex theory. Examples are the geocentric vs. helicentric theories of the solar system, and classical vs. modern physics. Similarly, since, as Dr. Mixter agrees, there is ample evidence that evolution has occured in some lines, it is in keeping with this principle to believe that evolution has occurred where the paleontological or comparitive morphologisal evidence is spotty or nonexistent. P. 1 par. 2 creates a false impression, that special creation is necessary to explain certain phenomena. As you can plainly see on further examination, the only thing he shows is a *lack of* necessity to believe in evolution in some particular cases, due to lack of evidence. Thus, on p. 23, par. 2 he says " . . . until bridges are found, one may hold that the groups so separated have arisen from independently created kinds". This is like saying that since I didn't see you go to Wheaton College, you must not have been there. I believe that a fair summary of Dr. Mixter's monograph would be: "Either evolution has occurred as the modern evolutionist says it has (with all his if's and but's), or that God has made the creatures of the world in such a way that we cannot distinguish the order from an evolutionary one". This is little consolation to the creationist unless he is willing to throw out evolution in its *entirety*.

As for the approach to unsolved problems, he does not discuss this further, and I fail to see how creationism can have anything to offer in this regard. If science is to make any progress, it must take as its basis the premise of being able to find mechanistic causes for natural phenomena. I don't see why the study of heredity and the bringing of order out of the chaotic array of plants and animals about us should be any exception to this.

This idea of there being a machinist (God) behind the complicated mechanisms that are living organisms was also found in the other pamphlet I referred to. It is a difficult idea to justify. Naturally there is something back of every natural phenomenon. But a watch is only evidence of a man in the sense that a hoof is evidence of a horse. If you have never seen a man make a watch, or have never seen a whole horse, you don't know the connection. There are a great many leaves and stems found as fossils that have been put into so-called "form genera" separately, because no one has found the leaves and stems attached. Conversely, because of the evidence that has been amassed (in the last hundred years, especially), evolution is conceived as the mechanism behind the diversity of life as we see it today, just as biochemistry gives mechanisms for many of the chemical events that occur in living things, (in contrast to the "vitalistic" theories of years past). If one wished to lay emphasis on this feature of the Bible, then one might say God *is* evolution.

I would like to give some examples of inconsistency in the argument presented in this pamphlet. Vide p. 16 par. 2 (X), p. 17 (X). If he believes that evolution can take place anywhere, anytime, why not in Hawaii in the recent past? Vide p. 27 last sentence. This statement is a direct denial that evolution has occurred at all, whereas Mixter admits that it has. The mutations only indirectly "cause" adaptation anyway, for the diversity must be present for natural selection to act upon it. The latter is the factor producing adaptation (see further discussion below). It would almost look as if Mixter was trying to confuse the reader.

## II. Regarding similarities

In pure genetic lines (in which brother-sister matings have gone on for many generations, i.e. laboratory mice, rats, etc.) the difference between parents and offspring is exceedingly minute, and that which occurs can usually be shown to be due to uncontrolled environmental influences. On the other hand, male and female Red-Wings (as well as males and females of any species) do not in fact receive the same chromosomes from their parents. This is the basis of sexual differentiation. In all natural populations which reproduce sexually there is a great deal of variation due to the fact that the genic material is not uniform and is being constantly shuffled. The point is that there is variation about a norm that is the "ideal" species. Now, when an evolutionist is talking about similarities he naturally takes this into account. As one goes to higher and higher categories (families, orders, etc.) the similarities he works with become more generalized. For example, all mammals and birds have a four chambered heart in contrast to all the lower animals, in spite of variations in size and shape of the details. One of the most important generalizations of the early biology (about a hundred years ago) is that all living things are composed of cells with nuclei. This is not quite true, but even bacteria, long thought to have no nuclei, have recently been shown to have them at least in some cases. The small size is a definite disadvantage to study in this instance. As Mixter says, however, (p. 9, par. 2) it is not individual likenesses that are compelling evidence, but the sum total of likenesses. He forgets this principle many times in his discussion, although it cannot be overemphasized. Believe me, if there is anything in biology that has been hashed and rehashed (and is still going on!) it is the probable relations of this group to that, for example: ideas of what is basicly simple and what is reduced from a more complex condition have undergone revision to the extent of completely inverting the accepted phylogeny in a number of cases.

This is indeed a difficult problem. I have one objection to this argument, however, that a mutation occurs means that the gene is no longer like that of its parent. Does this mean that it is not descended from its parent? Of course not. This insistence on almost complete similarity (in order to be able to demonstrate genetical relationship) is illogical. All automobiles are historically and functionally related, but one can't use a Ford carburetor in a Cadillac. The bolts won't fit, or the position of the air holes is wrong. These are features irrelevant to the *function*, but make it impossible to make a hybrid. Similarly with chromosomes and genes. Given enough time for random, nonessential changes to take place, and the "bolts don't fit" and there won't be any hybrid. So the ability to hybridize is just another of many features used to show relationships. Given sufficient complexity, I don't believe that convergent evolution occurs to such an extent as to confuse the evolutionary picture. That is, birds and bats are obviously different, even though they both fly. Similarly with whales and fish. With things like bacteria, it is a different story and little can be done in the way of phylogenies in this group, although people have tried.

Nor does the evolutionist say that *all* mammal-like reptiles became mammals. Again the judgement of evaluation of the degree of similarity is the critical matter, which depends at least in part on how carefully the subject is examined. Many of the fossils found which *might* be in the ancestral line of mammals, appear on close study to be more specialized in one respect or other than the mammals proper, and so are thought to be not in the direct line. Similarly, it is abvious that present day apes could not be the ancestors of ancient man. But in the judgement of those who have studied evidence, there does not appear to be any compelling necessity to invoke special creation in the ancestry of any line of organisms. The converse does appear to be true, however.

Regarding vestigial organs. The beginning of the first paragraph says that vestiges are evidence of a former perfection, but this is contradicted by the last part of the paragraph, both being views of creationists! What does it matter, whether it is vestigial or not, anyway, the crux of the matter is the business of degree of similarity. Both rabbits and man have a blind pouch off of the intestine, one big, one small. Does it matter if we can find no use for that in men?

Regarding adaptation. It is apparent from a few minutes thought that there is no reason to expect ill-adapted plants and animals to be found in any numbers, for they would soon lose out in the struggle for food, etc. This is natural selection. Read *Time's Arrow and Evolution* by Blum for an attempt to apply the second law of thermodynamics to evolution. Regarding the non-adaptive features in specific distinctions, this may be true of morphology, but I doubt it very much when it comes to physiology. Within the species, if it is at all widespread, there are bound to be adaptations to varying climatic conditions if nothing else.

The necessity to put labels on things (apparently dating back to Adam) is a source of much difficulty in understanding evolution . . .

### III. Regarding gaps in the record

A very good example of the slender thread on which hangs the evidence is that of the early history of birds. Archaeopterix and Archaeornis are the only birds known in the Jurassic (which lasted some 30 million years). The first represented is by one nearly perfect specimen (no skull) and one isolated feather, while the latter is known from one more perfect specimen. These are the only ones known, none at all having been found from 1877 until the present (at least until my reference book was written—1939). The number and distribution of avian fossils increase as we go up in the geological record, but we must consider ourselves very fortunate in the Jurassic specimens, as most of the later finds are fragmentary. I believe, therefore, that Mixter's argument that gaps=special creation, is based on the false premise that fossils are a likely occurrence. One has only to consider that all the $CO_2$ in the air would be used up in a few decades if there were no decay to realize that much less than one percent of all organisms can be preserved as fossils.

Granting this, a pretty good case can be presented that gaps in the record are probable. It has been shown that evolution among modern plants and animals (that is to say, formation of new species) goes on at a rate proportional to the smallness of the interbreeding group. That is, variations which would be swamped in a large population have a greater chance of expression in a small population. If speciation indeed may lead to formation of higher categories (genera, orders, etc.) as many geneticists believe, then the oldest population of a new Class (Birds) may indeed be quite small, say a few thousand. The chances of one or two of such to be preserved in 30 million years is small enough, let alone a set of specimens illustrating all the steps in the evolution.

So much for the gaps in the middle. I would like you to consider a few items about the earliest ages, also. It is certainly true that there is only very scanty morphological evidence of life in the Archaeozoic and the Proterozoic eras, but the chemical evidence is more abundant. Graphite, so cursorily dismissed in a footnote on p. 24, occurs in the Archaeozoic in beds as much as 13 feet thick. This is hardly comparable to "fine microscopic dust" in igneous rocks. The complete series from peat to graphite is known from later periods, but apparently all the carbonaceous deposits in the Archaeozoic were too far gone to yield any recognizable forms (even anthracite has none).

It is also apparent that any animals that were preserved ate to keep alive. What did they eat? Today the carbon cycle is founded on photosynthesis, carried on by plants. If we now look at the catalogue of species found in the early part of the Cambrian (according to a book by P. E. Raymond in 1939), we find 455 species, of which there were no protozoa, no algae except a few that formed calcareous matrices, 18.5% sponges, 2% jellyfish (etc.), 4.25% "worms", 27.5% brachiopods, 11.5% molluscs, 36.5% arthropods (note that all but 6.25% of these animals had hard skeletons of one sort or another). Are we to suppose that all of these animals lived on the few calcareous algae? It appears exceedingly unlikely from what we know of the feeding habits of modern animals of their ilk. For example, sponges feed on the microscopic plants and animals that they can strain out of the water, and have no means of locomotion. Does it not appear likely that the seas then, as now, were filled with minute plants and animals that formed the food for the larger forms? These animals also needed oxygen to respire and gain energy from their food. The only known source of oxygen in the atmosphere is that produced from green plants in photosynthesis.

It is also possible to outline in general terms how the origin of life probably took place. (theory according to Oparin, Huxley, Bernal, etc.) We know that before there were animals there were photosynthetic plants, but how did *they* get here? Since the photosynthetic plant is the most self sufficient organism existent, this means it has to synthesize all its complex proteins, enzymes, etc. from $CO_2$ and water. So this does not seem to be the beginning, but the climax of the evolution of the basic protoplasm. (most of the history of the earth went by before abundant fossils appeared!) So we start at the time when there was no oxygen and therefore, no oxone, in the atmosphere. Then the U-V from the sun could penetrate to the surface of the earth. Now it is known that U-V can produce all sorts of reactions, including making simple organic compounds (in the laboratory). In the eons that went by this organic matter accumulated, (there being no bacteria, etc. to eat it). More and more complicated molecules were formed until organic catalysts were formed which utilized the soup they were dissolved in. Somehow or other (it is very vague) these progressed to form organized complexes, and then cells with all their accouterments, living on this organic soup. The synthetic abilities of the early complexes or cells need not be formidable, since many complex molecules are present. The present complexity was then reached step by step as the complex molecules in the environmnt became used up one by one, working backwards down a chain of synthesis until the whole thing could be made from simple compounds. Then each step in the chain was useful when devised, which would not be true if one tried to build up a synthetic chain the other way. If all this seems very unlikely, remember that it might not have worked the first time through, in which case it could start all over (I don't believe that one can calculate how long it would take, for it depends on so many special circumstances, such as absorption on clay particles, concentrations, etc. that we know nothing about). Also remember that it would only take one individual (assuming no sex) to populate the world, and that this process was undoubtedly going on simultaneously in many places, if it was going on at all.

## REPLY TO CRITICISM OF "CREATION AND EVOLUTION"

### I. Regarding logic

The principle of parsimony requires that the simpler mechanism should be used to explain the facts of nature. This is a good principle and should be applied as long as it works. I maintain that evolution works within the groups which are separated by gaps in the fossil record and which are distinguished by structures not derived from previous ones by known genetic mechanisms. To illustrate: evolution explains the development of the first horse into the modern horse for the first horse is connected to the modern horse by a series of fossils, and genetic mutations could account for the modifications necessary to produce this amount of change. But the first horse is separated by a structural difference from other orders of animals and this difference is of the type that is not bridged by mutations. There is no fossil series connecting the horses to another order of mammals. Here, then, is where creation could have occurred, and the inability of mutations to produce new forms suggests that creation is needed to produce them. This reasoning does not give a proof but a possibility of creation. The real reason for believing in creation is that the Bible presents it and I take that to be a revelation that gives an explanation which is to be added to observations of living things.

Consider now the critic's statement that gaps are like his not seeing a person going to Wheaton College, so he must not have been there. I would say that if I had not seen someone at a college, he may not have been at that college. Notice the reserve in my actual conclusion, . . . "until bridges are found, *one may* hold that the groups so separated have arisen from independently created kinds". It isn't proof that creation did occur but it would be a place for its need if it is really needed.

My critic says, "A fair summary would be, 'Either evolution has occurred . . . or God has made the creatures of the world in such a way that we cannot distinguish the order from an evolutionary one." I think my paper shows that evolution does account for part of the variety of animal life but that the many breaks in the record (gaps between orders) permit one to believe that something else besides evolutionary forces are necessary for the complete account of what has happened. Members of an order of protozoa could have come from one species, but no mutations are known which could have added stinging cells containing nematocysts to protozoa so they could evolve into coelenterates of even the simplest sort. I say that until the gaps are bridged or a sufficient mechanism is described, one may be logical in his belief that evolution is not all that is at work. The evolutionist believes the gaps were not real, the creationist that many gaps were real; neither can prove his case, but neither can the evolutionist say that the creationist could not be right. He can only say, "He may not be correct."

The idea that science progresses by trying to find mechanistic causes for natural phenomena is good. But I object to saying that mechanistic causes can be the only causes. A scientist would have to be omniscient to say this.

Consider the argument that a machinist is necessary to explain machinery. The principle of parsimony would suggest this. We have seen men make watches; we have seen men construct levers and we know of nothing making levers except men. So when we see muscles work on the lever principle, we are entitled to assume that something with as much intelligence as a man designed the muscular lever. Then we read the Bible which says that God designed man. Revelation caps the argument.

Next my opponent claims I am inconsistent. His first example is a misunderstanding on his part probably caused by an ambiguous statement on my part. I do not believe that Hawaiian species were created in Hawaii; rather, I believe they were derived from mainland species probably South American. My statement about Hawaiian species was an attempt to show a creationist that he could not believe in the creation of species on the Hawaiian islands because the islands are not old enough. They did not exist when creation was occurring, which I assume was when the first members of the orders appeared. Modern orders of birds arose way back in the tertiary and the islands came out of the sea more recently than that. I apologize for giving the wrong impression and I appreciate very much the very careful reading my opponent has given to my manuscript.

The second inconsistency concerns the statement made by a world famous evolutionist who said "it would be a miracle that a mutation causing diversity would also cause convergence for an adaptive end." I did not use this statement to deny the possibility of any evolution but I used it to substantiate my claiming that the mutations which have occurred are not good enough to produce onward and upward evolution. The statement merely shows one of the inabilities of mutations. If I have confused the reader, I am sorry.

**II. Regarding similarities**

Both of us agree on the occurrence of similarities. I also recognize the use the evolutionist makes of the sum total of likenesses. I do, however, point out that not every similarity means kinship and the sum total need not be taken to mean kinship, unless the kinship is demonstrated by hybridization. When the kinship cannot be demonstrated, then the evolutionist infers it from the sum total of similarities. He could be right, but the sum totals could also be the result of creative activity - as are the resemblances between automobiles the result of the creative (constructive) activity of men.

The one who is quoted in my work as saying, "Design of creation is a psychological argument which cannot be definitely disproved" is the same world famous evolutionist who was quoted previously in regard to mutations producing adaptations. He was aware of this logic of creationist's viewpoint on similarities, but held that vestigial organs favored the evolutionist.

Vestigial organs may have functions. Apparently my critic does not object to my remarks on the appendix. Vestigial organs may be remnants of formerly more complete organs. The reduction may have been caused by mutation which is what I had in mind when I said that vestigial organs are gentic abnormalities. There is no contradiction here.

We agree regarding the "necessity to put labels on things being a source of much difficulty in understanding evolution." I went into this situation to some extent to try to show one kind of creationist, who believes all species were created, that he cannot hold such a position. That is, I was trying to convert some creationists to a limited form of evolution.

**III. Regarding gaps in the record**

My critic has helpful material here. Too many of us creationists do not appreciate how lucky we are to have any fossils and enough men to study them. The significant point about the gaps is that they are so consistent in the positions they occur, that is between orders, and classes and phyla. He says the links are missing because they occurred in small numbers. It is neither in the small nor the very large population that evolution is likely to occur but "probably the best chance for recombination is in a rather large population more or less broken up into local areas in which the frequencies of allelic genes are different." (Shull, *"Evolution"* 1951, p. 187) The size of a breeding population could be in the vicinity of 500 to 2000 individuals. It is reasonable to expect in 30,000,000 years that there would be a few specimens giving some hint of how feathers may have evolved —so far as I know, no such fossils have been found. The first birds had feathers which appear as complicated as the best of feathers today. Archaeopteryx is separated from their ancestral reptiles by a striking gap. The evolutionists (such as Heilmann in *"Origin of Birds"*) say that the real missing link is the proavis, which would be intermediate between reptiles and Archaeopteryx. So I hold that Archaeophteryx does not prove birds came from reptiles; it merely emphasizes the problem of gaps.

The reference to graphite being as much as 13 feet thick is a revelation to me. Apparently the footnote I quoted should be omitted. I am also grateful for the evidence that there was much plant life preceding the Cambrian animals.

I shall not comment on guesses as to the origin of life except to say that such works as du Nouy's "Human Destiny" and A. Cressy Morrison's "Man Does Not Stand Alone" show the extreme improbability of getting living things by chance happenings. The quotation at the end of my booklet is by a professor of botany at the University of Illinois. His statement admits that a belief in the creation of the first protoplasm is as logical as a faith in spontaneous generation.

R. L. Mixter

Wheaton, Illinois

# The Reconstruction or Cataclysmic Theory

PETER W. STONER, M. S.

FOREWORD

There are many different statements of the Reconstruction Theory. The statement in this article attempts to be somewhat inclusive of the various forms of the theory. I am sure that each proponent of the theory will find parts which he will consider as not belonging to the theory. This type of treatment seems necessary in order that the article may deal with the various parts of the theory held by different persons.

It should be clearly understood that no attempt has been made to criticise the authors of the Reconstruction Theory. The faith of those teaching the theory is not being attacked. I believe their faith to be as strong and sincere as the faith of those who see the great dangers of the theory.

The purpose of this treatment is not to offend any one but, if possible, to place our Christian faith on a higher scientific level, to save the faith of Christian young people entering college and to help open the door for evangelical work in the colleges of our country.

The reconstruction theory is very old and has been taught by the conservative church over a long period of time. It is an interpretation of the first chapter of Genesis and is something like this: In the beginning God created the heavens and the earth *perfect*, ready for the habitation of man. Life then developed on the earth very slowly. Very primitive forms came first, then more and more complex forms until all of the forms of plants and animals had evolved. Some believe they were created, others think they evolved by the evolutionary principle. The early forms left remains in the earliest and deepest strata while more and more recent forms occupied higher strata. This constituted all of the geological ages and accounts for all of the fossils found in the earth. Then Satan sinned and was cast out of heaven to the earth. The earth was destroyed and it *became* without form and void and dark, destroying all life. Then about 6,000 years ago God reformed the earth and recreated the life upon it in six 24-hour consecutive days. This last creation was so rapid that no remains can ever be found to record these creative acts of God.

The main arguments used to support this theory follow:

1. God is a perfect God and cannot do anything imperfectly. Therefore the first verse of Genesis means that "In the beginning God created the heavens and the earth *perfect*." Since God created the earth to be inhabited ("He formed it to be inhabited.") it came from his creative act ready for man.

2. Since the geological ages are very long, running into hundreds of millions of years, and since the days of Genesis are evidently 24-hour days, there can be no relation between the days of Genesis and the geological ages. Since Ussher worked out a very careful dating of the scriptural occurrences through the genealogies back to Adam and found it to be only about 6,000 years it can be only about 6,000 years back to the first day of Genesis. Since the first day of Genesis deals with the reformation of the earth we must place the geological ages between the first and second verse. Thus all of the geological ages occupy a long period of time between verse one and verse two.

3. Since the Bible says nothing as to how life came to the earth in this long period between verses 1 and 2, some are willing to accept the evolutionary explanation for the origin of all life before the first day of Genesis. A pre-Adamic race apparently existed before the first day of Genesis for God said to man "Be fruitful, and multiply, and replenish the earth, . . . " (Gen. 1:28) God would not have said "replenish" if man had not been here before. The same can be said for the other forms of life.

4. Satan sinned, lost his position as the covering cherub and was cast down to the earth and the earth was destroyed. This idea is supported by the following references: Jeremiah 4:23 26; Isaiah 24:1, 19-20; 45:18; 14:9-17; Ezekiel 28:13-19; and Luke 10:17-18. This justifies our changing the word "was" to "became" and making the second verse read "And the earth *became* without form, and void, . . . "

5. The days of Genesis are necessarily 24-hour days. There is such a close relation between plants and animals that it was necessary that they come into existence at about the same time. Certain types of plants cannot propagate without the bee, others require a special type of fly. God created plants for the food of animals and man. If he had created plants millions, or hundreds of millions of years before man and the animals, there would have resulted a terrible waste of this food. God is not wasteful.

6. The reconstruction theory by completely separating the days of Genesis from the geological ages has removed a great stumbling block from the minds of college students. They may now accept both the Scriptures and science and encounter no conflict.

Let us now consider these arguments for the reconstruction theory, item by item.

1. The introduction of the word "perfect" to the first verse of Genesis violates one of the most important rules of scriptural interpretation, namely: we must neither add nor subtract anything from

the scriptural statement. John emphasizes this principle in Revelation 22:18-19 when he writes: "If any man shall add unto these things, God shall add unto him the plagues that are written in this book: And if any man shall take away from the words of the book of this prophecy, God shall take away his part out of the book of life. . . . "

I would not object to the argument that God does everything perfectly. I do object to any man adding the word "perfect" to any scriptural statement and then interpreting the meaning of the word. This word "perfect" has given the church trouble before. Men believed that the circle was the only perfect curve and, therefore, every planet moved around the sun in a perfect circle. When there seemed to be trouble with the circle they had the planet move in a small circle whose center moved around a larger circle and the center of the larger circle moved around a still larger circle, etc. A planet moved around a very small circle and the center of this circle moved around a circle a little larger whose center in turn moved around a still larger circle, and finally, whose center moved on a circle around the sun. When the first astronomer suggested that the planets moved around the sun in an ellipse he was forced by the church to recant to save his life. Today we know that the ellipse is just as perfect a curve as is the circle.

We know today that a diffuse nebula, as described in the second verse, is just as perfect an astronomical body as was the earth when man was created. Furthermore God tells us just what steps the diffuse nebula went through in being prepared for the habitation of man. Those steps agree perfectly with modern science. Anyone can read them in the first chapter of Genesis and in books of science.

Should the church be guilty of adding to the very first verse in the Bible? God forbid!

There are two revelations of God. One is in the Scriptures and the other is in nature. The layers of the earth record the acts of God's creation and show their order. The chemist discovers the intricacies of the creation, in the way matter is built up from atoms and molecules, in the way atoms are built up from protons, electrons, neutrons, etc. The physicist discovers the laws governing the physical universe; laws of gravitation, of forces, of motion, etc. These are the laws which God laid down when the universe was created, laws to keep the universe orderly. The astronomer discovers the created universe, the composition of each part, how it moves, how it is held together, its probable origin and possible future. The biologist discovers something of the tremendous intricacies of living things and the tremendous wisdom involved in their creation.

These two revelations of God *must* agree. Any apparent disagreement demands, at once, a reconsideration of our scriptural interpretation and a very extensive re-examination of the scientific data involved.

2. The geological ages are very long. They represent great periods of time, some of them hundreds of millions of years. Radioactive materials in these layers reveal their ages or at least the general magnitude of the age. If we were to doubt the greatness of these ages we would doubt the working of the physical laws of God. We could as well doubt that hydrogen and oxygen can form water or that if you were to drop a heavy lead ball it would fall downward.

The datings of Ussher, although very carefully made, are no longer considered to be even approximately accurate by many anthropologists or by many careful students of the scriptural genealogies. Most of our students of genealogies believe that the word translated "begat" refers to a descendant, not necessarily a son or a daughter. This seems to be borne out by Deuteronomy 4:25 "When thou shalt beget children, and children's children, . . . " Many authorities are very sure that great gaps occur in all of the genealogies. This seems possible since Matthew 1:1 reads: "The book of the generation of Jesus Christ, the son of David, the son of Abraham." Certainly there are tremendous gaps here. I asked a prominent student of the genealogies what he would set as the lowest estimate of the age of man, taken only from the genealogies. He said he thought that 50,000 years was the absolute minimum, but it might be much longer. We do not have a definite statement of the age of man in the Scriptures. We should not be dogmatic about the 6,000 years.

The remains of man appear primarily in the upper sedimentary layers of the earth and cannot from the layers themselves be dated with any high degree of accuracy. The age of such remains are usually dated by the amount of sediment which has been deposited above them, but the rate at which this sediment was deposited over a period of 100,000 years has probably varied tremendously. The probable error here constitutes a sizable factor of the estimated age. However, there is a new method called the carbon-14 method of dating which has great promise. This method is being improved by Dr. Laurence Kulp of Columbia University, a member of the American Scientific Affiliation. It is expected that within a very few years it will be possible to date, with a fair degree of accuracy, all remains of man back to at least 100,000 years. Until this is accomplished, we can only say that all evidence, both scriptural and scientific point to an age of man far greater than 6,000 years.

If we place the geological ages between verse 1 and verse 2 we immediately move verse 2 to a recent period of time. Science cannot place verse 2, with the earth without form, void and dark within the last 6,000 years or even the last 60,000,000 years. Neither can you place the beginning of night and day in verses 4 and 5 or the separation of the oceans and the clouds in verses 6 to 8 only 6,000 years ago. Any science teacher, non-Christian or Christian, can show a recent date for verse 2 to be ridiculous. No, verse 2 is not recent.

Again suppose we try to place the geological layers between verses 1 and 2 and then let the earth become without form, void and dark. (Nathan Stone of Moody would add as basic meaning of the original words "of unmeasurable magnitude and in great commotion.") The only astronomical state of any body fulfilling this description is certainly a diffuse dark nebula. Here the atoms are separated from each other and many of the electrons are removed from the atoms. Now, suppose you place all of the geological ages with their layers between verses 1 and 2 and then let the earth become a diffuse nebula and again collect the nebula to form the earth, where would all of the fossils and layers be? No, the geological layers all followed verse 2. But I hear someone say that they believe that my interpretation of verse 2 is wrong. (Any astronomer will give the same interpretation.) They believe that it means that only the life was destroyed upon the earth. If so, it must have been all of the life on the earth or else no recreation would have been needed. If all of the life were destroyed on the earth, the layers of the earth would clearly record the event. There is no such record. We are forced to place the geological ages and their geological layers after verse 2, and in fact, the fossil-bearing layers after verse 10.

Are we going to place the geological layers between verses 1 and 2 and thereby make much of the remainder of this chapter contradict known facts of science? God forbid!

3. The original word translated "replenish" carries no meaning of filling again. It simply means, fill the earth. There is no scriptural evidence of a pre-Adamic race. Science indicates that the human race has been continuous since its first advent to this earth.

4. A second principle of scriptural interpretation is that we must not take a passage out of its context.

The idea that Satan sinned, was cast out of heaven and the earth was destroyed and it *became* without form and void can only be supported by violating the above principle.

Let us look at the references.

Jeremiah 4:23-26, "I beheld the earth, and lo, it was without form, and void; and the heavens and they had no light. I beheld the mountains and lo, they trembled, and all the hills moved lightly. I beheld, and lo, there was no man, and all the birds of the heavens were fled. I beheld, and lo, the fruitful place was a wilderness, and all the cities thereof were broken down at the presence of the Lord, and by his fierce anger."

What is the passage talking about? Look at the context. Verse 10 is talking about Jerusalem. Verse 31 is talking about the daughters of Zion. This passage is clearly talking about Jerusalem and the children of Israel. Jerusalem and the daughters of Zion, certainly, did not exist between verses 1 and 2. This passage is taken out of its context.

Isaiah 24:1, 19-20, "Behold the Lord maketh the earth empty, and maketh it waste, and turneth it upside down, and scattereth abroad the inhabitants thereof . . . The earth is utterly broken down, the earth is clean dissolved, the earth is moved exceedingly. The earth shall reel to and fro like a drunkard, and shall be removed like a cottage; and the transgression thereof shall be heavy upon it; and it shall fall and not rise again."

Verse 5 ". . . . because they have transgressed the laws, changed the ordinance, broken down the everlasting covenant." Certainly this is speaking about the Jewish people. Verse 15 "Wherefore glorify ye the Lord in the fires, even the name of the Lord God of Israel . . . " This also refers to the Jewish people. Verse 23 " . . . when the Lord of hosts shall reign in mount Zion and in Jerusalem . . . " Here we are not only speaking about the Jewish people, but about Christ's reigning in Jerusalem. Certainly the Jewish people did not exist between verses 1 and 2, nor did they have the everlasting covenant, nor did Christ reign on Mt. Zion between verses 1 and 2. This reference is absurdly taken out of its context.

Isaiah 45:18 "For thus saith the Lord that created the heavens; God himself that formed the earth and made it; he hath established it, he created it not in vain, he formed it to be inhabited; . . . "

Verse 19 " . . . I said not unto the seed of Jacob seek ye me in vain:"

Verse 18 is interpreted by the proponents of the reconstruction theory to mean that God created the earth ready to be inhabited as it came from his hand. Verse 18 like verse 19 implies a purpose. It appears to mean that God had a purpose in creating the earth and that purpose was that it should be inhabited. The condition of the earth is given in Genesis 1:2 and the steps that God took in preparing this earth for the habitation of man are related in the rest of the chapter.

Ezekiel 28:13-19 "Thou *hast been* in Eden the garden of God; every precious stone *was* thy covering, the sardius, topaz, and the diamond, the beryl, the onyx, and the jasper, the sapphire, the emerald, and the carbuncle, and gold; the workmanship of thy tabrets and of thy pipes *was p epared* in thee in the day that thou *wast* created. Thou *art* the anointed cherub that covereth; and I have *set* thee so: *thou wast* upon the holy mountain of God; *thou hast* walked up and down in the midst of the stones of fire. *Thou wast* perfect in thy ways from the day that thou *wast* created, till iniquity was found in thee. By the multitude of the merchandise they *have filled* the midst of thee with violence, and *thou hast sinned*: therefore *I will cast thee as profane* out of the mountain of God: and *I will destroy thee,* O covering cherub, from the midst of the stones of fire. Thine heart *was* lifted up because of thy beauty, thou *hast* corrupted thy wisdom by reason of thy brightness: I *will cast thee* to the ground, *I will lay thee* before kings, that they may behold thee. Thou *hast* defiled thy sanctuaries by the multitude of thine iniquities, by the iniquity of thy traffick; therefore *will I bring forth* a fire from the midst of thee, *it shall* devour thee, and *I will bring* thee to ashes upon the earth in the sight of all them that behold thee. All they that know thee among the people *shall be* astonished at thee: *thou shalt* be a terror, and never *shalt thou* be any more."

In general the tense in prophecy seems to have but little significance. Most prophecies are written in the past tense. (See Isaiah 53.) In this prophecy we have a remarkable case of change of tense. Here the past tense appears to have been used when the event was past and the future tense appears to definitely refer to future events.

Aside from naming Satan as the covering cherub, everything is past tense from the start of the thirteenth verse to the middle of the sixteenth verse, where the future tense starts with: "therefore I will cast thee as profane out of the mountain of God." Compare with this Revelation 12:7-12, particularly the ninth verse which reads: "And the great dragon was cast out, that old serpent, called the Devil, and Satan, which deceiveth the whole world: he was cast out into the earth, and his angels were cast out with him." The Ezekiel passage and the Revelation passage evidently refer to the same event. Certainly, the Revelation passage is future and is to take place at the time of the great tribulation period.

The next event in Ezekiel is: "I will destroy thee, O covering cherub, from the midst of the stones of fire." The destruction of Satan is forecast in Revelation 20, particularly the tenth verse. "And the devil that deceived them was cast into the lake of fire and brimstone, where the beast and the false prophet are, and shall be tormented day and night for ever and ever." This occurs at the end of the millenium and is still more future than the previous event.

At the seventeenth verse Ezekiel changes to the past tense again: "Thine heart was lifted up because of thy beauty, thou hast corrupted thy wisdom by reason of thy brightness:" This is referring to Satan's early sin.

Ezekiel again turns to the future: "I will cast thee to the ground, I will lay thee before Kings, that they may behold thee." Again Revelation 12 and 20 refer to the same future events.

Again the past tense is used: "Thou hast defiled thy sanctuaries by the multitude of thine iniquities, by the iniquity of thy traffick:"

The remainder of the Ezekiel passage is prophesied in Revelation 20. A part takes place before the millenium and the rest at its immediate end.

Isaiah 14:9-17 "Hell from beneath is moved for thee to meet thee at thy coming: it stirreth up the dead for thee, even all the chief ones of the earth; it hath raised up from their thrones all the kings of the nations. All they shall speak and say unto thee, Art thou also become weak as we? art thou become like unto us: Thy pomp is brought down to the grave, and the noise of thy viols: the worm is spread under thee, and the worms cover thee. How art thou fallen from heaven, O Lucifer, son of the morning! how art thou cut down to the ground, which didst weaken the nations! For thou hast said in thine heart, I will ascend into heaven, I will exalt my throne above the stars of God: I will sit also upon the mount of the congregation, in the sides of the north: I will ascend above the heights of the clouds; I will be like the most High. Yet thou shalt be brought down to hell, to the sides of the pit. They that see thee shall narrowly look upon thee, and consider thee, saying, Is this the man that made the earth to tremble, and did shake kingdoms; That made the world as a wilderness, and destroyed the cities thereof that opened not the house of his prisoners? . . . "

This passage describes Satan's sin. It describes very similar events to those of Ezekiel, but is addressed to Satan at the time of his fall, which time again is described in Revelation 12 and 20.

Luke 10:17-18 "And the seventy returned again with joy, saying, Lord, even the devils are subject unto us through thy name. And he said unto them, I beheld Satan as lightning fall from heaven."

The prophecies of Isaiah 53, regarding Christ are all written in the past tense, but refer to future events. I believe this is also a prophecy, made by Christ, referring to the future. It cannot refer to the same events referred to in Isaiah and Ezekiel. Satan is still our accuser in heaven as witnessed in the book of Job and in Revelation 12:10 " . . . for the accuser of our brethren is cast down, which accused them before our God day and night."

Satan continues to be our accuser up to the very last when he is cast out of heaven and starts the great tribulation period.

5. The days of Genesis.

The days of Genesis might be 24-hour days, but that is not necessary. Notice the following references where the same word, day, is used as that in Genesis.

| | |
|---|---|
| The day of destruction | Jb. 21:30 |
| The day of distress | Gen. 35:3 |
| The day of plague | Nu. 25:18 |
| The day of wrath | Jb. 20:28 |
| The day of temptation | Ps. 95:8 |
| The day of trouble | Ps. 102:2 |
| The day of vengeance | Prov. 6:24 |
| The day of adversity | Prov. 24:10 |
| The day of prosperity | Ec. 7:14 |
| The day of death | Ec. 8:8 |
| The day of Midian | Isa. 9:4 |
| The day of anger | Isa. 13:13 |
| The day of salvation | Isa. 49:8 |
| The day of evil | Jer. 17:17 |
| The day of Jerusalem | Jer. 38:28 |
| The day in which the walls were built | Mi. 7:11 |
| The day of Jesus Christ | Phil. 1:6 |

Your concordance will add many more.

These are certainly not all 24-hour days. The days

of Genesis could be either longer or shorter than a 24-hour day.

Genesis 2:4 is a very striking case. "These are the generations of the heavens and the earth when they were created, in the *day* that the Lord God made the earth and the heavens, and every plant of the field before it was in the earth . . . " Here all of the days of the first chapter are referred to as "the day." It is impossible for all of these days to be 24 hour days.

Some say that the use of morning and evening in the first chapter forces a 24-hour day, but look at Psalm 90:5-6 " . . . They are like grass which groweth up. In the morning it flourisheth, and groweth up; in the evening it is cut down, and withereth." This certainly is not a 24-hour day.

After some of the creations God told them to multiply and fill the earth. This practically requires a lapse of time between the acts of creation. Certainly, God gave them time to multiply and fill the earth. Plants in the sea necessarily became quite abundant before fish could live. Plants on the land necessarily became abundant before the creation of land animals.

The days of Genesis may have been very short periods of time, days of creation, then a geological age elapsed while that form of life multiplied and filled the earth. When the earth was ready for a new form of life, God made another creation.

These days of creation may have been 24-hour days, they may have been of less than a second duration, they may have been many years. They may have been geological ages, but I do not favor that interpretation. Psalm 33:6-9 "By the word of the Lord were the heavens made; and all the host of them by the breath of his mouth . . . For he spake, and it was done; he commanded, and it stood fast." This seems to imply short periods of creation.

The argument that plants could not exist without animals does not seem to be logical. The fossils in the earth certainly bear mute evidence that they did not appear at the same time. The fact that one type of plant requires a certain type of insect for its propagation is not proof that it always has had such a requirement. Nor is it proof that this particular type of plant was a part of the original creation of plants. Man has developed many new plants. Possibly insects have been responsible for the development of some types also.

The argument that God is not wasteful, and he would be wasteful if he created plants for food long before there were animals or man to use them, is a philosophical argument which is meaningless.

The great forests grew and fell one on top of another until our great coal beds were formed. Were they wasted? Were the materials that formed our great oil reservoirs, that formed our great graphite beds wasted? It is possible that more waste exists today with man in control than ever existed before he came on the scene.

6. Did the reconstruction theory remove a stumbling block from the minds of Christian students?

The Christian student is taught the reconstruction theory in the church; he enters college and has it proved to him that: there was no destruction of the earth after it was inhabited; life on the earth was not destroyed by a great cataclysm requiring a complete new creation; that light did not appear on this earth 6,000 years ago; that land did not rise from the water 6,000 years ago, or 60,000 or 60,000,000 years ago; that the age of life on this earth has been at least 2,500,-000,000 and in many cases hundreds of millions of years elapse between the advent of one general form of life and the next general form. A sad result follows. Nearly all of these students lose faith in the Bible. Certainly, they should question their teaching, to see if they are scriptural, but very few do. I have personally had the privilege of re-establishing the faith of several hundreds of students who were taught the reconstruction theory in a conservative church, and then lost their faith in college because of the reconstruction theory.

Accepted science, today, is literally at the door of the church asking the privilege of proving that the Bible is the inspired word of God. If we are willing to work with Christian men of science we can do great things for the faith of our Christian young people, and we can reach many non-Christian students in the colleges.

We of the church point the accusing finger at the state colleges and say that they are wrecking the faith of our young people. My own experiences indicate that the teaching of the reconstruction theory is wrecking the faith of more students than all of the atheistic professors put together. God forbid that we of the church shall so continue to give teachings which are unscientific and unscriptural. It would be very difficult for the church to change the teachings of these atheistic professors, but it could easily correct its own dangerous teachings.

May God prompt the ministers and Sunday School teachers of the conservative churches to carefully examine the scientific materials which they are teaching.

The findings of science today are such that they may be used by the church to prove, that of a certainty, the Bible is the inspired word of God.

The books *Modern Science and Christian Faith* and *Science Speaks* are both published by Van Kampen Press. They both constitute an important start at showing scientifically that the Bible is accurate and is God's word.

The American Scientific Affiliation is an organization of Christian men of Science. It stands ready to aid the church in this most important undertaking

## Letters

*Editor*:

As the author to the anonymous article which appeared in the March issue of your Journal under the heading, "A criticism of the A.S.A. monograph on 'Creation and evolution' ", I feel I owe an apology to the readers. I knew that Dr. Mixter wished to publish my comments, but put off revising them so long that he despaired of me and had them published as they stood. That is, as a hastily written epistle, poorly organized. One source of confusion is the omission of the page references in the margins to tie certain parts to corresponding sections of Dr. Mixter's pamphlet. The interested reader can, however, supply most of these.

I must admit to one inadvertent misstatement of fact. The 24 foot bed of graphite I refer to turns out to be mostly a metamorphosed limestone with numerous graphite inclusions. It is, nevertheless, a sedimentary formation, which may be evidence for organic origin. The unequivocal determination of the origin of ancient graphite by the carbon isotope ratio seems to be impossible, as the distinction between inorganic and organic carbon is not as distinct as once thought.

I feel I must reiterate three points. The first is stated clearly by Dr. Mixter (p. 2, par. 1), "The 'I don't know' of the Creationist is likely to soothe him into a complacency so that he does not search the natural sciences for explanations of 'things as they are'." The second is my opinion that the gaps in the geological record are not so remarkable nor so sharply defined as some would lead us to believe. The third is the matter of the machinist and the machine This common argument I tried to dispose of, but obviously did not convince Dr. Mixter. To counter his example muscles work on levers (bones) because it is the simplest way they can move a rigid structure. What alternatives are there? Pistons, inclined planes, gears and freely rotating axles are none of them readily adaptable to the mechanics of a body of flesh and bones. The soft bodied animals, too, use levers, but here the mechanism is not so elegantly simple. Dr. Mixter believes God invented the lever; I don't see how life could exist (aside from some simple microorganisms) otherwise. A predator with legs can move faster than his prey without them. To paraphrase Dr. Mixter: Evolution caps the argument.

All of which is not to say that I do not believe Dr. Mixter's monograph is not useful. He is, in the first place, much more radical and logical than some other advocates of special creation, and so serves to enlighten rather than obscure. On the other hand, he points up the deficiencies in the evolutionist's argument, and may help spur him to more rigid proof. And if he gives comfort to others in doubt, he has performed a service.

In conclusion, I would like to recommend that those interested in what modern scientists have to say about the possible (mechanistic) origin of life read the article by George Wald in the August (1954) issue of Scientific American, and the letters to the editor in the October issue. A much more detailed examination of the problem will be found in the April (1954) issue (no. 16) of New Biology (published by Penguin Books), where four eminent authors discuss (and disagree on) this knotty problem.

Sincerely yours,

John H. McClendon

48 W. Delaware Ave.
Newark, Delaware

# *The Impact of Geological Dating Upon the Interpretation of Biblical Chronology*

HERBERT FEELY
Columbia University

What impact has geologic dating upon the interpretation of Biblical Chronology? In the last few days many of you gentlemen, who should know much better than I, have indicated that it is slight; that aside from the lengthening of a few gaps in the chronologies, no alternations have been necessary. In so far as this is true it is indeed excellent. It is a sign of the accuracy of Evangelical theology. However a highly vocal, even if not large group, labelling themselves all things from theologists to geologists, have arisen to denounce geologic dating and geology itself as tools of the Devil which are seeking to destroy our faith in the Bible as the word of God. I haven't heard such a view expressed here so I won't waste your time trying to refute it.

Instead I would like to discuss briefly the theories and techniques of geologic dating. I do this not because the techniques are in themselves of especial importance to Christians, but because the results gotten from them seem to bother some. If this were not so I'm sure that this subject would not have been put on the program.

Let's consider first the reasons why the study of geology convinced men of the great antiquity of the Earth long before good quantitative methods of calculating its age were available. Then lets consider the nuclear physical data and geologic relationships upon which geochronology is based, and finally the degree of respect which theologians should give to geologic data when setting forth their interpretations of Biblical data.

The first thing that men had to learn, before they could appreciate the antiquity of the earth, is the fact that the earth, rather than being static, is undergoing continuous change. It is not a dark, dead sphere of rock. It is a pulsating body filled with an enormous amount of energy. This energy is slowly pushing forward processes of immense magnitude, changing continents into seas and seas into dry land. Earthquakes, volcanic eruptions, erosion, and sedimentation are manifestations of this energy.

Let's consider one example of this activity, a process involving great magnitudes of space, mass, and time. Vast quantities of material are being eroded by the Mississippi, its tributaries, and parallel streams. This material is ultimately carried down to the mouths of these rivers and deposited as sediment along the Gulf Coast. Measures of the rate of sedimentation off the shores of Texas, Louisiana, Mississippi, and the other Gulf Coast states indicate average rates of deposition of much less than a few inches per year. Yet exploration by oil companies has shown that the deposits are as much as 28,000 feet thick. The structures and textures within these rocks are the same in the deeply buried beds as in those at the surface indicating that all formed even as they are forming today. Great periods of time must have been required for their accumulation. The nature of sediments indicate that the floor of the basin has slowly and continuously sunk while the sediments have been piled on top of it. Yet the Gulf of Mexico is but one of many such basins which have existed at one time or another during the history of the Earth collecting millions of cubic miles of sediment.

Such rocks as these now seen forming along the Gulf Coast and in other sedimentary basins are found all around the earth. Some are still soft and uncompacted but most are consolidated into hard rock though still maintaining the same structures and textures as the soft sediments. Some are partially or largely recrystallized with loss of many of their original textures and some are completely recrystallized so that all, or almost all, primary structures are obliterated and their original sedimentary character is more inferred than observed.

Detailed studies of such masses of sedimentary rocks have shown parallel sequences in neighboring localities both in lithology, that is in the chemical composition and phyisical nature of the rock, and in their fossil content. These have permitted age correlations of beds in sections measured at different localities. Lithologic correlations are good only over short distances since they depend strongly on uniformity of environmental conditions during their deposition. However they do serve to substantiate the validity of parallel correlations made by use of certain fossils, or better yet, certain groups of fossils. Certainly not all age correlations made on these bases have proven to be correct, but experience shows stratigraphers which criteria are valid and which are not. Those which don't work are eliminated from stratigraphic methods. And let it not be thought that the validity of stratigraphic correlation by means of fossils in any way substantiates the theory of evolution any more than that of progressive creation or perhaps even modification of first creation. As with all scientific methods stratigraphic correlations and sequences cannot be applied without good judgment and adequate testing by the researchers.

Interspersed among the sedimentary rocks are igneous rocks which have formed by cooling of lava or magma, that is, fluid rock melts. Some have formed at the surface of the earth by extrusion from volcanoes or fissures, while others such as the feeders to volcanoes cooled and solidified below the surface.

The interrelations among the sediments and igneous rock units cutting across them or interbedded with them have been of great usefulness in working out the history of the rock sequences at many localities.

Deeper within the crust of the earth where higher temperature and pressure prevail, recrystallization of both sediments and igneous rocks has taken place. These processes are only relatively recently beginning to be understood in any detail.

Careful study of the rocks at many localities around the earth, using all methods of investigation available, has permitted the construction of a geologic history of the earth in general and of more detailed histories for multitudes of particular regions.

Our understanding of geologic history, to be sure, contains many gaps and uncertainties. Then, too future discoveries will doubtless necessitate many alterations and amendments. However the tremendous weight of factual evidence already available, and now serving as the foundation upon which the history has been built assures us that by and large the present history must be correct.

Studies of earth history show that tremendously long time periods must have been required for all that has taken place. An example is found in the petrified forests of Yellowstone Park where successive fossil forests are found, one above the other, where each has grown, been wiped out by volcanic ash, where the ash has weathered into soil at its upper surface, where another forest has grown only to be wiped out by another ash fall, and so on through several cycles. Yet these beds involved represent but a tiny fraction of the geologic column.

Thus a study of geology gives us the order of magnitude of earth history but doesn't give us good quantitative measurements. For such measurements geologists have turned to determinations of the radioactivity of rocks and minerals.

The phenomenon of radioactivity has been known and studied for almost 60 years but it has only been efficiently applied to geologic dating rather recently.

Radioactive decay results from an attempt by energetically unstable nuclei to reach a condition of greater stability. This is usually achieved by ejection from the nucleus of particles of matter or quanta of energy.

Much careful study by physicists has proven that the total activity of a radioactive isotope is directly proportional to the amount present. This decay can be expressed mathematically by a relatively simple differential equation containing one constant. The value of this constant, which may be expressed as the "half life" is different for each radioactive isotope and depends upon the particular energetics involved in its decay. The half life is the time required for half of any original amount of the species to decay. Thus Carbon-14 has a half life of 5580 years. That is, if I have one pound of C-14 today, in 5580 years only ½ pound will be left. The other half pound will have decayed to Nitrogen-14. Half lives range from those which are extremely long, such as the 62 billion years of Rubidium-87, to those which are extremely short, as the three ten millionths of a second of Polonium-212.

Typical decay schemes are shown by two radioactive isotopes important in geologic dating: Uranium-238 and Rubidium-87. Uranium-238, the most abundant isotope of uranium, has a half life of 4.5 billion years. It decays by the emission of an alpha particle to Thorium-234. This is itself unstable and has a half life of 24.5 days, decaying to Protactinium-234. The protactinium decays to Uranium-234, this is to Thorium-230, and so on until finally lead-206 is formed. This isotope is stable and will increase in amount as the Uranium-238 and the 16 intermediate products forming from it decay.

Rubidium-87 has a simpler decay scheme going by

Beta particle emission to strontium-87, a stable isotope. As previously stated this decay has a half life of about 60 billion years.

To measure the age of a rock or mineral one generally needs to know only the amount of a radioactive species in it, the amount of the daughter product which has been formed by its decay, and the rate of this decay.

The amount of the isotopes present may sometimes be measured by standard methods of chemical analysis but frequently mass spectrometers are used for determinations of isotope ratios of an element, and Geiger counters, scintillation counters, and other instruments for measuring radioactivity are used to determine the amount of a radioactive species present.

Ideally only minerals which contain none of the daughter product as a primary constituent should be dated. Daughter product initially present will make the mineral seem older than it really is. Thus Rubidium-87 ages are measured only for lepidolite and biotite micas which have negligible content of the strontium daughter. Potassium-40-Argon-40 ages and Uranium-Helium ages can be run on any mineral or rock since the argon and helium are gasses and are not included as primary constituents in rocks in measurable quantities. Similarly the best Uranium-Lead ages are gotten from minerals low in initial lead content.

Sometimes however mass spectrometer anlaysis permits corrections for initial daughter content and permits ages to be determined from minerals such as uranium ore containing primary lead. The isotopic composition of the primary lead may be gotten from associated lead minerals and this may be subtracted from the lead content of the uranium mineral. Ages from such minerals, however, seldom give as good precision as ages from lead-free minerals.

Another problem for the geologist measuring ages is the possibility of loss of daughter product during the history of the rock, giving ages which are too low. Thus argon formed from potassium decay and helium formed from uranium decay may leak out of a rock. Similarly either parent or daughter species may be leached out of the rock late in its history giving a spurious parent-daughter ratio, and hence an incorrect age. This is an especially important problem in lead-uranium age work.

These and similar problems must be overcome by the geochronologists. Their net result is to reduce the number of possible age methods which can be relied upon. Thus the Uranium-Helium method is practically defunct. The most recent trend in geochronology has been the measurement of single rock bodies by several methods to gain intercalibration. Results have been excellent, showing which methods are dependable and which have flaws, and what the nature of the flaws are.

At this time the uranium-lead method is considered the most reliable for dating rocks. Since uranium 235 decays to lead 207 and uranium 238 to lead-206, this method offers internal checks. The uranium-235, lead-207 ratio, the uranium-238, lead-206 ratio, and the lead-206, lead-207 ratio vary independently of each other. Rocks which give the same age by all three are considered reliably dated and are used as standards for the other methods. The potassium-argon method offers considerable promise for future work because it has shown good correlation with uranium-lead ages from standard localities and because potassium minerals make up a large proportion of the earth's crust. Thus geochronology is expanding and taking on new versatility.

Another age method, probably of greater popular interest than those thus far discussed, uses carbon-14. This isotope has a relatively short half-life compared to the age of the earth, and is, therefore, not useful in measurements on rocks, or, in fact, on any materials more than about 30,000 years old. However it is important in the dating of many archeological materials and very recent sediments. Because of its short half-life no carbon-14 formed at the creation of the universe, which apparently took place about 5 billion years ago, could now be detected. However cosmic ray bombardment of nitrogen in upper atmosphere produces carbon-14 as one if its products. This radioactive carbon is chemically the same as stable carbon and so is taken into the carbon cycle of the atmosphere, hydrosphere, and biosphere. All carbon in this cycle, such as carbon dioxide in the air, carbonate in near-surface sea water, and carbon in the organic compounds of plants and animals, has, within a per cent or so, the same content of this isotope.

When something is removed from the carbon cycle, such as when a plant or animal dies and is buried or when carbonate precipitates from sea water to form sediments, it ceases to receive carbon-14 into itself and that which it contains slowly disintegrates at the half-life of 5580 years. Thus the carbon-14 content of the carbon of an ancient tree-trunk or sea shell is a direct function of its age, and by measuring this content we may determine its age.

As with the other dating methods, so this one contains many sources of possible error. Contamination is a constant worry and a blunder in interpretation by a geologist or archaeologist can't be overcome by even the most expensive geiger counter. But these sources of error are known and constant care taken to eliminate them.

As we all know, geologic age determinations and even geology itself have been attacked by many of the hyperorthodox as invalid because they fail to uphold one or another interpretation of scripture. Numerous arguments have been brought forward to implement this attack. It is worthwhile, therefore, to discuss some of the possible errors in geochronology. First, however, it should be realized that attacks on the whole body of geology by the so-called "Flood Geologists" and those with similar ideas are pretty futile. Similarly theorizing by some that the laws of nuclear physics aren't really valid are rather hopeless. One might as well tell the engineer that his theories are all wrong and that automobiles can't possibly run or airplanes ever fly. Geology does work and nuclear physics does work. Eighty-five per cent of this country's geologists are employed by the oil industry not because that industry wishes to discredit somebody's interpretation of Genesis One, but because the geologist can tell the oil man where best to look for oil. He can do this because the principles of orthodox geology give right answers when properly applied. Similarly anyone doubting the validity of nuclear physics must consider the successful development of the atom and hydrogen bombs to be sheer luck. Such views seem to me little more than wishful thinking.

However, as with all scientific measurements, those of geochronology are subject to statistical errors and to errors of interpretation. The half-lives of carbon-14, uranium-235, and all other radioactive isotopes used in age determination must be measured. These are determined by nuclear physicists using the best techniques available, but the job is a difficult one and accuracies better than a per cent or two are hard to achieve. Similarly measurements of the radioactivity of a sample by standard counting techniques or of isotopic ratios by mass spectrometry may be accurate only to within a per cent or so of the true value. Similar statistical errors are unavoidable in other steps of the determinations. However these could only change a calculated age by a few per cent at most, and an age of 500 million years is scarcely different than one of 515 million.

More serious errors may be made in the interpretation and application of data. If uranium has been leached out of a mineral by circulating groundwater a uranium-lead age determined from it will be erroneous. Similarly if a proper correction is not made for primary lead deposited in a mineral when it crystallizes, even the most precise measurements made on the specimen will give the wrong answer. Many similar errors are possible, but workers in the field are well aware of them and are continually checking, rechecking, and crosschecking to be sure that they don't fall prey to them.

Another stimulus for the control or elimination of errors is the spirit of competition among the various laboratories engaged in geological age determinations. Each laboratory is only too eager to discover and pounce upon mistakes made by another lab. When something is agreed upon by all you can be quite sure that the evidence is very strong.

It may be safely stated then that the methods of geochronology have been well tested and have been found to work. Certainly they aren't perfect or inerrant but then no one claims that they are. They are a useful tool in the deciphering the history of the earth, and that's all that they are meant to be.

A few sages have chosen to argue against geochronology on the philosophical rather than the physical level, opining that extrapolation of present day physical laws back into the past is invalid, or that God created the earth with a history built into it. Certainly the first view leads logically to the second because of the orderliness of the geologic history discovered when backward extrapolation of today's physical laws is made.

Since this is primarily a philosophical question I won't attempt to discuss it here, but I do think that we must decide whether we believe that God, in letting us study His Creation, is trying to mislead us into unbelief or is showing us the grandeur of His Works. I think, that as Christians, we may only believe that He is showing us His Greatness and Majesty to increase our reverence and love for Him. "Speak to the earth, and it shall teach thee. Who knoweth not in all these, that the hand of Jehovah hath wrought this, in whose hand is the soul of every living thing, and the breath of all mankind."

How much then should the theologian be influenced by geochronological data relavent to his interpretations of scripture? I believe he should give it weight equal to that given to data from his own field of investigation. The exact amount of weight given, of course, depending on the precision of the measurements made and on the validity of the interpretations made from the measurements. Such judgments are not always easy to make of course, but careful study of reports almost always gives adequate information.

# Creation

John C. Sinclair
U.C.L.A. Medical School

Many things being taught our Sunday school pupils as God's Word are really only interpretations. Some of these can never be reconciled with the facts of science. Controversely many things being taught in our Universities as facts are really only generalizations from experimental observations. Some of these can never be reconciled with the Word of God. To my knowledge there is not fact of science or passage of Scripture that are contradictory to one another. The following account of creation is given from the best scientific evidence I know of today. Tomorrow parts of it may be obsolete in the light of new discoveries.

In the beginning God created the heavens and the earth. In a moment with explosive force the matter of the Universe came into existence. According to Gammow, the initial temperature and pressure was so great that the Universe was blown outward like an immense bomb. The force of this explosion was so great that the most distant galaxies visible to us are still traveling away from us at 1/5 the velocity of light (38,000 miles/second). That is when the light we see started on its way toward us, 360 million light years ago. In the first few minutes of its existence the elements of the Universe were formed from created neutrons, protons, and various forms of radiant energy. In this brief moment the present relative abundances of the elements were fixed, considering known transmutations such as the fusion of hydrogen to form helium, our source of solar energy. All matter is composed of atoms of the various elements. Water is composed of atoms of hydrogen and oxygen. Hydrogen is the smallest atom. It contains just one proton, or if it is heavy hydrogen it contains one proton and one neutron. The protons and neutrons are very dense and occupy the central nucleus of the atom. The electrons spin about the nucleus like the planets spin about the sun. Theories which require the continuous creation of matter in order to maintain a static universe, while still appearing to expand, can never be anything more than entertaining mathematical models until some process of neutron and proton creation is known.[5] Electron and positron pairs can unite to produce high energy light rays, and in turn can be formed by such rays; but there is no known way that neutrons and protons can be created or annihilated.

The earth and the whole Universe was a shapeless mass expanding with inconceivable speed. The initial flash of brilliant light had faded and it was dark and turbulent. It is difficult to imagine anything more desolate than an area devastated by an atomic blast. The word used to describe the Universe at this stage is void or a wreck. Isaiah 45:18 tells us that God created the earth not in vain, He created it to be inhabited. The matter out of which the earth will be formed is not inhabitable at this step of creation—but God is not finished! In the thinking of some, each stage of God's creative act must be perfect by *our* standards of perfection. When they read that the earth was shapeless, void and dark they conclude that it must be imperfect and hence something must have happened to destroy it. We would not consider a shapeless mass of pottery clay to be imperfect. A potter dug that clay and purified it not in vain. He intends to shape it into a beautiful piece. The earth of necessity would look desolate until it was clothed with grass and trees, and swarming with animal life.

Gradually the elements began to clump together and the larger clumps began to draw the smaller clumps into their gravitational fields. As the clumps grew to the size of stars the heat of contraction raised the temperature of the interior high enough for nuclear reactions to occur. And God said, "Let there be light", and there was light. The hot stars fueled by nuclear energy began to radiate light. Our sun (our star) was relatively recently formed from interstellar gas and dust. Our earth and the other planets in our solar system were probably forming at the same time our sun was taking shape. This step of creation (the formation of light) has reference primarily, I believe, to our solar system.

Did our sun begin to radiate in a moment of time? Or did God use the natural force of gravity and the natural forces of nuclear reactions to accomplish this result over the years and years of time such a process would require? We should not resent God's use of natural means when the laws of nature are God's laws. The force that parted the Red sea and froze a wall of ice on either side was, we are told, a wind. Let us conclude then, that the most natural means of accomplishing His purposes, is the method of God's choice. We may assume then that God used the forces of nature and the time these forces would require, to shape the Universe and our earth for habitation. Certain alternative explanations have led to impossible interpretations of other Scriptures and the facts of science in order to make them consistent, such as fossils created as fossils, and light created half-way to the earth. These are justified on the basis that God can do anything. But there are things that God cannot do. God cannot act contrary to His own nature. Titus 1:2 plainly tells us that God cannot lie. Fossils created as such and light created half-way to the earth would be a deceptive creation, giving the impression of great age that did not exist. Insisting on fantastic ideas such as this denies the Christian Faith to the scientist.

How then must we interpret the word day and the expression, "the evening and morning was the first day", if the time God took to form our sun into a light bearer was greater than our 24 hour day? The word "day" in Scripture, is used for periods of time longer than one solar day. It is used to describe the day of Salvation which has lasted now for thousands of years, II Cor. 6:2. "Evening and morning" could be a figure of speech, like when we speak of the sun rising and setting, when we know it is the earth that is moving. It could here convey the idea of finality; but I don't know that it means this. Once the sun began to radiate, it would continue to radiate until the creation of the new heavens and the new earth, II Peter 3:13. No sin of Satan could darken it until it had finished it's course. It is held by some that a literal interpretation of "evening and morning" necessitates a literal, 24 hour, interpretation for day. This is true. But if day is used figuratively then of course "evening and morning" will also be figurative, otherwise long periods of continual dark would be indicated, which we know could not have occurred.

If this Genesis account is a recreation account some explanation of the necessity of recreating the sun as a light emitting body must be found. Is it imagined that the sun was darkened in some way so that it could not give its light? Any process such as an explosion of the sun, which could darken it would also involve and destroy the earth as we know it, and our solar system would cease to exist. Bringing order out of such a chaos would look like an original creation because no remnant of a former order would survive it. The strata require a radiant sun for their formation. The principle strata are all the result of soil erosion by water; this water is drawn up into clouds from bodies of water by the sun, and dropped as rain. Darkening the sun is obviously inconsistent with the reconstruction theory for it provides no explanation of the strata and its fossils.

God separated the light from the darkness. The earth began to rotate on its axis, and to circle about the sun in a plane at an angle to its rotation to give us our seasons. Here also we could raise the question, was the rotation of the earth on its axis reinstituted about 6-12,000 years ago after a long period of absence? By every physical criteria, the radiant energy of the sun and the rotation of the earth on its axis have continued much as they are today since their first appearance.

The evidence against a recreation seems pretty strong, but is there any evidence for it? 1. The use of the word, *re*plenish, in Genesis 1:28. It is thought that this means that Adam is to *re*inhabit the world once inhabited by a pre-adamic race of men. The hebrew word *mal'a* simply means fill, or sometimes complete.[3]

2. The description of the, "at that time", future desolation of Palestine and Jerusalem, Jer. 4:16-29. Vs. 23-26, "I looked on the earth, and lo, it was waste and void; and to the heavens, and they had no light. I looked on the mountains, and lo, they were quaking, and all the hills moved to and fro. I looked, and lo, there was no man, and all the birds of the air had fled. I looked, and lo, the fruitful land was a desert, and all its cities were laid in ruins before the Lord, before his fierce anger". The wording here sounds a little like Gen. 1:2. Some Christians have interpreted this as a literal reference to Genesis, and have used the expression, there was no man, to infer the destruction of a preadamic race, which they believe explains the ancient human fossils, such as the Java man. Paleontology indicates modern man is a descendant of fossil man. The cultural objects, such as tools, pottery, fire, art and evidence of religion, proves fossil men were truly men in our sense of the word, they were men as we are, not super animals. Only the determmination to find proof for a recreation could lead a person to twist this description of a future desolation, to a literal reference to Genesis.

3. The possibility of using "became" in Gen. 1:2. It is possible to translate the verb "to be" as became, but this is not the usual meaning and nothing in the context suggests it. If it were to mean became, linguistic usage would require that the sentence begin with the verb followed by the preposition "lamedh"[3] The Septuigant uses *was*. After a brief statement that God created the record goes on to describe *how* God shaped this creation for man's habitation.

4. Why the seventh day should be hallowed when it corresponds to a seventh era. The seventh day was given as a day of rest to commemorate God's rest from all the things He created to make.[3] One day in the year commemorates the deliverance of Israel from Egypt, though the acutal time for all the plagues may have been more than a year. One day with the Lord is as a thousand years and a thousand years as one day. Setting one 24-hour day aside to commemorate the completion of God's creation, has nothing to do with the length of time it took.

I have tried to answer objections to the day-era concept before bringing plant and animal life on the scene, in order to avoid any confusion that it necessitates belief in a theory of Evolution (which has successive forms of all plants and animals coming into being by the natural selection of mutations). for it certainly does not!

How hot the original earth was we do not know, nor whether it is now cooling or heating up from radioactive energy in its interior. If it were originally quite hot the water would be in dense clouds of steam. As it cooled, water would condense into lakes and seas. "And God divided the waters which were under the firmament from the waters which were above the firmament."

The contraction of the earth and the buckling of its crust cannot account for the size of our continental masses and the height of our mountain ranges, nor can it explain how they are maintained when one would expect them to be eroded level. Heat convection currents in the mantel of the earth could explain our mountains. The mantel is composed of molten, viscous rock. The hotter interior expands it and causes it to flow toward the surface in periodic, rapid overturns. These could thrust the overlying crust up into mountain ranges. Our deep seas could be dug as the cooled rock flows back into the interior. And God said, "Let the waters under the heaven be gathered together unto one place, and let the dry land appear."

Now that land masses have been formed, the preparation of soil can begin. The water vapor and rains and ground water can chemically alter the rocks, carry them down and deposit them in gental slopes and valleys, ready for tiny roots of plants. And God said, "Let the earth bring forth grass." This grass is the first grass that ever grew. Gen. 2:5 says this is the record of how God created and made every plant of the field (fossil or otherwise), and every herb of the field before it grew. This event would probably correspond with the Cambrian strata of our earth. A person standing on an ancient Cambrian sea-shore would find it very similar to a modern sea-shore. It would abound with all forms of invertebrate animal life and the plants; algae, etc., that are in balance with them. The first appearance of life was rich and varied. And God said, "Let the waters bring forth abundantly". Why God waited until the fourth day of creation to designate the stars for signs and for seasons, I don't know. That some stars exist before our Solar system was formed we are reasonably sure.

There are local discontinuities in the fossil record and longrange cycles and trends, yet it tells but one unbroken account of life on this planet. There is no universal, simultaneous break in its continuity such as a recreation would be. Fern forests, dinosaurs and birds with teeth, blend imperceptibly with modern forms of life. As the fossil record is traced upward, vertebrate life makes its appearance, and the ancient seas begin to swarm with fish. Then mammals appear in terrestrial deposits, and finally man himself appears. And God said, "Let us make man in our image". In many ways man is similar to the animals and plants. We would expect God to use the same building blocks in us as in them to perform the same function.[2] Our similarity with them is certainly evidence of a common origin. For in God we both live and move and have our being. For His pleasure we are and were created.

This account is not intended to be an answer for everything, but a reasonable interpretation that fits the facts as I see them. Genesis, chapter one, is here considered to cover several billion years. Successive creative acts are separated by indeterminate periods of time, resulting in a sequence of events which seems natural to the scientist, but is the handiwork of our Creator. Widely held alternative theories have been found to be unscriptural and inconsistent with the facts. This period of time may seem long to us, but I am sure that it is not long to God. The Husbandman of the earth hath long patience over the precious fruit of the earth.

I have two purposes in presenting this paper:

1. To show how beautifully Genesis can fit what I consider good science.
2. To show how untenable the reconstruction theory is.

I do not attempt to prove the goodness of the science I use, but I give references to my source material, for those interested in this phase of it.

*Acknowledgement*: This paper is the result of a course of study made possible by the Los Angeles area A.S.A. The book covering the course material is "Modern Physics for the Engineer", Edited by, Louis Ridenour, Published by McGraw-Hill Book Co. 1954.

**BIBLIOGRAPHY**

1. D. Eggenberger, *J.A.S.A.*, *2-3*, 23(Sept. 1950).
2. W. Lammerts & J. Sinclair, *J.A.S.A. 5-3*, 7(Sept. 1953).
3. E. Harrison, Personal Comunication.
4. P. Stoner, *J.A.S.A.*, *6-3*,9(Sept. '54)
5. F. Hoyle, *The Nature of the Universe*, Harper Bros., N. Y. 1950

# *The Formation Of Living Organisms From Non-Living Systems**

WALTER R. HEARN

Assistant Professor of Chemistry, Iowa State College

In 1859, Charles Darwin cautiously stated, "Science as yet throws no light on the . . . problem of the essence or origin of life."[1] Actually, the possibility of observing the "spontaneous generation" of living things from inorganic matter was at that time a very controversial scientific question, as Pasteur had not yet completed his conclusive experiments along these lines[2]. However, instead of praising Darwin for this reluctance to extend his theory too far, critics of Darwin and of the theory of natural selection have often attacked at this very point, saying that the power of the theory to account for biological variations is vitiated by its failure to account for biological origins. Thus Clark, in 1950, stated that the influence of the theory of natural selection probably did biology more harm than good, by concentrating attention on small changes in living things and diverting it from the problem of their origin. He went on to say:

> "For the materialist, organisms are limited by the properties of atoms and molecules. Somehow or other, he must explain how chemical molecules of gigantic complexity came into existence and have been able to arrange themselves in increasingly complicated ways. This is the fundamental problem of evolution, yet it is generally ignored in modern books on the subject, nor, (at least to the author's knowledge) has the problem ever been fairly faced ."[3]

Oparin's book on the origin of life[4] was published in English translation in 1938 and certainly faced the problem fairly. The current scientific literature abounds with papers facing the problem, and today there is no reluctance at all to extend Darwin's principle of natural selection into the pre-biological realm, making the origin of living organisms merely a consequence of a still more fundamental "chemical evolution" of non-living molecules. The tendency today is to regard the original spontaneous development of living things from inanimate matter as not only highly probable, but actually inevitable[5]. What are the factors which have led to this recent extension of the evolutionary concept?

In the first place, the time available for a process of chemical evolution is now generally accepted to be about two billion years[6], which is regarded as adequate if the process is considered to be the over-all result of a large number of continuing small changes. Since Oparin[4], this gradualism has been a predominant feature of the theory, and has largely served to discredit arguments such as that of Lecomte du Nouy[7], who calculated that the formation of a single molecule of protein would require $10^{243}$ billions of years, if the probability of the chance collision of its constituent atoms were the only factor taken into consideration. If a series of stepwise reactions is postulated rather than a single all-or-none jump from simple molecules to complex ones, the probabilities dealt with are entirely different, and such calculations become meaningless.

A second factor, I think, has been an increased appreciation for the stability of biologically important molecules. Two previous papers in this symposium have described experiments in which mixtures of very simple compounds have been submitted to conditions simulating pre-biological conditions on the earth, and have produced more complex molecules identical to those common to living organisms today[8,9]. Undoubtedly such experiments demonstrate the intrinsic stability of amino acids and other such compounds and make their prebiological accumulation on the earth's surface highly probable. Another line of investigation is that of Abelson and others in the field of paleobiochemistry[10]. It has been found that recoverable amounts of the original organic matter are present in such fossil materials as shells, bones, petroleum, and coal, and possibly even in the Pre-Cambrian black shale. For example, it was possible to study the content of protein and amino acids in the shells of clams (*Mercenaria mercenaria*, the quahog of the Atlantic coast) now living and in those of the same species 1,000, 500,000, and 25 million years old. The 1,000 year old specimen had the same protein content and the same amino acid pattern as the modern specimen; the 500,000 year old specimen contained no protein and smaller amounts of amino acids, some of which were still in peptide chains; and the 25 million year old shell contained only a few of the original amino acids and no protein or even peptides. Laboratory experiments on the stability of alanine at various temperatures indicated that this amino acid would be sufficiently stable at a temperature of 20° C for half of the original amount present to remain after a period of about three billion years. Substantial quantities

* Paper presented at the Twelfth Annual Convention of the American Scientific Affiliation, Beverly Farms, Massachusetts, August, 1957.

of porphyrins, compounds of the same type as the chlorophylls or the heme of important respiratory pigments, have been found in petroleum samples at least tens of millions of years old. Such studies not only open up the exciting possibility of discovering something about the metabolism of early forms of life, but also emphasize in a dramatic way the inherent stability of these biologically important compounds. If their pre-biological synthesis was possible, as I think has been more or less clearly demonstrated, then they *must* have accumulated and set the stage for further developments.

Perhaps it should be pointed out here that our notions of the instability of large complex molecules rest partly on theoretical grounds and partly on the grounds of common experience. If chemical bonds have a certain probability of being broken, it stands to reason that under the same conditions a molecule with a large number of bonds will more likely be broken down to smaller molecules than will a molecule which has fewer susceptible bonds. However, organic chemists know that chemical bonds in a molecule may affect each other in different ways: in some cases a bond is "labilized" by the presence of other bonds (that is, made more susceptible to attack), but it is also possible for bonds to be "stabilized" by the influence of other bonds (for example, as by resonance in the case of the porphyrins already mentioned, or by the formation of zwitterionic dipoles in the case of the amino acids). In the case of very large molecules, such as the proteins and nucleic acids, steric factors as well come into the picture. Thus, with a molecule of sufficient size, forces such as those due to van der Waals attraction or hydrogen bonding may become *intra*-molecular, and their combined strength may impart considerable stability to a particular configuration of the molecule. Pauling and others have demonstrated this in the case of helical structures for certain proteins[11], and the Watson-Crick double helical structure for DNA (desoxyribose nucleic acid) is an even more striking example[12].

It is not so much these theoretical considerations, therefore, as it is our practical experience which leads us to believe intuitively in the instability of complex molecules. The phenomena of decay and decomposition are such commonplace experiences with us that I think we often fail to realize how much of this is actually biological. The earth's surface and atmosphere today is literally teeming with living organisms whose existence depends on their ability to degrade complex molecules and derive energy thereby. In Pasteur's day, people found it almost incredible that the air they breathed should be swarming with microorganisms and their spores[2], and sometimes even today we forget that we live in such a biologically crowded atmosphere. If we lived under circumstances in which food did *not* spoil at ordinary temperatures and mold did *not* grow on our protein solutions in the laboratory, we might not have such a strong feeling that complex molecules are intrinsically unstable. But, if the theory of chemical evolution is correct, complex organic compounds *were* synthesized and *did* accumulate in just such a "sterile" or "a-biological" atmosphere.

Now, the third factor which has led to the extension of the evolutionary concept into the pre-biological realm, and the one which I wish to emphasize in this paper, is our rapidly increasing understanding of biological processes at the molecular level. By "understanding" of such proccess I mean our ability to make precise enough measurements to lead us to mathematical generalizations, which may be tested for validity by comparison with quantitative data from other experiments. I do not mean to imply that we now fully "understand" life, or that we ever shall; however, it is true that whenever the complex inter-relationships of metabolism have been susceptible to quantitative investigation, the ordinary laws of thermodynamics and kinetics have proven adequate to account for the phenomena observed. For a profound discussion of the relationship between the laws of physics and biological phenomena. I recommend Schrodinger's essay, *What is Life?*[13], or for a slightly whimsical account, Gamow's *Mr. Tompkins Learns the Facts of Life*[14].

In general terms, what is it that we are seeking to understand about life in terms of the laws of chemistry and physics? It may seem strange that it is actually quite difficult even to define "life" or "living" satisfactorily. This difficulty has been emphasized by Pirie[15], who pointed out that any definition of life which would exclude all obviously non-living systems would probably also exclude some obviously living systems, and vice-versa. Pirie has set up what he considers generally acceptable criteria for a system to be regarded as "living" as a sort of ground-rule in lieu of a precise definition, in case someone claims to have produced such a system artificially [16]. If we cannot say precisely what a living thing *is*, we can at least specify some of the things it *does*. We might say in the most general terms that living systems are systems which are able to utilize a flow of energy through the system to maintain a higher degree of molecular order than their surroundings (or in thermodynamic terms, to "extract" negative entropy from the energy sources utilized by the system). In addition, although a system without the ability to reproduce may be regarded as living (a castrated animal, for example), at least the *possibility* of self-duplication must be present under certain conditions for a system to have much significance in evolutionary considerations. We should also keep in mind that if the Darwinian idea of variations plus natural selection is applicable to the pre-biological realm, variant systems of metabolism may have arisen and become "extinct" just as variant morphological forms seem to have done in the biological realm. In fact, the possibility exists of discovering such variant systems, possibly even as "vestigial remains" in modern organisms.

The standard procedure for studying any complex system is to resolve it into simpler components which can be studied separately, or perhaps to construct a model of similar but less complicated design and to study the model first before attempting the more difficult study of the system itself. Thus, we might resolve living things into systems for obtaining energy and systems for utilizing this energy (or the entropy deficit accompanying it) for synthetic purposes; in addition we might investigate the way these two systems are "coupled," and ways in which the entire complex is held together or reduplicated. In practice this is what biochemistry has been doing in recent years, with varying degrees of success in elucidating the details of the different systems. It must be admitted that we do not as yet have a complete picture of any of the isolated systems, but we do have a partial picture of each of them. What is most significant here is the fact that it has been possible to study such isolated systems in this way at all. Now, let me sketch in a broad outline of those metabolic processes which may be said to have had the greatest "survival value," and then we shall look at some possible ways in which they might have developed from less complex processes.

There seem to be three major processes by which living things obtain energy. These are oxidation, fermentation, and photosynthesis. Actually, if we limit ourselves for the moment to energy-yielding processes, the latter should really be called "photodecomposition," since the synthetic part of the process actually utilizes the energy thus obtained.

Oxidative metabolism makes use of molecular oxygen to give final products of carbon dioxide and water in the usual scheme in which organic compounds are oxidized. In a series of steps, hydrogen atoms are removed from the substrate, with or without a preliminary addition of a molecule of water, and eventually made to react with oxygen. The carbon dioxide is liberated when the substrate has been rearranged to form a carboxylic acid with unsaturation on the beta-carbon atom, a type of compound which readily decarboxylates. To oxidize a two-carbon compound such as acetic acid, the compound is first condensed to a "carrier" molecule in order to have enough carbon atoms to form the beta-keto acid analog; in the course of the subsequent oxidations and decarboxylations, the carrier molecule is regenerated, giving a cyclic process. The energy-yielding part of such a cycle, the Krebs citric acid for example, is the oxidation of the hydrogen to water. In all cases of oxidative metabolism investigated, this is a stepwise process in which the hydrogen atom, or an electron accompanied by a hydrogen ion, is transferred from one oxidation-reduction system to another, each of higher oxidation potential, until finally two hydrogens combine with an atom of oxygen. There are some strictly anaerobic microorganisms which apparently do not make use of this process at all, and there are also autotrophic chemosynthetic bacteria which oxidize sulfur, iron, or hydrogen principally rather than carbon compounds[17].

Fermentation, an anaerobic process, consists basically of the oxidation of one compound by reducing another compound other than oxygen, so that the presence of oxygen is not required. In some cases, this oxidation-reduction may be intra-molecular, as in the dismutation of glucose to lactic acid in the familiar Embden-Meyerhof scheme. In this case, which also involves a cleavage into two smaller molecules, it might be said that one end of the molecule has been oxidized (the carboxyl group of the lactic acid) while the other end has been reduced (the methyl group). Anaerobic processes are often found in tissues whose primary mechanism for obtaining energy is oxidative, such as mammalian muscle.

In photosynthesis, the energy involved is derived from sunlight, which is absorbed by certain pigments and used to break down water to hydrogen and oxygen, the hydrogen then being used to reduce carbon dioxide to yield compounds of the state of oxidation of carbohydrates. The oxygen may be released as a gas, as in higher plants, or may oxidize other compounds, as in the photosynthetic bacteria[18]. It is interesting that the green plants apparently make use of all three types of metabolic scheme, synthesizing carbohydrate by photosynthesis, rearranging it by anaerobic glycolysis, and then oxidizing the glycolysis product (lactic or pyruvic acid) to carbon dioxide and water.

Of synthetic pathways in various types of organisms we now have a considerable body of knowledge, and a general similarity of these pathways is evident even in the most widely differing species. We can trace the chemical steps involved in the conversion of simple one-, two-, and three-carbon compounds into porphyrins, steroids, amino acids, carbohydrates, nucleotides, etc. The chief interest at the moment is in the biosynthesis of the most complex and specific compounds of all, the proteins and nucleic acids. Indications are that the mechanisms will probably be similar to those involved in the synthesis of the simpler compounds, but the way in which order or "information" is built into the molecules is a complicating factor. This "coding" of biological macromolecules has become a quantitative study itself[19].

Of the mechanisms involved in coupling these synthetic processes to the exergonic ones, we have varying degrees of knowledge. We have come to realize that the only possible way such coupling can take place is through the raising of a specific chemical compound to a higher energy state by the rearranging of the electronic configuration of its constituent atoms.

In some cases we know what the compound is, and how the rearrangement takes place. In many cases there is an intermediate compound which can serve as an energy "carrier" because it is thermodynamically unstable but kinetically stable; such a compound transfers its energy only when a suitable catalyst is present to speed up the reaction. Perhaps the most common example is that of ATP (adenosine triphosphate), which has at one end of the molecule two pyrophosphate groups which show this property of thermodynamic instability (*i.e.*, "high energy") but which are not hydrolyzed at an appreciable rate at the ordinary temperatures and pH's of living cells and tissues. When a suitable catalyst is present, ATP is rapidly hydrolyzed, and we have a rather definite picture of how the catalyst accomplishes this[20]. ATP seems to play a role in all types of metabolic processes, and in some cases we know exactly how it works. For example, in the anaerobic fermentation of glucose we know exactly how the energy is used to synthesize ATP, and exactly how many molecules are synthesized. In the case of oxidative metabolism, we know that ATP is synthesized, and we know approximately how many molecules must be synthesized for every hydrogen atom that eventually combines with oxygen, but we do not know in this case the actual chemical details of the synthesis[21].

As for mechanisms by which metabolic interrelationships are preserved and duplicated in reproduction, it must be said that we have probably just discovered the most important clues but have not yet had time to solve the mystery. The identification of the molecule that actually carries the hereditary pattern as DNA[22], and the elucidation of the major features of its structure[12] have certainly given us hope for the first time of understanding heredity at the molecular level. It is interesting to note that DNA is made up entirely of molecules of the same general type as ATP minus the high-energy pyrophosphate, and that the structural features of the molecule already account beautifully for many of the properties of the hereditary carrier, such as stability in thread-like form and ease of exact duplication.

Now, extrapolating backwards, what can we say about the sequence of events leading to the accumulation of these systems into what we know as living things? In the first place, if the original atmosphere was really a reducing one rather than an oxidizing one, it is highly probable that oxidative metabolism was *not* the original exergonic process utilized. A number of authors have regarded anaerobic fermentation as the most likely original pathway; Kirkwood points out that it may be regarded as possibly more "primitive" than oxidation or photosynthesis because the anaerobic enzymes are usually soluble, while the others are usually intimately connected with cell structure[23]. If the atmosphere gradually became more oxidizing, due to the photolysis of atmospheric water and the loss of the lighter hydrogen resulting, then oxidative metabolism may have developed. Oxidation is capable of yielding much more energy than fermentation; with the large quantity of organic compounds available from pre-biological synthesis, this would have been a very advantageous scheme—until, of course, the supply of organic compounds ran out! The oxidation of iron and sulfur may have been resorted to, but such schemes would also be heading up a blind alley. There may have occurred crisis after crisis, but eventually the atmosphere would have become too oxidizing to replenish the supply of reduced molecules; some other scheme must have developed, in this case, photosynthesis. The energy of sunlight could then be used to reduce carbon dioxide resulting from oxidative metabolism, and the familiar carbon cycle of nature as we know it today was underway.

Is it reasonable that metabolizing systems should arise spontaneously and persist? The systems we know today are much too complicated to have appeared suddenly, but there is yet the possibility that the system might have evolved by the continued modification of only slightly advantageous chemical reactions. Several suggestions have been made along these lines. In a general sense, any chemical reaction which produced a compound with even slight catalytic power to stimulate the reaction by which it was produced would have "survival value" over competing reactions. Calvin[6] has pointed out that iron-porphyrins are catalysts which will promote certain steps in the synthesis of porphyrins from succinic acid and glycine, two compounds already known to be synthesized under model "pre-biological" conditions; furthermore, since iron itself has very slight catalytic power to do the same thing, we have a good example of a synthetic reaction which would be expected to "evolve" from very simple precursors. It might be mentioned that the further combination of the iron-porphyrin compound with a specific protein increases its catalytic power many thousand-fold, but this elaborate refinement probably came much later in the development of this system.

It has also been pointed out that inorganic pyrophosphate itself has the same properties of thermodynamic instability plus kinetic stability as ATP, which may be regarded merely as pyrophosphate with an organic "handle" on it. Therefore, pyrophosphate could have served as one of the first molecules capable of transferring energy from exergonic reactions to endergonic synthetic reactions. Kirkwood[23] suggests that pyrophosphate probably accumulated from thermal reactions along with the pre-biological supply of organic compounds, and Calvin[6] has even suggested a very rudimentary coupling reaction for the synthesis of pyrophosphate from orthophosphate, linking the dehydration of orthophosphate to the squeezing out of a molecule of water from the coordination sphere of iron or vanadium ions when they are oxidized. It is thus possible to devise fairly reasonable models for "primitive" metabolic schemes, a stimulating hobby of growing popularity among biochemists.

We do know from recent studies on enzymes and co-enzymes that simple molecules may often have enough catalytic power without combination with a specific protein to make such primitive metabolizing systems seem reasonable. Some attempts have already been made to design a "mathematical model" of the evolution of primitive metabolizing systems by making various assumptions about possible effects the products might have on the catalysis of their synthesis[24]. For example, a compound would not have to serve as a specific catalyst in order to promote its own synthesis over that of other compounds: if it were a macromolecule it might serve just as well by adsorbing the reacting molecules or even by forming a prototype of a cell wall to hold the system together. It is probably too soon to assess the validity of the various mechanisms proposed[5,23,25,26,27] but the general principle does seem valid in the light of our present knowledge of biochemistry[28]. The difficulty of obtaining conclusive data will no doubt produce a great deal of speculation, such as the argument over whether the assembling of metabolizing systems most probably began by adsorption on inorganic clays, as proposed by Bernal[29], by aggregation in quiet waters of ocean depths, as proposed by Pringle[30], or in water-logged soil or layers adsorbed at air-water interfaces, as proposed by Haldane[31]. However, the concept of pre-biological evolution results not only in speculation but also in the stimulation of specific lines of research[32], such as the investigation of "primitive" types of organisms, of variant pathways of metabolism, and of model enzyme systems.

Continuing biochemical research on "sub-vital" systems may be expected to lead to further explanation of biological mechanisms at the molecular level. Further research on cell nuclei[33,34], on crystalline DNA which "transforms" bacteria to other genetic types[35,36], on bacteriophage which inject DNA into bacteria and bring about replication of the phage[37,38], on TMV (tobacco mosaic virus) protein and RNA (ribose nucleic acid) which recombine to form an infective virus particle [39,40], on the "electron transport particle" of mitochondria[41,42], on chloroplasts which photosynthesize[43,44], on protoplasts which synthesize induced enzymes[45], and so on, will probably intensify the conviction that the living cell is an already highly developed system, a milestone perhaps, but not the starting point, of evolutionary processes.

## Bibliography


1 C. Darwin, *The Origin of Species by Means of Natural Selection*, with additions and corrections from sixth and last English edition. D. Appleton & Co., New York, 1920, p. 294.

2 J. B. Conant, Ed., *Pasteur's and Tyndall's Study of Spontaneous Generation. Case 7. Harvard Case Histories in Experimental Science*. Harvard University Press, Cambridge, Mass., 1953.

3 R. E. D. Clark, *Darwin, Before and After*. The Paternoster Press, London, 1950, pp. 126-127.

4 A. I. Oparin, *The Origin of Life*, translation with annotations by S. Morgulis. McMillan Co., New York, 1938. 2nd Edition, Dover Publications, New York, 1953.

5 G. Wald, The origin of life. *Scientific American*, 191, No. 2, 44 (Aug., 1954); see also letters to the editor in subsequent issues.

6 M. Calvin, Chemical evolution and the origin of life. *American Scientist*, 44, 248 (1956).

7 P. Lecomte du Nouy, *Human Destiny*. Longmans, Green & Co., New York, 1947. Signet Books, New American Library, New York, 1949, pp. 30-39.

8 S. L. Miller, Production of some organic compounds under possible primitive earth conditions. *J. Am. Chem. Soc.*, 77, 2351 (1955); for references to other such syntheses, see S. W. Fox, Origin of protein, in "Letters", *Chem. Eng. News* 35, No. 25, x (June 24, 1957).

9 S. L. Miller, The mechanism of synthesis of amino acids by electric discharges. *Biochim. Biophys. Acta*, 23, 480 (1957).

10 P. H. Abelson, Paleobiochemistry. *Scientific American*, 195, No. 1, 83 (July, 1956).

11 L. Pauling, R. B. Corey and R. Hayward, The structure of protein molecules. *Scientific American*, 191, No. 1, 51 (July, 1954)

12 F. H. C. Crick, The structure of the hereditary material, *Scientific American*, 191, No. 4, 54 (Oct., 1954).

13 E. Schrodinger, *What is Life?* Cambridge University Press, Cambridge, 1944.

14 G. Gamow, *Mr. Tompkins Learns the Facts of Life*. Cambridge University Press, Cambridge, 1953, especially pp. 69-79.

15 N. W. Pirie, The meaninglessness of the terms life and living. In J. Needham and D. E. Green, *Perspectives in Biochemistry*. Cambridge University Press, Cambridge, 1937, pp. 11-22.

16 N. W. Pirie, on making and recognizing life. *New Biology*, No. 16. Penguin Books, Ltd., London, 1954, pp. 41-53.

17 J. R. Porter, *Bacterial Chemistry and Physiology*. John Wiley & Sons, New York, 1946, pp. 635-661.

18 C. B. van Niel, The comparative biochemistry of photosynthesis. In J. Franck and W. E. Loomis, *Photosynthesis in Plants*. Iowa State College Press, Ames, 1949, pp. 437-495.

19 G. Gamow, Information transfer in the living cell. *Scientific American*, 193, No. 4, 70 (Oct., 1955).

20 B. Axelrod, Enzymatic phosphate transfer. *Advances in Enzymol.*, 17, 159 (1956).

21 B. Chance and G. R. Williams. The respiratory chain and oxidative phosphorylation. *Advances in Enzymol.*, 17, 65 (1956).

22 R. L. Sinsheimer, First steps toward a genetic chemistry. *Science*, 125, 1123 (1957).

23 S. Kirkwood, The origin of life. *Chemistry in Canada*, 8, No. 2, 25 (Feb., 1956).

24 G. Allen, Reflexive catalysis, a possible mechanism of molecular duplication in prebiological evolution. American Naturalist. 91, 65 (1957).

25 A. Gulick, Phosphorus as a factor in the origin of life. *American Scientist*, 43, 479 (1955).

26 H. F. Blum, Perspectives in evolution. *American Scientist*, 43, 595 (1955).

27 S. W. Fox. Evolution of protein molecules and thermal synthesis of biochemical substances. *American Scientist*, 44, 347 (1956).

28 M. Ycas, A note on the origin of life. *Proc. Natl. Acad. Sci. U. S.*, 41, 714 (1955).

29 J. D. Bernal, The origin of life. *New Biology*, No. 16, Penguin Books, Ltd., London, 1954, pp. 28-40.

30 J. W. S. Pringle, The evolution of living matter. *New Biology*, No. 16. Penguin Books, Ltd., London. 1954, pp. 54-67.

31 J. B. S. Haldane, The origins of life. *New Biology*, No. 16. Penguin Books Ltd., London, 1954. pp. 12-27.

32 W. P. Woodring, Conference on biochemistry, paleo-

ecology, and evolution. *Proc. Natl. Acad. Sci. U. S.*, 40, 219 (1954).

33 V. G. Allfrey, A. E. Mirsky and H. Stern, The chemistry of the cell nucleus. *Advances in Enzymol.*, 16, 411 (1955).

34 N. H. Horowitz, The gene. *Scientific American*, 195, No. 4, 78 (Oct., 1956).

35 R. D. Hotchkiss and J. Marmur, Double marker transformations as evidence of linked factors in desoxyribonucleate transforming agents. *Proc. Natl. Acad. Sci. U.S.*, 40, 55 (1954).

36 R. D. Hotchkiss and E. Weiss, Transformed bacteria. *Scientific American*, 195, No. 5, 48 (Nov., 1956).

37 A. D. Hershey, Chemistry and viral growth. In D. E. Green, *Currents in Biochemical Research*, 1956. Interscience Publishers, Inc., New York, 1956, pp. 1-28.

38 A. Lwoff, The life cycle of a virus. *Scientific American*, 190, No. 3, 34 (Mar., 1954).

39 H. Fraenkel-Conrat and R. C. Williams, Reconstitution of active tobacco mosaic virus from its inactive protein and nucleic acid components. *Proc. Natl. Acad. Sci. U. S.*, 41, 690 (1955).

40 H. Fraenkel-Conrat, Rebuilding a virus. *Scientific American*, 194, No. 6, 42 (June, 1956).

41 B. Mackler and N. Penn, Studies on the electron transport system. IX. Fragmentation of DPNH oxidase. *Biochim. Biophys. Acta*, 24, 294 (1957).

42 P. Siekevitz, Powerhouse of the cell. *Scientific American*, 197, No. 1, 131 (July, 1957).

43 J. A. Bassham and M. Calvin, Photosynthesis. In D. E. Green, *Currents in Biochemical Research*, 1956. Interscience Publishers, Inc., New York, 1956, pp. 29-69.

44 E. I. Rabinowitch, Progress in photosynthesis. *Scientific American*, 189, No. 5, 80 (Nov., 1953).

45 O. E. Landman and S. Spiegelman, Enzyme formation in protoplasts of *Bacillus megaterium*. *Proc. Natl. Acad. Sci. U. S.*, 41, 698 (1955).

# *An Evaluation Of The Fossil Record**

RUSSELL L. MIXTER

If you have ever sat in the shade of western yellow pines and tried to write about fossils you can feel how I felt when I wrote this. A vireo is singing his monotonous phrases, a flycatcher has failed to find the insects buzzing around me and a kingfisher rattles cross-lots on his way to the creek. All I see is alive except the pre-Cambrian rocks on which I sit. They contain no fossil animals, or at least none that everybody agrees is a real fossil, although I read that "there is no serious dispute that algæ do occur from at least the middle part of the long pre-Cambrian span onward."[1]

About a half mile from here is a cliff of redish brown standstone of Cambrian age. In this same Cambrian in various parts of the world are found fossils of all the major groups of animals, except backboned ones. There are single celled animals, brachiopods (which resemble clams which exchanged their side shells for top and bottom ones), starfish-like creatures, mollusks, and the jointed-leggers, of which the trilobites are an outstanding example. Do not think any of these appeared suddenly for the Cambrian period lasted a long time and these groups, which we call Phyla, came in at various times over some millions of years. It is striking that the jointed legged ones (called arthopods) are found in the same stratum as the single celled ones (which are called protozoa).

You readily notice that we have an abundance of fossils when we get any at all. We know nothing from the record of the rocks about the supposed progression from the first living thing up to the protozoan and this change "was probably the most complex that has occurred in evolution, and it may well have taken as long as the change from protozoan to man."[2] But we have in the fossil record much information about what has been here after the first members of the major groups, or phyla, were fossilized.

Because you wish more detail than I can give you in this paper, you will read, "The Meaning of Evolution" by George Gaylord Simpson, who was chairman of the department of paleontology at the American Museum of Natural History and professor of vertebrate paleontology at Columbia University and is now at the Harvard Museum. He is a frequent contributor to the popular and technical literature dealing with the significance of fossils and he is obviously honest, for if there is any doubt about what he is saying, he will suggest the qualification of his belief on the spot. For example, he wrote, "I am trying to pursue a science that is beginning to have a good many practitioners but that has no name: the science of four-dimensional biology or of time and life. Fossils are pertinent to this field when they are treated as historical records (paleontologists do not always treat them so.")[3] Another example:

> Some idea of probable rates of evolution in gaps in the record that correspond with major changes in adaptation can be gained by comparing the possible lengths of the gaps with the length of the relatively continuous record in the sixteen orders of mammals for which there is fair knowledge. These estimates are highly unreliable, because a large element of personal opinion is involved and no fully objective method of measurement exists as yet. The averages and consistant tendencies... give as good an indication of probabilities as can now be achieved.[4]

You realize that he knows his own limitations as well as those of his co-workers.

As you read some of Simpson's books you will be impressed with two ideas. One is how much we do know about fossils and the other is how much we really do not know after all. Consider what seem to be fairly certain conclusions from paleontological observations.

First, species can change into other species. I quote Simpson and he can list for you many examples. "Among the examples are many in which, beyond the slightest doubt, a species or a genus has been gradually transformed into another."[5] Also he writes, "Splitting and subsequent gradual divergence of species is also exemplified..."[6] Some of you may still doubt this. If you wade through one of his books you will at least be less sure that your doubt is justified.

Second, the members of families and orders had their origin in a common ancestor. Consider the record of the horse along with the ass and the zebra. It belongs to the family called Equidae. It is a reasonable belief that all these sprang from the same ancestral stock. If you do not accept this belief, you are holding to one which is less simple, that is, one which would demand more specific creations, and you would be saying that hereditary transformations cannot produce the changes required. Let us admit that no one actually knows what the hereditary sequence in the line-

*This paper was first presented at the 1957 Convention of the American Scientific Affiliation but publication was delayed since the program chairman was holding it in an attempt to obtain a cross-section of critiques to accompany it. Not realizing this objective in full, it is felt desirable to publish it now to coincide with the A.S.A. Centennial Volume on Darwin.

age of horses really was. We are dealing, not with precise facts, but with greater or lesser probabilities. Nevertheless, we should be willing to admit the possibility that the ancient horse (Eohippus) was the remote grandparent of the modern horse (Equus). We should admit it because there are *so many* stages and *such gradual* stages between Eohippus and Equus and the best explanation for slightly different stages is that they are related to one another by descent.

As more specimens accumulate, and more missing links are found, it may be more and more evident that whole families do arise from species.

Here is Simpson's summary of the second point. "Gradual transformation is also fairly well exemplified for subfamilies and occassionally for families, as the groups are commonly ranked."[7]

Third, as Simpson states,

> In spite of these examples, it remains true, as every paleontologist knows, that most new species, genera, and families and that nearly all new categories above the level of families appear in the record suddenly and are not led up to by known, gradual, completely continuous transitional sequences. When paleontological collecting was still in its infancy and no clear examples of transitional origin had been found, most paleontologists were anti-evolutionists. Darwin (1859) recognized the fact that paleontology then seemed to provide evidence against rather than for evolution in general or the gradual origin of taxonomic categories in particular. Now we do have many examples of transitional sequences. Almost all paleontologists recognize that the discovery of a complete transition is in any case unlikely. *Most* of them find it logical, if not scientifically required, to assume that the sudden appearance of new systematic groups is not evidence for special creation or for saltation, but simply means that a full transitional sequence more or less like those that are known did occur and simply has not been found in this instance.[8]

One view of the meaning of the gaps in the record is as follows. The gaps set the boundaries between groups of species which have descended from created kinds. Because the gaps were "systematic defiiciencies of record" and occurred so regularly between orders, it was held that each order had a start with a created being. To illustrate. Between the first horse and the supposed ancestor of it (the condylarthra) there is a definite break in the continuity of successive forms. So it may be concluded that the first condylarthra are created, and that the first horse was created.[9]

But on the other hand the breaks may represent links which *did live* but were not fossilized, or forms which did live but have not been found or recorded in the literature. For instance, "Eohippus is more like some condylarths than it is like *Equus.*" But in other words, there is not as much difference between the supposed ancestor of the horse and the horse as there is between the first horse and the latest horse. So, if there was a transition from the horse of ages ago to the one of today, which was a considerable change, there could well have been animals which were stages between the horse and its ancestors, which is a lesser amount of change. If there already is much change on the record, why believe that a little which is not documented did not occur?

In the 1944 treatise on *Tempo and Mode in Evolution*, Simpson said that the first horse was "an equid, which is a classificatory way of saying that the vast majority of its multitude of morphological characters were already the same as those preserved in *Equus* and in all equids as well as in many other more or less related animals."[10] That is, the first horse was a genuine horse. In 1953 in *The Major Features of* Evolution, Simpson remarks, "eohippus is more like some condylarths than it is like Equus."[11] Also he states,

> "For instance, Matthew (1926) pointed out, but later students have mostly ignored, the fact that eohippus was *not* a horse, that it is about as good an ancestor for *Rhinoceros* as for *Equus*. In effect, there was no family Equidae when eohippus lived. The family and all its distinctive characters developed gradually as time went on. Eohippus is referred to the Equidae because we happen to have more nearly complete lines back to it from later members of this family than from other families. There is no particular time at which the Equidae became a family rather than a genus or a species; the whole process is gradual and we assign the categorical rank after the result is before us.[12]

Simpson has probably more lucidly and substantially stated the case for transformations from one order to another than any other recent writer.

It is not possible to study much of evolution experimentally.

> When a genus in a steadily moving lineage like that of *Equus* has an average duration of 7½ million years, nothing we can do to speed up experimental evolution is going to bring such events, let alone the longer spans of families and still higher categories, down to a period men can hope to follow in experimentation. Cross-breeding, essential for most genetical analyses, is almost never satisfactorily possible at the level of genera, and absolutely never above that level. Here, then, is a domain in which the observational approach and, when available, paleontological materials are the only ones possible.[13]

We should recall that it was possible to breed backwards to a horse that is now extinct, the tarpan. But we can never prove by producing eohippus again, that it was in the direct line of horse ancestry. We merely accept the plausible statement about its relationship to the horse.

What does Genesis really say? I take my clue from John Ockenga's statement in a book that discusses the first woman.[14] Ockenga says that Genesis says there were three creations as shown by the use of the Hebrew word bara: the heavens and the earth; animal life; and man. "Between these stages," he writes, "there is room for evolution in our thinking when it is taken to mean under the power of God."

I do believe in creation because of the Bible statement—and because of the argument from design. Arthur Holly Compton, Nobel prize winner in physics said, "The argument on the basis of design, through trite, has never been adequately refuted."[15] In recent years it has been admirably presented by the Moody Science films. No one need apologize for believing in creation. But the question is, "How much was directly created and how much was left to hereditary processes, which were also created?"


References

1. Simpson, *The Meaning of Evolution,* p. 17.
2. *Ibid.,* p. 16.
3. *Ibid.,* p. xii.
4. *Ibid.,* pp. 351-352.
5. *Ibid.,* p. 360.
6. *Ibid.,* p. 360.
7. *Ibid.,* p. 360.
8. *Ibid.,* p. 360.
9. Mixter, *Creation and Evolution.*
10. Simpson, *Tempo and Mode In Evolution,* p. 159.
11. Simpson, *The Major Features of Evolution,* p. 352.
12. *Ibid.,* p. 345.
13. *Ibid.,* p. 340.
14. Ockenga, *Have You Met These Women?* pp. 80-83.
15. Compton, *The Freedom of Man,* p. 73.


## Comments On Dr. Mixter's Paper


Wilbur L. Bullock

Zoology Department, University of New Hampshire


There are several basic problems indicated in Dr. Mixter's paper that have too often gone unanswered by evangelical Christians. The first of these is that "species can change into other species". To a Christian that has been brought up to defend the concept of fixity of species this may sound like a shocking statement. Actually, as we study the living world around us and the interpretation of that world in scientific writing it should be obvious that: 1) we are not really sure of exactly what a species is, and 2) no matter what definition we accept, we can find considerable evidence for a common origin of at least closely related species.

The common ancestor of members of a family or order is not so directly observable but on the basis of the tendency of organisms to vary, the amount of time involved from the geological record, and the probable changes that have taken place in the environments during that time, it is not unreasonable to interpolate to the recognition for the common ancestry of the members of these groups.

The origin of the major groups—phyla and classes—constitutes a more serious problem as any honest evolutionist will readily point out. The evidence for common ancestry at this level is based on analogy, interpolation, and an extremely incomplete fossil record. However, just as we criticize the materialistic evolutionist for reading too much into the inadequate record we too jump to hasty and ill advised conclusions every time we insist that such and such a gap "proves" a supernatural creative act. Then, as our gap gets filled in we are forced to contract the role of God. Our God is too great to become a synonym for our ignorance. Is He not the author of any natural laws that may be behind our puny theorizing? Let's not try to outdogmatize the materialistic evolutionist. As Christians I think we could more profitably consider the role of God the Creator throughout the entire course of evolution instead of defending the God of the gaps. I personally am not convinced of the "entire course of evolution" but in answering the question "*How* did God create the physical and biological world"? I am not averse to considering the evolutionary hypothesis as a *tentative* working hypothesis until we can come up with something better.

It seems to me that this is the direction of Dr. Mixter's arguments. Certainly neither his paper nor my comments imply a blanket endorsement of "Theistic evolution" but the discussion would seem to indicate that we have arrived at a point where the differences between "Theistic evolution" and "progressive or step-wise creation" need to be calmly and thoughtfully considered.

## The Evolution of Evangelical Thinking on Evolution

J. Frank Cassel

North Dakota State College, Fargo, North Dakota

Over twenty years ago a group of zoology majors at a Christian college agreed in all seriousness that as part of his life's work each would take a certain phase of evolution, explore it carefully, and derive therefrom inherent data to refute the evolutionary concept. At that time these men had no question as to the refutability of the theory, but were dissatisfied with the kind and use of data (if any) then presented in refutation.

The viewpoint of this group was probably representative of that of many Christians of this period who were students of biology and was, no doubt, a reflection of the tenor of the times. The conflict between Fundamentalism and Modernism was at a whiteheat. Separationist groups were even dividing within themselves. To be right doctrinally was more to be desired than to show the love of the indwelling Christ in a regenerated life. One's status in a Christian fellowship seemed to depend as much upon the method, the place, or the person involved in one's conversion as upon the fact of it. In this climate much contending for the faith had become contentious and much standing fast in the faith had become reaction.

"Evolution" was a dirty word, and hence to be dealt with by invective and shunned as sin. Under the guise of scholarship, a few crusaders stumped the country building up straw men to knock down as they "warned true Christians" of the evils perpetrated in the name of science. And small wonder, for had not the deism of Darwin been replaced by the mechanism of Spencer and Haeckel as bioliogists strove to grasp the significance of nature's laws? And was this not the sperm which fertilized the egg of Natural Philosophy giving rise to "higher criticism"—the father of Modernism?

The basic ("fundamental") beliefs of the Bible-believing Christian were being challenged from within and from without by scholarship, and orthodox scholars were slow to meet the challenge on its own ground. But this group of well-meaning college kids is an example of the awakening to the need for Christian scholarship. As the recognition of this need developed, a few Christian men of science realized that effectiveness of scholarship can be enhanced by an organization providing for the stimulation of the exchange of views and for the publication of results of investigation; and the A. S. A. was born.

Although the contribution of an A. S. A.-inspired scholar to either his discipline or to the synthesis of the Christian view has only rarely been startling, certain trends can be seen to have been developing. Perhaps the lack of original contributions can be attributed (at least to some extent) to the personal stress on each individual who has dug at all deeply into the data of evolution. First to be overcome was the onus of dealing with a "verboten" term and in a "non-existent" area. Then, as each made an honest and objective consideration of the data, he was struck with the validity and undeniability of datum after datum. As he strove to incorporate each of these facts into his Biblico-scientific frame of reference, he found that—while the frame became more complete and satisfying—he began to question first the feasibility and then the desirability of an effort to refute the total evolutionary concept, and finally he became impressed by its impossibility on the basis of existing data. This has been a heart-rending, soul-searching experience for the committed Christian as he has seen what he had long considered the *raison d'etre* of God's call for his life endeavor fade away, and as he has struggled to release strongly held convictions as to the close limitations of Creationism.

This struggle is made no easier by the lack of approbation (much less acceptance) of some of his less well-informed colleagues, some of whom seem to question motives or even to imply heresy. To watch the still eager acceptance of views he has been forced to reject and the stolid rejections of views he feels clearly supported by evidence while having his own character brought into question at the same time often makes him question the worth of the whole effort.

Some have reacted by lessening their activities or at least their pronouncements in the field, others by becoming less active in the Affiliation, while a few hardy souls (to which the paper under discussion gives witness) continue to give forth their God-given convictions and with them perhaps a simple challenge —"If you have a better approach please tell me quickly for this one of mine has been bought at great price to my peace of mind."

My acceptance or rejection of Dr. Mixter's conclusions depends upon (indeed, demands) my honest exploration and consideration of the data and references he presents. Whether or not I agree or disagree with him has no bearing upon his right to be heard. His contribution is of real significance to A.S.A.; and his presentation needs to be cloaked in the dignity commensurate with the significance of his pronouncement —a pronouncement in essence that:

> I, an evangelical Christian, can accept the basic concepts of evolution. Although not exclusively demanded by the data involved, it is certainly allowed, and in fact I can see no better or more logical way to handle the data. I believe in Creation, and simply affirm that in the light of the evidence now available, I think some evolution—that is, development of present-day forms by differentiation of previously existing forms—the most likely way God accomplished much of His Creation.

Thus, in fifteen years we have seen develop within A. S. A. a spectrum of belief in evolution that would have shocked all of us at the inception of our organization. Many still reserve judgment but few, I believe, are able to meet Dr. Mixter's challenge of, "Show me a better explanation." Some may see in this developing view the demise of our organization, but it seems to me that we only now are ready to move into the field of real potential of contribution—that in releasing Truth from the restrictions we have been prone to place upon it, we can really view it in the true fullness which the Christian perspective gives us.

# "A Critique of Evolution"*

W. R. THOMPSON

When I was asked by the publishers of this new edition of *The Origin of Species* to write an introduction replacing the one prepared a quarter of a century ago by the distinguished Darwinian, Sir Arthur Keith, I felt extremely hesitant to accept the invitation. I admire, as all biologists must, the immense scientific labours of Charles Darwin and his lifelong, singlehearted devotion to his theory of evolution. I agree that although, as he himself readily admitted, he did not invent the doctrine of organic evolution, or even the idea of natural selection, his arguments, and especially the arguments in *The Origin of Species,* convinced the world that he had discovered the true explanation of biological diversity, and had shown how the intricate adaptations of living things develop by a simple, inevitable process which even the most simple minded and unlearned can understand. But I am not satisfied that Darwin proved his point or that his influence in scientific and public thinking has been beneficial.

I therefore felt obliged to explain to the editors of Everyman's Library, that my introduction would be very different from that of Sir Arthur Keith, and that I could not content myself with mere variations on the hymn to Darwin and Darwinism that introduces so many text-books on biology and evolution, and might well be expected to precede a reprinting of the *Origin.* They raised no objection, so my main difficulty was removed. I am of course well aware that my views will be regarded by many biologists as heretical and reactionary. However, I happen to believe that in science heresy is a virtue and reaction often a necessity, and that in no field of science are heresy and reaction more desirable than in evolutionary theory. I have written what I think should be written; but the responsibility of the editors of the library is not involved.

I have said that it was mainly *The Origin of Species* that converted the majority of men to the evolutionary doctrine. Sir Arthur Keith emphatically agreed. 'No book,' he said, 'has appeared to replace it; *The Origin of Species* is still the book which contains the most complete demonstration that the law of evolution is true.' But the more strongly we insist on this point, the more necessary it is to scrutinize the proofs given in the *Origin*. Of course, we may be induced to accept a statement that is true, by agreements that are fallacious or inadequate. Still, no one would seriously maintain that it is good to do the right thing for the wrong reasons. If arguments fail to resist analysis, assent should be withheld, and a wholesale conversion due to unsound argument must be regarded as deplorable.

For Sir Arthur Keith, Darwin as a writer may be classed among the 'small select group of great Englishmen which holds Shakespeare.' The literary critics, apparently, did not agree with him. Though he has often been regarded as an obscure writer, Darwin usually expresses himself clearly enough. He was not interested in philosophical considerations or in the exact definition of the terms he used. In the final chapter of the first edition of *Origin,* where he recapitulates his arguments, the word evolution is not even mentioned; yet the proposition he is defending can easily be defined. This is, that all the organisms that exist or have existed have developed from a few extremely simple forms or from one alone, by a process of descent with modification. The mechanism of these transformations though infinitely complex in its detailed working, is very simple in principle. For reasons not fully understood organisms tend to vary slightly in their various characteristics. These variations must be called random in the sense that they have no predestined relation to the well-being of the organism. Nevertheless since they occur continually in many directions, an individual in which a particular variation has occurred will have a slight advantage over its competitors in a particular environment. The advantage will be transmitted to its progeny in which, owing to varation, it will be manifested in different degrees, and thus there will occur through successive generations, a progressive adaptation to the environment from which the inadequately equipped competitors will disappear either through extinction or by adaptation to a different environment. We must, says Darwin, admit the truth of the following propositions: 'that gradations in the perfection of any organ or instinct, which we may consider, either do now exist or could have existed, each good of its kind —that all organs and instincts are, in ever so slight a degree, variable—and, lastly, that there is a struggle for existence leading to the preservation of each profitable deviation of structure or instinct.' These truths being admited, the theory of descent with modification through natural selection, must be accepted. This explanation has universal value. It enables us to understand that every mental power and capacity has been a gradual but necessary acquirement and thus the origin and history of man become scientifically comprehensible. And as the past has been, so will the future. We may look with some confidence, says Darwin, 'to a secure future of equally inappreciable length. And as na-

*This is the introduction to Charles Darwin's "Origin of Species" New York, E. P. Dutton & Co. Inc. 1956 and is published with the permission of Dr. Thompson, Everyman's Library, and E. P. Dutton & Co., Inc.

tural selection works solely by and for the good of each being, all corporeal and mental endowments will tend to progress towards perfection.'

The view that natural selection, leading to the survival of the fittest, in populations of individuals of varying charateristics and competing among themselves, has produced in the course of geological time gradual transformations leading from a simple primitive organism to the highest forms of life, without the intervention of any directive agency or force, is thus the essence of the Darwinian position. Purposeless and undirected evolution, says J. S. Huxley, eventually produced, in man, a being capable of purpose and of directing evolutionary change. This, it appears to me, remains the view of the most representative modern Darwinians. It is true that Darwin himself admitted a Lamarckian element, the effects of use and disuse, and Sir Arthur Keith defended him against those who accused him of relying exclusively on natural selection. But this, in the modern view, would be a virtue of Darwin's theory since the inheritance of acquired characters is now generally denied by biologists.

We must now examine the arguments in the 'demonstration that the law of evolution is true.'

Darwin's first argument, to which he devoted a great deal of labour, is that there is great variation among the individuals of many species. This variation is particularly evident among domesticated animals and plants. From these undeniable facts Darwin drew several conclusions. One was that species are not strictly immutable as biologists commonly maintained. The difference between the various types of domesticated species is often much greater than that which exists between wild species, and even in these it is often extremely difficult to decide whether a particular form is a species or a variety. The great difference in the forms of domesticated species shows, on one hand, that variation can be stimulated by particular conditions and that the artificial selection made by breeders has produced forms with extremely distinctive characteristics. The differences between the various species of violets or between the species of the hymenopterous genus *Mesoleius,* for example, are clearly far less striking than the differences between a pekinese and an Irish setter, or between a snow apple and a russet. Darwin points out that under certain conditions abnormal individuals are produced, and he maintains that it is impossible to draw a line between such monstrosities and the individuals regarded as normal. These converging arguments indicate that what we call a species is just a transitional stage in a genealogical succession which cannot at any time be regarded as having a permanent definable essence or nature. There is therefore no intrinsic obstacle to unlimited evolution and the extrinsic conditions for it exist.

That natural selection directs the course of evolution Darwin could not prove by an appeal to facts. However, he felt certain that all organisms tend to increase in geometrical ratio, that each lives by a struggle for its requirements at some period in its life and that among individuals differing even to a slight degree, the fittest must survive and transmit their characteristics to their offspring and, since these will continue to vary, natural selection will progressively improve the adaptations and equipment of each species. 'What checks the natural tendency of each species to increase in number,' said Darwin, 'is most obscure . . . ' 'We know not exactly what the checks are even in one single instance.' He was able to show from factual examples that there is a great destruction of individuals in nature and to indicate some of the causes of this destruction; but he had little detailed evidence to offer concerning the action of natural selection.

Whether or not natural selection has produced the existing and past diversity of organic forms, this diversity exists, not only in space but in time. Such facts as the presence of different species of the same genus in different islands in the same area are consonant with the idea of descent with modification from a common ancestor as is the absence in isolated islands of organisms without active powers of migration and the presence of others such as bats and birds, taxonomically related to those of mainland areas.

Other supporting arguments were advanced by Darwin: the slow change and apparent progression of organic forms in the geological strata, the evidence of the existence in the past of a great variety of organisms now extinct; the similarity between the embryonic stages of organisms quite distinct in the adult condition; the existence of rudimentary organs; and the fact that a natural classification of organisms is possible, since this indicates real blood relationship and is therefore in a sense a mirror of the genealogical system by which they arose.

I have tried to include in a necessarily brief summary the most important points in Darwin's argument and have not designedly attempted to weaken the presentation. If Darwin convinced the world that species had originated through evolution by natural selection, it was, I think, on the basis of the arguments I have mentioned

But in a matter of this kind a great deal depends on the manner in which arguments are presented. Darwin considered that the doctrine of the origin of living forms by descent with modification, even if well founded, would be unsatisfactory unless the causes at work were correctly identified, so his theory of modification by natural selection was, for him, of absolutely major importance. Since he had at the time the *Origin* was published no body of experimental evidence to support his theory, he fell back on speculative arguments The argumentation used by evolutionists, said de Quatrefages, makes the discussion of their ideas extremely difficult. Personal convictions, simple possibilities, are presented as if they were proofs, or at least valid arguments in favor of the theory. As an example de Quatrefages cited Darwin's explanation of the manner in which the titmouse might become transformed into the nutcracker, by the accumulation of small changes in structure and instinct owing to the effect of natural selection; and then proceeded to show that it is just as easy to transform the nutcracker into the titmouse. The demonstration can be modified without difficulty to fit any conceivable case. It is without scientific value, since it cannot be verified; but since the imagination has free rein, it is easy to convey the impression that a concrete example of real transmutation has been given. This is the more appealing because of the extreme fundamental simplicity of the the Darwinian explanation. The reader may be completely ignorant of biological processes yet he feels that he really understands and in a sense dominates the machinery by which the marvellous variety of living forms has been produced.

This was certainly a major reason for the success of the *Origin*. Another is the elusive character of the Darwinian argument. Every characteristic of organisms is maintained in existence because it has survival value. But this value relates to the struggle for existence. Therefore we are not obliged to commit ourselves in regard to the meaning of difference between individuals or species since the possessor of a particular modification may be, in the race for life, moving up or falling behind. On the other hand, we can commit ourselves if we like, since it is impossible to disprove our statement. The plausibility of the argument eliminates the need for proof and its very nature gives it a kind of immunity to disproof. Darwin did not show in the *Origin* that species had originated by natural selection, he merely showed, on the basis of certain facts and assumptions, how this might have happened, and as he had convinced himself he was able to convince others.

But the facts and interpretations on which Darwin relied have now ceased to convince. The long-continued investigations on heredity and variation have undermined the Darwinian position. We now know that the variations determined by environmental changes—the individual differences regarded by Darwin as the material on which natural selection acts—are not hereditary. We can, by selection, sort out from a natural population a number of pure lines or genotypes, each possessing with respect to a given character its special curve of variability; but we cannot change this curve by selection within the genotype. For example, in a certain pure line of the house-fly, those with the longest wings may conceivably have an advantage—though I cannot see how this could be demonstrated. But we cannot, by choosing and mating these long-winger flies, produce a progressive increase in the proportion of long-winged flies, or a progressive increase in wing length.

It is true that some variations are hereditary. These are the so-called mutations which do not develop gradually but appear suddenly and remain as they appeared. The varieties of domesticated plants and animals are the result of mutations. But such forms must be eliminated in nature, which would other-wise present a spectacle entirely different from the reality. This is partly due to the fact that mutations are not adaptive. If we say that it is only by chance that they are useful, we are still speaking too leniently. In general, they are useless, detrimental, or lethal. Darwin himself did not think that the races of domesticated animals were capable of surviving in nature, but the modern Darwinians are obliged to explain evolution as the result of mutations. If we minimize or at least limit the survival value of characters in general, we can agree that certain distinctive morphological dispositions may well be the the result of mutations. But the neo-Darwinians hold firmly to the belief that every specific character has survival value. This to my mind puts them in a very awkward position.

To realize how unconvincing their position is, we have only to consider the fact of organic correlation. Strangely enough, though Darwin was evidently well acquainted with the work of Cuvier he pays practically no attention, in the *Origin*, to Cuvier's principle of adaptive correlation. For him correlation is merely a concurrence of characters like 'the relation between blue eyes and deafness in cats, and the tortoise-shell colour with the female sex, the feathered felt and skin between the outer toes of pigeons, and the presence of more or less down on the young birds when first hatched, with the future colour of their plumage; or, again, the relation between the hair and teeth in the naked Turkish dog.' Indeed Darwin's remarks suggest that he thinks of correlation as a material connection between malformations rather than as an adaptation. His modern disciples in general simply ignore the problem of correlation. However, to ignore it is easier than to solve it. As Emile Guyenot has said, mutations are powerless to explain the general adaptation which is the basis of organization. 'It is impossible to produce the world of life where the dominant note is functional organization, correlated variation and progression, from a series of random events.' The position therefore is that while the modern Darwinians have retained the essentials of Darwin's evolutionary machinery, to wit, natural selection, acting on random hereditary variations, their explanation, plausible in Darwin's day, is not plausible now.

It has been said that the substitution of particulate for blending inheritance removed what was a serious difficulty in Darwin's own position. The interference with progressive evolution resulting from blending inheritance was certainly a weakness in the argument of the *Origin* but, as I have said, particulate inheritance has introduced other difficulties.

An important point in Darwin's doctrine, as set out in the *Origin*, was the conviction that evolution is a progressive process. We may look forward, he said, to a secure future of inappreciable length. 'And as natural selection works solely by and for the good of each being, all corporeal and mental endowments will tend to progress towards perfection.' The Victorians accepted this idea with enthusiasm. Here I need only to say that on this point Darwin was inconsistent since, in his view, natural selection acts not only by the survival of the fittest but also by the exterminiation of the less fit and may produce anatomical degradation as well as improvement.

That owing to the existence of different genotypes within a species and the somewhat different adaptive-characters of these genotypes, samples of a widespread population taken at different points may be reconizably different in various ways, or a population of this kind spreading from a centre (as in the case of an introduced insect) may develop local varieties sufficiently marked to be regarded as species by a taxonomist, may be freely acknowledged. Furthermore, when we consider the development of a complex organism from the structurally simple germ cell, we must recognize that in this field, at least, evolution, in the classical sense, is a fact accessible to direct observation. But it is a far cry from these facts to the speculations of the *Origin* and the Victorian concept of evolution.

It is hardly necessary to dwell at length on all the minor arguments advanced by Darwin. These consist essentially in a translation of certain facts in terms of evolutionary theory, or, in other words, on an historical basis. If an organism possesses a structure having no assignable function, but looking like a reduced specimen of a functional structure existing in some other form, it was regarded as a 'rudiment' whose existence is explicable only as a relic that has gradually degenerated in coming down from a remote ancestor, where it was well developed and functional.

It is clear that this supposition has no demonstrative value. It itself requires demonstration. Unless one adopts the Darwinian postulate that all characteristics have survival value, it is not necessary to assume that they have, or ever had, definite functions. Some so-called rudiments, such as the homologues of the mammary glands in man cannot, so far as any plausible evidence goes, have been inherited from an ancestor in which they were functional. Others, once believed to be useless, have definite functions. The existence in whales of transitory teeth and of small bones buried in the flesh, but corresponding to the pelvis, the femur, and the tibia, is commonly regarded as a proof of their descent from ancestors of the tetrapod type with functional teeth; but in the first place some anatomists consider that these structures have an impotant role in the developmental process; in the second place, we have no proof of a descent from ancestors in which these structures were more strongly developed; in the third place, it is clear that if they exist now, this is not primarily because they existed in the past, but because actual present causes now operate to produce them. What such cases like those of anatomical 'convergence' and general homology actually demonstrate is that there are large numbers of organisms, differing considerably in the details of structure but constructed on the same fundamental plan. However, this is no proof of descent from one original ancestor of this anatomical type. This itself requires proof. It may be said that unless we admit this, we must make the much more difficult supposition that many complex types originated independently. This, it will be remembered, was a point Darwin made against Lamarck. But I, for my part, do not see that I am obliged to express a view on such matters. Darwin himself considered that the idea of evolution is unsatisfactory unless its mechanism can be explained. I agree, but since no one has explained to my satisfaction how evolution could happen I do not feel impelled to say that it has happened. I prefer to say that on this matter our information is inadequate.

Darwin suggested in the *Origin* that embryological development provides evidence for evolution. He postulated that characteristics appear in the embryo at the stage in which they developed in the ancestor, so that new developments may be tacked on, so to speak, to a phase representing the ancestral development, since Darwin also held that the slight variations on which, in his view, evolution depends, 'generally appear at a not very early period of life.' This idea, elaborated by other workers, eventually became in the hands of Haeckel the 'great biogenetic law,' according to which the ontogeny repeats the phylogeny, or, as propagandists have put it, the developing animal 'climbs up its family tree.'

A natural law can only be established as an induction from facts. Haeckel was of course unable to do this. What he did was to arrange existing forms of animal life in a series proceeding from the simple to the complex, intercalating imaginary entities where discontinuity existed and then giving the embryonic phases names corresponding to the stages in his so-called evolutionary series. Cases in which this parallelism did not exist were dealt with by the simple expedient of saying that the embryological development had been falsified. When the 'convergence' of embryos was not entirely satisfactory, Haeckel altered the illustrations of them to fit his theory. The alterations were slight but significant. The 'biogenetic law' as a proof of evolution is valueless.

A more important argument in the opinion of Darwin himself was the possibility of classifying organisms. All true classification, he said is genealogical. Community of descent 'is the hidden bond which naturalists have been unconsciously seeking.' The arrangement of the groups within each class, 'in due subordination and relation to the other groups, must be strictly genealogical in order to be natural.' And again, 'the natural system is genealogical in its arrangement, like a pedigree; but the degrees of modification which the different groups have undergone have to be expressed by ranking them under so-called different genera, subfamilies, sections, orders, and classes.' What we call the natural system of classification is a proof of evolution since it can only be explained as a result of evolution.

The plausibility of this argument is obvious. Yet it is not so convincing as it may appear at first sight. On the Darwinian theory, evolution is essentially undirected, being the result of natural selection, acting on small fortuitous variations. The argument specifically implies that nothing is exempt from this evolutionary process. Therefore, the last thing we should expect on Darwinian principles is the persistence of a few common fundamental structural plans. Yet this is what we find. The animal world, for example, can be divided into some ten great groups or phyla, all of which are not morphologically as coherent and clear-cut as we might wish for convenience in classification, but nevertheless are stable and definable entities from the taxonomic standpoint. All identifiable animals that ever have existed can be placed in these groups. Generally speaking, the subordinate groups are equally well defined. We can tell at a glance to what Order or Family a particular insect belongs. As I have already noted there is often controversy and uncertainty about the definitions of genera, species, and varieties; but taking the taxonomic system as a whole, it appears as an orderly arrangement of clear-cut entities which are clear-cut because they are separated by gaps. These gaps Darwin explained by the hypothesis that the intermediates are constantly eliminated by natural selection. I do not think we can be expected to accept his unproved supposition as an argument for Darwinism. But in any case it has no bearing on the persistence throughout geological time, in spite of the fortuitous variation and natural selection, on the persistence of the fundamental anatomical plans exhibited by the great groups. Darwin insisted on several occasions that characteristics long inherited become stabilized and perhaps he considered that the persistence of morphological types can be explained in this way. But without introducing considerations quite foreign to his system, we cannot explain why the anatomical type of the Echinoderm or the Insect continued to be inherited.

Because all organisms we know are generated by other organisms, it is natural to interpret biological classification in terms of genealogy, But not all the thing that can be classified are connected by generation. The arrangement of the chemical elements and their compounds is a true classification and so is the arrangement of geometric forms; yet no genealogical considerations are involved. Looking at the matter from this angle, we can easily see that in actual fact the system of biological classification is simply based on the characteristics of organisms as they are here and now. The basis of these characteristics here and now is the physicochemical constitution. If we wish to erect a genealogical classification we cannot do so with a collection of abstractions drawn from our arrangement of existing organisms—we must discover through what forms the existing organisms have actually descended. If these historical facts cannot be ascertained, then it is useless to seek for substitutes, and from the fact that a classification is possible we certainly cannot infer that it is genealogical and is in any sense a proof of evolution.

Evolution, if it has occurred, can in a rather loose sense be called a historical process; and therefore to show that it has occurred historical evidence is required. History in the strict sense is dependent on human testimony. Since this is not available with respect to the development of the world of life we must be satisfied with something less satisfactory. The only evidence available is that provided by the fossils. It has been pointed out by both supporters and opponents of the evolutionary doctrine, that even if we can demonstrate the chronological succession of certain organisms, this is not proof of descent. This may seem like a quibble. If we put a pair of houseflies in a cage and let them breed, we do not doubt that the live flies we find there in a month's time are the descendants of the original pair. Similarly, if in an apparently undisturbed geological formation we find snail shells at an upper level very similar to those at a lower level, we may reasonably conclude that there is some genealogical connection between the two groups, though we cannot trace the descent from individual to individual as is required in a true family tree. Therefore, if we found in the geological strata a series of fossils showing a gradual transition from simple to complex forms, and could be sure that they correspond to a true time-sequence, then we should be inclined to feel that Darwinian evolution has occured, even though its mechanism remained unknown. This is certainly what Darwin would have liked to report but of course he was unable to do so. What the available data indicated was a remarkable absence of the many intermediate forms required by the theory; the absence of the primitive types that should have existed in the strata regarded as the most ancient; and the sudden appearance of the principle taxonomic groups. Against these difficulties he could only suggest that the geological record is imperfect, but that if it had been perfect it would have provided evidence for his views. It is clear therefore that the palaeontological evidence at his disposal, since it had not led competent naturalists acquainted with it to a belief in evolution, could only justify a suspense of judgment. The condition of fossil material is, of course, unsatisfactory since soft tissues usually disappear, leaving only skeletal structures, frequently much distorted. The fossil insects of the group with which I am best acquainted cannot be accurately determined, even to genera. It is evident that many organisms now extinct existed in the past, but we can never know them as we know living forms. The chronological succession of the fossils is also open to doubt, for it appears, generally speaking, that the age of the rocks is not determined by their intrinsic characteristics but by the fossils they contain; while the succession of the fossils is determined by the succession of the strata. It was thought also that the fossils should appear in a certain order, corresponding roughly to the stages in embryological development. In fact the strata, and therefore the fossils they contain, do not always occur in the accepted order. In some areas of the world for example, the Cambrian strata, which are regarded as the oldest fossiliferous formations, rest on the Cretaceous which are regarded as relatively recent; in other, Cretaceous or Tertiary beds appear, instead of the Cambrian, on the granite. Sometimes the character of the deposits would lead to the belief that they were chronologically continuous since they can be separated only by the fossils they contain. Various hypotheses have been proposed to explain these departures from accepted theory, and though they are often the subject of controversy among geologists I do not suggest that the problems to which they relate are insoluble.

On the other hand, it does appear to me, in the first place, that Darwin in the *Origin* was not able to produce palaeontological evidence sufficient to prove his views but that the evidence he did produce was adverse to them; and I may note that the position is not notably different to-day. The modern Darwinian palaeontologists are obliged, just like their predecessors and like Darwin, to water down the facts with subsidiary hypotheses which, however plausible, are in the nature of things unverifiable.

It has been said that though we do not find in the geological deposits the intermediates required by Darwinian theory, some very striking intermediates have been found of which the classical oft-cited example is *Archaeopteryx*. To me, however, it appears that since the geological strata probably represent environmental conditions very different from those of the present, collections made in them may be regarded something like those made on the continent of Europe or in the tropics, with respect to the fauna and flora of the British Isles. As the range of our collections extends, so we invariably enrich our representation of various groups, and this necessarily and inevitably entails the appearance of intermediate between the forms in the collection from the restricted area in which we started. The recognition of this fact, with respect to the collections of organisms existing here and now, does not necessarily commit us to any particular view of the origin of species; and the same thing is true of the collection of fossil material.

*The Origin of Species* converted the majority of its readers to a belief in Darwinian evolution. We must now ask whether this was an unadulterated benefit to biology and to mankind. Sir Arthur Keith, as we have seen, had no doubts about this point. Some of the Darwinian propagandists were even more positive.

Writing in his Anthropogeny of the evolutionary controversy, Haeckel asserted, that in this intellectual battle, which excites all the thinking sections of humanity, and prepares for the future a truly humane society, we see on one side, under the splendid banner of science, the liberation of the mind, truth, reason, civilization, development, and progress. In the other camp are ranged, under the banner of the hierarchy, intellectual servitude, error, irrationality, barbarous ways of life, superstition, and decadence. Quite recently an evolutionary propagandist has said, that without the evolutionary doctrine, biology, except in certain restricted fields, becomes unintelligible.

I find myself unable to agree with these views. I do not contest the fact that the advent of the evolutionary idea, due mainly to the *Origin*, very greatly stimulated biological research. But it appears to me that owing precisely to the nature of the stimulus, a great deal of this work was directed into unprofitable channels or devoted to the pursuit of will-o'-the-wisps. I am not the only biologist of this opinion. Darwin's conviction that evolution is the result of natural selection, acting on small foruitous variations, says Guyenot, was to delay the progress of investigations on evolution by half a century. Really fruitful researches on heredity did not begin until the rediscovery in 1900 of the fundamental work of Mendel, published in 1865 and owing nothing to the work of Darwin. In his great work *Growth and Form*, D'Arcy Thompson remarked on the stultifying effect of Darwinian theory. 'So long and so far as "fortuitous variation" and the "survival of the fittest" remain engrained as fundamental and satisfactory hypotheses in the philosophy of biology, so long will these "satisfactory and specious causes" tend to stay "severe and diligent inquiry," "to the great arrest and prejudice of future discovery." ' Much time was wasted in the production of unverifiable family trees. For example, by plausible but unconvincing arguments zoologists have 'demonstrated' the descent of the Vertebrates from almost every group of the Invertebrates. During the thirty years from 1870 to 1900, there was an immense concentration of effort on embryology, inspired by the 'biogenetic law.' Here again the main objective was the tracing of ancestries. The attempt of his to explain development in terms of actual physical causes was rejected with contempt by authors like Haeckel. 'We have better things to do in embryology,' said one of them, 'than to discuss tensions of germinal layers and similar questions, since all explanations must of necessity be of a phylogenetic nature.' Gradually it was realized that the objective was unattainable. Embryology then ceased to be fashionable. Taxonomists also followed the trend, constructing hypothetical ancestors for their groups and explaining the derivation of existing forms from these imaginary entities. I do not of course deny that a great amount of valuable information was gathered in these studies, but I think it could have been obtained more effectively on a purely objective basis. My impression is, also, that though it was unproductive from the Darwinian standpoint, this was not usually admitted. The deficiencies of the data were patched up with hypotheses, and the reader is left with the feeling that if the data do not support the theory they really ought to.

A long-enduring and regrettable effect of the success of the *Origin* was the addiction of biologists to unverifiable speculation. 'Explanations' of the origin of structures, instincts, and mental aptitudes of all kinds, in terms of Darwinian principles, marked with the Darwinian plausibility but hopelessly unverifiable, poured out from every research centre. The speculations on the origin and significance of the resemblances between animals, or between animals and their environment and of the striking colour patterns they often exhibit, constitute one of the best-known examples. In the article on 'Mimicry' in the 14th edition of the *Encyclopedia Britannica* we find a remarkable explanation of the form of tropical insect belonging to the group of the 'lantern-flies.' The head of this insect, which is not very large, resembles, in miniature, the head of an alligator, being prolonged into a snout at the base of which is a protuberance resembling an eye, while along the side are formations resembling minute teeth. Curious though the resemblance is, it is obviously a mere coincidence. The insect as a whole does not look anything like an alligator. However, for the Darwinian author of the article we have here an example of the development of protective resemblance by natural selection. The similarity of the head of the insect to the head of an alligator is a protection against monkeys. The monkey does not actually mistake the insect for an alligator but the sight of its head recalls to him the occasion on which an alligator almost seized him when he was drinking from a stream. Such is the effect of Darwinian fantasy on biological thinking.

The success of Darwinism was accompanied by a decline in scientific integrity. This is already evident in the reckless statements of Haeckel and in the shifting, devious, and histrionic argumentation of T. H. Huxley. A striking example, which has only recently come to light, is the alteration of the Piltdown skull so that it could be used as evidence for the descent of man from the apes; but even before this a similar instance of tinkering with evidence was finally revealed by the discoverer of Pithecanthropus, who admitted, many years after his sensational report, that he had found in the same deposits bones that are definitely human. Though these facts are now well known, a work published in 1943 still accepts the diagnosis of Pithecanthropus given by Dubois, as a creature with a femur of human form permitting an erect posture. Not long ago (1947), an exhibit in London, designed for public instruction, presented human development in such a way as to insinuate the truth of the 'biogenetic law'; and in the same exhibit were problematic reconstructions indicating the descent of man and including the Piltdown type.

As we know, there is a great divergence of opinion among biologists, not only about the causes of evolution but even about the actual process. This divergence exists because the evidence is unsatisfactory and does not permit any certain conclusion. It is therefore right and proper to draw the attention of the non-scientific public to the disagreements about evolution. But some recent remarks of evolutionists show that they think this unreasonable. This situation, where scientific men rally to the defence of a doctrine they are unable to define scientifically, much less demonstrate with scientific rigour, attempting to maintain its credit with the public by the suppression of criticism and the elimination of difficulties, is abnormal and undesirable in science.

It is difficult to assess the effect of the *Origin* on the public mentality. It must be considered in conjunction with Darwins' later work: *The Descent of Man* and the writings of the supporters of Darwin in several countries. However, Sir Arthur Keith said that Darwin himself had done more than anyone to lift 'the pall of superstition' from mankind and, in another place, that Darwinism is a 'basal doctrine in the rationalistic liturgy.' These remarks suggest that in his opinion the decline of belief in the supernatural, and probably the decline of Christinity, is largely due to the influence of Darwin. I think there is much to be said for this view. It is true that in the *Origin* Darwin speaks of life 'having been originally breathed into a few forms or into one'; and refers to a Creator. Furthermore, he objected to the spontaneous generations for which Lamarck argued. But I think this objection was merely to an idea that would have made his own theory less comprehensively explanatory.

Although the *Origin* contains no direct attack on the Christian concept of the universe, it is, on a number of crucial points, opposed to this concept. The biblical account of the creation of living things can be, and often has been interpreted in a manner more or less compatible with the doctrine of evolution. Propagandists like T. H. Huxley, however, made every effort to minimize this possibility, and to prove that Christian orthodoxy implies a literal interpretation of Genesis which is irreconcilable with the evolutionary idea. Darwin himself though he once held some rather vaguely Christian views, abandoned them quite rapidly and soon ceased to believe in the Christian revelation.

The doctrine of evolution by natural selection as Darwin formulated, and as his followers still explain it, has a strong anti-religious flavour. This is due to the fact that the intricate adaptations and co-ordinations we see in living things, naturally evoking the idea of finality and design and, therefore, of an intelligent providence, are explained, with what seems to be a rigorous argument, as the result of chance. It may be said, and the most orthodox theologians indeed hold, that God controls and guides even the events due to chance; but this proposition the Darwinians emphatically reject, and it is clear that in the *Origin* evolution is presented as an essentially undirected process. For the majority of its readers, therefore, the *Origin* effectively dissipated the evidence of providential control. It might be said that this was their own fault. Nevertheles the failure of Darwin and his successors to attempt an equitable assessment of the religious issues at stake indicates a regrettable obtuseness and lack of responsibility. Furthermore, on the pure philosophical plane, the Darwinian doctrine of evolution involves some difficulties which Darwin and Huxley were unable to appreciate. Between the organism that simply lives, the organism that lives and feels, and the organism that lives, feels, and reasons, there are, in the opinion of respectable philosophers, abrupt transitions corresponding to an ascent in the scale of being, and they hold that the agencies of the material world cannot produce transitions of this kind. I shall not attempt to discuss this difficult question here. Nevertheless it is clear that the view just mentioned has been that of mankind in general. That plants, animals, and man can be distinguished because they are radically different is the common-sense conviction, or was, at least until the time of Darwin. Biologists still agree on the separation of plants and animals, but the idea that man and animals differ only in degree is now so general among them, that even psychologists no longer attempt to use words like 'reason' or 'intelligence' in an exact sense.

This general tendency to eliminate, by means of unverifiable speculations, the limits of the categories Nature presents to us, is the inheritance of biology from *The Origin of Species*. To establish the continuity required by theory, historical arguments are invoked, even though historical evidence is lacking. Thus are engendered those fragile towers of hypothesis based on hypothesis, where fact and fiction intermingle in an inextricable confusion. That these constructions correspond to a natural appetite, there can be no doubt. It is certain also that in the *Origin* Darwin established what may be called the classical method of satisfying this appetite. We are beginning to realize now that the method is unsound and the satisfaction illusory. But to understand our own thinking, to see what fallacies we must eradicate in order to establish general biology on a scientific basis, we can still return with profit to the source-book which is *The Origin of Species*.

# *Species Concepts and Definitions**

J. FRANK CASSEL
North Dakota State College, Fargo, North Dakota

The purpose of the American Scientific Affiliation as it is usually stated on our convention programs is that we are—

> A group of Christian scientific men, devoting themselves to the task of reviewing, preparing, and distributing information on the authenticity, historicity, and scientific aspects of the Holy Scriptures in order that the faith of many in Jesus Christ may be firmly established.

Reviewing, preparing, and distributing are sometimes hard to separate. Part of our objective, and our purpose in these discussions, is to determine just what the Bible says and just what it allows. Sometimes some of us who are in the scientific fields feel that perhaps the theologians are able to give us better ideas along these lines than we are able to derive for ourselves. I think this is part of the reason for our getting together with the ETS periodically.

We also want to know the facts of the created world. We believe that God, who has revealed Himself in various ways, will not be contradictory. Indeed He *cannot* be contradictory in His various revelations of Himself and hence the facts of science can in no way conflict or contradict the revelations of God by other means, such as in His Word and in Jesus Christ. What we *think* He reveals in the written Word and what we *think* He reveals in creation may be different from what is actually there. That is, there is the truth, and then, there is our view of the truth. And this is where the conflicts arise and this is the reason for setting up panels of this nature—discussions—so that we can gain as much of an insight into the facts—find out as many of the facts as possible. There are some things that we just have not found out yet and we want to know the areas in which we lack information. There can be much profitable discussion as to the meaning of the Scriptures or as to what the Scriptures allow. We want to be as informed as possible as to the possibilities and learn as much of God Himself (as He has revealed Himself in His various ways) as we can.

The *species problem* I think of as being similar to the problems some of us biologists have faced when we speak on science and the Bible. I at least tend to speak more about science and don't say very much about Genesis. People say, "You have told me these things, but you haven't said anything about how this fits in with Genesis." My answer often is, "If you tell me what you think Genesis says, then I can tell you how I think that it fits." There are more views of what the first chapter of Genesis says than there are possibilities of the interpretation of the origin of the organisms that we have today.

This holds also for the species. You can ask, "What is the kind?" "What does the kind mean in Genesis? Does it mean a species?" And we must answer, "You tell me what you mean by species and I'll tell you whether 'kind' means 'species.'" I think as you follow this discussion, you will gather that the biologists who are working in this area in the various fields that we are going to be discussing, and in many that we do not have time to discuss, are not in total agreement as to just what this group is.

What are the biologist's kinds? We have a plant kind and an animal kind. These we call kingdoms. Now we're not even in agreement up in this realm. Take the little organism Euglena. It has chloroplasts and it is also mobile. Is this a plant or an animal? The botanist teaches that it's a plant, and the zoologist teaches that it's an animal. There are numbers of organisms that are not as well defined as Euglena. Many of us have been taught that there are two kingdoms—the plant kingdom and the animal kingdom. Some taxonomists—people who work at classifying organisms—say that to classify simply as "plant" and "animal" is not satisfactory. Some suggest as many as six kingdoms now. In the kingdom Protista, for instance, some include viruses, bacteria, protozoa, and algae. There isn't total agreement, then, as to how many major categories or groups there should be, nor as to the category into which some organisms fall.

Classification is based (as we will emphasize

* Edited from a tape of an extemporaneous presentation at the Thirteenth Annual Convention of the American Scientific Affiliation, Ames, Iowa, August, 1958. Introduction to a symposium on *What Is a Kind—The Species Problem.*

| Phylum | Some Distinctive Characteristics | Common Name | Representative Forms |
|---|---|---|---|
| Protozoa | Forms self-sufficient in nature as single cells | Protozoans | Amoeba, paramecium, euglena |
| Porifera | Body having many pores | Sponges | Bath sponge |
| Coelenterata | Saclike digestive tract, nematocycts for protection and food getting | Coelenterates | Hydra, jellyfish, coral |
| Platyhelminthes | Flat and soft, often parasitic | Flatworms | Planaria, liver fluke, tapeworm |
| Aschelminthes | Body slender, digestive tract complete | Roundworms and others | Rotifers, hookworm, trichina worm |
| Echinodermata | Radial symmetry, tube feet | Echinoderms | Starfish, sea urchin |
| Mollusca | Limy external shell in most | Mollusks | Clam, snail, octopus |
| Annelida | Body with numerous similar segments, setae | Segmented worms | Earthworm, leech |
| Arthropoda | Chitinous exoskeleton segmented, jointed appendages | Joint-footed animals | Crayfish, lobster, insect, spider |
| Chordata | Notochord, dorsal, hollow nerve cord, gill arches | Chordates | Fish, reptile, bird, mammal, man |

later) on degree of similarity and degree of difference. Animals are grouped according to their similarities; they are divided according to their differences. We shall be discussing the animal kingdom for the most part because it is the one on which the panel members are best informed. It is divided into "kinds" which are known as phyla, the major ones of which are listed at bottom of preceding page.

The phyla are divided into "kinds," groups called classes. The classes of the Chordata, for instance, include cartilaginous fishes, bony fishes, amphibians, reptiles, birds, mammals. Each of these may be divided still further into groups (kinds) known as orders, the order into families, the family into genera, and the genus into species. Each kind is grouped with others of its kind because of similarity and separated from other kinds by differences. To illustrate, a robin is classified as follows:

| *Category (Kind)* | *Scientific Name* | *Common Name* |
|---|---|---|
| Kingdom | Animalia | Animals |
| Phylum | Chordata | Chordates |
| Class | Aves | Birds |
| Order | Passeriformes | Perching birds |
| Family | Turidae | Thrushes |
| Genus | Turdus | Typical thrushes |
| Species | *Turdus migratorius* | Robin |

A species has been described broadly as a group of individual organisms which are as much alike (similar) as the offspring of the same parents. We are humans. There are dogs, there are cats, there are lions, there are tigers. And there are robins.

It is the classification category level of Species which we will be discussing. When considered in the above relationship, the category of Species may seem rather clear-cut. May we examine it more closely? Just what is a species? Since this category is considered the basic unit of classification, we have to understand what it is before we can understand fully any of the other categories. And we must understand the nature of each category before we can determine its relationship to the use of the word "kind" in Genesis. That the idea of species is not clear-cut is demonstrated by the fact that at its annual meeting in 1955, the American Association for the Advancement of Science sponsored a symposium on "The Species Problem." This symposium was under the chairmanship (and editorship) of Ernst Mayr (*Publ. 50, A.A.A.S.*, Washington, 1957). Most of our remarks will be based upon this symposium as being representative of views now held by people working in this field and concerned with the problem of the species.

In his introduction Mayr points out that several approaches have been used in defining species. One is like that mentioned above which has to do with a group of animals that are very similar in their structure. It is known as a "Typological-Morphological Species Concept" (p. 11-14). According to this view all animals having certain characteristics (usually as represented by certain specimens, "type specimens," on file in a designated museum) constitute a certain species. This can be quite restrictive. Sometimes from a single specimen, or a very small series, a taxonomist will describe a new species, designating one specimen as his "type"—that is, as representing his best concept of the characteristics of the species he is describing. The next worker then has to determine whether what he has is enough like the first man's description or type to be the same thing or whether it is different enough to give it another name. This is taxonomy in action. Do we need another name or can we use the same name? Upon what basis shall the decision be made? How much alike are the specimens in question? How do they differ? How much do they differ? After considering the data amassed from such inquiries, the investigator then must decide whether the later form can be grouped with the former according to the restrictions of the description. If not, he will probably go through the same process to describe and name his form, so that future investigators will have a name to use in discussing this kind of organism. This is one aspect of a species, organisms that are like certain other organisms previously described and named.

There is another aspect of species which Mayr points out. When you see or hear the word "robin," you get a mental picture of a robin. You may think of robins in your back yard, of robins flying around Ames, or wherever else you've seen them. You get a certain idea of *Robin* as a group of animals actually living in nature. I want to designate this as a "Conceptual View" (p. 14-16). (Mayr gives it no name, actually.) What is your *idea*? It is not based upon a type, an actual organism, but on what you think of when the name comes to mind. This "definition" is somewhat broader in its limits than the typological one.

The third definition, one emphasized by Mayr and others in the symposium, is called the "Biological Species Concept" (p. 16-19). This view pays attention to how a group of animals come to be, to how it is maintained, and to what it will be in the future, as well as to what it actually looks like now. It presents the species as being a group of animals or plants which are interchanging their genes. The form of animals or plants is determined by structures within the nucleus of the cell known as chromosomes. In these chromosomes are genes—the hereditary factors. These factors, to be effective upon the offspring of any mating, must be passed on to that offspring. The particular combination of genes occurring in the offspring determine its form. Its actual structure—what the animals look like—is perpetuated by the passing on of these genetic characteristics from

(Continued on page 5)

## SPECIES CONCEPTS AND DEFINITIONS

(From page 3)

one generation to another. This passing on of characteristics is very important in the concept of the species. If there is going to be any change in the species it must come through a change in the genetic make-up of the organisms which are further produced. This is the reason there is a great deal of emphasis being given to this particular concept at the present time. If all organisms, all the various different ones, have indeed evolved from ancestors in time past—then this is the only known mechanism whereby such differentiation could have been accomplished.

But how much change—how much differentiation is possible? Are there any limits to this process? If so, what might they be? As we discuss this matter of organisms containing hereditary characteristics which they pass on to their offspring, we are discussing, as many authors have pointed out, "evolution in action." Evolution is defined by some as any change in plants or animals which can be passed on to their offspring. By this definition any inheritable change is a form of evolution. Such change would tend in some cases perhaps to develop new species.

We want also to examine "creation in action." One of the points that we need to consider is, "Was God's creative activity greater at one time or another?" Is it true, as Carl Henry ("Science and Religion" in *Contemporary Evangelical Thought*, New York: Harper, 1957) says in one of his recent books, that one of the essential criteria of creationism is "that God created *ex nihilo* by divine fiat." Or, if I may put it this way, magical development—the appearance of a form where there was none before. This is considered by some, as I say, to be a necessary postulate of creationism. Is it?

These are some of the things, then, that we want to explore—that we want to consider—as we look at this problem of "What is a species?" How does it apply to, "kind," as the word is used in Genesis, in particular, as well as elsewhere in the Bible? Several papers have been presented on this subject in past conventions. Today, as I said before, Dr. Marsh will read his paper on "Genesis Kinds in Our Modern World," which we have characterized as a conservative creationist viewpoint of this problem, and then the other panel members will suggest some of the viewpoints that have been brought out either in the A.A.A.S. symposium or in other realms. They are by no means restricted to the symposium. We use this as a basis because it gives an excellent summary of what present-day scientists are thinking along this line. Then we'll have further discussion.

REFERENCES

Henry, Carl F. H. "Science and Religion" in *Contemporary Evangelical Thought*. New York, Harper, 1957. (See also his "Science and the Bible," 1958, *Christianity Today* II [23]:20-22.)

Mayr, Ernst. "Species Concepts and Definitions in the Species Problem." *Publication No. 50* of the *American Association for the Advancement of Science*, Washington, D.C., 1957.

# *The Genesis Kinds in Our Modern World**

FRANK L. MARSH
Professor of Science and Religion, Andrews University (Berrien Springs, Mich.)

In the first chapter of Genesis we read that the basic types of plants and animals appeared upon our earth through an act of special fiat creation. These basic types are described as not only being formed each after its specific morphological pattern, but also with a reproductive mechanism which caused each type to produce new individuals like itself.

The briefness of this Genesis account of origins gives opportunity for the development of at least two schools of interpretation with regard to the degree of fixity in nature indicated by this terse record. During the Middle Ages, or medieval period of history, in round numbers from about A.D. 400 to A.D. 1400, the opinion prevailed among scientists that the statements of Genesis declared that in reproduction the new individuals of a kind were as like as pennies from a mint. The only institutions of learning of the Middle Ages were those controlled by the church, and in these centers of thought the deductive type of reasoning was used. With regard to origins, the general premise was always the assertion of extreme fixity said to be taught in Genesis. In certain theological centers this idea of extreme fixity even resisted the changes of the Renaissance and the shift to the inductive method of reasoning. And it was still taught as dogma to the students of theology at Cambridge when Charles Darwin was graduated from the Department of Theology in that university in 1831.

At Cambridge, along with the idea of an extreme fixity which did not permit the development of new varieties since creation, Darwin was also taught that all modern forms of plants and animals had been created and set down in the very pattern of geographical distribution in which we find them today. These two bits of dogma were presented to the students in theology at Cambridge as the only possible teaching of Genesis on these points. Accoutered with these extreme views of special creation, Darwin went forth on his five-year circumnavigation of the globe as a sincere creationistic naturalist.

There is small wonder that as Darwin proceeded from region to region on that voyage and observed abundant concrete evidences that organisms had spread over the surface of the earth and had commonly varied somewhat as groups became more or less isolated from their relatives, he became more and more troubled in his mind over the fixity of the kinds which he had been told was the teaching of Genesis. He pondered the apparent head-on collision between what he had been told that Genesis said and what he could see was really happening. We wish that Charles Darwin had studied Genesis for himself and seen the actual harmony between the Bible and nature. But he accepted the teaching of Genesis as extremists had interpreted it to him, and after battling with the problem for years after his return from his voyage, he finally reached a tragic decision. This decision was to refuse the idea of the fiat creation of basic types of organisms.

This decision was reached in the year 1844. At that time, in a letter written to his friend, the botanist Hooker, he said:

"I have read heaps of agricultural and horticultural books and have never ceased collecting facts. At last gleams of light have come, and I am almost convinced (quite contrary to the opinion I started with) that species are not (it is like confessing a murder) immutable." (See Erik Nordenskiold, *The History of Biology*, 1928, page 463.)

The second school of interpretation with regard to the degree of fixity within the kinds indicated by the statement of Genesis is based upon the opinion that the book of nature and the written Word shed light upon each other. Correctly interpreted these two sources of truth *do* agree. They have the same Author. The Bible itself directs us to go to nature for confirmation of profound verities. In Job 12:7-11 we read:

"But ask now the beasts, and they shall teach thee; and the fowls of the air, and they shall tell thee: or speak to the earth, and it shall teach thee: and the fishes of the sea shall declare unto thee. Who knoweth not in all these that the hand of the Lord hath wrought this? In whose hand is the soul of every living thing, and the breath of all mankind."

Therefore, the members of this second school of interpretation go first to the Scriptures and learn that the statements of Genesis neither exclude the possibilities of variation within the kinds, nor do they assert that plants and animals were created in their present details and set down in the areas where we find them today. Then turning to nature these students find that Darwin was entirely correct in his observation of migration over the earth accompanied with variation. What Darwin failed to observe was that variation is not without bounds and is definitely limited in each case to the locus of its basic type or Genesis kind. All individuals of even abundantly variable forms, such as men and dogs, are unquestionably in every instance bona fide members of their respective basic types.

Because of his outstanding ability and because of his great contributions to the basic science of taxonomy, believers in special creation are always glad to recall that the Swedish botanist, Carolus Linnaeus,

* A talk given at the Thirteenth Annual Convention of the American Scientific Affiliation at Ames, Iowa, August, 1958.

was a creationist. Interestingly it is not unusual even in our day to find people who are of the opinion that he was especially endowed by heaven in his ability to point out the created units or Genesis kinds among living forms. However, an endeavor to learn just what classification groups in nature were considered by Linnaeus to be the Genesis kinds is likely to end in some confusion because during his life he published at least two opinions on the extent of the basic created unit. During the most active period of his life we find in the various editions of his *Systema Naturae* the following assertion:

"We count as many species as have been created from the beginning; the individual creatures are reproduced from eggs, and each egg produces a progeny in all respects like the parents."

Linnaeus realized the difficulty of determining natural affinities and did, in my opinion, make many mistakes in his endeavor to distinguish the created kinds in nature. Illustrations of this would be his assignment of different species names to the American bison and the European bison, and to spring wheat and winter wheat.

In his later life, after a great deal of observation of the bordering of some species on one another, and particularly as a result of his own experiments in hybridization, he changed his opinion of the created unit. From his twelfth and last edition of *Systema Naturae* he omitted the statement, "No new species arise." Then in his *Systema Vegetabilium*, published in 1774, four years before his death, we read the following interesting opinion regarding the original created unit:

"Let us suppose that the Divine Being in the beginning progressed from the simpler to the complex; from few to many; similarly that He in the beginning of the plant kingdom created as many plants as there were natural orders. These plant orders He Himself, therefrom producing, mixed among themselves until from them originated those plants which today exist as genera.

"Nature then mixed up these plant genera among themselves through generations of double origin (hybrids) and multiplied them into existing species, as many as possible (whereby the flower structures were not changed) excluding from the number of species the almost sterile hybrids, which are produced by the same mode of origin."

Because Linnaeus used a purely artificial system of classification and recognized only the four taxonomic categories, Class, Order, Genus, and Species, it is not easy from the above statement to secure a clear picture of what was his mature concept of the created unit. It may be helpful, in an effort to understand his mature opinion here, to select his order *Gymnospermia* as an example. Today our taxonomists use the name *Gymnospermae* for a class of plants made up of cycads, ginkgo, and conifers. However, Linnaeus' *Gymnospermia* consisted largely of mints and snapdragons.

Thus in Linnaeus' opinion God spoke into being parent forms of such groups as the mints and snapdragons and then by His own controlled hybridization developed among these, additional plant groups which Linnaeus called genera. Then as the centuries passed He built up the modern groups which we call biological species—groups which, to continue with our example, are illustrated by such plants as skullcap, catnip, motherwort, sage, horsemint, mullein, toadflax, and painted cup. It is possible that not all special creationists of today would be willing to agree that plants as varied as mullein and foxglove had evolved from a single created unit. However, we would stress the fact that we believe Linnaeus was certainly on the right track when he judged that any forms which would hybridize had sprung from a common ancestor. This would be a limited form of change, but certainly not evolution of new basic types. Possibly it would be more accurate to designate such change as mere variation within the original basic units.

It is obvious from the wording in Genesis that the expression "after his kind," includes both morphological and physiological characteristics. That is to say, when the plants and animals appeared upon the earth the individuals of each basic type were distinctly different in the details of their form and structure from the individuals of all other basic types. With regard to their physiology we read in Genesis 1:12 that not only were the members of a basic type alike in their form and structure, but when they reproduced, new individuals appeared which were like their parents. The Scriptures state that each of these forms yielded "seed after his kind." To express it mildly, in the light of this text it is difficult to understand how a basic type could transmute into a new basic type, or could give rise to a new basic type, if its reproductive performance was such as to bring forth additional individuals of the same kind as their parents.

Such is the letter of the written record. Should we conclude from this record that Genesis does teach the extreme fixity in nature which the creationists of the Middle Ages averred it did? The creationist of today believes that the Bible and nature are complemental, each helping to explain the other. Therefore, we turn to nature to discover the degree of fixity indicated by Genesis. In speaking of this situation in nature, Dr. Theodosius Dobzhansky, Professor of Zoology, Columbia University, says:

"Organic diversity is an observational fact more or less familiar to everyone. . . . If we assemble as many individuals living at a given time as we can, we notice at once that the observed variation does not form any kind of continuous distribution. Instead, a multitude of separate, discrete distributions are found. In other words, the living world is not a single array of individuals in which any two variants are connected by unbroken series of intergrades, but an array of more or less distinctly separate arrays, intermediates between which are absent or at least rare." —*Genetics and the Origin of Species*, second edition, pp. 3, 4.

This discontinuity is one of the most familiar characteristics of the living world as we recognize men, horses, cows, dogs, and cats, and roses, petunias, marigolds, zinnias, and water lilies. This same discontinuity is also one of the most striking features of the fossil world.

The explanation of this very real existence of gaps between the basic types of organisms is one of the great problems of the evolutionist. If all modern forms have evolved from one or a few primeval protoplasmic blobs, why should both the fossils and the living world present us with this striking discontinuity just as if the different kinds had originated as Genesis declares they did?

This problem was one of the topics in a series of letter discussions which I had with Dr. Dobzhansky a few years ago. Dr. Dobzhansky is today one of the leading American disciples of the theory of evolution. In our discussion I pressed him to give me just one instance in our living world where evolution of a new basic type was known to occur. His reply was as follows:

"When one says that evolution is established beyond reasonable doubt, one obviously does not mean that one can see evolution happen and reproduce it in a test tube, but this is the evidence which you escape by your device of saying that it is all change within a 'kind.' What you are after is evidently evidence for the thing which is called by this rather unfortunate term 'macroevolution,' Now, this is a process taking place in geological time, hence it, as any other historical process (human or natural), can be proven or disproved only by inference from the available evidence."

Dobzhansky's admission of the impossibility to demonstrate the evolution of new basic types among living forms is typical of the testimony of all evolutionists who are really conversant with the pertinent facts. After having admitted that evolution cannot be demonstrated among living forms, Dobzhansky passed the burden of demonstration over to the paleontologists who, in his opinion, could demonstrate that evolution of new basic types had occurred during geological time. He referred me to the then new work of George Gaylord Simpson, famous paleontologist of the American Museum of Natural History. This book had just come from the press under the title, *Tempo and Mode in Evolution*. Of this book, Dobzhansky remarked, "To me at least this is a most lucid explanation of paleontological evidence."

I secured a copy of Simpson's book and among much interesting material found the following assertions:

"On still higher levels, those of what is here called 'mega-evolution,' the inferences might still apply, but caution is enjoined, because here essentially continuous transitional sequences are not merely rare, but are virtually absent. These large discontinuities are less numerous, so that paleontological examples of their origin should also be less numerous; but their absence is so nearly universal that it cannot, offhand, be imputed entirely to chance and does require some attempt at special explanation, as has been felt by most paleontologists." *Id.*, pp. 105, 106.

On this same point of gaps between the various types of fossil forms, D. Dwight Davis, Curator, Divison of Vertebrate Anatomy, Chicago Natural History Museum, says on pages 74 and 77 of *Genetics, Paleontology, and Evolution:*

"The sudden emergence of major adaptive types, as seen in the abrupt appearance in the fossil record of families and orders, continued to give trouble. The phenomenon lay in the genetical no man's land beyond the limits of experimentation. A few paleontologists even today cling to the idea that these gaps will be closed by further collecting, i.e., that they are accidents of sampling; but most regard the observed discontinuity as real and have sought an explanation for them."

"But the facts of paleontology conform equally well with other interpretations that have been discredited by neobiological work, e.g., divine creation, etc., and paleontology by itself can neither prove nor refute such ideas."

We will agree with Dr. Davis that it is correct that divine creation of basic types cannot be demonstrated by the fossil record, but we cannot refrain from saying that the distinctness of the basic types in the fossil record with the absence of intergrading forms is completely in harmony with the creation of plants and animals after their kinds as portrayed in Genesis. The fossil record constitutes the only natural record we have of what occurred before the dawn of secular history. In the light of the fossil record, the theory of evolution which asserts that all modern types have evolved gradually from one or more simple blobs of protoplasm requires more faith for its acceptance than does the theory of special creation which asserts that God created the basic types instantaneously in all their morphological differences. We hear every now and then of "the missing link." Actually among both fossil and living forms great chains of links are everywhere absent between the basic types.

Thus it is that by observation of the living world and of the natural record of the past we find that there is no disagreement between Genesis and nature. Organisms appear always to have existed and still do live in a discontinuous pattern. We return to the question which concerns the degree of fixity that is referred to in Genesis. Were new individuals as like one another and their parents as coins from the mint are alike and are identical to the die which stamped them? The creationistic schoolmen of the Middle Ages declared that such was the doctrine of Genesis.

A study of the fossil record reveals to us that groups of organisms have maintained their individuality all the way down to our time. Austin H. Clark of the United States National Museum refers to this fact in the following statement from pp. 100 and 101 of his book, *The New Evolution: Zoogenesis:*

"Strange as it may seem, the animals of the very earliest fauna of which our knowledge is sufficient to enable us to speak with confidence, the fauna of the Cambrian period, were singularly similar to the animals of the present day. In the Cambrian period crustaceans were crustaceans, echinoderms were echinoderms, arrow worms were arrow worms, and mollusks were mollusks just as unmistakably as they are now."

Here is the sort of fixity referred to in Genesis, and behold nature shows us that the fixity is that of group characters and not a fixity of all individual characters. Each individual bears the distinguishing marks of his kind but is not necessarily identical with other individuals of his kind. Dr. Clark refers to this fact in the following statement from p. 100 of his book:

"In the details of their structure these fossils are not necessarily like the crustaceans, starfishes, brachiopods, annelids, or other creatures living in the present seas. Nevertheless, if they are sufficiently well preserved, we have no difficulty in recognizing at once the group to which each and every fossil animal belongs."

The testimony of *living* nature with regard to the extent of fixity indicated in Genesis is all about us in most intriguing forms. The processes of variation furnish us with many interesting breeds of plants and animals. Individuals often vary considerably within some groups. We have over 500 varieties of the sweetly scented sweet pea and over two hundred breeds of dogs. One author (see Griffith Taylor in *Environment and Race*) has divided human beings into as many as 160 breeds. Evolutionists love to call our attention to all this variation that is going on and insist that here is evolution before our very eyes. We all observe that variation *does* occur, but evolutionists fail to perceive that after all that the process of variation can accomplish has been accomplished, we unquestionably still have sweet peas, dogs, and men. The sort of evolution that the theory of evolution requires is the natural development of *new basic types.* But every additional case of variation that is studied, be it among the fossils or living forms, merely brings additional evidence that there is a law in nature which declares that every organism can produce only individuals which are unquestionably of the same basic type as the parents.

The evolutionist makes a creator out of Father Time by affirming that if we will just allow enough duration then processes of variation will produce new basic types. The plea that time will do it is no more reasonable here than it would be should we invoke it in trying to lift ourselves. If we see a lad trying to lift himself by his bootstraps, we would be incorrect if we were to say to him, "Just keep trying long enough, sonny, and finally you will be able to do it!" Such a feat can never be accomplished because there is a law in nature which says that just as hard as you pull up just that hard you push down. In the same way time cannot accomplish the appearance of new basic types because there is no mechanism in existence which can accomplish changes of sufficient magnitude to produce one new basic type. Every additional case of variation studied adds one more bit of evidence further to clarify this principle.

Not infrequently the creationistic biologist is asked, "In our present system of classification of plants and animals is there any category which is an equivalent of the Genesis kind or created unit?" The answer to this question seems to be "No." At the time of creation the kinds of basic types were each created after a distinguishing pattern in form and structure, and they were able to produce other individuals like themselves. As we look into nature today we find that man, *Homo sapiens,* can cross with no other animal. So in his case the species is the created unit. In other instances we find that the dog, *Canis familiaris,* will cross with the gray wolf, *Canis nubilis,* and the horse, *Equus caballus,* will cross with the ass, *Equus asinus.* Here the genus is the created unit. Again the common goat, genus *Capra,* will cross with the common sheep, genus *Ovis* to the extent of producing fetuses which will live until just before the time for birth. A more successful generic hybrid is the case of the genus *Bibos* which will cross with the Brahma cow, genus *Bos,* making the family the created unit. (See *Mammalian Hybrids,* 1954, by Annie F. Gray, published by Commonwealth Agricultural Bureaux, Farnham Royal, Bucks, England.) Yet again the domestic hen, family *Phasionidae,* has been crossed with the turkey, family *Meleagrididae.* Thus the order becomes the created unit. (See *Handbook of Biological Data,* 1956. Edited by W. S. Spector. Published by Saunders. ) In the naming of plants we find the same lack of harmony with the Genesis kinds. Very commonly species of the same genus will cross, as the bur oak with the swamp white oak. Genera not infrequently cross, for example, rye with wheat, and field corn with Teosinte and gama grass. One of the most interesting crosses in plants probably is that of radish with cabbage, both representing genera of the mustard family. To my knowledge, among plants, members of two different families have not been crossed.

If we accept the ability to hybridize as the principal characteristic of the groups within the Genesis kind, we apparently have no single category in our present-day taxonomic system which is equivalent to the Genesis kind. It was in the light of these facts that in 1941 I suggested the new name *baramin* (Hebrew *bara*—created, *min*—kind) for the Genesis kind. (See F. L. Marsh, *Evolution, Creation, and Science,* p. 174.) This would be a physiological group into which all forms would be placed where even so much as true fertilization of the egg occurred. Even though the embryo might develop no farther than the earlier stages, its parents would still qualify as members of one basic type. This would give us the man baramin, the dog baramin, the horse baramin, the oak baramin, the maize baramin, and so on.

*Deductively,* the idea of the baramin springs from Genesis 1:12 where we are told that plants not only were made after their kinds but also brought forth after their kinds. This reproductive isolation would depend largely upon the chemistry of the different protoplasms. If two protoplasms were compatible, then true fertilization would occur. Morphology arises from the chemistry of the hereditary determiners in the protoplasm. We may believe that forms which could reproduce would be similar chemically and would be expected to be quite similar morphologically.

*Inductively,* in every known instance in living nature where true fertilization can occur, the parents are sufficiently similar morphologically to be considered members of a single kind, such as the man kind, the dog kind, the cow kind, the oak kind, the corn kind, the apple kind, and so on.

It is sometimes objected that the baramin concept is weak in that many of the crosses obtained have occurred in captivity and probably would not take place in undisturbed nature. This objection indicates that I have failed to clarify the baramin in the mind of the objector. Actually animal psychology does not enter into the baramin concept. Rather it is a physiological, that is, chemical, test and still applies whether occurring naturally in the aisles of the forest or artificially in vitro in the laboratory. The essential assumption is that the chemistry of the members of the Genesis kind are identical enough to cause them to produce germ cells which will be compatible and able to unite in true fertilization. Artificial pollination and artificial insemination would be the best tools for the discovery of the limits of the baramin.

We realize that the processes of variation, namely mutation, recombination, and chromosomal aberrations, have been working in these basic types since creation and have produced physiological incompatibilities within the Genesis kinds so that ability to interbreed may not now exist among all members of the baramin. In such instances morphological characters will have to be used to determine membership. An illustration here would be the two groups of the

(Continued on page 11)

## THE GENESIS KINDS

(From page 8)

fruit fly, *Drosophila pseudoobscura,* which were formerly called Race A and Race B of *D. pseudoobscura.* Because complete sterility was found between some members of Race B when placed with Race A, Dr. Dobzhansky assigned to Race B the new species name, *D. persimilis.* The individuals of *D. pseudoobscura* and *D. persimilis* appear identical in external characters but may be completely sterile when mated. In such cases the morphological similarity of adults is sufficient to show that they belong to the same baramin.

Sometimes the question is asked, "Is the modern biological species identical with the Genesis kind?" I would answer that such may occasionally be the case. An example would be the biological species *man* which is also a Genesis kind. To be true members of the same biological species the individuals must be fertile interse. If within a biological species a group arises whose members are sterile when mated with others of the group, a new biological species would have arisen. The fruit fly mentioned above illustrates such a case. *D. persimilis* would be a new biological species arising within the older biological species *D. pseudoobscura.* Obviously all biological species are not originally created units. The growing popularity of the biological species concept among evolutionists is evidenced by the fact that, except for one, all eight contributors to the book, *The Species Problem,* a symposium edited by Ernst Mayr and published late in 1957 by the American Association for the Advancement of Science, accept the biological species and are rather enthusiastic about it. In recognizing the biological species as a natural unit, biologists are becoming less artificial in their classification and are making progress in the discovery of the Genesis kinds in nature.

Of course there are many forms in nature where the fertility test cannot be applied to determine either the biological species or the baramin. This situation would exist where new individuals are produced by such asexual processes as simple fission, budding, formation of spores, and even by the sexual process of hermaphroditism. The fertilization of their own eggs is quite common in higher plants and in a few animals. However, in these forms it is clearly evident that each is following closely the law of Genesis which says that basic types bring forth after their kinds.

The scientist reads in Genesis of the fiat creation and instantaneous appearance in the beginning of basic types of plants and animals which were made and which reproduced according to a certain fixity. The book of nature, through its fossil record, and in the world of living things, reveals that an actual fixity has ever existed and still does exist among these forms. The fixity is not one which produces identical

(Continued on page 13)

THE GENESIS KINDS (From page 11)

individuals, but rather is one which produces groups which enjoy considerable variation within their boundaries. These original groups demonstrate that they have no power to produce any new basic types. In this complete verification in nature of the assertions of Genesis, the Christian man of science receives added assurance that the Bible is indeed a book breathed by the God of Truth.

JUNE, 1960

# *Some Implications of Evolution for A.S.A.*

IRVING A. COWPERTHWAITE

At two recent annual conventions of our American Scientific Affiliation (at Gordon College and Divinity School, 1957, and at Iowa State College, 1958) there appeared to be a growing conviction that inexorable pressure of expanding knowledge is about to force us to accept some formulation of the theory of evolution, including the evolutionary origin of man, and that we must adjust our thinking in accordance with this eventuality. Naturally, if evolution has become established beyond a reasonable doubt, we must accept it even though it means that we have to make an "agonizing reappraisal" of our interpretation of Scripture.

There are a number of difficult, if not insoluble, problems involved in trying to harmonize the theory of the evolutionary origin of man with the account of the creation of man as found in the Bible.

In the first place, the Biblical account clearly implies that man was a special creation of God, climaxing His creative work in which He made the heavens and the earth. To be sure, Genesis 2:7 reads: "Then the Lord God formed man of dust from the ground, and breathed into his nostrils the breath of life; and man became a living being." Some seem to feel that the words, "formed man of dust from the ground," can be interpreted in terms of an evolutionary process spanning millions of years. Whether or not this is regarded as an acceptable interpretation, there is a still greater difficulty involved in the account of the creation of Eve. This story is told in Genesis 2:21, 22 in the following words: "So the Lord God caused a deep sleep to fall upon the man, and while he slept took one of his ribs and closed up its place with flesh; and the rib which the Lord God had taken from the man he made into a woman and brought her to the man." As Dr. R. Laird Harris pointed out at the convention at Gordon, these words quite clearly describe a specific act of creation in an apparently brief moment of time. The wording used in describing the creation of Eve would make harmonization with the theory of evolution most difficult.

A second difficulty is the Biblical teaching that the whole human race is descended from one man and one woman—Adam and Eve. It is difficult to imagine an evolutionary process that would culminate in the development of one viable pair. It would be much more reasonable to expect such a process to result in the production of a more or less widespread population.

In the third place there is the matter of the antiquity of man. The working of the genealogy in the fifth chapter of Genesis as presented in the English translation would certainly justify Ussher in assuming that he could determine the date of the creation of Adam by figuring back to the event. This genealogy is expressed in terms of unbroken succession of father and son from Adam to Noah. Possibly the original Hebrew language permits other interpretations. In any case, there are conservative scholars who believe that there are gaps of unknown duration in this genealogy and that these gaps push the date of Adam back far earlier than Ussher's estimate. However, it is hard to see how any intellectually tenable gaps could push the date of Adam back 100,000 and more years to a time when anthropologists and evolutionists claim that early man lived upon earth.

The foregoing paragraphs indicate three apparent discrepancies between the theory of the evolutionary development of man and the Scriptural account of his origin in a special creative act of God. These points are briefly outlined and there has been no attempt to present an exhaustive study of the problem. However, perhaps enough has been pointed out to indicate that very serious disagreement exists between the two. The following discussion is based upon two crucial "ifs." If evolution is now a solidly established scientific law which must be accepted and if evolution cannot be harmonized with the Biblical story then what is to be our attitude toward the Bible?

Can we dismiss the problem by assuming that the early chapters of Genesis are somehow of a lower order of inspiration than other parts of the Bible? If there are any reasons for assigning inferior authority to these passages they certainly are not generally known to reasonably well-informed laymen. However, the problem would not be solved by such an assumption because other passages of Scripture base teachings upon the story of the creation of man and by inference place their endorsement upon it. The following are a few examples of such teachings. The genealogy of Jesus as given in the third chapter of Luke traces His ancestry back to Adam. "The son of Enos, the son of Seth, the son of Adam, the son of God" (Luke 3:38). A number of passages teach that the sin of one man, Adam, passed a heritage of sin upon the entire human race. The following quotations are typical of this teaching. "Therefore as sin came into the world through one man and death through sin, and so death spread to all men because all men sinned—sin indeed was in the world before the law was given, but sin is not counted where there is no law. Yet death reigned from Adam to Moses, even over those whose sins were not like the transgression of Adam, who was a type of one who was to come" (Romans 5:12-14). "For as in Adam all die, so also in Christ shall all be made alive" (I Corinthians 15:22). The creation of Eve subsequent to the creation of Adam is also the basis of a New Testament teaching as found in I Timothy 2:12, 13: "I permit no woman to teach or to have authority over men; she is to keep silent. For Adam was formed first, then Eve." Thus it would appear that any doubts cast upon the authority of the story of creation would, by inference, also be cast upon many other passages of Scripture.

Perhaps we can resolve the difficulty by assuming that the story of the creation was written in picturesque figurative language and was never meant to be taken literally. It was, perhaps, a poetic "Psalm of the Creation." This solution to the problem would seem to set a dangerous precedent of "explaining away" troublesome passages of Scripture. What passages are meant to be interpreted literally and what figuratively?

Another possibility would be to consider that the Biblical writers wrote their messages in terms of the scientific theories and superstitions of their days and that any such scientific allusions can be ignored. Such a proposition could logically be extended to a similar assumption that these same writers also expressed their messages in terms of the moral and cultural patterns of their times, and that many problems dealt with have no application to us in this modern day and age. These suggestions would be equivalent to saying that "the Bible contains the Word of God" instead of "the Bible is the Word of God." This would mean that God has revealed Himself to man in a book written in terms of discredited science and outmoded cultural patterns. Acceptance of such a proposition would put a great responsibility on the Biblical scholar to search out the eternal revelation of God from among the temporal encumbrances.

On the basis of the two big "ifs"; if evolution is established and must be accepted, and if it cannot be harmonized with the Scriptures, some consequences in the field of Biblical interpretation have been suggested. All of the suggestions are apparently contrary to A.S.A. traditions and are doubtless contrary to our doctrinal statement. No solution to the problem of fitting acceptance of evolution into the framework of traditional A.S.A. beliefs regarding the harmony of science and the Scriptures has been found in the course of this study. It is therefore necessary to conclude this discussion with a question instead of an affirmation. Is harmonization possible, and if so, how?

# *The Origin of Man and the Bible*

J. Frank Cassel
North Dakota State College, Fargo, North Dakota

**This paper is in reply to the paper by Cowperthwaite on page 12.—*Editor*.**

As evangelicals give closer consideration to the claims of science, particularly in the area of evolution, questions arise as to the assessment of the claims in the light of the Scriptures, and eventually vice versa. Although knowledge is expanding, I would judge that it is not the new information as much as the better understanding through more honest consideration of the old that accounts for the shift of emphasis by the A.S.A. with respect to evolution. When I was in college, "evolution" was a dirty word. When A.S.A. began, those of us who questioned this position and insisted that to study it we had to say it (and that study it we must if we were to combat it) were read out of fundamentalism—remember the comment on Mixter's first draft of his "Creation and Evolution"[1] after it had been circulated to the membership. As you will see from my comments[2] on Mixter's Gordon paper,[3] that as we began seriously, critically, but objectively to examine the data of evolution, we became less and less convinced of its refutability. That, then, which you find no way to refute, demands a measure of positive consideration. This we've been doing—with the caution of one who extends his hand toward a fire, to be sure. But we are beginning to approach the problem positively. The first results of such an approach are reflected in the Gordon and Ames meetings and led, of course, to the principal topic for discussion at the Chicago meeting—that is, if our approach is changing, does this mean our philosophy has changed? Do our present studies stem from, or lead to, a different philosophy of science and/or a different or modified theology? The discussions of Chicago have made it abundantly clear, I think, that we scientists not only do not know the answer to this question, but that the philosophers and theologians aren't too sure either. Had they been sure, I think they would have carefully, logically, and kindly led us to their point of view instead of pouring on us the tremendous heckle which may be good pedagogy for freshman philosophy classes but is hardly productive in solving the serious problems we were approaching. I thought when I left Chicago that perhaps I was just oversensitive, but after my experience at the N.S.F. Institute,[4]

1. Mixter, Russell L., "Creation and Evolution" Monograph 2, Am. Sci. Affil., Goshen, Ind., 1951.
2. Cassel, J. Frank, "The Evolution of Evangelical Thinking on Evolution." *J. Am. Sci. Affil.* 11:26-27, 1959.
3. Mixter, Russell L., "An Evaluation of the Fossil Record." *J. Am. Sci. Affil.* 11:24-26, 1959.
4. National Science Foundation Institute on History and Philosophy of Science and Mathematics, American University, Washington, D.C., June 15–July 24, 1959.

I am convinced this heckle was more an evidence of lack of a positive contribution to offer. I am also convinced that if an acceptable, satisfying, and productive philosophy of science is to be developed it will have to be done by scientists—the philosophers will make fine and much-needed critics—but the basic formulation must be done by those who are immediately involved in the implications of induction—and of that sacred cow, "the scientific method."

But back to the immediate problems. Remember, what we're interested in is how God accomplished creation. We begin with the acceptance by faith of Genesis 1:1 (and elsewhere) that God is *Creator* and *Sustainer* (Colossians 1:16, 17). God has revealed Himself in creation as well as in His Word, in Jesus Christ, in the Holy Spirit, in the believer, etc. Hence, there can be no real contradictions in His various revelations. If such appear, they must be only apparent and lie in the fallacy of our interpretations of one or the other or of both revelations under question. You pose the problem of contradictions between the creation and the Bible, but the basic assumption you seem to be making is that our interpretation of creation must be in error and the phase in which active harmonization must take place rather than vice versa. This is our basic fundamentalist heresy and the hardest to overcome, since we are products of the vicious battles of the twenties and thirties during which time we were assured, and not without reason, that any hint of fallacy in our *interpretation* of the Bible was an attack upon the Bible itself and upon its God.

The infallability of Scripture and the truth it reveals need no vindication, intellectual or otherwise. Being what they are, they will stand. Let's look, then, to our views concerning them, which may really leave something to be desired here and there. What I'm trying to say is that if we're to attempt a harmony, let's glean the facts where they are, put them together as best we can as we prayerfully work under the Spirit's guidance, realizing that He may not see fit to give us the whole picture at this time. I am becoming more and more convinced that we as Christian men of science should approach our scientific endeavors with the same reverence and devotion with which we approach any other spiritual office and that we can look for God's enlightment in our study of His revelation in creation just as we can in our study of His revelation in Scripture. Somehow, by force of habit and the climate of the times, my faith is weaker here —but it shouldn't be!

But to Genesis. The current insistence by Henry[5] and many others upon *fiat creation* as an evangelical position, I do not see demanded by Genesis, at least with the connotation always made, namely, the accompanying insistence upon *ex nihilo* creation. If "fiat" simply means at God's command, well and good, for God is not restricted by our time-space continuum. But the connotation I always feel in these discussions is that of "immediately—out of nothing." These are human terms by which God may or may not choose to restrict Himself. As to the *ex nihilo*, this bothers me, perhaps because I'm human and can't comprehend it, but perhaps by intuition. God Himself has always been; so there was never *nothing*. God had Himself to draw on for creation. Brackbill at Harrisonburg helped me with this when he said he liked to think of God as *omnienergic*. This is certainly compatible with almighty and all powerful, but ties in more denotatively with current scientific thinking. It makes me wonder whether $e=mc^2$ may not be a *creative* as well as a destructive formula.

I also "bump" at our current usage of "special" in *special creation*. Is not any creation *special?* That man is more "special" in God's sight in that He desires fellowship with him and sent His Son to die for him is undeniable. Any way that God did really achieve this end of His creation is *special*, whether it has been by gradual, well-planned development over a course of time or has been by instantaneous command. He could have done it as He desired. Why do we insist in the name of orthodoxy to restrict Him before He reveals it to us in His own good time? "But He has!" I hear the cry, "You're just too blind to see it." True, but aren't we all blind? Do we have trouble ascertaining God's will for man in Jesus Christ? Do we hassle over the sinfulness of man or the righteousness of God? Why should God, who reveals Himself so clearly in this way, be so obscure that only a favored few of the chosen may know how He made us? In other words, I don't think the Bible is crystal clear as to the processes involved in the creation of man, and so we do it no insult to look elsewhere for evidences. But before we leave Genesis 2:7, the principle that stands out to me in its very strange way of putting it, is the principle of "process." In fact, the only real "fiat" act in Genesis that I can see by any connotation is in 1:3. The rest of the account as in 2:7 is an account of making, of "process."

With respect to the account of the making of Eve, here the process seems well and explicitly described. Here, as in the Virgin Birth, there seems to be certain process without scientific precedent or explanation. In the case of the Virgin Birth, it is not easy to explain genetically the birth of a *male* except through the act of a father contributing the male sex chromosome. Neither is it to be expected that from cells containing male sex chromosomes, should be formed a creature not containing any of these, but rather another in its place. Herein, then, God did not act according to the usual pattern by which He operates today. There are numerous instances of this. We call them miracles. But this does not make the other acts—the sustaining acts of God in nature—any less miraculous. We are so bound in our thinking, however, by the doctrine of uniformitarianism that we

5. Henry, Carl F. H., "Science and the Bible." *Christianity Today* 2[23]:20-22, 1958.

have trouble believing that God may not so restrict Himself, or that if He does, it's in a broader way than we are able to comprehend. We assume only that He will be what He is—consistent with Himself—but this may not always seem consistent from our finite viewpoint.

I see then at least two possible conclusions that I can draw from Gen. 2:18-25. One is that it happened just exactly as it says it happened. This position demands no change in my views of Genesis 1 or of 2:7 as expressed above. Here I have detail which I accept; there the detail is lacking. If man evolved, the mutant that became man because of God's breathing into him the breath of life and endowing him with powers far above other creatures would not by natural process have been likely to have had a mate available with these same endowments. Hence the need of "flesh of my flesh."

On the other hand, I can say that this is a beautifully poetic account of the birth of mankind—that "Adam" means not "a man" but "man, the species," and that "woman" is of the same nature and "flesh" as man. The distinction is here made to set the stage for chapter 3. God has acted marvelously in creation. He has given man all of His bounty culminated by an understanding of love in the love of man and wife—and still man sinned! Repugnant as this sort of interpretation is to me because of my background and training, I cannot deny the possibility of this sort of interpretation because this is the way I've been taught to interpret Revelation 2 and 3 for instance.

But must I choose? Must I say that it's an "either-or"? Might it be a "both-and"? Or might there be but a dim picture of reality here—a picture so dim as to reflect little of what truly happened? Whom, indeed, did Cain marry? I conclude—I don't know! But I'll keep on washing, lest I throw out the baby with the bath.

I was at a conference last fall at which Davie Napier of Yale Divinity School spoke. He spoke on Genesis and pictured Genesis 1-11 as "mythological" in import. Not, he assured us, in the sense of "myth" as a fairy tale but rather as the essence of man's experience in prehistory—as he encountered God. It shows us, he said, the eternal plan of God in operation from the beginning of time—*creation, sin, grace, redemption*. What more do you ask of the first chapters of Genesis—in essence?

As I read on, it seems that I have in my meanderings above taken care of both alternatives in paragraph 4, your second difficulty. You again pose the question in an "either . . . or" fashion, as if God had no hand in evolution. If this indeed was His creative process, could He not guide it to His ends?

The age of man is a real problem. Here, of course, lies the area which perhaps more than any other has brought us face to face with the irrefutability of some of the scientific claims—God's revelation in nature. Times were originally estimated from the sedimentary bed-fossil correlation. Geologists examining processes in action today and applying uniformitarian thinking to the interpretation of the geological sedimentary formations estimated the time necessary to lay down and consolidate these beds. Obviously a great margin of error must be allowed. But when, with carbon-14 and fluorine dating, their estimations are refined, but broadly confirmed, then we feel that we must account for this time in creation itself. In other words, our interpretation of this revelation is verified, sending us to re-examine our interpretation of other revelation at this point. As to the validity of Ussher's and others' exegeses, I cannot speak; for I am not critically acquainted with the facts or the methods. But I am left with the problem of accounting for human remains and artifacts which, by using various interverifying methods, show an antiquity far greater than 6,000 years. My alternatives? One, my methods, assuming uniform process through the ages, may be off because of a fallacy in the basic assumption. This is a live question to be asked of most of our scientific conclusions in many fields. But when I have several independent methods of verification which complement each other, this becomes less likely—although there is still the possibility of the discontinuity affecting each line of evidence equally and thus obscuring the discrepancy.

Another alternative is that the remains and artifacts identified as human, are so only in the taxonomic, morphological, and archaeological sense, and do indeed represent something prior to the man of Genesis 2:7. That is, God's *man*, the God-conscious man, may have come long after the brute man was differentiated morphologically and even mentally—long after he learned the use of tools and fire, etc. Spiritually, men may be but 6,000 years old, but such a man is essentially undefinable by any presently recognized scientific terminology. Hence the matter of the age of such a man becomes one of faith (as indeed does any other view), but is not verifiable by present methods.

A third alternative is that there really is more time allowed by the Biblical context than is obvious on first reading. Naturally when I see the need of more time, and some Biblical scholar says, "Sure, you may have more time," I tend to jump on his band wagon very quickly. But I still wonder whether he's given me more time because I demanded it, or because it's really there. The same question applies to the days of Genesis 1 or the universality of the Noahic flood. Just because one interpretation makes better sense to me than another right now, doesn't mean it's right, but as Ramm says about his "progressive creation," it is "that theory of the relationship of God's works and God's Holy Word which makes most sense to the author—and upon what other basis can he make up his mind?"[6] I'm not always sure I have to make up

6. Ramm, B., *The Christian View of Science and the Scriptures*. Grand Rapids: Eerdmans, 1955:293.

my mind right now—but I tend, of course, toward that combination of views which seem most consistent and is hence most satisfying to me. (They're no less satisfying to me, incidently, if someone disagrees with me. He may be able to show me a still more satisfying correlation—or I, him. But why should we fight about it? Particularly so when we both are working from the same basic presupposition of God—the Creator, Sustainer, Self-Revealer, and Redeemer.)

Let me note, before passing on, that though the antiquity of man becomes solidly established at ten or even one hundred times that allowed by Ussher, this in itself does not establish the fact of the evolution of man, either mechanistically or creatively. Time is one problem; change is another. With respect to change, we must also remember that there *has* been evolution within the species *Homo sapiens* since the time of Adam as is evidenced by the various races now living in various parts of the globe. But we still have not established the limits of evolution—nor can we at the present time. Anthropologists are still pretty well agreed that they do not as yet have a clear picture of the origin of man. There are numerous possibilities suggested, but no sequence of forms have yet been found that come near being as complete or as correlated as that of the horse or of some other forms. The "missing link" is still missing here, and there is even disagreement as to what it might link man to, were it found.[7]

But what influence has all this on our view of the Scriptures? What is the role of inspiration? At this point, I feel our basic philosophy of science and our basic theology as well as our personal relationship with Christ is most important. Our relationship with Christ transcends, it seems to me, both exegetical and scientific problems. In Jesus Christ we have found reality and whatever other mistakes we make in our interpretation, He remains our solid fact! In the light of this knowledge, and by His grace, then, I feel that I can question some of my long-cherished interpretations of the Bible without being damned by doing so or being led astray by Satan in the process.

Certainly the idea of a lower and higher order of inspiration gives little satisfaction to our resolution of the problem—but what do we mean by inspiration, anyway? God certainly does not reveal Himself in falsehood, nor is He a God of confusion. Hence I look for consistency in the Bible as I do in truth wherever it is found. And with truth goes authority. I think I have shown that there are alternative interpretations possible in Genesis without, in my opinion, altering the basic truth and authority of the Word. The New Testament commentaries which you quote do not necessarily alter the basic alternatives but have their relevance in each framework. Some may not be as satisfying as others, to be sure, but to the person who holds to Adam as "man," Paul's "one man" would be to make his point, not to vindicate Genesis, and would negate neither. But rather, inspired by the account in Genesis, the Spirit guides him to explain the effect of Christ's sacrifice. Again, the Adam-Eve sequence can be explained as spiritual. Whether this is true or a dodge is, of course, an academic question, for is it not the spiritual message which God seeks most to impart to us? Then why worry about what passages are to be interpreted literally and which figuratively. Look, rather, to God to reveal Himself more fully and most directly to you from each passage according to your need. Do I sound neo-orthodox here? I can't help it, because it seems to me that, practically speaking, the Bible is the Word of God to you as He uses it to reveal Himself to you. But the responsibility placed upon the Biblical scholar is not to *"search out* the eternal revelation of God from among the temporal encumbrances" but rather, as he studies, to have prayerful faith that God will make clear to him from His Word those eternal truths which he needs for this time. Let God be true and every man a liar—but let His truth be ours in Jesus Christ according to His will.

So without accepting your "ifs" categorically, I still find no problem in doing so academically without feeling that I must alter my basic view of the nature, authority, and inspiration of the Scriptures. To the man of faith, *God is* and whatsoever is of Him is truth. The scientist, philosopher, and theologian alike are seeking to learn more of this truth; and as they do, they learn of God. There can be no inherent con-

(Continued on page 28)

---

**7. See Leakey, L. S. B. "From Taungo Skull to 'Nutcracker Man.'" *Illus. London News,* Jan. 9, 1960. Recent finds in Africa of *Zinjanthropus* need to be carefully assessed in this light as some are proclaiming this form as the "missing link."**

**ORIGIN OF MAN** (from page 16)

tradictions in God or in His revelations of Himself. We must be careful alike of our affirmations and of our negations and know that when harmony seems lacking it is because our eyes are blinded. "Is harmonization possible?" you ask. Anything else is impossible. "How?" By faith, for it is here that the Christian must both begin and end. This is always the ultimate resolution of our "How?"

## *Letter*

This is a letter in reply to the paper by Cowperthwaite on page 12.—*Editor.*

Thank you very much for your fine letter of February 1 together with the enclosure, "Some Implications of Evolution for A.S.A." Please do not feel that I am ever so busy that I could not be bothered with your problems. I am much concerned over the problem which you present, namely that of the attitude of the A.S.A. toward the subject of evolution.

You are not the only one who was deeply disturbed at the apparent drift toward evolution at the Gordon meeting in 1957. On the other hand it has always been the policy of the A.S.A. to attempt to think through difficult problems in the general area of the subject of science and the Bible. So it has been my policy not to be too disturbed by the approach of any individual papers on this subject of evolution. There are many unsolved problems in the area, and I am sure that the A.S.A. will not bring forth a final solution. Rather, we should do as we have been doing in the past—investigate possible solutions. However, we do need to keep in mind that we accept the inspiration and divine authority of the Scriptures. The problem, of course, boils down to this: "What is our interpretation of any given passage?"

I have read your paper over carefully and have done considerable thinking on it since that reading. I have been attempting to find a mature Christian person in this vicinity with whom I might discuss your paper. I must say, as of now, I have not found such a person.

I certainly do agree with you that the Biblical account clearly implies man is a special creation of God: and furthermore, that this climaxes His creative work. Personally, I hold to the point of view that evolution does not have the answer here. However, we must reserve final judgment on this point until we have more evidence.

Concerning the matter of the antiquity of man, I must say that I do not see any possible difficulty between the Biblical account and the theory of evolution. Since the Biblical account itself skips as many as six generations when you compare one genealogy with the other, I see no reason for not putting in some similar gaps in early genealogical tables—that is, before the time of Noah.

Concerning the two crucial "ifs"—first, I wonder whether scientifically we have the evidence to state that evolution is now a fully established scientific law which must be accepted. I realize that this is almost universally accepted in the scientific world, but it always struck me that we have a good illustration of the scientist exhibiting his faith in a law rather than in divine revelation. Now, of course, it is true that in science we must exercise a faith—otherwise we could make no headway. On the other hand, is it reasonable to assume that we should have more faith in scientific principles than in divine revelation? Your second "if" suggests that the acceptance of evolution can not be harmonized with the Biblical story. Now, of course, there are many who say that this is simply not so. This has been brought out in several A.S.A. discussions. So you see, as far as I am concerned, this whole subject does not vitally effect my faith in the Bible. Your suggestion that perhaps the early chapters of Genesis are of lower order of inspiration than other parts of the Bible, of course, is firmly rejected by me. I assume that most of the members of the A.S.A. would agree with me on this point. Your other suggestion that the story of the creation was written in picturesque, figurative language and also that it was never meant to be taken literally is accepted by many Christians. I am quite sure that this is the attitude of a number of A.S.A. members. Here I agree with you that such a solution would seem to set a dangerous precedent of "explaining away" troublesome passages of Scripture. However, here again we must be very cautious and indeed quite charitable with those who take this point of view. We do realize that many parts of Scripture are quite figurative and it may be that this is another case.

Personally, I tend to take the story of the creation rather literally and I still am interested in attempting to correlate the Scripture passages and scientific facts and theory which may bear upon the same. One of our dangers in this area, of course, is that we tend to form final conclusions. We should realize that our interpretation of both scientific principles and of Scriptural passages must only be tentative.

Perhaps I have not helped you very much in your thinking, but I wish you to realize fully that I am deeply sympathetic with your point of view. I would be most happy if you could put together the various answers you get from different people and circulate them among your friends. Perhaps we could all be edified by such a procedure.

Yours in Christian service,
H. Harold Hartzler.
President

# *Neutron-Induced Variation of Roses**

WALTER E. LAMMERTS**

## INTRODUCTION

Curiosity over the origin of a naturally occurring albino bud sport on a plant of Audie Murphy, on the part of my physicist friends, Philip Livdahl and Harlan Zodtner, led to the consideration of experiments to induce mutation. Since studies on the effect of gamma radiation were already underway at Brookhaven, neutron radiation was an obvious line of attack. The following experiments, begun in the fall of 1957 and continuing until 1959, have given such unusual results as to be worth reporting.

### Materials and Methods

Rosebuds of the vigorous Grandiflora rose, Queen Elizabeth, a red Hybrid Tea H56024/39, and Rosa *multiflora* were cut from bud sticks and placed on wet filter paper, 50 in each petri dish. Except for a control group of about 100 buds left on bud sticks in Experiment 1, this method of exposing buds was used throughout. The following five experiments were performed:

1. In the fall of 1957, Philip Livdahl, then at Lawrence Radiation Laboratory in Livermore, California (now at Argonne Laboratory near Chicago), used the A-48 linear accelerator bombarding a copper target with 7.5 Mev deuterons. The approximate radiation intensity at the main axis of the beam was estimated to be about 400 rad, with neutrons emitted at a wide angle. Two petri dishes, each with 50 Queen Elizabeth buds were placed one immediately behind the other. A small bundle of bud sticks having about 100 buds of Queen Elizabeth was placed as near as possible to these petri dishes as a control, to study possible shielding action of bud stick tissue. All the exposed buds and 50 control unexposed buds were budded into R. *multiflora* understock the same afternoon immediately following exposure.

2. On March 14, 1958, 50 buds of Golden Rapture #5 were exposed to thermal neutrons from a "water boiler" reactor receiving approximate 1000 rad and budded into R. *multiflora* understock in the greenhouse.

3. On March 18, 1958, four petri dishes, each with 25 buds of Golden Rapture #5 were exposed to 400, 600, 1500, and 3000 rads respectively of 14 Mev neutrons from a Cockcroft-Walton accelerator. These were neutrons resulting from deuterons accelerated to 500 kev bombarding a tritium target. The buds were immediately budded into R. *multiflora* understock in 4" pots in the greenhouse.

4. On April 23, 1959, 50 buds of a red Hybrid Tea H55024/39 were placed 11 centimeters from the target of a Van de Graaff accelerator and exposed to about 90 rads of 15 Mev neutrons obtained by the accelerated deuterons impinging on a tritium target. Similarly 4 other petri dishes of 50 Queen Elizabeth buds each, were exposed to 90 and 120 rads, and 2 at 180 rads. As a check to test the effect of placing one petri dish behind the other, an upper dish at 9 centimeters away was exposed to 120 rad. All were budded in R. *multiflora* understock soon after exposure; i.e., the same afternoon.

5. Finally on July 9, 1959, 3 petri dishes (150 buds) of H56024/39, 1 petri dish (50 buds) of Queen Elizabeth, and 50 buds of the single 5-petaled white R. *multiflora* understock were exposed at 9½ cm from the target to 15 Mev neutron radiation obtained by using a Van de Graaff accelerator. As a check on the "protective effect" of petri dishes a "triple piggy back" group of 3 petri dishes, each with 50 buds of Queen Elizabeth, were placed one above the other so that the dish closest to the target (8 cm) received 500 rad of neutron radiation. These were all budded into R. *multiflora* understock the same afternoon following radiation exposure. The R. *multiflora,* single petri dish of Queen Elizabeth, and all 3 of the H56024/39 petri dishes at 9½ cm received about 350 rad of neutron radiation. The third petri dish of the "triple piggy back" group received about 280 rad. One of the H56024/39 dishes was removed after 180 rad of neutron radiation.

### Results

Before giving results it should be explained that Queen Elizabeth is a hybrid of the Hybrid Tea rose, Charlotte Armstrong x Floradora, a Florabunda.

*Experiment 1*—The most striking result of this experiment was the almost complete lack of mutations or even bud retardation from buds left on bud sticks when exposed. By comparison the 50 buds in the petri dish nearest the target were *all* retarded and slow to start and very deformed as to foliage and leaf appearance. Gradually many of them became more normal by growth from buds on deformed stems, but even these shoots were abnormal, having "strap" leaves and segments of heavily pigmented tissue. Finally by the spring of 1958, most of the plants had normal shoots which "took over" the growth. The original deformed branches continued to grow, and buds taken from them showed a wide range of variation. About 20 of the plants, however, remained completely changed, both as to color, form, number of flower petals, and appearance of plant and leaves. A thick large rugose leaf type was obtained so unlike

*Paper presented at the Fifteenth Annual Convention of the American Scientific Affiliation, Seattle, Washington, August, 1960.

**Dr. Lammerts received his Ph.D. in Genetics from the University of California and is Director of Research for Germains Seed Company, Livermore, California.

Queen Elizabeth that one would never suspect its relationship by induced mutation. Equally unusual was a very narrow strap leaf type, having thin long, pointed buds. Several had flowers about the size of Floradora with 50-60 petals as compared to 20-25 in Queen Elizabeth and were light scarlet in color. All were *weaker* in growth than Queen Elizabeth and one was so similar, except for its dwarf habit, as to be of possible commercial value, since some gardeners object to the remarkably vigorous growth of Queen Elizabeth.

Even more interesting was the unusual fact that plants from buds in the petri dish above the one next to the target were normal, except for one possible color variation to light scarlet. The buds were but little retarded, as compared to the control buds not exposed to any radiation.

*Experiment 2* may briefly be stated as negative, in that no mutations were observed following exposure to thermal neutron radiation.

*Experiment 3* was discouraging in that the two most desired mutations, i.e., increased petal number and darker yellow color, were not obtained. Occasional changes, such as reduction of petal number to 5-6 from the usual 20-25 (2 plants), and one plant with slightly narrower leaves were observed in those radiated at 3000 rad. These changes are valueless commercially. Evidently the genetic variability potential of Golden Rapture #5 was not great. Possibly this may be due to the fact that it carries at least 2 factors for pale yellow (YY). Hence, though genetically carrying recessive factors for deep yellow (yy), both dominant factors would have to be inactivated to allow expression of recessive deep yellow as hoped for.

*Experiment 4*—As the buds of those exposed to lower dosages of the high energy neutron radiation of this experiment began growing, it seemed at first that, even at this low level, we had variability comparable to that of Experiment 1. The following is a summary of observations made June 9, 1959, followed by summary of final results, as of August 18, 1960.

1. H56024/39—20-25 petaled, rose-red hybrid were radiated at 89 rad. 15 plants were normal in height, i.e., 16"-20". 7 were 6"-8", 5 were 2"-4", and 8 buds were just starting. Some were of normal height with lighter anthocyanin pigmentation. One plant has flowers with only 7 petals.

8/18/60. 34 surviving plants. One very weak plant died after transplanting. Final results were: 1 plant has 8-10 petals and 2-3 petaloids, and 1 plant is very slender and weak and 12" high compared to normal pf 26"-30".

2. Queen Elizabeth—89 rads exposure. Nine were normal 16"-20", 13 were 6"-8", 20 were 1"-4", 3 buds were just starting, 2 were dormant, and 3 were dead. Some normal plants were darker in anthocyanin pigmentation and all showed malformation areas.

8/18/60. 22 surviving plants. All were normal Queen Elizabeth plants both as to flower and plant habit.

3. Queen Elizabeth—178 rads exposure. Ten normal plants were 16"-20", 20 were 6"-8", 16 were 1"-4", 3 buds were just starting, 1 was dormant, and 2 buds were dead. Much greater deformity was present, even in those plants normal in height, than among the group receiving only 90 rads. One plant was almost black with anthocyanin. About a dozen were sectoral chimaeras for 50-60 petals; several of these also were light scarlet.

8/18/60. 28 surviving plants. Six were much weaker than normal, of which 2 had abnormally small leaves and flowers.

4. Queen Elizabeth—178 rads exposure. Four were normal 16"-20", 22 were 6"-8", 19 were 1"-4" high, 3 were dormant, and 4 buds were dead. Like #3, all plants were much deformed, even those normal in height.

8/18/60. 24 surviving plants. Four were much weaker than normal.

5. Queen Elizabeth—118 rads exposure. Four were normal 16"-20", 18 were 6"-8", 18 were 1"-4", 4 buds were just starting, 4 were dormant, and 1 was dead. Much deformity was present in all, seemingly more than in #3 or #4. Some small plants with leaves were not at all like Queen Elizabeth.

8/18/60. 34 surviving plants. All quite normal in appearance.

6. Queen Elizabeth—177 rads exposure. The bottom petri dish was nearest target at 9 centimeters. These generally were more affected than any. None were normal in height: 15 were 6"-8" high, 24 were 1"-4", 3 were just starting, 3 were dormant, and 2 were dead. Many, as they developed, showed sectoral chimaeras for increased petal number, light scarlet, and small-sized flower.

8/18/60. 37 surviving plants. All were quite normal in appearance.

7. Queen Elizabeth—118 rads exposure. Upper dish, 11 centimeters from target, was protected by first petri dish. Eight were normal height 16"-20", 26 were 6"-8", 9 were 1"-4", and 7 were dormant buds. Relatively little deformity noted on 4-5 plants affected at all.

8/18/60. 33 surviving plants. One plant had most of its leaves smaller and yellowish green. Flowers definitely were more scarlet. A normal shoot is now growing vigorously from the base. One plant had a thicker, leathery, dark green foliage and a light scarlet flower. Two very weak plants had narrow, dark green foliage. Three plants had sectoral chimaeras, i.e., stems having flowers much more scarlet than typical of Queen Elizabeth; one of these had over 60 petals, beautifully imbricated. All three stems were weaker than the rest of the plant.

At low-dosage rates evidently only a small percentage of cells in each bud mutates. As the plants continue to grow, normal shoots "take over" and except as noted the final result was a normal Queen Elizabeth plant.

Generally then the many changes were transitory, i.e., chimaeral. However, by budding from these, some were stabilized and may have horticultural value, even though they are not as vigorous as Queen Elizabeth.

*Experiment 5,* which was started on July 9, 1959, was comparable to Experiment 4, as regards Queen Elizabeth and H56024/39. In addition, 50 buds of R. *multiflora* were exposed to 356 rads of neutrons. Since the results are rather similar, they may briefly be summarized as follows:

1. Queen Elizabeth—50 buds in petri dish 8 cm from target receiving 502 rads of neutrons. Only 4 plants survived which are now quite normal. Ten of the exposed buds started and the plants at first were very delicate and prostrate. But buds from these more normal and finally basal shoots were completely normal. The original prostrate shoots are being budded to study variability.

2. Queen Elizabeth—50 buds in petri dish 9.5 cm from target, i.e., just above #1 and receiving 356 rads of neutron radiation. 41 plants survived of which 5 remained much weaker than normal. One originally prostrate plant now has normal basal shoot. There were many sectoral chimaeras for such qualities as:

(a) Scarlet.
(b) Small flowers similar to Floradora, having many petals.
(c) Flowers the size of Queen Elizabeth with many petals.

3. Queen Elizabeth—50 buds in petri dish 11 cm from target, i.e., above #3 and receiving 265 rads of neutron radiation. 42 plants survived, one with chlorotic leaves and 7 weak plants. Relatively few sectoral chimaeras appeared compared to #2.

Samples 4, 5, and 6 each had 50 buds of H56024/39 and received 356 rads of neutron radiation. Sample #7, also H56024/39 received only 178 rads. There were 49, 50, 46, and 43 surviving plants, all discouragingly normal in appearance. Occasionally flowers with only 10-12 petals were found, but unlike Queen Elizabeth radiated buds, no flowers with increased petal number were found.

The last experiment—petri dish #8—was 9.5 cm from the target and received 356 rads of neutron radiation. It contained 50 buds of a basic species, R. *multiflora,* having small white flowers and only 5 petals.

Evidently this species was highly resistant to change as there were no variations in flower color, petal number, or size. Of the 47 surviving plants, 6 are very definitely much weaker than normal and one of these is very weak. These plants will be budded to see if this reduced vigor is a stable condition.

### Discussion

Quite evidently by neutron radiation of buds at about 4000 rad we can tremendously increase the rate of mutation almost to the 100 per cent level in terms of mutation per bud radiated, since all buds showed at least sectoral chimaeras for changed tissue.

Horticulturally, the results were encouraging in that we did get several selection with increased petal number, having good bud form and a pleasing change to a light scarlet or what is popularly known as "coral pink" color. Also a dwarf type identical to Queen Elizabeth in bud, open flower form, and color was obtained which may have commercial value in that some gardeners object to the extremely vigorous growth of Queen Elizabeth (to 8 feet high in California gardens). This "buds true" in tests so far made.

Horticulturally, the results were discouraging in that we did not get color variations at least theoretically possible genetically. Thus, though we did get a white variation, the leaves were adversely affected, being semi-chlorotic, i.e., yellowish green in appearance. Possibly bud selection may segregate the color change from the leaf change should the cell tissue still be variable. Perhaps the situation may best be stated by listing the variations we did *not* get as follows:

1. Change to dominant $R^1$ or crimson red color from the recessive r or light scarlet range.

2. Change to dominant S or dark scarlet such as is characteristic of Floradora. Evidently all variations were to the recessive s or salmon to coral pink (light scarlet of British Horticultural Color Chart).

3. Change to the large-flower and large petal characteristics of Charlotte Armstrong. All the variations had petals either slightly smaller to much smaller than Queen Elizabeth.

4. Change to the long bud characteristic of Charlotte Armstrong.[2] Long bud is dominant to short bud due to a series of dominant or semi-dominant factors.

5. Change to climbing type or more vigorous pillar type.

6. Changes to long cutting stem with flowers occurring singly as in Hybrid Teas. All changes were toward cluster behavior of Floradora, the Floribunda parent.

7. Increase in leaf size. Queen Elizabeth has a large-sized leaflet fully as large as Charlotte Armstrong. Most variations, even the large rugose one, were toward smaller-sized leaflets.

Quite evidently most mutations were toward the recessive type as exemplified by:

1. Decrease in size of flower coincident with increase in petal number, i.e., resulting in many smaller petals.

2. Change to the recessive s or salmon (light scarlet appearance). These mutations were very frequent, i.e., in ratio of about 100 to 1 for change to white.

It would seem that the dominant M factor[3] was frequently inactivated, thus allowing the recessive s factor to show, or simultaneously the dominant S factor was also changed to s.

Compared to Queen Elizabeth, Golden Rapture is much more difficult to affect by radiation. Genetics tests show definitely that it carries recessive factors (yy) for dark yellow, a much desired color. Accordingly it was hoped that neutron radiation would result in mutation to dark yellow. However, quite evidently, much higher dosage rates must be used and probably more buds radiated in order to inactivate both of the dominant YY factors present. Contrariwise, Queen Elizabeth has only one dominant M factor which was relatively easy to inactivate, as judged by the frequency of sectoral chimaeras for light scarlet.

Both the red H56024/39 and Golden Rapture are of Hybrid Tea ancestry. Unlike Queen Elizabeth they are not Fl hybrids between a many-petaled, small-flowered Floribunda and a large flowered Hybrid Tea. Accordingly they do not carry factors for many petals. Hence their variability potential in this respect is low. We know that the single, 5-petal con-

---

1. Mehlquist, G. A. L., "Inheritance in the Carnation, Dianthus Caryophyllus—1. Inheritance of Flower Color," *Proc. Amer. Soc. Horticultural Science, 37*: 1019-1021 (1940).

2. Lammerts, Walter E., "The Scientific Basis of Rose Breeding," *American Rose Annual,* 1945, pp. 71-79.

3. Lammerts, Walter E., "Inheritance of Magenta Red Color in Roses," *American Rose Annual,* 1960, pp. 119-125.

dition is recessive. As noted above, both H56024/39 and Golden Rapture mutations to 5 petals were found, as well as intermediate conditions. Mutations increasing the number of dominant factors for increased petalage simply did not occur.

Finally the stability of the R. *multiflora* species is highly significant. Again mutations to the dominant, many-petaled condition simply do not occur. Likewise mutations to the A factor, basic to pigment formation, the dominant R factor for crimson red, or the S factor for dark scarlet did not occur. Again it is significant that there were no variations to increased flower size. R. *multiflora* is a *semi-trailing*, climbing type of plant. Though the sample is small, it is also significant that no mutations to upright bush type of growth, or recurrent flowering occurred. The conclusion is inescapable that mutations to dominant traits are difficult to effect by radiation.

In other words the success of a radiation experiment depends, not only on dosage rate, but the variability potential of the variety used.

Biologically stated, *all* changes were toward weaker less viable types, certainly not comparable to Queen Elizabeth in vigor or survival ability. In fact, it was quite clear that as the plants developed, the most *viable* tissue combination "took over" in that if any normal unchanged cells were present, eventually buds formed from them and they became the main plant. In the case of the 1957 experiment, evidently the 20 plants which remained changed and transmitted their changed traits by budding, were of homogeneous cell type. However, as these grew from buds, one could observe segregation of tissues and the plants grew progressively stronger as the cells with best gene combinations "took over" and became buds growing into uniform-appearing but changed plant type.

Some of these obviously differ from Queen Elizabeth by many factcrs. Since *all* are either completely sterile or semi-sterile, it may never be possible to determine their exact genetic make-up. Crosses of four, which showed a low percentage of apparently good pollen back to Queen Elizabeth, do not give much promise of setting seed.

Though it is recognized that many of the mutated types may well be "abnormalities" involving chromosome disintegration and so not be pertinent to the following discussion, many obviously are rather simple changes to the recessive condition. Certainly we have here a technique by which we can observe more "mutations" from one group of 50 neutron-radiated buds than usually is possible in a lifetime of looking at several million plants per year in the field run of plants from unradiated buds.

Are then these changes the sort of thing leading to evolutionary divergence? In this respect the results of neutron radiation of rosebuds checks closely with reports of other radiation work. By far the greater majority of all mutations found are defective. Occasionally as in the rose experiments, some are found which *horticulturally* are desirable. Even certain ones are recorded in the literature as showing increased resistance to some plant disease. But so far careful inquiry has brought out that as in roses, *associated* defects make the *survival* value of the mutant type in a natural environment highly questionable.

It would seem then that the idea that evolution can occur through the accumulation of mutations is simply not borne out by the ever-increasing array of experimental evidence. In fact, it becomes more and more clear that species and even varieties are very complexly integrated units tolerating very little "tinkering" or change. Quite obviously some species such as the chromosomally basic R. *multiflora* species are particularly resistant to change. Others have a greater variability potential.

Quite obviously the ability of species and varieties to change as they spread over the surface of the earth has been pushed out of all proportion into the idea of evolution. The inability of older naturalists such as Darwin to evaluate this variability potential of species and varieties properly and so to mistake it for evolution or change of one species into another is quite understandable. After all, they had no clear concepts of genetics and even set up genetic postulates completely at variance to what we now know occurs. But the continued interest and belief in evolution, at least in some modified form, on the part of so many modern biologists is puzzling, since the facts of genetic variability and mutation so clearly show that species and varieties have such clearly defined boundaries of variability.

Contrariwise, the idea of evolution has had a deadening effect on the minds of many, especially younger and less imaginative research workers, since the very concept tends to make one feel that changes come slowly. Actually, plant breeding experience shows that within the limits of species or variety-variability potentials changes can be effected very rapidly. In fact, with the development of modern tools of radiation research, we can in a few months probe the variability range of even such normally slow reproducing species as apples, pears, and peaches. Thus by neutron radiating buds of peach varieties and growing the budlings under continuous light,[4] we can get a crop in one year from radiated buds from which we can see the entire range of variability potential. However, we also must be clearly alert to the fact that once the range of variation is reached, further attempts to increase variability are a waste of time.

## Conclusion

1. Data are given showing technique and dosages use to get maximum percentage of mutations from rosebuds by neutron radiation.

---

4. Lammerts, Walter E., "Effect of Photoperiod and Temperature on Growth of Embryo-Cultured Peach Seedlings," *American Journal of Botany 30*, No. 9: 707-711.

2. Varieties differ strikingly in their variability potential. Thus variations in Queen Elizabeth toward increased petal number, smaller flowers, and change in color from carmine rose (presence of M factor) to coral pink (m) with s factor for light scarlet occur frequently. Change to white occurred only once.

3. Horticulturally, neutron radiation of buds shows great promise, since it so greatly speeds up mutation.

4. Biologically, all mutations are defectives, hence any postulate of evolution by mutation is contrary to evidence.

5. It is observed that the concept of evolution has had a bad influence on research in that it tends to make the scientist work on the theory that nature acts slowly and mutational changes occur slowly. Hence he tends to limit himself as to possibilities of getting results rapidly.

### ACKNOWLEDGMENT

The author wishes to acknowledge the advice and help of Philip Livdahl, Howard Tewes, and Harlan Zodtner in planning and executing the radiation of the rosebuds in these experiments.

# *Creation, a Finished Work**

WILLIAM J. TINKLE**

There is agreement among Christians that God should be credited with the existence of the universe. If there were no God, there would be no universe, for His plan and activity have brought it into being.

Since the universe has immense size and complexity, however, the methods and details of the formative process are not all known. One need not fear to plead ignorance regarding some of the steps in creation. Indeed the person who has all the answers is often suspect for that very reason.

The purpose of this paper is to shed light on one embarrassing controversy among Christians regarding the general nature of creation. Is it a process which was finished, or is it being accomplished gradually at the present time? In speaking of the formation of the universe, should one use the past or present tense? It is agreed among Christians that God had the power to do either, but considering His revealed Word and the record of nature, which method was used?

### Definition

The word creation is understood to mean bringing something into being, whereas neither it nor its forebears had being before. Primary creation means the formation of substances out of nothing. Secondary creation is the ordering and combining of substances in such a way that a new entity is formed which is more than the product of pre-existing factors.

Creation is not the sole type of divine activity. The continuous maintaining and ordering of the universe is called providence, and excites one's admiration in a manner similar to the contemplation of the original bringing into being. When one observes a landscape and is religiously moved, he should not say he has evidence that God is still creating.

Growth and reproduction of living things should be classified as providence rather than creation. The new plant or animal is built up according to a plan which is not new at all but which existed in its ancestors. The plan is embodied in a minute blueprint composed of genes, passed on to the new individual in the egg and sperm of the parents.

### Does It Matter?

Since it is admitted that God could finish His creative work or prolong it endlessly, some say that the question does not matter or that it is not more than academic. A minister of my acquaintance said that one guess is as good as another. He simply refused to consider the question. He and his kin are so interested in building a tall superstructure that they neglect the foundation.

Another minister, a student in Juniata College, approaches the question through reference to the nonconformist group of young people commonly called the Beat Generation. After discussing their rebellion against organized authority, he continues: "A second characteristic or expression of the Beat Generation is their denial of meaning and purpose in life. To them, life is blatantly full of question marks but starkly void of satisfactory answers. We stare about us in search of meaning or purpose but we see only the reflections of our own anxiety. It is as if life were a blank slate on which the only meanings to be found are those which we ourselves in desperation write to ease the pain of meaningless existence. We cry out for meaning and purpose, and the world cries back: 'Confusion and frustration. . . .' The conformist may be content with his unreal set of meanings, but the Beat chooses despairing sincerity rather than superficial contentment."[1]

Need we wonder at this denial of all purpose, since evolutionists have been crying out against teleology for one hundred years? Evolution is not the only support of the idea that the world has no purpose, but it definitely is one support. Natural selection based upon struggle for existence presupposes an arena where the violent and crafty ones win, whether they deserve to win or not. It denies the providence of God in order to account for creation without God. Those who give only lip service to this philosophy are not greatly affected, but the Beat actually accepts the outlook and yields to despair.

At this point someone objects that he does not believe in materialistic evolution but in development guided by God at each step. Such a belief, however, is at variance with that of the scientific evolutionists, and thus does not resolve any conflict. This desire to compromise with science and avoid an argument is the chief reason for formulating theistic evolution. Since it does not agree with the conventional evolutionists after all, one reason for accepting theistic evolution is lost.

Another reason why it is important to consider God's method of creation is the Biblical statement that God made man like Himself.[2] Why should God find it necessary to perfect His creature by eons of

---

*Paper presented at the Fifteenth Annual Convention of the American Scientific Affiliation, Seattle, Washington, August 22-25, 1960.

**Dr. Tinkle is a biologist and for many years was a professor at Taylor University. Following this, he taught at Ball State Teachers College and is now retired. He has published a number of papers as well as a textbook in biology.

1. R. B. Gardner, *Brethren Life and Thought,* Summer 1960, p. 53.
2. Genesis 1:26, 27.

brutal struggle in order to make him in His own image? And what does such a process suggest about the nature of God? It might teach that He is a selfish, relentless warrior rather than a God of love.

Concerning the fall of man, it may be that some theistic evolutionists accept it because it is in the Bible; yet there is no logical place for it in that system of thought. Evil is accounted for by man's incomplete development, and time alone may make up the deficiency. Evolution does not point out the need for a Saviour.

As stated above, not all evolutionists hold these anti-Christian ideas, but they do if they have accepted evolution wholeheartedly. Such ideas are inherent in the philosophy of evolution.

Another reason for holding the original meaning of creation is that some people reject the Bible rather than compromise. They do not appreciate the circumlocution which is necessary to harmonize the Bible with evolution.

### Compromise Seems Necessary

Let us consider the situation which has made a revision of the meaning of creation seem necessary. Science has been very successful in improving machinery and human health, thus winning the acclaim and gratitude of the public. One branch of science receives accolades as well as another, even though its pronouncements are only theories. When it is stated in the name of science that all life has developed gradually from lifeless matter, people feel that it is futile to deny it. It must be that we have not understood the Bible, and it is necessary now to interpret it in such a way that it will agree with this new truth.

To aggravate the situation, students in some Christian institutions have been advised not to study the data on which evolution is based. Such was not the experience of the author of this paper. In college I was not told what I must believe about creation, either by the college or by the church. The textbooks taught evolution in the same arguments that are used now, while the teachers made no definite commitments. Plodding and vacillating through my biology major, studying geology also, I finally decided that nature says, "No."

While some evolutionists browbeat their opponents by classifying them as either uninformed or prejudiced, others admit that gradual development has not been demonstrated. Newman wrote, "Reluctant as he may be to admit it, honesty compels the evolutionist to admit that there is no absolute proof of organic evolution."[3] S. W. Fox writes in much the same vein in the present year: "Darwin's theory of evolution has, for example, been judged and has proved to be intellectually useful on the basis of its consistency with much general knowledge, rather than on that of any single dramatic experiment."[4] While it agrees with the temper and thoughts of the modern age, it has not been clearly demonstrated.

On the other hand, most scientific conclusions have been demonstrated. Photographs of the earth taken from great height show its surface to be shaped like an arc. The earth's rotation is demonstrated by Foucault's pendulum, one of which is located at the Museum of Science and Industry in Chicago.

### Biblical Testimony

Since all Christians accept the Bible, although they do not agree as to its interpretation, we should seek to learn what is says. The first chapter of Genesis relates the formation of the earth and the different types of plant and animal life, concluding with man, and God is pleased with the success of His work. This seems to be a series of completed acts and the idea of completion is confirmed in Genesis 2:1: "Thus the heavens and the earth were finished, and all the host of them." Edwin Monsma comments, "This rules out any idea of a continuous creation after the sixth day, an idea inherent in evolutionary thinking."[5]

Later Scriptures likewise speak of creation as having been accomplished in the past. "Thou, even thou, art Lord alone; thou hast made heaven, the heaven of heavens, with all their host, the earth, and all things that are therein, the seas, and all that is therein, and thou preservest them all; and the host of heaven worshippeth thee."[6] This agrees with the Genesis account, using the perfect tense in speaking of the creation and the present tense in speaking of His providence, the preservation of these things.

Further confirmation is given in Psalms: "Let them praise the name of the Lord: for he commanded, and they were created."[7]

These Scriptures do not rule out the possibility of change, and certainly do not make it impossible for God to create a new heaven and a new earth, but they give no hint of a slow, upward, developmental process. We see in the Biblical account idea, plan, purpose, and command, rather than development by struggle, with elimination of the ones that fail to make the grade.

### Testimony of Nature

Since God is the Author of nature, we should see in it the same kind of creation which is described in God's Word. Many scientists have not noted the similarity, however, but have observed individuality and change. When some Christian protagonists denied changes in living things they made a mistake. The truth is, however, that the changes which can be observed do not add up in such a way as to make permanent progress, except where man guides the process.

---

3. H. H. Newman, *Readings in Evolution, Genetics and Eugenics*, University of Chicago, 1921, p. 57.
4. S. W. Fox, *Science 132*, 22, July 1960, p. 207.
5. E. Y. Monsma, *If Not Evolution, What Then*, privately published.
6. Nehemiah 9:6.
7. Psalm 148:5.

Much change is cyclical in nature. Growth in a living thing is progressive, but the next generation starts back at the base line, with no advantage from the environment of the parents. There is agreement among geneticists that "acquired characters" are not inherited.[8] Since this is true, most of the changes in plants and animals are for one generation only and do not influence future generations.

A tremendous amount of data has been accumulated to show that the characters of living things are determined by standard factors known as genes. They are located in the rod-shaped chromosomes in the nucleus of each cell, and had been seen by only a few men a hundred years ago, but now are observed and drawn by college freshmen. Genes are in no sense hypothetical or theoretical, but are entities proved by experiment.

If creation took place gradually by development, we might expect that a gene would develop gradually, making its character more pronounced with each generation. But Mendel, the father of genetics, held a very different idea, as reported by Nordenskiold: "Much surprise has been expressed over the fact that Mendel's brilliant observations did not attract greater attention, and the blame has been laid upon the unknown journal in which they were published. One might with greater justification ask oneself whether any of the more important publications of the time would have undertaken to print results of research so utterly at variance with the prevailing conception of biology. We have only to remember that Mendel denies variability in those characters that he observed, whereas all the biologists were just at the time seeking after variations as material in proof of natural selection; and then come these assertions as to absolutely constant or constantly divisible characters from the pen of a monk in a monastery! It would certainly have been a miracle if they had found support from the generation that had been brought up on Haeckel's *Natural History of Creation*. Nevertheless, the Mendelian principle of cleavage now forms the basis of all hybrid research."[9]

Modern geneticists are agreed that a gene does not change at all unless it makes a sudden and fortuitous change called a mutation. This change may be small or great, but takes place rarely. Snyder and David make the following statement: "Most genes are exceedingly stable. . . . The natural mutation rate is very low. Many species have remained much the same for long geologic ages. The brachiopods among animals and the seaweeds and others among plants are examples of groups of organisms in which almost no changes are observed in present-day species as compared with fossils. Even in the laboratory among organisms chosen for their capacity to produce mutations a high stability of genes is found. Muller, by means of cleverly devised experiments in Drosophila, has recently estimated that the mean life of a gene (that is, the average time elapsing without change in any particular gene and its descendants) approximates 100,000 years."[10]

Moreover, a very large percentage of changes due to mutation are harmful to the plant or animal. This is true to the extent that beneficial mutations are seldom named. If mammals developed from reptiles we should expect to find incipient organs developing in reptiles such as mammary glands or umbilical cords. If birds developed from reptiles, we might expect to observe a pin feather on a lizard now and then. Such advances have not been reported. Under the scrutiny of many observers, mutation should be caught in the act of originating some advanced character, of which the above are examples.

At the A.A.A.S. Convention at Indianapolis, T. Dobzhansky spoke on evolution and asked for written questions following the address. I asked him to name several mutations which have conferred an advantage to the organism. His reply was that there are a number which give an advantage under changed conditions, but one should not look for a mutation which gives an advantage where the environment has not changed.

I appreciated his frank reply which makes creation by development a devious and long-drawn-out process, if it exists at all.

Much could be written about abnormal numbers of chromosomes and abnormal positions of genes within them. But since no change in a gene itself is involved, the change in the plant or animal usually is slight, unstable, and not of an advanced nature. Such details are interesting to students of heredity, but they would make this paper too long.

## Conclusion

Every Christian who is also a scientist is confronted with conflicting systems of interpreting the universe. He has God's revealed Word and also the explanations formulated by men who have thought but little about God, yet have some interpretations which are helpful.

For one who is not averse to recognizing the power of God, it would seem that the most direct interpretation is that God created the genes of plants, animals, and men, some of which have reproduced themselves down to the present, but others have met with accident. Theistic evolution is an unnecessary compromise, not well supported by observation.

---

8. W. R. Breneman, *Anatomical Form and Function*, Ginn, 1954, p. 413.
9. E. Nordenskiold, *History of Biology*, Knopf, 1928, p. 591.
10. Snyder and David, *Principles of Heredity*, Heath, 1957, p. 349 f.

# *Origin of Life**

WALTER R. HEARN**

Some of you may feel that I have a lot of gall to say some of the things I'm going to say this afternoon, so I may as well begin by rendering unto Caesar that emulation which is due him and divide *my* gall, like his, into three parts. (Now, Dr. Kraakevik told me this talk should be *semi*technical, and I assumed that the "semi-" part would cover puns as well as misinformation, but maybe I was mistaken.)

In the first part of my talk I want to discuss some recent technical developments pertaining to the origin of life; in the second part, a little about the attitude of scientists toward this kind of investigation—the "climate" in which this work is being done, as it were; and finally, I want to comment on the reactions of Christians to these matters. All of my remarks will be rather sketchy, a few of them will be impertinent (to stir up more lively response from you during the discussion later on), and what I say will of course bear the marks of my own personal orientation both as a biochemist and as an evangelical Christian.

You may be aware that some reviewers of *Evolution and Christian Thought Today* in the conservative Christian press have felt obliged to put that word, "evangelical," in quotation marks when referring to authors of some of the chapters, or to prefix it with the adjective, "so-called." I shall have an opportunity at the banquet tonight to speak more fully of my Christian experience, so may I suggest that you leave those quotation marks only tentative until after this evening, anyway? Now then, since none of the reviewers felt obliged to call me a "so-called biochemist," I shall proceed now to the technical discussion and you can all relax, knowing it is safe to follow me *that* far at least. (As a matter of fact, I hope most of you *can* stand the suspense of leaving me inadequately classified for a while—you're likely to get a lot more out of the talk that way. I've noticed that when a speaker addresses a Christian audience and they don't follow him at *all*, if they think he's a good guy they'll say "Boy, that guy is profound!"—and if they think he is one of the bad guys they'll say, "Well, he sounded sorta liberal to me!" In either case, if they've got him pegged, they don't have to bother to try to understand what he says. Well, this will be an *adult*-western-type talk, so hold on to your seats: it'll be hard to tell the good guys from the bad guys. I'll try not to spoil the plot by giving myself away).

## Part I. (Technical)

The goal of *all* basic biochemical and biophysical research is increased understanding of the mechanisms of life processes: i.e., the translation of biology into the language of chemistry and physics. In a sense then, *any* advance in biochemistry is likely to shed some light on the scientific problem of the origin of life. If one concedes that the origin of at least *some* of what we think of as "life processes" may have preceded, or been concomitant with, the origin of life itself on earth, then learning the chemical details of *life* processes shows us what to look for in the *non*living realm so we can extrapolate intelligently back to the *pre*-living realm. Now, many exciting things are going on in fields such as intermediary metabolism, the study of viruses, and biochemical gentics particularly, which have a bearing on our ideas of the origin of life. Some of these have been spotlighted in the news, as when Severo Ochoa and Arthur Kornberg won the Nobel Prize for their work on the enzymatic synthesis of nucleic acids—work which should certainly make us ponder a bit before saying dogmatically that no possibility exists for the synthesis of new genetic material (which was said in a letter to the editor of *Christianity Today* a few weeks ago). In fact, I understand that in the discussion at the Darwin Centennial at the University of Chicago, H. J. Muller said that if the artificial "creation of life" could be defined as putting together a strictly chemical system that can synthesize new genetic material—well, then Kornberg has already done it! (I think this was regarded as a radical opinion, even in that environment.)

Another recent news story dealt with the complete determination of the sequence of amino acids in the protein of the tobacco mosaic virus—work which may be of particular interest to this audience because at least two of the men who did it are devout evangelical Christians, C. A. Knight of the Virus Lab at Berkeley and Duane Gish, now at Upjohn and Co. One of the most fascinating and still most frustrating problems of current biochemistry is learning how the genetic information, encoded in the sequence of four repeating bases in the nucleic acid structure, is "translated" into the particular sequence of twenty repeating amino acids

*Presented at the Wheaton College Science Symposium on "Origins and Christian Thought Today," February 17, 1961.

**Dr. Hearn is Associate Professor of Biochemistry at Iowa State University, Ames, Iowa.

which gives each protein its specific biological characteristics. The TMV work is important because in this case a single protein is presumably associated with a single nucleic acid molecule, and now we know the "code" on the protein part. We still cannot read the code on the nucleic acid part, which will require a tremendous amount of chemical work to unravel—and we may still be frustrated even then if it turns out that the nucleic acid code gets translated in the infected plant, not into *virus* protein, but into a *different,* still unknown protein, specifically the enzyme or enzymes which put together the virus protein. I emphasize this difficulty for the benefit of students in the audience who may get the impression that the really interesting problems in biochemistry are just about all solved, so you might as well go into some other field. STOP! DON'T JUMP! We are *looking* for new students sharp enough to come up with a way of cracking this nucleic acid code, of studying what goes *on* in an infected cell when virus particles are being synthesized, and of attacking a "host" of other challenging problems. In particular, *I'm* looking for some sharp graduate students, and I'll be happy to give you some literature to lure you into *our* department for your graduate training if you're qualified. If I should happen to miss contact with you, by all means write to me if you are the least bit interested.

I think the work you really want me to discuss, however, is the work aimed *directly* at the problem of the origin of life, and there is a considerable amount of this. I shall discuss some research not unrelated to the subjects I have just mentioned; as you can see we now have protein molecules which can catalyze the synthesis of nucleic acids, and there is plenty of evidence that the pattern for synthesizing specific proteins can be built into the structure of a nucleic acid molecule. There is still a huge gap in our understanding of the origin of *metabolizing* systems, i.e., complex enzyme systems in which endergonic (energy-requiring) reactions are coupled to exergonic (energy-yielding) reactions to give us a steady state not at thermodynamic equilibrium. However, I think you can see that if we *can* show that under conditions likely to have existed on the pre-biological earth, proteins and nucleic acids are formed chemically, then we have taken a big step in our understanding of the origin of life.

The significant work of Harold Urey and his student, Stanley Miller, in actually trying out some postulated primitive atmospheres to see what might happen, has been cited in my chapter in *Evolution and Christian Thought Today.* You will recall that a number of amino acids, the building blocks of protein, were found in the reaction mixture after those experiments. Since that time, a number of other investigators have continued along that line, and have shown that at least the simplest amino acids can be produced under a wide variety of relatively mild conditions. One effect of these findings has been to remove some of the restrictions, from the biochemist's point of view, on possibilities for pre-biological environments, still an area of some disagreement among geologists and geochemists. In a sense, whatever the environment, the origin of the building blocks of protein molecules has been settled.

With regard to the polymerization of these amino acids into proteins, there have also been some interesting developments. After many years of feeling that thermodynamic considerations would make this step essentially impossible, there is a rather sudden shift to thinking of it as highly probable. In fact, it has been shown that with HCN as an intermediate in the formation of amino acids (as in Miller's 1957 experiments), the *amides* of the amino acids are also intermediates, and we know that the energy barrier for polymerizing amino acid amides should be much lower than that for polymerizing the amino acids themselves. One investigator I know personally, John Oro of the U. of Houston, has shown that the amides can be polymerized readily under quite reasonable "primitive earth conditions." This might even make the amino acids *products* of protein-like materials via subsequent hydrolysis rather than their *precursors* under some conditions. (To put it crudely, you get to the chicken first, and then the chicken lays the egg instead of growing the chicken from the egg. That's about as crude as it can be put!)

Another investigator, Sidney Fox of the Oceanographic Institute of Florida State, has, however, also made a "breakthrough" in the polymerization of amino acids themselves into what he calls "proteinoids." (Incidentally, Urey and Miller are now attached to the Oceanographic Institute of the U. of *California,* so you can imagine what sort of rivalry goes on in this field between Tallahassee and La Jolla—but a public lecture is no place for gossip.) Fox tried thermal polymerization—i.e., just heating up a mixture of amino acids. Now generally, as all of us know who've ever left something on the steam bath too long by mistake, such a mixture is liable to give you a mess—"intractable tars" is the way you describe it in the literature (if you have to admit a reaction of yours went wrong somewhere). I'm not sure whether Fox actually looked this up *before* he planned the experiment, or whether one experiment worked and *then* he looked it up (and I was polite enough not to ask him when I talked to him about it this fall), but at any rate he noted that many proteins from widely varying biological sources contain relatively large amounts of the two acidic amino acids, aspartic and glutamic acids, so he loaded his amino acid mixture with these two and fired away. Whether to his surprise or not, this time he came out with a nice clean product, not crystallizable because it represents a heterogeneous mixture of polymeric molecules, but certainly much more pleasant to behold than the red-brown gunks he and everybody else had always obtained before. Recently he has been trying variations on this theme, tossing in phosphoric acid as a catalyst, etc., and has also been analyzing the polymers which he can make in this way. They deserve the term "protein-like" in almost every sense of the word, having molecular weights of around 10,000, containing all the amino acids, being hydrolyzed by proteases, etc. One feature of Fox's work, or rather of his description of it, that annoys some people in the field is his claim for "morphogenicity" of his proteinoids. In his paper in *Science* last July, Fox showed a photograph of "microspherules" which were formed by his proteinoid products and described these as models of primitive cells; you don't have to be from California to think that might be stretching the point a bit for effect. However, I think Fox's experimental work is important in spite of some annoyance at the way he writes: he has essentially proved the point that protein structures *do* form at temperatures under 100°C from mixtures of amino acids. There is even some indication in his work that the sequence of amino acids in these thermal co-polymers is not entirely random—i.e., preferred sequence patterns may show up even this early in the game.

With regard to the pre-biological synthesis of nucleic acid precursors, there is also some interesting work by the same investigators. While heating up things, Fox put together some malic acid and urea, both reasonable components of "pre-biological soup," and found he had synthesized ureidosuccinic acid with unexpected ease; this compound is a known biological precursor of the pyrimidine bases of nucleic acids. More significantly, Oro actually has produced adenine, one of the two purine bases, merely by heating ammonium cyanide at 70° C for half a day. Furthermore, from the same mixture he also isolated 4-aminoimidazole-5-carboxamide, and imidazole groups are now generally conceded to do a lot of the business in biochemical catalysis. The carbohydrate components of nucleic acids can be obtained, everybody agrees, if you have a little formaldehyde in the soup. So . . . we are on our way, and more and more biochemists are getting into the soup—I mean, into the act.

## Part II. (Less technical—wake up!)

The kind of *psychological* atmosphere in which this work is going on is pretty obvious, even if the nature of the ancient earth's atmosphere is still being haggled over. There is an air of exhilaration from the results of the few experiments already done that is bound to lead to new theoretical thought and subsequently to accelerated experimentation. This is the excitement of a "new frontier" (if I may use such a Democratic phrase on a protestant campus)—and the assurance that "thar's gold in that thar soup!" This enthusiasm over "pre-biological evolution" is getting more biochemists interested in taking a closer look at mechanisms for biological evolution—and biology is liable to perk up considerably as a result. If you were under the impression that evolution was about to blow over, I think you've misread the weather signs!

The extent to which these ideas permeate the thinking of biochemists not actually working in the field was brought home to me only a few weeks ago when a manuscript I had submitted to a journal was returned to me with a referee's comment. The paper was on a laboratory synthesis of delta-aminolevulinic acid, a known biological precursor of the porphyrins (such as the heme of hemoglobin), and we mentioned that even though the yield was poor our method had some interesting analogies to the known biosynthetic pathway. The referee commented that our method did throw some light on a possible *pre*-biological pathway to porphyrins, but why didn't we use succinic anhydride instead of the acyl chloride? The anhydride was much more likely to occur in a pre-biological soup! Well, we wrote back to the editor that we hadn't had pre-biological evolution in mind at all when doing the work; and besides, we *had* tried the anhydride first and it didn't work—gave us only "intractable tars"!

## Part III. (Danger—this is philosophical!)

Well, I've used up my time, so maybe we just won't have a third part and you'll never know whether the speaker was a bad-good guy or a good-bad guy—whether he belongs in quotation marks or not. I *would* like to say a few words about some of the people I've mentioned who are working in this field, in case you have *them* pegged as bad guys, a snarling bunch of vicious anti-Christian types plotting to do away with the necessity for God, or something. Well, they are certainly not vicious, and hardly plotting anything except how to do tomorrow's experiment. They know some interesting questions, they've found out how to get some good answers, and they're having fun. If you think of this as the Lord's universe, then you should realize that these are the men who are having the fun of exploring it for you. You ought to get acquainted, you know? Last year at the Federation meetings I met Stanley Miller through John Oro and spent a marvelous afternoon talking about these things; he is a delightful chap. I don't know whether he has any strong feelings about Christianity or not; he may not even be aware of it yet as a live option, although I did get a chance to identify myself as a Christian in our conversation. John Oro has at least come in contact with evangelical Christians—I was on his graduate committee. I have given copies of *Evolution and Christian Thought Today* to both John Oro and Sidney Fox. Fox is probably a thoroughgoing logical positivist with no concept of the supernatural at all, but as far as I could tell from a long discussion with him he has little interest in philosophical argument and none at all in undermining anyone else's position. He did say he would like to get people to *think*, but you can hardly blame him for *that*.

I know that Dr. Fox continually comes in contact with Christians, however—at least by correspondence. His work has been widely publicized in the popular press, and he told me he has a huge scrapbook full of letters, some highly emotional, some denouncing him for monkeying around with God's Word (he receives Bibles in the mail regularly), and some calling him a fool for denying God by trying to do something *anybody* knows only God can do.

Now I ask you, if *you* are an evangelical Christian, is that the way *you* would go about trying to witness to Sid Fox?

I hope not. That's not the way I'm trying to go about it, anyway. Thank you.

REFERENCES

Hearn, W. R., and Hendry, R. A., The Origin of Life, Chapter 3 of *Evolution and Christian Thought Today*, Edited by Russell L. Mixter; Eerdmans, Grand Rapids, 1959.

* * *

Fox, S. W., The Chemical Problem of Spontaneous Generation. *J. Chem. Ed. 34*, 472-9 (October, 1957).

Harada, K., and Fox, S. W., The Thermal Copolymerization of Aspartic Acid and Glutamic Acid. *Arch. Biochem. Biophys. 86*, 274-80 (February, 1960).

Fox, S. W., and Harada, K., Thermal Copolymerization of Amino Acids in the Presence of Phosphoric Acid. *Arch. Biochem. Biophys. 86*, 281-5 (February, 1960).

Fox, S. W., and Harada, K., The Thermal Copolymerization of Amino Acids Common to Protein. *J. Am. Chem. Soc. 82*, 3745-51 (20 July, 1960).

Fox, S. W., How Did Life Begin? *Science 132*, 200-208 (22 July, 1960).

* * *

Oro', J., Kimball, A., Fritz, R., and Master, F., Amino Acid Synthesis from Formaldehyde and Hydroxylamine. *Arch. Biochem. Biophys. 85*, 115-30 (November, 1959).

Oro', J., and Guidry, C. L., A Novel Synthesis of Polypeptides. *Nature 186*, 156-7 (9 April, 1960).

Oro', J., and Guidry, C. L., Direct Synthesis of Polypeptides. I. Polycondensation of Glycine in Aqueous Ammonia. *Arch. Biochem. Biophys.* (In press).

Oro', J., Synthesis of Adenine from Ammonium Cyanide. *Biochem. Biophys. Research Commun. 2*, 407-12 (June, 1960).

Oro', J., Synthesis of Purines under Possible Primitive Earth Conditions. I. Hydrogen Cyanide Condensation in Aqueous Ammonia. Determination of Adenine after Acid Treatment of the Reaction Product. *Arch. Biochem. Biophys.* (In press).

## Summary of Comments by Walter R. Hearn, Wheaton Science Symposium Panel Discussion

February 18, 1961

In my talk on the "Origin of Life" I tried to point out that the success of recent chemical experimentation in this field has had a profound effect on scientific thinking: events interpretable as filling in the gap between nonliving and living systems are now regarded as being highly probable. At both the pre-biological and biological levels, there is great interest in *The Molecular Basis of Evolution* (see book with this title by C. B. Anfinsen, published by John Wiley & Sons, 1959).

Christians have sometimes taken refuge in the Second Law of Thermodynamics to argue against the evolution of complex living systems from simpler systems (R.E.D. Clark, for example). I consider this type of argument entirely invalid, and generally based on a misunderstanding of the Second Law. There is abundant evidence that local order can be produced in a system in which the general trend is toward disorder, as required by the Second Law. (For entertaining discussions of the "order from disorder" principle, see H. von Foerster, in *Self-Organizing Systems*, edited by M. C. Yovits and S. Cameron; Pergamon, New York, 1960.) Christians need not be upset by this idea if they are willing to focus their attention on God's universe as a whole, instead of only on minute parts of it, such as themselves. There is no less majesty in the view that the marvelous complexity of our bodies and minds was actually structured into the original creation, than in the view that there are limits to what God's matter can accomplish under His steady direction. I think there is also no reason in this view to think that God is any less personally interested in us, or that the Bible is any less His Word revealed to us.

However, many Christians have gotten into the habit of thinking of God's direct action in nature as *always* of a sudden, instantaneous type—*never* a steady type involving processes which could be studied by the scientific method. "Processes" are considered "natural" and instantaneous events "supernatural." This sort of thinking inevitably leads one to a "God of the gaps" philosophy, no matter how sophisticated he may be about the nature of the present gaps. I am for a "Science of the gaps"—that's exactly what science is for, to fill in gaps. I am also for a "God of Creation," who is involved directly in all natural processes (See W. B. Pollard, *Chance and Providence*, Scribners). I am baffled by the idea that God is "in" some events more than in others. An illustration in the spiritual realm: In what way was God involved in our becoming a Christian—in our own "New Birth"? Only at the "instant" of that birth? What a strange view! Surely He guided us to Himself through the influence of others, the testimony of His Word, etc. And if so, then surely in the chain of events that led to our physical birth (and what a chain of events, stretching back centuries while God protected our germ plasm in the bodies of our ancestors—and even beyond that), in the preservation of the Bible, and in all the other events that eventually brought us together in Christ. Why pick out only the event that is particularly dramatic to us personally and say "*This* was God's doing!" Why not get in the habit of seeing God's creative handiwork in *everything* that happens? Only then can we honestly call ourselves His creatures, and Him our Creator, for surely we know that processes have been involved in bringing *us* into existence. Why shudder, then, at the idea that processes were involved in bringing Adam into existence? Granted that we do not yet know details of the processes, why may we not assume that God *did* use processes?

This outlook probably makes me a "General Creationist" rather than a "Special Creationist" (see my letter to *Christianity Today*, "Evolution Revisited," September 29, 1958), but gives me no less reverence for Genesis as the revealed Word of God. I cannot understand why people are disturbed at my statement that "The expressions in Scripture regarding the creation of life are sufficiently figurative to imply little or no limitation on possible mechanisms" (Chapter 3 of *Evolution and Christian Thought Today*). Genesis says that God breathed into Adam the breath of life, and he became a living soul; surely no one would question that "God breathed" is highly figurative, anthropomorphic, symbolic language (and I hope no one would be so immodest as to say he knows exactly what the figure of speech means!). To say that I don't take this passage literally—i.e., that I don't think the Creator exhaled through mouth or nostrils—does *not* imply that I don't take the passage *seriously;* I regard it as true, as God's Word, as intended to reveal something about God to me. In addition, I think it is a beautifully poetic expression and I value it for this reason as well. Biochemical descriptions of God's creative activity are almost invariably more complicated and less beautiful.

# *Comments on the Origin of Species* *

J. FRANK CASSEL**

### Introduction

As we move in our thinking from the origin of the universe—to the origin of life—to the origin of species—to the origin of man, we move into a more and more charged atmosphere. We are moving into the areas encompassed by what is usually referred to as *evolution*. "The lover of paradox," says Bertalanffy (1952), "could say that the main objection to the selection theory is that it cannot be disproved." This statement gives us a hint as to the nature of the data we are discussing.

In the fall of 1959 the centennial of the publication of Darwin's *Origin of Species* (1859) was celebrated at the University of Chicago. Only recently the proceedings of these meetings, including lectures, papers, panels, and general discussions, have been published in three large volumes (Tax, 1960; Tax and Callender, 1960). These volumes summarize what is known about the subject of evolution and what some of the greatest minds of our time are thinking about it. Volume I, the largest, in 629 pages and 20 papers, deals formally with the subjects Dr. Hearn and I are discussing. Obviously we cannot do the field justice in one hour.

I have not had these volumes long enough to be able to summarize them for you. It was only this morning, in fact, that I came across a paper (Olson, 1960: 523-545), which questioned what I had felt was the generally if not universally accepted trend of the field. If then my presentation seems somewhat disjointed, it is partially because I have not yet adjusted my own thinking to these new ideas—I have seen my paper for this afternoon—as it were—go up in steam.

What *is* our purpose? What *can* we do here? It seems to me that the advantage of a gathering such as this is to find out what is being thought—to assess the facts that we have—to assess the interpretations of these facts not only in the light of themselves but in the light of the presuppositions that we make. I don't want to stand here today—in fact I'm not able to stand here and tell you what you should believe. But I think that here we can exchange ideas and thus come closer to a recognition of reality.

Over a hundred years ago (1831-36) Darwin sailed around the world in the Beagle. A short time after that, he tells us in his introduction to the *Origin of Species* (1859 (1909:21)) he began to

*Condensed (and supplemented) from a tape recording of an address given from notes 17 Feb. 1961 as part of the Wheaton College Symposium on "Origins and Christian Thought Today."

**Dr. Cassel is President of the American Scientific Affiliation and is Chairman of the Department of Zoology, North Dakota State University, Fargo, North Dakota.

keep notes—and after five years, he says, he allowed himself to think about what some of these observations might mean and two years later (1844) sketched some conclusions. It was some 15 years after that that the *Origin of Species* was published. I beleive the *Origin of Species* had the influence upon the biological and the rest of the world that it had, because of the intricate documentation that Darwin gave, showing the background of his ideas—stimulating other people to take this data and see what they could get out of it—stimulating others to find out more data and see how it would correlate. And so for a hundred years in various disciplines this has been done. Thus a great body of data has been accumulated, which formed the basis of the Chicago discussions and forms the basis of our discussions today.

## Definitions

*Origin*

"Origin" or "origins" may be self-explanatory. This is not the key word in our discussion. Origin refers to the beginning. Where did something start?

*Species*

"Species" is defined in various ways. Perhaps the problem is best summarized by Mayr (1956) when he points out that there are three different species concepts.

1. Species can be thought of from the standpoint of a *type*. A biologist describes an animal. He says, "Hereafter this species will consist of all animals which are like this animal which I am describing which I will deposit in this museum." Thereafter a decision must be made with respect to any unknown animal as to whether it is enough like the one on deposit at the museum and like the original description to be called the same species, or whether it's different. This is the classical idea of species.

2. A second idea of species is based on *concept*. (This is my own term, not Mayr's.) It suggests that if we say "robin," for instance, each one of you in your mind gets a picture of a robin. It isn't a particularly detailed technical picture, but at least you know what a robin is.

3. The third view of species we call *biological*. This one is most useful in the discussions we are having. The biological species is conceived as a group of animals having a certain genetic make-up, the elements of which can be passed on to their offspring maintaining essentially the same genetic make-up. Hence, a species is a group of naturally occurring populations which freely interbreed at the periphery of their ranges (where they come together), or are potentially capable of doing so. Or, briefly, a species consists of those animals which freely breed with one another.

The division between species comes then when interbreeding cannot occur for some reason or other. This would seem rather easy to observe. But we find that there are some organisms which are partially fertile with other organisms. We also find organisms which are identical in their morphology and their external appearance, but cannot interbreed. We find other organisms which look very different in their outward appearance, but which do interbreed freely. We find all the spectrum in between these. So this is not an easy criterion.

*Kind*

What does "kind" mean as it is used in Genesis? I am not posing as an authority on this but in discussions that I have seen of this word it is taken often to refer to a certain category of organisms (Mixter, 1951:3; Bullock, 1952; Payne, 1957), but not consistently to any specific biological category. Oftentimes, I think, we create a problem by saying (or thinking) that "kind" *must* refer to "species." Undoubtedly "kind" sometimes *does* refer to species. But "kind" also may refer to other groups, perhaps to genera, perhaps to families, perhaps to higher categories. So that it is not accurate, it seems to me, to restrict the use of "kind" to *species*.

Origin of species—how could it take place? Darwin suggested, arguing from indirect or circumstantial evidence, that it seems as if it has in some cases by natural means. We know a great deal more about the possible mechanisms by which species might be formed than did Darwin. Our knowledge of chemistry is much greater (Hearn, 1961). What are these mechanisms whereby species could be formed? Is there a mechanism whereby organisms which can at one time freely interbreed might at another point of time not be able to interbreed? If such a situation can be shown to obtain, then we can say that *species* have been formed and that *speciation* has taken place. And the word that we use in the field to talk about this is *evolution*.

*Evolution*

Evolution means a number of things to a number of people. I raised my eyebrows when I saw that this symposium was going to be on "Origins and Christian Thought Today." We're talking about the book, *"Evolution* and Christian Thought Today" (Mixter, 1960). Perhaps "origins" is a much better word, because we all are in fair agreement as to what it means. On the other hand, practically every one of us has a different idea of what the word "evolution" means.

Indeed the word does have several meanings which are currently used in scientific or other literature. Some of the main ones follow.

1. Evolution is the process whereby all life has developed by mechanistic means *without God* from single cells or primordial protoplasm to all the different forms we have today. But with such a definition (equating evolution with atheism), if this be the only one allowed—as many affirm—we as evangelicals are left with no satisfactory word to use to

describe the phenomena relating to species formation, phylogeny, and similar scientific data.

2. Evolution is the general theory that life as we know it on earth today has over a long period of time been developed by differentiation from a single or several primordial cells; descent with modification.

3. Evolution is any change in plants or animals which can be passed on to their offspring. Many writers say or infer that such changes must take place "in natural populations" to get away from the discussion of the significance of changes due to controlled breeding practices. Mayr (1959) puts it, "Organic evolution . . . refers to a change in genetic properties from generation to generation owing to reproduction." It is this connotation of the word which is most commonly intended, when modern writers discuss evolution in action. According to this concept evolution can be either progressive or retrogressive. In other words, the change may go toward the complex or toward the the simple.

4. One thing that evolution does not mean is that man descended from apes. The word may be used to refer to the idea that man and apes may have descended from some common ancestor. But straight line descent from anthropoid ape to man (or any other straight line descent) is no longer inferred.

I have often seen the word "evolution" used in the same paragraph in the same article several times with a different connotation each time. Therefore it is very difficult to talk about "evolution" and to understand just exactly what we mean when we're talking about it.

### Speciation

What do we know more than Darwin knew about the formation of species? We know about genetics. We know something of the intricacies of cellular structure. We know much of the biochemistry of the cell, particularly of the nucleus. We know something of the operation of the particular factors which seem to influence the heredity of the organism.

*Mutation*

Although not all heredity is due to nuclear components, there being some cytoplasmic genes, "Synthetic Evolution" (Olson, 1960:527) is based mainly upon the concept of hereditary characteristics which are governed by those factors in the nucleus known as "genes." These genes can and do change. These changes we call "mutations." A mutation is a change in a factor governing the heredity of an organism. This can basically be explained as change in chemical structure.

*Isolation*

In a population of freely interbreeding organisms in nature a change within one organism is not likely to affect the population unless something happens to favor the maintainence of this particular change. There are so many other genes in any group, or gene pool as we call it, that one mutation in and of itself has no effect upon the population, because the mutant gene does not often even come into enough abundance to be seen, that is to effect an observable change in any organism. In artificial breeding, if we find a characteristic in a plant or animal which we want to keep, we then separate or segregate this individual and breed it only to other individuals which have other characteristics which we want to keep. In other words, these individuals have to be *isolated*. In the process of speciation (species formation) in nature, this is another necessary factor. The population must be so small that mutations will be maintained because the mutant genes form a significant portion of the genes in the gene pool.

Natural populations of organisms may be isolated or separated from other such populations by geographical barriers such as mountains or rivers, by habitat preference, by climatic responses, and eventually by morphological or physiological differences. Species, however, are not formed until reproductive isolation is developed, that is until the separate populations are no longer even potentially capable of interbreeding. There are numerous examples of this occurring in nature. Perhaps the most striking one is the circumpolar distribution of overlapping populations of gulls of the Herring Gull-Brown Gull complex. In Scandinavia the Brown Gull does not interbreed with (is reproductively isolated from) the Herring Gull and yet each population does freely interbreed with adjacent populations all round the pole in a continuous ring broken only where the Brown and Herring Gulls maintain distinct populations where they overlap in northern Scandinavia. (See Mayr, 1940).

*Selection*

In artificial breeding, we select for those factors which we want. Darwin suggested that this same thing happens in nature. With the proper isolation, selection can be achieved. He suggested that if any change in the organisms was of particular advantage to that organism, this would give that organism greater strength to carry on its life than if this change had not occurred. Actually it does not appear that great advantage is a necessity for the effective selection and development of a characteristic in an isolated population. It is obvious that if any change occurs in a plant or animal which is of disadvantage to it, that individual is not likely to exist or reproduce as well as it would otherwise. (The human species is an exception, because much of our medical work maintains life in those individuals that are not as well adapted to the environment as others). On the other hand, in nature, if there is a change within a small population which is sufficiently isolated, this change can be maintained without any great advantageous selection. But if there is a factor which is of advantage to the existence of the organism so that it breeds more freely or produces more offspring which then contain these same genetic characteristics, this particular genetic characteristic is likely to be developed with a greater frequency than those which do not have such an advantage. It is usually held that such changes are quite slight (microevolution) and that it may take several to effect much visible change in the individuals of a population. It should be remembered, too, that speciation is not accomplished until reproductive isolation is established no matter how great or how little the other changes involved. This then very briefly, and sketchily, delineates the thinking at the present time of those who hold to the "synthetic theory." I have pointed out elements or mechanisms which must be considered by anyone contemplating on how changes might come about in populations of organisms.

In the paper I came across this morning, the author makes a point which is seldom emphasized. In our present state of knowledge he suggests that it is presumptuous in the face of a *variety* of interpretations of the data to hold that there is only *one* interpretation possible. "The statement is made, in effect, that those who do *not* agree with the synthetic theory do *not* understand evolution and are incapable of so doing. . . . Regardless of the apparent merit and strength of the synthetic theory, it seems to me that the more cautious and thoughtful attitudes . . . are more appropriate" (Olson 1960:526, 527). He then gives the several theories which must be considered.

In addition to the (1) *Synthetic theory,* he lists: " (2) *Saltation theories,* involving major, abrupt reorganizations . . . ," which hold that many changes in plants and animals are not minute but are rather large—large jumps can take place, changes which would be of great magnitude — macroevolution (Goldschmidt, 1940); (3) *"Metaphysical theories,"* calling upon some force outside or inherent in the organism which will govern and guide the changes which take place; and (4) *"Lamarckian* or *Neo-Lamarckian* theories, involving inheritance in one way or another of acquired characteristics"—the concepts involving the idea of use and disuse being controlling factors in the development of certain characteristics.

Olson includes the alternative of a *Metaphysical Theory of Evolution*. The discussion which we have here at this symposium wll be along the lines, I hope, of developing in a sound way some real contributions to this alternative. This, ladies and gentlemen, I suggest is our challenge as evangelical Christians. May we stop shadow-boxing. May we stop finger-pointing. May we be careful not to fall into the same fallacy for which Olson chides the synthetic evolutionists. From our Christian presuppositions of the creating and sustaining God, may we work eagerly to develop without apology an explanation for the biological phenomena which we see in the universe. Particularly with regard to the phenomenon of speciation, or the origin of species, should we do some of our most careful observation and our best thinking so that we can prepare as accurate and true a description of the reality as presently possible in the light of presently discernible revelation.

## LITERATURE CITED


Bertalanffy, L. von, *Problems of Life.* Wiley, New York. 1952.

Bullock, W. L., The "kind" of Genesis and the "species" of biology. *J. Am. Sci. Affil. 4*:5-7. 1952.

Darwin, C., *On the Origin of Species by Means of Natural Selection.* London. 1859. Reprint: The Harvard Classics, C. W. Eliot. 1909. Collier, New York.

Goldschmidt, R., *The Material Basis of Evolution.* Yale, New Haven. 1940.

Hearn, W., *Origin of Life. J. Am. Sci. Affil. 13*:38-42. 1961.

Mayr, E., Speciation Phenomena in Birds. *Am. Nat. 74*: 249-278. 1940. Ecological Factors in Speciation. *Evolution 1*:263-288. 1947. Species Concepts and Definitions. *AAAS Publ. 510*:1-22. 1957. Darwin and Evolutionary Thought, p. 1-10. In *Evolution and Anthropology, a Centennial Appraisal.* Gaus', Brooklyn. 1959.

Mixter, R., Creation and Evolution. *Am. Sci. Affil. Mono. 2*:1-31. 1951. *Evolution and Christian Thought Today.* Eerdmans, Grand Rapids. 1959.

Olson, E. C., Morphology, Paleontology and Evolution. Pp. 5231545. In Tax (ed.) *Evolution after Darwin.* Chicago U., Chicago. 1960.

Payne, J. B., The concept of "kinds" in Scripture. *J. Am. Sci. Affil. 10*:17-20. 1957.

Tax, S., ed., *Evolution after Darwin.* Vol. 1. The Evolution of Life. Chicago U., Chicago. 1960. Vol. 2. The Evolution of Man. Chicago U., Chicago. 1960.

Tax, S. and C. Callender, ed., *Evolution after Darwin.* Vol. 3. Issues in Evolution. Chicago U., Chicago. 1960.

# *Christian Thought Today—On Origin of Species**

J. FRANK CASSEL

I find myself this morning in a more accustomed role—a role I'm more at home in—that of a heckler. (Let me say in passing that some have objected to my use of the term "heckler" on previous occasions. In my use of the term I connote nothing evil, no criticism. I think this is the way we stimulate thinking—at least it's the way I'm stimulated. So

*Prepared from notes of introductory statement made during panel discussion on "Origins and Christian Thought Today"at Wheaton College, February 18, 1961.

when I use the word, I mean "thought provoking.")

Yesterday I had a chance to establish the positive side of what we know about the origin of species and last night a positive approach to a Christian philosophy of science. I don't think I need to expand this morning upon this aspect. Rather I would like to ask some questions—questions which I feel we as Christians must approach and deal with positively in our consideration of evolution—or admit we don't have enough data to answer and indicate that we are withholding judgment.

1. Speciation can be seen in action—in Drosophila, the fruit fly in both the lab and in nature, in moths around smoky cities in the British Isles, in frogs, in herring, gulls, etc. What are its limits? Some are content to point out that changes are only at the species level or at the family category at the most. Is this really true, or is our data simply too limited by virtue of time restrictions? But even if it is so at present, what are the implications? Does it not indicate at least that this is one of God's methods of creation? Might it be the only one?

2. Is a time difference necessary between creating and sustaining? Some hold (e.g., Tinkle, *J. Am. Sci. Affil.* 13:15-17, 1961) that God's creative work is finished—that He is now in His seventh day of rest. Is this what the Scripture really teaches? If so, then speciation cannot be held to be creation.

3. In understanding the first chapter of Genesis, what of the sequence problem? If we hold to the creative days being a period of geological time, then we have particular trouble explaining the sequence of events on days 3 and 4. To review, the creation sequence briefly is as follows:

Day I —Light
Day II —Firmament
Day III —Earth, and plants (including Angiosperms - "seed bearing")
Day IV —Sun, moon, and stars
Day V —Aquatic life and *birds!*
Day VI —Cattle, creeping things, beasts, and man

It can be seen that plants, including the most complex forms were created before the sun. To be sure there was light on the first day—but what was its source? How strong was it? And in a heliocentric solar system, what of the earth before the sun? On day five, compared with the fossil record, how do birds fit before "creeping things" for instance? I do not mean to indicate, let me emphasize, by asking these questions that they are unanswerable. Nor do I mean to throw doubt on the Scriptural account. I simply mean that these are questions which have to be faced and handled in a sound and scholarly manner. Let me say again, as I did yesterday, that I am not questioning the authority of the Scripture, nor that God is speaking in it—our question is "Just what is God really saying"—but I'm getting ahead of myself!

4. In understanding the fossil record, what of the sequence problem, particularly in the vertebrates? In the Cambrian, the first fossil-bearing beds of any significance as far as number of organisms are concerned, are found representatives of all the major phyla—except chordates! Then briefly the sequence in time shows fish being the earliest vertebrates found, later amphibians, still later reptiles, and then (although there is some discussion just when) mammals and then birds. And the earliest birds have teeth and tail vertebrae—like a feathered lizard's tail. Some more recent birds had teeth but no tail vertebrae, while modern birds have no teeth and no tail vertebrae. Why this sequence of creation or how *did* it come about? To be sure we have trouble affixing ancestory, particularly for classes as in mammals—but we do have the sequence to explain.

It should be recognized also that there are many problems inherent in our present methods of dating and of assigning and defining taxonomic categories.

5. Natural processes, as we pointed out last night, are God's processes, hence in a manner of speaking "supernatural." Is the miracle of birth (and of the life that follows) any less miraculous than the second birth? Does God not use His own methods? I'm more and more impressed by the wonder-working hand of God in all things—why pick out only the rare occurrences to give Him the glory? And because these occurrences are rare does not necessitate a different method of operation or sustaining—simply a point of time occurrence and perhaps a speeding up or acceleration of process. In other words, I expect the wine at Cana was alcoholic, and that it could have been chemically analyzed as such—and because molecules of water changed so that there were molecules of alcohol—God working with this material—His creation—to achieve His ends. Is this not the daily course of things as well?

I've always promised myself that I would work up an opening lecture in zoology in which when I discussed biology as the study of life, I would include eternal life in the consideration. I haven't yet figured just how to do it well, but I think it illustrates the point.

6. Let me reiterate what we said last night. The Bible reveals God. The universe reveals God. Each can be used to interpret the other. They cannot, as far as I can see, be contradictory. This use of creation to interpret Scripture, does not question the authority of the revelation—simply the authority of the *interpretation*, for example Scofield's much-used notes. These are often very helpful, but they are not Scripture, and they do not deserve

being accorded the reverence, authority, or infallibility of the Word itself. God speaks in the Bible. God speaks in nature. What is He saying?

7. Someone else has mentioned Carl Henry. In writing on this subject several places, Carl has said that to be a Creationist one must believe in three principles,

a. Creation in divinely graded stages of living organisms
b. Creation by fiat command
c. Creation *ex nihilo*—out of nothing (Henry, *Christianity Today* 2 [23]: 20-22, 1958)

I suppose I must go along with the first in the light of what I've been saying. If there are levels or stages—categories of living organisms—and there are—then certainly God did it.

But I do not see at present the necessity of "fiat command" in the sense that "Let there be" demands instantaneous, point of time appearance. Can it not possibly mean "Let there become"? This is a question to be answered from the context and connotation of the original language to be sure. Does this really restrict the connotation to "fiat"?

Nor do I see the necessity for *ex nihilo*. This seems a problem perhaps in logic, or as Dr. Buswell points out in realms of theology in which I am not versed. It has seemed to me that God possibly created out of Himself. Col. 1:17. God the Almighty, All Powerful—is a God of all energy. Might $e = mc^2$ be a *creative* formula?

Here I need the Hebrew scholar and the theologian to point out to me where I'm off base and this I think is the advantage of symposia such as this one. Might I suggest further that in the light of these discussions, the next step might be (and it might logically be sponsored by Wheaton) to get together a smaller group of authorities by invitation, to sit down together in a retreat situation, for two or three days or more, toss these ideas back and forth, come to such understanding as is possible at present and summarize their findings for publication and hence wider and more extended discussion. Missouri Lutherans have been using this technique effectively, I believe—for example, "What, Then, Is Man" by Paul Meehl, *et al.* (Concordia, 1958).

Well, that's enough of a heckle for now—but seriously I feel these are all questions which have to be faced if we are to demand the ear of a thinking world.

# *The Origin of Man, and the Bio-cultural Gap**

JAMES O. BUSWELL, III**

## I.

It is indeed significant that the planners of this symposium have directed the focus of attention upon origins. They are to be congratulated for thus precluding continuing confusion over the details of evolutionary process. For it is largely the genetic and geological processes of life and earth history with which evolutionists have dealt. Thus it is within these areas that the most scientific progress has been made, and concerning which, consequently, there has been a decreasing basis for argument. But with matters of origin, evolutionists have been admittedly hesitant to deal. The noted physical anthropologist William Howells wrote sixteen years ago,

> We are totally bewildered, of course, about the beginning of life and the reasons for our existence, and these are questions which have been grist to the mills of philosophers and mythmakers alike. But we know, roughly, what happened along the way, and that is the story of human evolution.[1]

Recently, however, there has been increased interest shown in the problems of origins, and, as one might expect, it has served to bring into unprecedented prominence the significance of one's philosophy of science. In another context, but with far-reaching application, Dr. J. O. Buswell, Jr., speaking here before the recent annual philosophy conference, pointed out that

> . . . Whereas 35 years ago Christian thought was most seriously challenged by the natural sciences, the crucial problem today is philosophy.

*Presented at Wheaton College Science Symposium on "Origins and Christian Thought Today," February 17, 1961. Slightly revised to incorporate ideas stimulated by the panel discussion held on February 18, and by new material previously unavailable to the author. Credit is due my colleague, Mr. Donald Wilson of Covenant College (St. Louis) for able criticism and enlightening discussion on problems basic to this paper.

**Mr. Buswell is Associate Professor of Anthropology at Wheaton College and is presently completing work toward a Ph.D. degree in Anthropology at Columbia University.

1. Howells, 1944, p. 3.

This is, in part, due to the fact that large segments of evangelicalism have paid increasing attention to what may be called true scientific progress, and have broken away from the straight jacket of certain interpretations of Scripture which ran headlong into conflict with factual scientific data. It is also due to the fact that, with Bible-believing scientists taking an active part in this scientific progress, the focal points of concern have shifted from a broad, over-all question of science vs. the Bible, or evolution vs. creation, which led inevitably to a high degree of fruitless controversy, to a much finer focus upon problems of positive interpretation with basic creationist presuppositions. Thus in the volume which has stimulated this symposium,[2] it is clear that Schweitzer is not attempting to debate *whether* God created the universe but *how* God created it; and that Hearn and Hendry are not examining *whether* God created life, but *how* God created it. Thirty-five years ago, and less, it would not have been likely that a fundamentalist, doctrinally speaking, would have found himself able to state, as Schweitzer does, that there is no conflict between the best scientific theories as to *how* the universe came into being, and the Biblical view (p. 50). Nor could it have been admitted, as Hearn and Hendry stated, that "The expressions in Scripture regarding the creation of life (are) sufficiently figurative to imply little or no limitation on possible mechanisms" (p. 69). Thus it is not the natural sciences themselves which challenge Christian thought today, but it is the underlying naturalistic and mechanistic philosophy of their leading practitioners.

With reference to considerations of the origin of man, the situation is found to be the same. It will be unnecessary to review here the differences between the interpretations of man's origin from a naturalistic point of view, as contrasted with the creationist, or supernaturalistic approach. Suffice it to say that it is not necessary to quibble with the evolutionist over any of the data pertaining to prehistoric man, in order to maintain a sound Scriptural position which does not jeopardize any of the conservative doctrines such as the creation of a single pair of humans, their original perfection and subsequence fall, and the unity of the human race. One's interpretation of this data will depend upon one's underlying philosophical presuppositions. Indeed, there is perhaps more active debate today over the exact interpretation of man's origin *within* creationist circles than there is elsewhere.

II.

Turning now to an examination of some of these problems, it will be remembered that in the chapter on prehistoric man[2a] in the Mixter volume, we came to certain tentative conclusions regarding the interpretation of the Australopithecinae, the so-called South African man-apes, in the context of a consideration of the question, What is man? These conclusions were that "so far no definite indication of any cultural assemblage has been identified" in connection with these types, and that, "should such a cultural assemblage be identified for the Australopithecines, it will necessitate perhaps a drastic revision of what we are used to considering 'human' but nothing more as far as the creationist position is concerned" (p. 187).

Even then mention should have been made concerning stone tools which had indeed been discovered *in situ* in deposits which also contained the remains of Australopithecus. These had been discovered by C. K. Brain in 1956 at Sterkfontein and were reported by the British paleontologist Kenneth Oakley the following year as proving "that pebble tools were made at the very site where Australopithecus occurred."[3]

There remained, however, some doubt as to the likelihood that the man-apes were the manufacturers of these tools. Brain himself reported in 1958 that no trace of the stone culture existed in the lower levels of the deposits where most bones including Australopithecus were accumulated.[4] Oakley felt that since the pebble tools thus ". . . have no background . . . at this or any other Australopithecine site" that judgment must be reserved. Although more stone pebble tools have been discovered subsequently at two other sites bearing Australopithecus remains, there is no proof that these animals made them. Oakley's expression regarding the 1956 discovery, which he considered "possibly the most important discovery in the field of paleo-anthropology since the finding of implements with Peking man," was almost identical to our own conclusion, as he stated that, "If in fact Australopithecus was the maker of the Sterkfontein tools, it would involve almost a revolution in our conception of 'man.' "[5]

Parenthetically, it should be emphasized at this point, as the British anatomist LeGros Clark has pointed out, that in using the terms "man" and "human" we tend to think only of modern man. But the terms must be taken to include extinct races such as the Pithecanthropus and Neandertal as well. As LeGros Clark puts it, "the terms 'man' and 'human' have come to assume, by common usage, a much narrower and more rigid connotation, which for most of us (however we may try to persuade ourselves otherwise) also involves a real emotional element."[6] If we could divorce from our

2. Mixter, R. L. (ed.), *Evolution and Christian Thought Today*. Grand Rapids: Eerdmans, 1959. London: The Paternoster Press, 1960.
2a. Buswell, 1959.
3. Oakley, 1957, p. 441.
4. Sabels, 1959, p. 248.
5. Oakley, 1957, p. 443.
6. LeGros Clark, 1955, p. 6.

thinking any preconceived notions of what man *should look* like, discussions of human paleontology might involve less strain on our powers of interpretation. As I mentioned in the chapter under review, "I believe that the question of human or nonhuman cannot be answered categorically upon morphological grounds. The question must be answered on spiritual grounds, which, I presume, are only indicated by cultural remains" (p. 187). The noted paleontologist Franz Weidenreich some years ago made the same point in a remarkably parallel fashion, although his reference to "spiritual life" applied merely to his German conception of man's mentality and culture. He wrote that

> . . . studies made on skeletons alone will never enable us to make statements about either the mentality of the individuals concerned or about mental change or progress over a period of time. Cultural objects are the only guide so far as spiritual life is concerned. They may be fallacious guides, but we are completely lost if those objects are missing.[7]

Raymond Dart, discoverer of the first Australopithecus fossil, has noted recently in an article describing the rather extensive use of bone weapons,

> It is important to recognize that one of the outstanding effects of the South African man-apes upon physical anthropology has been to display its limitations in assisting us to define what is man and what is ape anatomically.[8]

That is why the consideration of the Australopithecine weapons and tools is a crucial one. To the anthropologist the manufacture of tools in distinguishable traditions becomes the archaeological hallmark of mankind. It must be remembered, however, that these pebble tools which are found elsewhere widely distributed over Africa at this early pleistocene horizon do not show any distinct variation of tradition or style or method of manufacturing such as we have become accustomed to identifying in connection with the undisputed prehistoric culture complexes elsewhere in the world beginning only slightly later. As J. Desmond Clark, paleontologist from Southern Rhodesia has pointed out, "The stone tools of this time show no significant typological variation from one end of the continent to the other."[9] This, together with the total absence of any more conclusive cultural associations in the comparatively extensive recovery of Australopithecus remains, such as the use of fire, would seem to warrant a certain degree of caution in assigning them a completely human cultural status.

The recent discovery by Dr. L. S. B. Leakey of an Australopithecine skull in the summer of 1959 in Tanganyika, however, prompts us not to make too permanent a place for the South African man-apes on the nonhuman shelf. Here, about 22 feet below the surface of Bed I, the earliest cultural horizon exposed in the famous Olduvai Gorge, was uncovered a part of a living floor of Oldowan culture, stone tools of that culture, and the nearly complete skull of a manlike primate which Leakey and others described as "Australopithecine." With the tools was a hammer stone and waste flakes from the manufacture of them. These pebble tools had long been know at this site, but there had been no evidence of their makers. Clark Howell of the University of Chicago, who examined the find shortly after discovery has written that "it is clear that an Austarlopithecine group was responsible."[10]

The Australopithecines have not generally been considered directly ancestral to later Hominids despite their marked physical resemblance such as dentition and bipedal locomotion, and despite their certain classification, morphologically, with the Hominids rather than with the Pongids, or great apes. The reason for this has been that they occur too late in time. Pithecanthropoid man is found soon afterward in geological time, if not contemporary with some of the Australopiths. Thus it has been held that processes of racial or evolutionary differentiation could not have produced the former, with its yet marked morphological differences, from the latter. This position on the Australopiths is expressed by Ashley Montagu when he concludes,

> In the present state of our knowledge one can only point to the Australopithecines and say, that while no one of them may have been directly ancestral to man, a type very like them must have been.[11]

III.

In view of these recent, fast-moving, and important developments in African prehistory, activity in another and vastly more important area of concern has been stimulated as never before in anthropology and related sciences.

Man has been the central subject of anthropological concern. Since the underlying basic assumptions of man's origin which have held sway for most anthropologists have been those of organic evolution, it has followed that physical anthropology has traditionally carried this study to its greatest refinement in its attention to the variations of man's body in time and space, and to corresponding refinements in the related sciences of genetics and geology. Early attempts of cultural anthropology to view man's culture in evolutionary

---

7. Weidenreich, 1948, in Washburn and Wolffson, 1949, p. 23.
8. Dart, 1959, p. 87.
9. J. D. Clark, 1960, p. 312.
10. Howell, 1960a. p. 76.
11. Montagu, 1960, p. 147.

constructs, lacked historical and ethnological validity and were destined to be repulsed by the refinement of its own methods. Not only were the data, by comparison, more imponderable than the data of physical anthropology, but the anthropologists were more prone to indulge in independent speculation since they were relatively bereft of any stabilizing body of theoretical antecedents. But, as Professor Hallowell of the University of Pennsylvania has pointed out, when the constructs of *cultural* evolution were rejected by most 20th-century anthropologists, rejected as well was their basic notion that present-day primitives represented stages of physical evolutionary development.[12]

Origins of the various aspects of man's culture were no longer sought among the remotest tribes. With the advent of Franz Boas and the rigorous methodological house cleaning of anthropological theory undertaken by him and his followers, who established the foundations of American Anthropology, the mind of the savage was seen to be the potential equal of every man's, and all human culture behavior, primitive and civilized, was seen to be explainable in socio-historical, or cultural terms rather than biological or evolutionary ones. "The conclusion was drawn," Hallowell points out, "that culture change and development in Homo sapiens is not primarily linked with an evolution of mentality."[13]

The consequence which Hallowell deplores, however, was that this left the pursuit of evolution largely in the hands of the biologists, and, as he regretfully observed, "the psychological dimension of evolution, which to Darwin himself was an integral part of the total evolutionary process and of vital significance for our comprehension of man's place in nature, fell upon evil days."[14] Physical anthropologists concerning themselves with problems of fossil man, growth, race, and later, population genetics, were nevertheless devoting their entire attention to the evolution of man's body. Cultural anthropologists, concerned now almost wholly with culture as such, and cultures in synchronic instead of diachronic perspective, were paying no attention to the evolution of man's behavior.

Furthermore, most significant was the result that, whereas development of evolutionary theory had served to close the vast *biological* gap between man and his nearest relatives, the living primates, at least by the assumption of their respective derivation from a common ancestor, the cultural anthropologists by their refinement of the concept of culture and the distinctive human possession of language were driving a fundamental conceptual wedge deeper and deeper *between* man and ape. "In effect," Hallowell explains, "this preoccupation with culture led to a re-creation of the old gap between man and the other primates which it was once thought the adoption of an evolutionary frame of reference would serve to bridge."[15]

The biological continuum has been postulated for so long, and the behavioral continuum so subconsciously assumed, that the present investigators are faced with a highly sophisticated body of culture theory which has man's distinctive and unique characteristics as its foundation and major supports. "All these characterizations stress man's difference from other living creatures," declares Hallowell. "Like the criteria of culture and speech, they emphasize discontinuity rather than the continuity that is likewise inherent in the evolutionary process."[16]

There are two bodies of data available, according to the evolutionary perspective, for the investigation of man's behavioral evolution, that is, for the attempt to get at its original cultural development from a nonhuman ancestry. These are the fossil evidence for early man, and the study of the complex social behavior of the various species of living primates. Since the cultural remains of fossil man are so limited in what they can reveal about behavior and mentality, the accumulation of data on the man-apes has been bringing about the focus of a very high degree of research upon the study of the living primates. Their social organization, their communication, their aesthetics, their use of tools, and their response to varied social changes are being scrutinized for some new light on the original development of man's kinship, language, art, manufacturing of artifacts, and social dynamics.

One of the pioneers in primatology, C. R. Carpenter, observed a few months ago that "for thirty years field studies of monkeys and apes have been the lonely fate of a few hybrid scientists. Now, upsurges of interest and effort are occurring. . . ."[17] For example, the January issue of *Natural History* includes a report of recent research on "Primate's Aesthetics."[18] The subtitle reads, "An ape provides clues to the origins of artistic activities." A chimpanzee named Congo had been provided with opportunity and facilities for drawing and painting, and the results had been analyzed with the aim of "probing the biological origins of aesthetics." Naturally the resulting "art" provided a great deal of unique data on the mentality of chimpanzees, as well as for the analysis of nonhuman projective visual representations as such. However, in line with the primary aims, the author reports:

> The early results reveal that in the chimpanzee there is what we might call the germ of visual

---

12. Hallowell, 1959, p. 37.
13. *Ibid.*
14. *Ibid.*, pp. 37, 38.
15. *Ibid.*, p. 38.
16. *Ibid.*, p. 39.
17. Carpenter, 1960, p. 403.
18. Morris, 1961.

> composition . . . even now it is . . . quite clear that the basic visual rules that control composition and design in painting by human beings are shown in a rudimentary form in the work done by Congo.[19]

Charles F. Hockett, distinguished linguist and anthropologist at Cornell, in a recent article on "The Origin of Speech" stated that "with this comparative method it may be possible to reconstruct the communicative habits of the remote ancestors of the hominoid line, which may be called the protohominoids. The task, then," he continues, "is to work out the sequence by which that ancestral system became language, as the hominids—the man-apes and ancient men—became man."[20]

One thing which characterizes all such efforts to be found in the literature is the readiness of the investigator to admit that his investigations are being undertaken in a theoretical context of pure speculation. Marshall Sahlins of Michigan, in a recent article on "The Origin of Society" states plainly that,

> This discussion of the early phases of human society considers events that occurred a million years ago, in places not specificially determined, under circumstances known only by informed speculation. It will therefore be an exercise in inference, not in observation. This means juxtaposing the social life of man's closest relations —monkeys and apes—on the one side, with the organization of known primitive societies on the other. The gap that remains is then bridged by the mind.[21]

Among other statements embodying the same admission regularly occurs the assumption that "deductions from comparative behavior are as methodologically legitimate as those from comparative anatomy."[22] It is plain to see that the logic involved is only valid within a framework of evolutionary presuppositions.

At least one other candid statement of the rationale for nonhuman social investigations may actually underlie existing theory more widely than is often admitted. That is the point made by N. C. Tappen in a recent article on the distribution of African monkeys:

> Assuming that there was nothing miraculous about the evolution of man from primate ancestors, it follows that the potential for paralleling the process is still present and could be realized, given the right circumstances.[23]

Perhaps the physical anthropologist with the least reservations on the matter in his writings is Dr. S. L. Washburn of the University of California, who states in no uncertain terms:

> Complex and technical society evolved from the sporadic tool-using of an ape, through the simple pebble tools of the man-ape and the complex tool-making traditions of ancient men to the hugely complicated culture of modern man.[24]

According to Washburn this may be "seen in the scanty fossil record and can be inferred from the study of the living forms."[25] Such a generalization is just as naive and invalid, from the creationist standpoint, as the more sweeping unilineal evolutionary generalizations of Edward B. Tylor in 1871 and Lewis H. Morgan in 1877 were to Boas and Lowie in the 1920's. Some scientists of today who would not claim to be creationists still see the fallacies of this comparative method for achieving its stated goals. Marston Bates, for example, in a volume published this month, deplores the basic theory behind the method,[26] and further states very simply, that "The method of comparing different kinds of living organisms is not, in itself, an adequate basis for reconstructing a plausible evolutionary story. . . ."[27]

Modern anthropology is seeking desperately to bring the study of behavioral evolution to the status of settled certainty which organic evolution is assumed to have, by seeking to close the gap between human culture and nonhuman behavior and re-establish the evidence for a behavioral continuity equal to the evidence for the assumed morphological continuity. One of the most remarkable things that consistently occurs, however, is that every attempt to get at human cultural origins, no matter whether it uncovers new information about primate behavior and social structure or no, *always* serves to sharpen and reinforce the gap between them more explicitly. Long ago Thomas Huxley in his famous work *Man's Place in Nature* in which he established the incontrovertible fact that anatomically, man is more similar to the great apes than the apes are to the monkeys, mentioned ". . . the great gulf which intervenes between the lowest man and the highest ape in intellectual power," and " . . . the vast intellectual chasm between the ape and man."[28] But, whereas Huxley was only speaking of brain power, as it were, the modern investigators, by the most detailed and illuminating descriptions of behavioral distinctions, sharpen the understanding of the very nature of the differences between learned, cultural, human behavior, and instinctive, genetically controlled, animal behavior. The late A. L. Kroeber in his famous text, *Anthropology,* pointed to the relative insignificance of the anatomical differences which he considered merely "differences of detail and degree, and mostly of no very great degree. But the difference as re-

19. *Ibid.,* pp. 27, 28.
20. Hockett, 1960, p. 4.
21. Sahlins, 1960, p. 2.
22. Hallowell, 1959, p. 44; Etkin, 1954, p. 140.
23. Tappen, 1960, p. 116.
24. Washburn, 1960, p. 15.
25. *Ibid.*
26. Bates, 1961, pp. 162, 163.
27. *Ibid.,* p. 168.
28. Huxley, 1863 (1959), pp. 120-122.

gards culture is one of so enormous a degree . . . as to become virtually equivalent to a difference in kind".[29]

Now the exact reason behind this vast difference between man's culture and the behavior of the nonhuman primates, and the bearing it has upon the attempts to derive human culture from pre-existing primate heredity is so fundamental that its importance can hardly be overestimated. The crux of the matter is that man's culture is not genetically inherited nor genetically controlled. Virtually all of man's behavior is learned after birth, while typically, though learning plays a part, nonhuman behavior for each particular species is "given" at birth and pre-determined so that each member of each species behaves in essentially the same restricted patterns of his species, without having any choice in the matter. Thus the attempted comparative studies of living primates and the entire preoccupation with closing this bio-cultural gap is based upon the contradiction of deriving something cultural from something noncultural; of producing something nongenetic and nonphysiological from something that is genetically controlled. Sahlins puts it very precisely as follows:

> There is quantum difference, at points a complete opposition, between even the most rudimentary human society and the most advanced subhuman one. The discontinuity implies that the emergence of human society required some suppression, rather than a direct expression of man's primate nature.[30]

Thus a suppression of instinctively controlled behavior is seen as a necessary step in cultural origins. Sahlins refers to this "overthrow of human primate nature" as "the greatest reform in history."[31]

Loren Eiseley, the distinguished anthropologist, author, and Provost of the University of Pennsylvania, speaks of "the 'new' brain, denuded of precise instinctive responses . . ."[32] and of man's "societal universe, with its institutions supplanting his lost instincts . . ."[33] and of his supreme and characteristic capability of symbolic language. Eiseley also points out that

> What must have been the frightening withdrawal of instinct in man and its replacement by the culture-building brain is a passage that the Darwinian world failed to grasp or appreciate clearly.[34]

He, too, refers to this event as "of the nature of a quantum step,"[35] and Bartholomew and Birdsell refer to the transition in the same terms as they presume that the first bipedal tool-using primates "were entering a period of rapid change leading to a new kind of adaptedness. In the terminology of Simpson (1944) they were a group undergoing quantum evolution."[36]

It is at this point that the question of the origin of man is lifted from the level of mere concern with bones and morphology to the level of phenomena of a very different order. Creationists have too long entered into heated controversy among themselves as well as with evolutionists over various aspects of the fossil record, to the exclusion of the consideration of the very area where the modern evolutionary explanation is totally at a loss. This admitted perplexity on the part of anthropologists is being expressed as never before in the context of primate studies as the renewed preoccupation with origins forces recognition of the reality of this elusive increment.

Thus Chance and Mead in their article "Social Behavior and Primate Evolution" state that,

> The anatomical features which differentiate man from the other primates, and the fossil evidence make clear the major changes which have led to his emergence from the primitive mammalian state to his present taxonomic position. Yet we are still without an adequate theory to explain this process in terms of adaptive evolution.[37]

Kroeber refers to man's original acquisition of language and culture as "an event of unusual novelty on this planet,"[38] and, in speculating as to whether or not it could have been due to some "supermutation—in the genes" he remarks that if it was such a mutation "this one was different in that the genetic change set something going outside of heredity also."[39]

Regarding the problem of the identification of culture with a biological base, Kroeber writes:

> There is no new organ, no new layer, no new chemical substance that we know of, peculiar to the human cortex.[40]

There are those who hold that the supposed transition was purely due to genetic processes. Dobzhansky and Ashley Montagu stated in 1947 that

> The biologist insists that the evolutionary changes that occurred before the prehuman could become human, as well as those which supervened since the attainment of the human estate, can be described casually only in terms of mutation, selection, genetic drift, and hybridization—familiar processes throughout the living world.[41]

Washburn likewise would pass completely over the question of the nongenetic nature of culture ori-

---

29. Kroeber, 1948, p. 70.
30. Sahlins, 1960, p. 3.
31. *Ibid.*, p. 12.
32. Eiseley, 1955, p. 73.
33. *Ibid.*, p. 74.
34. *Ibid.*, p. 75.
35. *Ibid.*, p. 67.
36. Bartholomew and Birdsell, 1953, p. 492.
37. Chance and Mead, 1953, p. 395.
38. Kroeber, 1948, p. 58.
39. *Ibid.*, p. 71.
40. *Ibid.*, p. 70.
41. Dobzhansky and Montagu, 1947, p. 587.

gins, preferring to think of it as a unified and inseparable process with organic changes. Thus he asserts that

> Selection produced new systems of child care, maturation and sex, just as it did alterations in the skull and the teeth. Tools, hunting, fire, (sic) complex social life, speech, the human way and the brain evolved together to produce ancient man of the genus *Homo* . . .[42]

Perhaps the most candidly qualified explanation is that of William Etkin in his study of "Social Behavior and the Evolution of Man's Mental Faculties." He states that "If this uniqueness of man is to be understood in terms of evolutionary biology it can only be as the resultant of a biological history that includes unique conditions."[43] Then, after making the observation that animals do not develop behaviors beyond their functional requirements any more than structures or physiological capacities, he offers the following conclusions to his examination of possible preadaptive pressures toward the development of culture:

> On this basis we expect selection pressure to push language development only to the point where it serves a function of identification of concrete objects and of socialization but not to the level of its use in abstract thought. Similarly the evolution of co-operative behavior can be explained to the point where it permits a degree of stabilization of the male into the family and pack but no further. In this view the origin of abstract thought . . . and of truly ethical behavior . . . are not explicable in the biological terms developed here.[44]

He calls this "the limitation of the biological explanation" only claims that "These biological factors are held to account for only the first steps toward a culture-capable organism."[45]

## IV.

Our final consideration involves the origin of the human brain, and will take us back once more to the African man-apes.

Two of the difficulties with any present attempt to put the Australopiths into the status *or* ancestry of man, are (a) the time factor mentioned above, and (b) his brain. The span of time in which the man-apes seem to have lived runs from the end of the Pliocene through the earliest phases of the Pleistocene. Thus they meet, or even possibly overlap men whose cultural status is unquestioned. It is felt by some (although not by all authorities) that in order for them to have been ancestral to man they would have had to live much further back in Pliocene times. My colleague Don Wilson has reminded me in a recent letter that "the most vital period for the understanding of the development of man (from an evolutionary standpoint) is the Pliocene and in this 10 million years fossil primates are presently almost unknown." A number of authorities, however, feel that the earlier Australopithecines could have been ancestral to Pithecanthropus-type man.[46]

The *brain* of the man-apes is the single most outstanding morphological characteristic which does not compare with the hominid proportions. It is only about one half as large, and clearly within the range of the modern anthropoids. With these considerations in mind let us follow Eiseley in his presentation of Alfred Russell Wallace's remarkable anticipation of this crucial aspect of the problem of man's origin. In 1876 Wallace said that either man would be found very early " 'spread in dense waves of population over all suitable portions of the great continent—for this on Mr. Darwin's hypothesis is essential to rapid developmental progress . . .' or, on the contrary, if 'continued researches in all parts of Europe and Asia fail to bring to light any proof of his presence (during the Pliocene and before), it will be at least a presumption that he came into existence at a much later date and by a more rapid process of development.' In that case, Wallace continued, it will be a reasonable argument that man's origin 'is due to distinct and higher agencies.' "[47] Eiseley adds that "It should now be apparent, through these propositions of Wallace, where the rearrangement of our remaining human fossils is leading us. It is leading us straight toward Wallace's second proposition . . ."[48] though Eiseley would not accept Wallace's idea of a supernatural agency. Nevertheless he does add:

> If our briefly sketched confinement of the major rise of the human brain to the Pleistocene is even approximately correct, it would appear to demand some other evolutionary mechanism beyond that of the old Darwinian struggle of man with man or group with group. The movement would appear much too fast[49] . . . some other more rapid process of evolution . . . must have been at work in the production of man.[50]

Washburn, too, stresses the fact that by the time of the first known man who used fire and had clearly defined tool traditions, "the brain had doubled in size" over that of the man-apes.[51] He then points out that, "It then appears to have increased much more slowly; there is no substantial change in gross size (of the brain) during the last 100,000 years."[52] Chance and Mead in their study quoted above come to the conclusion that "No adequate explana-

42. Washburn, 1960, p. 3.
43. Etkin, 1954, p. 129.
44. *Ibid.*, p. 140.
45. *Ibid.*, p. 141.
46. Washburn and Howell, 1960; Washburn and Avis, 1958; Washburn, 1960; and Howell, 1959 and 1960b.
47. Wallace, 1876, pp. 64, 65; quoted by Eiseley, 1955, p. 67.
48. Eiseley, 1955, p. 67.
49. *Ibid.*
50. *Ibid.*, p. 69.
51. Washburn, 1960, p. 9.
52. *Ibid.*, p. 11.

tion has been put forward . . . to account for the development of so large a cerebrum as that found in man."[53]

The question should also be raised, Why are mankind's mental traits and spiritual traits so uniform the world over when racial and cultural traits are so exceedingly varied. With reference to the latter two, natural selection and heredity control the one, and freedom from these has permitted the other. But concerning the uniform nature of man's mind which, Eiseley reminds us, could not have resulted from a long, slow evolutionary process of Darwinian struggle, he concludes, "Something—some other factor—has excaped our scientific attention."[54]

V.

Conclusions:

1. If the Australopithecinae have any relevance for the consideration of man's organic ancestry, it is at present within an evolutionary frame of reference which presumes that there must have been such a stage because it fits perfectly the present state of theory.

2. The tendency to refer to the known man-apes as representing the earliest known stage in a continuum of human cultural development occurs frequently in the literature but is dependent upon the interpretation of the factors of time, and brain size in comparison to other examples of early man.

3. The whole question of theistic evolution, as defined in the chapter under review, should seem to hinge on the matter of the bio-cultural gap. As we have seen, evolutionary science approaches it via the comparative examination of ape and man, fossil and living, but when all is said and done, the gap remains clearer than ever. Loren Eiseley, himself an evolutionist, examining the problem of man's origin meticulously from the historical as well as the contemporary point of view, finds that "the key to the secret doorway by which he came into this world is still unknown."[55]

(a) If one wishes to handle this in the evolutionary frame of reference, but with supernaturalistic assumptions, he may interpret God's creation of man as itself constituting the crucial transition—mental, cultural, spiritual. This would be the position of theistic evolution held by many Christian people, and the position most widely held by Roman Catholic scholars,[56] allowed by the Church, if God's intervention is safeguarded.

(b) If, on the other hand, one wishes to handle the problem not as a transition, but as a definite origin—physical, mental, cultural, spiritual—one may interpret God's creation of man as constituting the origin of man's body as well as his culture, mind and soul.

Genesis has been interpreted to include both of these positions by conservative Bible scholars. The facts of the fossil record do not at present demand the adoption of either position necessarily; but neither do they conflict with either one.

(c) The alternative, of course, would be the common evolutionary interpretation based upon naturalistic assumptions.

Summarizing: the problems of the nature and antiquity of the fossil bones are vastly overshadowed by the importance of the intangible problems of the spiritual and the cultural spheres in the consideration of the origin of man.

---

53. Chance and Mead, *loc. cit.* Also quoted by Eiseley, 1955, p. 67.
54. Eiseley, 1957, p. 91.
55. Eiseley, 1955, p. 75.
56. E.g., Hauret, 1955, p. 97: "God drew the human body . . . from an animal organism, which He transformed and adapted so as to receive a human soul."


BIBLIOGRAPHY

Bartholomew, G. A., and J. B. Birdsell
1953 "Ecology and the Protohominids," *American Anthropologist*, Vol. 55, No. 4, pp. 481-498.

Bates, Marston
1961 *The Forest and the Sea: A Look at the Economy of Nature and the Ecology of Man.* New York: The New American Library (Mentor).

Beadle, G. W.
1957 "The Uniqueness of Man," *Science*, Vol. 125, No. 3236, pp. 9-11.

Bonner, J. T.
1960 Review of *Evolution Above the Species Level* by B. Rensch. *American Scientist*, Vol. 48, No. 4, pp. 370A-372A.

Buswell, J. O., III
1959 "A Creationist Interpretation of Prehistoric Man," in Mixter (ed.), 1959.

Carpenter, C. R.
1960 "Comments" on "Social Organization of Subhuman Primates in Their Natural Habitat," by K. Imanishi. *Current Anthropology*, Vol. I, Nos. 5-6 (Sept.-Nov.), pp. 402, 403.

Chance, M. R. A., and A. P. Mead
1953 "Social Behavior and Primate Evolution," *Symposia of the Society for Experimental Biology, No. VII*, "Evolution." New York: Academic Press.

Clark, J. Desmond
1960 "Human Ecology During Pleistocene and Later Times in Africa South of the Sahara," *Current Anthropology*, Vol. I, No. 4, pp. 307-321.

Clark, W. LeGros
1955 *The Fossil Evidence for Human Evolution: An Introduction to the Study of Pleoanthropology.* Chicago: University of Chicago Press.

Dart, Raymond
1959 "Further Light on Australopithecine Humeral and Femoral Weapons," *American Journal of Physical Anthropology*, Vol. 17, No. 2, pp. 87-93.

Dobzhansky, Th., and M. F. Ashley Montagu
1947 "Natural Selection and the Mental Capacities of Mankind," *Science*, Vol. 105, pp. 587-590.

Eiseley, Loren
1955 "Fossil Man and Human Evolution" in *Yearbook of Anthropology—1955.* Republished in part as *Current Anthropology: A Supplement to Anthropology Today*, edited by W. L. Thomas, Jr. Chicago: University of Chicago Press, 1956.
1957 *The Immense Journey.* New York: Random House.

Etkin, Wm.
1954 "Social Behavior and the Evolution of Man's Mental Faculties," *The American Naturalist*, Vol. LXXXVIII, No. 840, pp. 129-142.

Hallowell, A. I.
1959 "Behavioral Evolution and the Emergence of the Self," in Meggars (ed.), 1959.

Hauret, Charles

1955 *Beginnings: Genesis and Modern Science.* Dubuque: Priory Press.
Hockett, Charles D.
1960 "The Origin of Speech," *Scientific American*, September (Reprint).
Howell, F. Clark
1959 "The Villafranchian and Human Origins," *Science*, Vol. 130, pp. 831-844.
1960a "Howell's Comments" on "The Discovery by L. S. B. Leakey of *Zinjanthropus boisei*," *Current Anthropology*, Vol. I, No. 1, pp. 76, 77.
1960b "European and Northwest African Middle Pleistocene Hominids," *Current Anthropology*, Vol. I, No. 3, pp. 195-232.
Howells, W. W.
1944 *Mankind So Far.* New York: Doubleday.
1959 *Mankind in the Making: The Story of Human Evolution.* New York: Doubleday.
Huxley, Thomas H.
1863 *Evidence as to Man's Place in Nature.* Reprinted, 1959, Ann Arbor: University of Michigan Press.
Kroeber, Alfred L.
1948 *Anthropology.* New York: Harcourt, Brace.
Meggars, Betty J. (ed.)
1959 *Evolution and Anthropology: A Centennial Appraisal.* Washington, D.C.: The Anthropological Society of Washington.
Mixter, Russell L. (ed.)
1959 *Evolution and Christian Thought Today.* Grand Rapids: Eerdmans.
Montagu, M. F. Ashley
1960 *An Introduction to Physical Anthropology*, Third Edition. Springfield, Ill.: C. C. Thomas.
Morris, Desmond
1961 "Primate's Aesthetics," *Natural History*, Vol. LXX, No. 1, pp. 22-29.
Oakley, Kenneth
1957 "Tools Makyth Man," Antiquity, Vol. 31; reprinted in *Annual Report . . . of the Smithsonian Institution, 1958*, pp. 431-445. Washington, D.C.
Roe, A., and G. G. Simpson (eds.)
1958 *Behavior and Evolution.* New Haven: Yale University Press.
Sabels, Bruno E.
1959 Review of "The Transvaal Ape-Man-Bearing Cave Deposits," by C. K. Brain (Pretoria, 1958), *American Journal of Physical Anthropology*, Vol. 17, No. 3, pp. 247-249.
Sahlins, M. D.
1960 "The Origin of Society," *Scientific American*, September (Reprint).
Simpson, G. G.
1944 *Tempo and Mode in Evolution.* New York: Columbia University Press.
Tappen, N. C.
1960 "Problems of Distribution and Adaptation of the African Monkeys," *Current Anthropology*, Vol. 1, No. 2, pp. 91-120.
Tax, S. (ed.)
1960 *Evolution After Darwin*, Vol. II *The Evolution of Man.* Chicago: University of Chicago Press.
Wallace, A. R.
1876 "Difficulties of Development as Applied to Man," *Popular Science Monthly*, Vol. 10, pp. 60-72.
Washburn. S. L.
1960 "Tools and Human Evolution," *Scientific American*, September (Reprint).
Washburn, S. L., and V. Avis
1958 "Evolution of Human Behavior," in Roe and Simpson (eds.), 1958.
Washburn, S. L., and F. C. Howell
1960 "Human Evolution and Culture," in Tax, S. (ed.), 1960.
Washburn, S. L., and D. Wolffson (eds.)
1949 *The Shorter Anthropological Papers of Franz Weidenreich Published in the Period 1939-1948. A Memorial Volume.* New York: The Viking Fund.
Weidenreich, Franz
1948 "The Human Brain in Light of Its Phylogenetic Development," *The Scientific Monthly*, Vol. LXVII, No. 2; reprinted in Washburn and Wolffson, 1949.

# *Geology and the Great Flood*

WILLIAM F. TANNER*

Many Christians accept the idea that a single flood swept, in the day of Noah, across the entire surface of the earth, submerging even such lofty peaks as Mt. Everest. Many scientists have doubted this story, for two reasons: (1) there is neither a known source from which to draw such fantastically large volumes of water, nor a place to store it after the flood has subsided, and (2) there has not seemed to be any particularly convincing geological evidence in favor of such a flood.

Christian authors have sought scientific support for a strict, literal, interpretation of the Bible account, among the facts of geology. This search has led to rather ludicrous results. Various writers have cited, as proof of Noah's flood, Pennsylvanian beds (deposited about 3 x $10^8$ years ago), as well as rocks which were formed at other times, both earlier and later than that.

Scientists have, in many cases, used their scorn for the flood story as a means of casting doubt on the authenticity of the entire Bible.

Although scientists and theologians have been able to resolve their differences in many fields, the flood narrative has presented an apparently insoluble problem.

Recent results in the earth sciences, however, now raise the hope that perhaps here, again, the historians who wrote the early books of the Bible will be vindicated. It may well be that some Christians will not be pleased with the result; on the other hand, the present agreement beween the observations of geology and the Bible story confirms the latter as well as the nature of the Genesis account will permit.

Many geologists may not be pleased with the results, either. Traditionally, geologists have thought that earth history, over the last 25,000 years or so, was sufficiently well documented to preclude anything remotely resembling the Noachian deluge. The present report is based on a modification of each of these extreme points of view. However, it is thought that in no sense is either the Bible account, or the geological record, undermined or diluted.

The science-oriented Christian must admit, in the first place, that God is quite capable of performing miracles. It is not necessary, therefore, that the total amount of water on and in the earth be conserved. God can create any necessary additional amount, and likewise can later dispose of the surplus. If, however, we can see both a science-based mechanism, and a miraculous mechanism (i.e., one not understood or apparently mechanically contradictory), we should choose the former. This is because God is *not* capricious; He *does* operate within the framework of His own laws; and science is, in the final analysis, merely a study of the behavior of those God-ordained laws which affect the physical universe. A decision to accept the scientifically plausible explanation, rather than the scientifically illogical one, does not prove anything one way or the other; it merely shows that we think that the same God who guided the writing of the Bible account also established the laws which govern the day-to-day operation of the universe. In other words, the *best* answers which we are able to obtain are those that fit into both realms of thought: the Biblical, and the scientific.

If the Bible states flatly, beyond any possible argument, that 100% of the presently-known earth was inundated, we must face the alternative: a miraculous act of some kind. The Genesis account, in translation, does use the word "all." Our interpretation of this word must be colored, however, by other uses of the same word. For example, in Luke 2:1-3 we read that Caesar ordered "all" the world to be enrolled (that is, counted for census and tax purposes), and that "all" the world was, consequently, enrolled. It should be fairly clear that the Incas, Aztecs, and Eskimos were not affected by this order, and that probably large numbers of people in Australia, Asia and Africa never heard of it either. The usage here is essentially that of "all the world that is pertinent to the discussion," which is roughly, but not precisely, the same thing as saying "all the known or civilized world." Such a usage of "all" is standard in spoken English today. When we say, "Everybody's doing it" (whatever "it" may be), we do not mean "100%." The same thing is true in other languages, and particularly true in languages which have not (or

*Dr. Tanner is Associate Professor of Geology at Florida State University, Tallahassee.

had not) been influenced by scientific concepts. The notion of "100%" (mathematically pure and complete) is largely a scientific idea, and should not be forced onto other types of communication. The statement, "All have sinned, and come short of the glory of God," is something else again. This deals with a spiritual principle, is reiterated many times in various forms, and does not admit of an exception.

In the case of Noah's flood, we must choose between the two uses. The physical, non-mathematical one (i.e., "all of the world pertinent to the discussion") is to be preferred.

The foregoing treatment of the meaning of "all" would be pointless if there were no other evidence bearing on the problem. Fortunately, there *is* other evidence, and the notion of "all of the world pertinent to the discussion" is therefore significant.

The Bible does not provide precise dates for the events in the early chapters of Genesis. Bishop Ussher's famous dates are only *minimum* dates; that is, they represent a summation of the recorded events, and hence the least possible time in which those events could have happened. They do not allow for events which Bible writers thought unimportant or not germane, and therefore may, or may not, be accurate dates. It is interesting to note that Ussher's chronology places the Noachian flood at about $5 \times 10^3$ years ago, or even further back. Any important flood which occurred appreciably later than about $5 \times 10^3$ years ago is, apparently, not the flood discussed in Genesis.

One other general fact emerges from the Bible account. Man was, at the time of the flood, practicing agriculture. This development occurred, according to the most reliable archeological estimates, between 7,000 and 9,000 B. C.—that is, 9 to $11 \times 10^3$ years ago (Braidwood, 1960). The Ussher chronology, plus the limit placed by the advent of agriculture, restrict us to a period between $11 \times 10^3$ and $5 \times 10^3$ years ago. Of course these dates are not precise, but they may prove useful.

General information obtained from the Bible story includes the fact of severe, prolonged, wet, stormy weather. This, also, may prove useful.

One important fact is not provided. The Genesis writer states that the flood receded, but we do not know whether or not the water level actually returned to its original position. It is quite possible that, after Noah and his family disembarked, sea level stood higher than it had before. The Bible does not clarify this point.

Recent geological research contributes a considerable body of information to the discussion. Radiocarbon dates from the Gulf of Mexico area (including the Mississippi River delta) show that an extremely rapid rise in sea level began about $12 \times 10^3$ years ago (Fairbridge, 1960). This rise continued until roughly $9 \times 10^3$ years ago. During this period, sea level was changed by about 400 feet. The average rate of rise was close to 1.5 inches per year (about 12 feet per century). At the end of this period sea level stood about 50 feet below its present position. The water for the rise was provided by partial melting of the major ice caps and glaciers of the world.

A reversal of the climate trend (back toward more snow and ice) lowered sea level temporarily, until it reached a position approximately 100 feet below the present level, roughly $8 \times 10^3$ years ago. Between that date and about $6 \times 10^3$ years ago a warming trend sent sea level up to within 15 or 20 feet of the modern position. This was, likewise, a rapid rise, and an erratic one. The average rate was only about five feet per century, but much of this rise may have been concentrated in relatively short periods of time at rates perhaps as rapid as one or more feet per year. No precise geological figures are available on this point.

At approximately $2 \times 10^3$ years ago sea level stood at minus five feet. It has risen slowly and somewhat irregularly since then, and apparently is still rising. If estimates of present ice volume are correct, a complete melting would result in a rise to about 300 or 350 feet above the present level. Such a rise, if fairly rapid, could be quite disastrous.

A summary of the geological data shows two rapid rises, separated by a fall of about 50 feet, with the second rapid rise followed by a slow rise up to the present. Of the two rapid rises, the second appears to coincide with the general interval available for the Great Flood, as established by Bible chronology and general archaeology. This agreement between the two lines of evidence may make a solution possible.

| Geological Events | Thousands of Yrs. ago | Biblical and Archeological events |
|---|---|---|
| Rapid rise begins | 12 | |
| | 11 | ) |
| Rise ends; fall begins | 10 | ) *First agriculture?* |
| | 9 | ) |
| *Rapid rise begins* | 8 | |
| | 7 | |
| *Rise slows down* | 6 | |
| | 5 | *Chronology limit?* |
| Slow rise | 4 | Abraham |
| | 3 | |
| Slow rise | 2 | Birth of Christ |
| | 1 | |
| Slow rise; levelling off? | 0 | |

The accompanying table compares the two sequences of events, as obtained from geological evidence, on the one hand, and from archeological and Biblical evidence, on the other. Noah's flood apparently must be fitted between the italicized items. The agreement is excellent, considering the limitations of the method of study.

This preliminary comparison allows examination of a possible geological setting for the Great Flood. Sea level was rising rapidly from about minus 100 feet to about minus 20 feet. The climate was becoming warmer, and the moisture capacity of the atmosphere correspondingly greater. The latter suggests, for non-glacial regions such as Asia Minor, the possibility of more rain, perhaps (temporarily) extended periods of extremely heavy rain. Wet periods, covering very much of the world, coupled with a warming trend, should accelerate glacial melting, and produce short, fairly sharp, rises of sea level.

The pre-flood civilization described in the Bible was probably centered in the valley of the Tigris-Euphrates river system. There the relief of the land is low, largely because much of the area is constructional, built up of river sediment. A low, smooth, nearly flat alluvial plain of this kind is easily covered, to relatively large depths, by a slight rise in water level. A typical alluvial or delta-plain gradient is only one or two inches per mile at the lower end. A water level rise (for example, during a severe storm) of 15 feet would translate the zero contour to a point about 100 miles inland. If the gradient of the plain were much steeper, perhaps as much as four inches per mile, a 15-foot rise would move the shore line about 50 miles inland.

The entire low-land length of the drainage system is only about 400 miles (straight-line measurement; *not* winding river length) at the present time. A disastrous river flood, coupled with a rise and advance of storm-driven sea-water, could inundate the entire valley, sweep everything man-made away, leave nothing (within an area of roughly $10^5$ sq. mi.) visible above the water surface, and drown the entire population of the area (with the exception of people who were already on or near well-provisioned boats). Combine this with an extended torrential down-pour, and we have a pretty fair picture of what may have been the flood story.

When the waters receded, the face of the plain would have been unrecognizable, for several reasons: (1) Storm tides, along the Persian Gulf, would have severely eroded the coastal areas; (2) Various rivers, during the high-water stages, would have changed their courses (a well-known geological fact); and (3) Much of the plain would have been plastered with new alluvial deposits, possibly several feet thick. Furthermore, by the time man could re-inhabit the plain, rising sea-level would have buried relatively large areas which previously had been cultivated.

Similar catastrophes must have occurred in the earlier, longer, sea-level rise (from 12 to 10 x $10^3$ years ago). Man was not, at that time, a farmer. Since he was more or less nomadic, he would not have been anchored to cultivated fields, probably would not have preferred the deltaic end of the valley, and would not think of the sea level rise as a disaster.

The incident of the birds, and the tree leaves, as recounted in Genesis, is revealing. Burial of leaves under water, for long periods of time, is fatal. The green branch was either a direct miracle, or the bird found it on an unharmed tree growing on adjacent hills, perhaps 100 or more miles away.

We do not know the path taken by the drifting ark. One would infer that it, eventually, drifted sea-ward, rather than upstream against the flood-time river current. Where it finally struck land we do not know. The traditional identification of the spot with a mountain in Turkey does not, at the present moment, seem to have any validity.

A combination of sea-level rise, rapid glacial melting, torrential rains, and widespread floods would affect, primarily, coastal areas and the down-stream ends of river valleys. The entire world would have been so affected. Man probably had not spread over the entire world; the Bible account does not clarify this point, but geological and glaciological evidence do. However, wherever he was, along the waterways and coasts (which provided the only good means of transportation), he must have suffered the same fate.

The hypothesis advanced here distorts neither the geological, nor the Biblical, picture. The result is not a 100% universal flood, with water 30,000 or more feet deep over the modern coasts. The Bible, however, does not require such a flood. Falling back on minimum Bible requirements, we find a reasonable agreement with the best evidence provided by geology. The relatively simple, straight-forward picture which emerges is not miraculous. But it does tend to show that Bible history is completely reliable. That, in itself, is a worth-while result.

### References

Braidwood, Robert J., 1960: "The Agricultural Revolution," *Scientific American*, September, pp. 130-148.

Fairbridge, Rhodes W., 1960: "The Changing Level of the Sea," *Scientific American*, May, pp. 70-79.

# Acknowledgments

American Scientific Affiliation. *The Story of the American Scientific Affiliation.* (1942): 1–11. Courtesy of the American Scientific Affiliation.

Stoner, Peter W. "Introduction, Preface and Chapter One." In Peter W. Stoner, ed. *From Science to Souls* (Chicago: Moody Press, 1944): 14–61. Courtesy of Biola University.

Gedney, Edwin K. "Geology and the Bible." In *Modern Science and Christian Faith* (Wheaton: Van Kampen Press, 1948): 39–73. Courtesy of Yale University Divinity Library.

Tinkle, William J. and Walter E. Lammerts. "Biology and Creation." In *Modern Science and Christian Faith* (Wheaton: Van Kampen Press, 1948): 75–107. Courtesy of Yale University Divinity Library.

Stoner, Peter W. "Astronomy and the First Chapter of Genesis." In *Modern Science and Christian Faith* (Wheaton: Van Kampen Press, 1948): 23–37. Courtesy of Yale University Divinity Library.

American Scientific Affiliation. *A Symposium on "The Age of the Earth."* (1948): 1–22. Courtesy of the American Scientific Affiliation.

Mixter, Russell L. *Creation and Evolution.* (1950): 1–31. Courtesy of the American Scientific Affiliation.

Allen, Frank. "The Eye as an Optical Instrument." *Journal of the American Scientific Affiliation* 1 (1949): 9–20. Courtesy of the American Scientific Affiliation.

Mixter, Russell L. "The Science of Heredity and the Source of Species." *Journal of the American Scientific Affiliation* 1 (1949): 1–9. Courtesy of the American Scientific Affiliation.

Ramm, Bernard. "The Scientifico-Logical Structure of the Theory of Evolution." *Journal of the American Scientific Affiliation* 1 (1949): 10–15. Courtesy of the American Scientific Affiliation.

Monsma, E. Y. "Some Presuppositions in Evolutionary Thinking." *Journal of the American Scientific Affiliation* 1 (1949): 15–10. Courtesy of the American Scientific Affiliation.

Mixter, Russell L. "Biology and Christian Fundamentals." *Journal of the American Scientific Affiliation* 2 (1950): 20–23. Courtesy of the American Scientific Affiliation.

Kulp, J. Laurence. "Deluge Geology." *Journal of the American Scientific Affiliation* 2 (1950): 1–15. Courtesy of the American Scientific Affiliation.

[Morris, Henry] "Comment on the 'Deluge Geology.'" *Journal of the American Scientific Affiliation* 2 (1950): 2. Courtesy of the American Scientific Affiliation.

Fulwood, R. T. "The Plan in the First Chapter of Genesis." *Journal of the American Scientific Affiliation* 2 (1950): 1–13. Courtesy of the American Scientific Affiliation.

Tinkle, William J. "Why God Called His Creation Good." *Journal of the American Scientific Affiliation* 2 (1950): 20–25. Courtesy of the American Scientific Affiliation.

Erdman, Cordelia. "The Paleontology of the Horse." *Journal of the American Scientific Affiliation* 2 (1950): 25–36. Courtesy of the American Scientific Affiliation.

Eggenberger, Delbert. "Methods of Dating the Earth and the Universe." *Journal of the American Scientific Affiliation* 3 (1951):1–3. Courtesy of the American Scientific Affiliation.

Everest, F. Alton. "The American Scientific Affiliation—The First Decade." *Journal of the American Scientific Affiliation* 3 (1951): 33–38. Courtesy of the American Scientific Affiliation.

Young, G. Douglas. "Biblical Chronology and Genesis 1:1–3." *Journal of the American Scientific Affiliation* 3 (1951): 19–23. Courtesy of the American Scientific Affiliation.

Buswell, J. Oliver, Jr. "Creation Days." *Journal of the American Scientific Affiliation* 4 (1952): 10–15. Courtesy of the American Scientific Affiliation.

Bullock, Wilbur L. "The 'Kind of Genesis and the Species' of Biology." *Journal of the American Scientific Affiliation* 4 (1952): 5–6. Courtesy of the American Scientific Affiliation.

Allen, Roy M. "The Evaluation of Radioactive Evidence on the Age of the Earth." *Journal of the American Scientific Affiliation* 4 (1952): 11–20. Courtesy of the American Scientific Affiliation.

Erdman, Cordelia. "Stratigraphy and Paleontology." *Journal of the American Scientific Affiliation* 5 (1953): 3–6, 11. Courtesy of the American Scientific Affiliation.

Lammerts, Walter E. and John C. Sinclair. "Creation in Terms of Modern Concepts of Genetics and Physics." *Journal of the American Scientific Affiliation* 5 (1953): 7–10. Courtesy of the American Scientific Affiliation.

[McClendon, John H.] "A Criticism of the A.S.A. Monograph on 'Creation and Evolution.'" and "Reply to Criticism of 'Creation and Evolution' by R.L. Mixter. *Journal of the American Scientific Affiliation* 6 (1954): 24–28. Courtesy of the American Scientific Affiliation.

Stoner, Peter W. "The Reconstruction or Cataclysmic Theory." *Journal of the American Scientific Affiliation* 6 (1954):9–13. Courtesy of the American Scientific Affiliation.

McClendon, John H. [Letter] "Response to Mixter regarding A.S.A. Monograph on 'Creation and Evolution.'" *Journal of the American Scientific Affiliation* 6 (1954): 19–20 Courtesy of the American Scientific Affiliation.

Feely, Herbert. "The Impact of Geological Dating Upon the Interpretation of Biblical Chronology." *Journal of the American Scientific Affiliation* 7 (1955): 46–49. Courtesy of the American Scientific Affiliation.

Sinclair, John C. "Creation." *Journal of the American Scientific Affiliation* 8 (1956): 16–18. Courtesy of the American Scientific Affiliation.

Hearn, Walter R. "The Formation of Living Organisms from Non-Living Systems." *Journal of the American Scientific Affiliation* 10 (1958): 3–8. Courtesy of the American Scientific Affiliation.

Mixter, Russell L. "An Evaluation of the Fossil Record." and "Comments on Dr. Mixter's Paper" by Wilbur L. Bullock. *Journal of the American Scientific Affiliation* 11 (1959): 24–26. Courtesy of the American Scientific Affiliation.

Cassel, J. Frank. "The Evolution of Evangelical Thinking on Evolution." *Journal of the American Scientific Affiliation* 11 (1959): 26–27. Courtesy of the American Scientific Affiliation.

Thompson, W.R. "A Critique of Evolution." *Journal of the American Scientific Affiliation* 12 (1960): 2–9. Courtesy of the American Scientific Affiliation.

Cassel, J. Frank. "Species Concepts and Definitions." *Journal of the American Scientific Affiliation* 12 (1960): 2–3, 5. Courtesy of the American Scientific Affiliation.

Marsh, Frank L. "The Genesis Kinds in Our Modern World." *Journal of the American Scientific Affiliation* 12 (1960): 4–8, 11, 13. Courtesy of the American Scientific Affiliation.

Cowperthwaite, Irving A. "Some Implications of Evolution for A.S.A." *Journal of the American Scientific Affiliation* 12 (1960): 12–13. Courtesy of the American Scientific Affiliation.

Cassel, J. Frank. "The Origin of Man and the Bible." *Journal of the American Scientific Affiliation* 12 (1960): 13–16, 28. Courtesy of the American Scientific Affiliation.

Hartzler, H. Harold. "Letter." *Journal of the American Scientific Affiliation* 12 (1960): 16–17. Courtesy of the American Scientific Affiliation.

Lammerts, Walter E . "Neutron-Induced Variation of Roses." *Journal of the American Scientific Affiliation* 13 (1961): 2–6. Courtesy of the American Scientific Affiliation.

Tinkle, William J . "Creation, a Finished Work." *Journal of the American Scientific Affiliation* 13 (1961): 15–17. Courtesy of the American Scientific Affiliation.

Hearn, Walter R. "Origin of Life." *Journal of the American Scientific Affiliation* 13 (1961): 38–42. Courtesy of the American Scientific Affiliation.

Cassel, J. Frank. "Comments on the Origin of Species." *Journal of the American Scientific Affiliation* 13 (1961): 42–45. Courtesy of the American Scientific Affiliation.

Cassel, J. Frank. "Christian Thought Today—On Origin of Species." *Journal of the American Scientific Affiliation* 13 (1961): 45–47. Courtesy of the American Scientific Affiliation.

Buswell, James O., III. "The Origin of Man, and the Bio-cultural Gap." *Journal of the American Scientific Affiliation* 13 (1961): 47–55. Courtesy of the American Scientific Affiliation.

Tanner, William F. "Geology and the Great Flood." *Journal of the American Scientific Affiliation* 13 (1961): 117–119. Courtesy of the American Scientific Affiliation.

American Scientific Affiliation. *"The Story of the American Scientific Affiliation",* Journal of the American Scientific Affiliation *(1942):* 1-11

American Scientific Affiliation. *"A Symposium on "The Age of the Earth"* Journal of the American Scientific Affiliation *(1948)*: 1-22

Mixter, Russell L. "Creation and Evolution". Journal of the American Scientific Affiliation (1950): 1-31.

Allen, Frank. "The Eye as an Optical Instrument" Journal of the American Scientific Affiliation 1, (1949): 9-20

Mixter, Russell L. "The Science of Heredity and the Source of Species". Journal of the American Scientific Affiliation 1 (1949): 1-9

Ramm, Bernard. "The Scientifico-Logical Structure of the Theory of Evolution" Journal of the American Scientific Affiliation 1, (1949): 10-15

Monsma, E.Y. "Some Presuppositions in Evolutionary Thinking" Journal of the American Scientific Affiliation 1 (1949): 15-10

Mixter, Russell L. "Biology and Christian Fundamentals." Journal of the American Scientific Affiliation 2 (1950): 20-23

Kulp, J. Laurence, "Deluge Geology." Journal of the American Scientific Affiliation 2 (1950): 2.

Morris, Henry "Comment on the 'Deluge Geology.'" Journal of the American Scientific Affiliation 2 (1950): 2

Fulwood, R.T. "The Plan in the First Chapter of Genesis." Journal of the American Scientific Affiliation 2 (1950): 2.

Tinkle, William J. "Why God Called His Creation Good." Journal of the American Scientific Affiliation 2 (1950): 20-25

Erdman, Cordelia. "The Palaeontology of the Horse." Journal of the American Scientific Affiliation (1950): 25-36

Eggenberger, Delbert. "Methods of Dating the Earth and the Universe." Journal of the American Scientific Affiliation 3 (1951): 1-3

Young, G. Douglas. "Biblical Chronology and Genesis 1:1-3." Journal of the American Scientific Affiliation 3 (1951): 19-23.

Buswell J. Oliver, Jr. "Creation Days." Journal of the American Scientific Affiliation 4 (1952): 10-15.

Bullock, Wilbur L. "The 'Kind of Genesis and the Species' of Biology." Journal of the American Scientific Affiliation 4 (1952): 5-6

Allen, Roy M. "The Evaluation of Radioactive Evidence on the Age of the Earth." Journal of the American Scientific Affiliation 4 (1952): 11-20.

Erdman, Cordelia. "Stratigraphy and Paleontology." Journal of the American Scientific Affiliation 5 (1953): 3-6, 11

Lammerts, Walter E. and John C. Sinclair. "Creation in Terms of Modern Concepts of Genetic and Physics." Journal of the American Scientific Affiliation 5 (1953): 7-10.

McClendon, John H. "A Criticism of the A.S.A. Monograph on 'Creation and Evolution.'" And "Reply to Criticism of 'Creation and Evolution' by R.L. Mixter. Journal of the American Scientific Affiliation 6 (1954): 24-28

Stoner, Peter W. "The Reconstruction or Cataclysmic Theory." Journal of the American Scientific Affiliation 6 (1954): 9-13.

McClendon, John H. [Letter] "Response to Mixter regarding A.S.A. Monograph on 'Creation and Evolution.'" Journal of the American Scientific Affiliation 6 (1954): 19-20

Feely, Herbert. "The Impact of Geological Dating Upon the Interpretation of Biblical Chronology." Journal of the American Scientific Affiliation 7 (1955): 46-49

Sinclair, John C. "Creation." Journal of the American Scientific Affiliation 8 (1956): 16-18

Hearn, Walter R. "The Formation of Living Organisms from Non-Living Systems." Journal of the American Scientific Affiliation 10 (1958): 3-8

Mixter, Russell L. " An Evaluation of the Fossil Record." And "Comments on Dr. Mixter's Paper" by Wilbur L. Bullock. Journal of the American Scientific Affiliation 11 (1959): 24-26

Cassel, J. Frank. "The Evolution of Evangelical Thinking on Evolution." Journal of the American Scientific Affiliation 11 (1959): 26-27

Thompson, W.R. "A Critique of Evolution." Journal of the American Scientific Affiliation 12 (1960): 2-9

Cassel, J. Frank. "Species Concepts and Definitions." Journal of the American Scientific Affiliation 12 (1960): 2-3,5

Marsh, Frank L. "The Genesis Kinds in Our Modern World." Journal of the American Scientific Affiliation 12 (1960): 4-8, 11, 13

Cowperthwaite, Irving A. "Some Implications of Evolution for A.S.A" Journal of the American Scientific Affiliation 12 (1960): 12-13

Cassell, J. Frank. "The Origin of Man and the Bible." Journal of the American Scientific Affiliation 12 (1960): 13-16, 28

Hartzler, H. Harold. "Letter." Journal of the American Scientific Affiliation 12 (1960): 16-17

Lammerts, Walter E. "Neutron-Induced Variation of Roses." Journal of the American Scientific Affiliation 13 (1961): 2-6.

Tinkle, William J. "Creation, a Finished Work." Journal of the American Scientific Affiliation 13 (1961): 15-17

Hearn, Walter R. “Origin of Life.” Journal of the American Scientific Affiliation 13 (1961): 38-42

Cassel J. Frank. “Comments on the Origin of Species.” Journal of the American Scientific Affiliation 13 (1961): 42-45

Cassell J. Frank “Christian Thought Today – On Origin of Species.” Journal of the American Scientific Affiliation 13 (1961): 47-55

Buswell, James O., III. “The Origin of Man, and the Bio-Cultural Gap.” Journal of the American Scientific Affiliation 13 (1961): 47-55

Tanner, William F. “Geology and the Great Flood.” 13 (1961): 117-119